International Conference on Mathematics, Science, and Education (ICoMSE 2017)

Journal of Physics: Conference Series Volume 1093

Malang, Indonesia
29 - 30 August 2017

Editor:

Nandang Mufti

ISBN: 978-1-5108-7340-7
ISSN: 1742-6588

Printed from e-media with permission by:

Curran Associates, Inc.
57 Morehouse Lane
Red Hook, NY 12571

CURRAN ASSOCIATES INC.
proceedings
.com

Some format issues inherent in the e-media version may also appear in this print version.

Printed by Curran Associates, Inc. (2018)

Institute of Physics
Dirac House, Temple Back
Bristol BS1 6BE UK

Phone: 44 1 17 929 7481
Fax: 44 1 17 920 0979

techtracking@iop.org

Additional copies of this publication are available from:

Curran Associates, Inc.
57 Morehouse Lane
Red Hook, NY 12571 USA
Phone: 845-758-0400
Fax: 845-758-2633
Email: curran@proceedings.com
Web: www.proceedings.com

Table of contents

Volume 1093

The 2017 International Conference on Mathematics, Science, and Education

29–30 August 2017, Malang, East Java, Indonesia

Accepted papers received: 24 August 2018
Published online: 10 October 2018

Preface

Papers

IOP Conf. Series: Journal of Physics: Conf. Series **1093** (2018) 011001 doi:10.1088/1742-6596/1093/1/011001

Preface: The 2017 International Conference on Mathematics, Science, and Education (ICoMSE 2017)

The 2017 International Conference on Mathematics, Science, and Education (ICoMSE 2017) was held in Malang, East Java Indonesia on 29-30 August 2017. It is an international conference covering a wide range of subjects in mathematics and natural sciences. This Conference was organized by the Faculty of Mathematics and Natural Sciences, State University of Malang (UM) in collaboration with the Indonesian Chemical Society, Indonesian Physical Society, and the Indonesian Biology Consortium.

Many people have enthusiastically participated in the preparation of this conference, including the Organizing Committees, the speakers, technicians, students, etc. Thank you all for your collaboration and extraordinary support. Thanks to all invited speakers: Prof. Dr. Hadi Nur (Universiti Teknologi Malaysia, Malaysia), Dr. Siriwat Soontaranon (Synchrotron Light Research Institute, Thailand), Prof. Dr. Oenardi Lawanto (Utah State Univesity, USA), Prof. Chokchai Yuenyong, Ph.D. (Khon Kaen University, Thailand), Prof. Dr. rer. nat. Gereon Elbers (FH Aachen University of Applied Sciences, Germany), Prof. Akhmaloka (Institut Teknologi Bandung), that you accepted our invitation.

The financial support from the Faculty of Mathematics and Natural Sciences, State University of Malang is gratefully acknowledged. ICoMSE 2017 is aimed at providing the scientific forum to encourage communication and interaction among distinguished scientists working in related fields in order to construct research collaborations and extending international research networking for the future. More than 500 authors from six countries have submitted their work to the conference. ICoMSE 2017 finally accepted 51 original research papers after a peer review process at least by two reviewers. During the conference, 10 parallel sessions were held in order to advance and contribute to specific research area in Mathematics, Science, and Education. Finally, special thanks to you, the delegates, for supplying the input needed for successful scientific conferences. We wholeheartedly welcome you and hope to find you in the next conference that is organized by our faculty.

Chief Editors

Nandang Mufti, MT. Ph.D

Photographs of ICOMSE 2017

Registration of participant on ICOMSE 2017

Registration of participant on ICOMSE 2017

Traditional dancing performance by Dance Art Group of Universitas Negeri Malang on ICOMSE 2017

Opening Speech by Rector of Universitas Negeri Malang on ICOMSE 2017

IOP Conf. Series: Journal of Physics: Conf. Series **1093** (2018) 011002 doi:10.1088/1742-6596/1093/1/011002

Opening Speech by Chairman of ICOMSE 2017

Group Pictures of Keynote Speakers with Rector and Principle Committee ICOMSE 2017

IOP Conf. Series: Journal of Physics: Conf. Series **1093** (2018) 011002 doi:10.1088/1742-6596/1093/1/011002

Presentation by Prof. Dr. Hadi Nur on Plenary Session of ICOMSE 2017

Giving Souvenir from the Dean of Faculty of Mathematics & Natural Sciences, Universitas Negeri Malang to Keynote Speakers (Prof. Dr. Hadi Nur from Universiti Teknologi Malaysia, and Prof. Dr. Akhmaloka from Institut Teknologi Bandung, Indonesia)

IOP Conf. Series: Journal of Physics: Conf. Series **1093** (2018) 011002 doi:10.1088/1742-6596/1093/1/011002

Presentation by Prof. Dr. Gereon Elbers (Germany) on Plenary Session of ICOMSE 2017

Giving Souvenir from the Chairman of ICOMSE 2017, Universitas Negeri Malang to Keynote Speakers (Prof. Dr. Gereon Elbers from Germany)

IOP Conf. Series: Journal of Physics: Conf. Series **1093** (2018) 011002 doi:10.1088/1742-6596/1093/1/011002

The Participants on Plenary Session of ICOMSE 2017

The Partisipants on Plenary Session of ICOMSE 2017

IOP Conf. Series: Journal of Physics: Conf. Series **1093** (2018) 011002 doi:10.1088/1742-6596/1093/1/011002

Discussion Session on Plenary Session of ICOMSE 2017

Discussion Session on Plenary Session of ICOMSE 2017

IOP Conf. Series: Journal of Physics: Conf. Series **1093** (2018) 011002 doi:10.1088/1742-6596/1093/1/011002

Presentation by a Presenter on Parallel Session of ICOMSE 2017

Presentation by a Presenter on Parallel Session of ICOMSE 2017

IOP Conf. Series: Journal of Physics: Conf. Series **1093** (2018) 011002 doi:10.1088/1742-6596/1093/1/011002

Presentation by a Presenter on Parallel Session of ICOMSE 2017

Presentation by a Presenter on Parallel Session of ICOMSE 2017

Presentation by a Presenter on Posters Session of ICOMSE 2017

Presentation by a Presenter on Posters Session of ICOMSE 2017

Peer review statement

All papers published in this volume of *Journal of Physics: Conference Series* have been peer reviewed through processes administered by the proceedings Editors. Reviews were conducted by expert referees to the professional and scientific standards expected of a proceedings journal published by IOP Publishing.

Structural, Optical, and Antifungal Characters of Zinc Oxide Nanoparticles Prepared by Sol-gel Method

Ahmad Taufiq[1,*], Habibatun Nurul Ulya[1], Joko Utomo[1], Sunaryono[1], N Hidayat[1], Hendra Susanto[2], Nandang Mufti[1], Munasir[3], Siriwat Soontaranon[4]

[1]Department of Physics, Faculty of Mathematics and Sciences, Universitas Negeri Malang, Jl. Semarang 5 Malang 65145, Indonesia.
[2]Department of Biology, Faculty of Mathematics and Sciences, Universitas Negeri Malang, Jl. Semarang 5 Malang 65145, Indonesia.
[3]Department of Physics, Faculty of Mathematics and Sciences, Universitas Negeri Surabaya, Jl. Ketintang, Surabaya 60231, Indonesia.
[4]Synchrotron Light Research Institute, Muang, Nakhon Ratchasima 30000, Thailand

*Corresponding author's email: ahmad.taufiq.fmipa@um.ac.id

Abstract. The preparation of zinc oxide nanoparticles has been successfully done by employing a sol-gel route. The x-ray diffraction data analysis presented that the zinc oxide particles crystallized as hexagonal wurtzite structure and sized in nanometric scale of 35.8 nm. The scanning electron microscopy image showed that the sample had agglomeration pattern with the particles size of 38.2 nm. Such results were confirmed by a synchrotron small-angle x-ray scattering that the prepared sample agglomerated in 3-dimensional structure with a fractal dimension of about 3. The functional group of the Zn-O as the main component of zinc oxide were observed at the wavelengths of about 454 and 523 cm^{-1}. Moreover, the optical band gap energy of the zinc oxide nanoparticles was of 3.53 eV. Excitingly, the powdered sample in this work had an average inhibition zone diameter of 4.59 mm for *Candida albicans* fungus. Therefore, the prepared zinc oxide nanoparticles exhibited imperative materials for antifungal agent.

Keywords: Zinc oxide, nanoparticle, structure, optical band gap energy, antifungal agent.

1. Introduction

A nanomaterials study to develop medical applications including antibacterial and antifungal applications has been an interest of many experts in the nanoscience and nanotechnology fields. Previous report shows that the nanomaterials having the ability to kill microorganism in the large quantity is known as "wonder of modern medicine" [1]. The high performance of nanomaterials in the application is generally related to the nanomaterials characteristics such as particle size, distribution, and morphology. Furthermore, several previous studies show that nanomaterials have been developed in recent years are able to show a prime performance for various applications especially related to the uniqueness of the catalytic [2], optical [3], electronic [4] and antimicrobial [5] properties.

Of the many nanomaterials being developed by the researchers, zinc oxide has become one of the materials that is interesting to be studied, particularly related to the biomedical application [6]. A number of several previous research showed that zinc oxide nanoparticles gave the highest catalytic efficiency, strong adsorption, and act as an antibacterial agent [7] and antifungal agent in the wastewater treatment [8,9]. Interestingly, the particle is designated as one of the many materials that are recognized as safe [10]. In relation to the development of zinc oxide particles for an antifungal application, the fundamental study that covers the nanostructural hierarchy and the optical properties become necessary to be performed, including its correlation with the antifungal performance.

2. Materials and Methods

The materials used were zinc acetate dihydrate, NaOH, H_2O, methanol, DMSO, *Sabouraud Dextrose Agar* (SDA), dan dextrose. The instruments used were the magnetic stirrer, beaker glass, pipette, filter paper, petri dish, pH paper, spatula, mortar, and oven. The synthesis of zinc oxide nanoparticles was done by a sol-gel method began by a dissolution process of *zinc acetate dihydrate* in H_2O using a magnetic stirrer, and then proceeded with a titration process with NaOH until reaching pH 13 to obtain a gel solution. Next, the gel solution was heated at the temperature of 90 °C for 30 minutes which then followed by a washing process with methanol until reaching pH 7. The next method is the filtering which was followed by the heating process at the temperature of 100 °C for 60 minutes to generate zinc oxide particle powder.

The zinc oxide particle powder was then characterized using an x-ray diffractometer (XRD), small angle x-ray scattering (SAXS), scanning electron microscopy (SEM), Fourier transform infra-red (FTIR) spectroscopy, as well as antifungal testing. The XRD, SAXS, SEM, and FTIR characterizations were performed to investigate the structural and optical characters of the sample. Moreover, the antifungal performance on the sample using *Candida albicans* fungus was investigated by means of a well-diffusion method.

3. Results and Discussion

Figure 1 presents the diffraction pattern of an x-ray for the zinc oxide nanoparticles was fitted through a refinement analysis using Rietica software. In the refinement process, the AMCSD data no 0005203 was used as the zinc oxide data model. The refinement analysis results showed that the experimental data have the highest match with the data model indicated by the statistical parameter value $Rp = 14.7$, $Rwp = 20.9$ and *chi-square* 1.8. Based on Figure 1, visually, the zinc oxide phase peaks were identified at 2-theta at angles of $(2\theta) = 31.7°$, 34.4°, 36.2°, 47.5°, 56.6°, 62.8°, 66.3°, 67.9°, 69.1°, 72.5°, and 76.9°. The diffraction patterns have similar patterns to other research that have been reported previously [11]. Generally, the zinc oxide particles have a diffraction pattern with the characteristics that tend to be high and sharp indicating that the sample has a high crystallinity. Furthermore, from the data analysis, it was not found a new phase or impurity indicating that the sample has a high purity level.

The results of the X-ray diffraction data analysis quantitatively showed that zinc oxide particles have a hexagonal wurtzite crystal structure with a lattice parameter value of $a = b = 3.251$ Å and $c = 5.211$ Å, with $\alpha = \beta = 90° \neq \gamma = 120°$. The sample of synthesis result in this research also has the crystal volume of 47.7 Å^3 with the crystallite size of 35.8 nm. This crystal structure has a *P63mc* space group indicated by two sublattices interconnected including Zn^{2+} and O^{2-}, thus each Zn ion is surrounded by tetrahedral O ions and vice versa. In addition, such *wurtzite* crystal has the value of the axial ratio of $c/a = 1.6029$ and parameter of $u = 0.379$ whose results are nearing the ideal atomic structure [12]. Based on the data analysis, the values of crystal lattice distortion and bond-length degree of Zn-O on zinc oxide particles are 1.0186 and 1.9783 respectively. Crystallographically, the crystal distortion on zinc oxide particles caused a change in the electronic bond between two nearest atoms which affected the length of Zn-O bond.

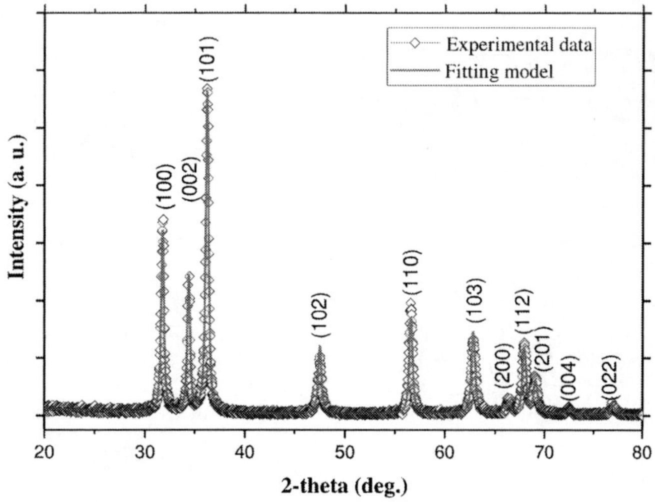

Figure 1. Experimental x-ray diffraction data and fitting model of the zinc oxide nanoparticles

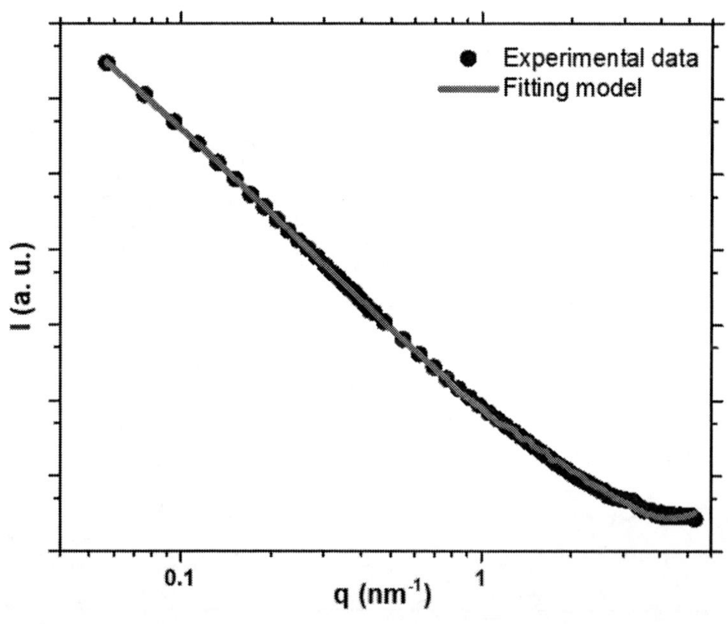

Figure 2. Scattering profile of the zinc oxide nanoparticles

A scattering was employed to determine the size, shape, distribution, and their correlation in a nanometre scale. Based on Figure 2, generally, the zinc oxide nanoparticles were consisted of primary particles forming an aggregation in fractal structure. Therefore, we applied a lognormal distribution combining with mass fractal model to fit the scattering data. The mathematical model for fitting the scattering data was modified from the mathematical model reported in previous work [13].

$$I(q) \approx \int_0^\infty N(R)F_N^2(q,R)dR\, S(q,\varepsilon,D,R) \quad (1)$$

where F is the amplitude of scattering, N is the total of particle density, $S(q,\varepsilon,D,R)$ is the mass fractal structure factor, q is the scattering vector, ε is the correlation length, D and R are the fractal dimension and particle size.

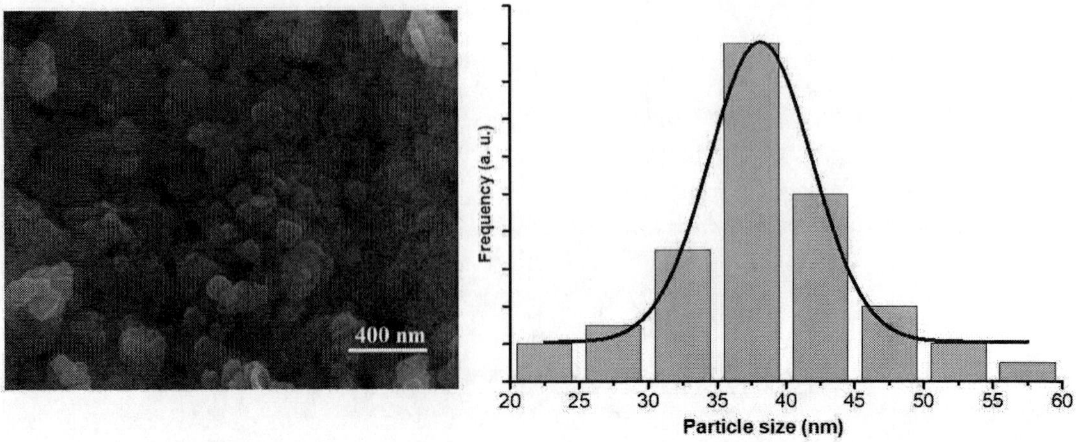

(a) **(b)**

Figure 3. (a) SEM image and (b) particle size distribution of the zinc oxide nanoparticles

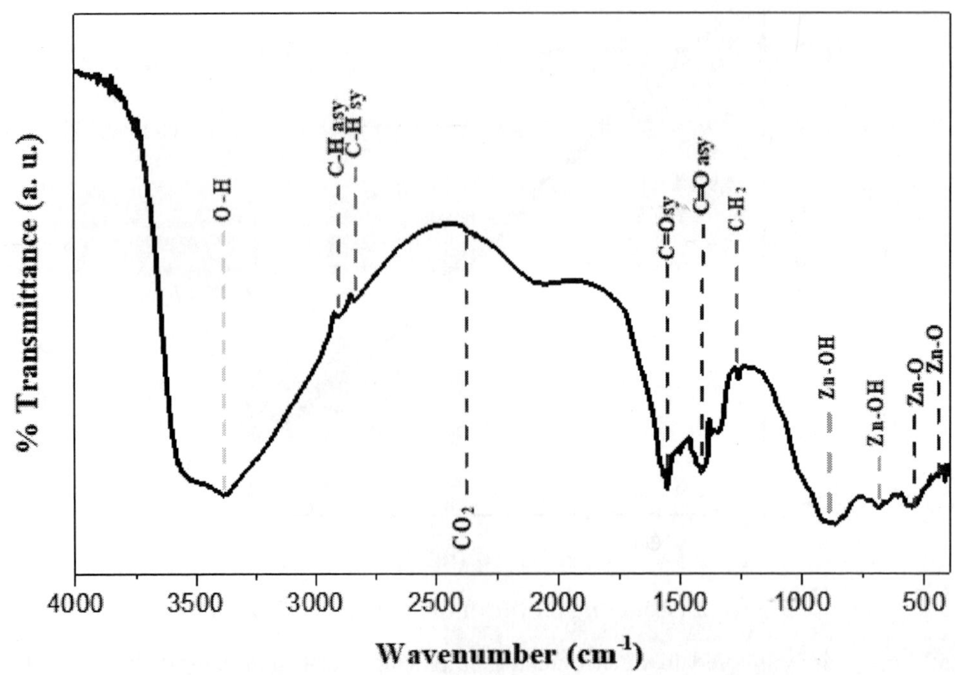

Figure 4. FTIR spectrum of the zinc oxide nanoparticles

The result of data analysis for the scattering data shows that the zinc oxide nanoparticles has a primary particles of approximately 9.14 nm constructing a bigger cluster with the size of 34.39 nm. Furthermore, the fractal dimension of the zinc oxide nanoparticles is about 2.85 which represents that the particles constructed in 3-dimensional structure. These result is in line with the results of data analysis obtained from XRD and SEM characterizations. The morphology of the zinc oxide nanoparticles visually can be seen in Figure 3. The visual results of SEM data analysis on the zinc oxide nanoparticles sample indicated that the shape of the particle tends to be spherical and slightly monodisperse, both for primary and secondary particles. However, generally, zinc oxide nanoparticles still experienced a slight agglomeration with the particle size of 38.16 nm. Interestingly, the particle size from SEM data analysis results corresponded/approached the results generated from the X-ray diffraction data analysis.

Table 1. Functional groups of the synthesized the zinc oxide nanoparticles

Wavenumber (cm^{-1}) Experimental data	**Functional groups**	**Wavenumber (cm^{-1}) References**	**References**
454 and 523	Zn-O	454 and 523	[14], [15]
690 and ~900	Zn-OH	717 and ~ 900	[16,17]
3443	O-H stretching	3400	[18]
1384 and 1630	C=O asymetris and C=O symetris	1384 and 1630	[19,20]
2376	CO_2	2376	[21]
2885 and 2937	C-H symetris and C-H asymetris	2885, 2937, 2860 and 2950	[11,22]
1268	CH_2	1050	[16]

In general, the agglomeration process was caused by the high surface energy and specific surface area on the particle of zinc oxide nanoparticles. Besides that, the agglomeration also enabled a crystal nucleation dominantly due to the higher NaOH saturation, thus producing a fast conversion of the Zn^{2+} ion to be $[Zn(OH)_4]^{2-}$ with low agglomeration directing to the smaller particle size [23]. Furthermore, the crystal growth in zinc oxide nanoparticles was caused by the chemical reaction rate and the Zn^{2+} and OH^- ion concentration. When noted on SEM figure closely, it is very apparent that there is crystal growth. The important roles which caused the crystal growth and nucleation among them are the role of ion OH^- in the solution and during the process of sample heating.

The functional group formed on the zinc oxide nanoparticles sample was reviewed using FTIR spectroscopy as shown in Figure 4. Based on Figure 4 and Table 1, it is apparent that the functional group of zinc oxide nanoparticles from the synthesis results in this research showed a similar result with the previous research. At the low wavenumber, the Zn-O functional group was formed at the wavenumbers of 454 and 523 cm^{-1}. These results showed that the zinc oxide particles have been well-configured as presented in the x-ray diffraction data. Besides that, the Zn-OH functional group appeared at the wavenumbers of 690 cm^{-1} and approximately 900 cm^{-1}. Furthermore, at the wavenumber of 3443 cm^{-1}, the hydroxyl stretching vibration of O-H often occurred. The appearance of the O-H bond showed the existence of water absorbed in the zinc oxide particles surface [24].

The carboxyl C-O bond was also identified at the wavenumber of 1384 cm^{-1} and 1630 cm^{-1} which possibly came from the reactive carbon during the synthesis process [20]. The appearance of $C\text{-}H_2$ and the symmetrical and asymmetrical C-H functional group bonds at the wavenumbers 1268, 2885, and 2937 cm^{-1} identified the existence of the monoacetate group as the intermediate product [22]. Meanwhile, the functional group of CO_2 at the wavenumber of 2376 cm^{-1} was predicted coming from the outside air. To study the other important characteristics related to the optical properties of band gap energy, the UV-Vis characterization was performed at room temperature. The characterization results showed the absorbance relation with the wavelength in the range of 200-800 nm as presented in Figure 4.

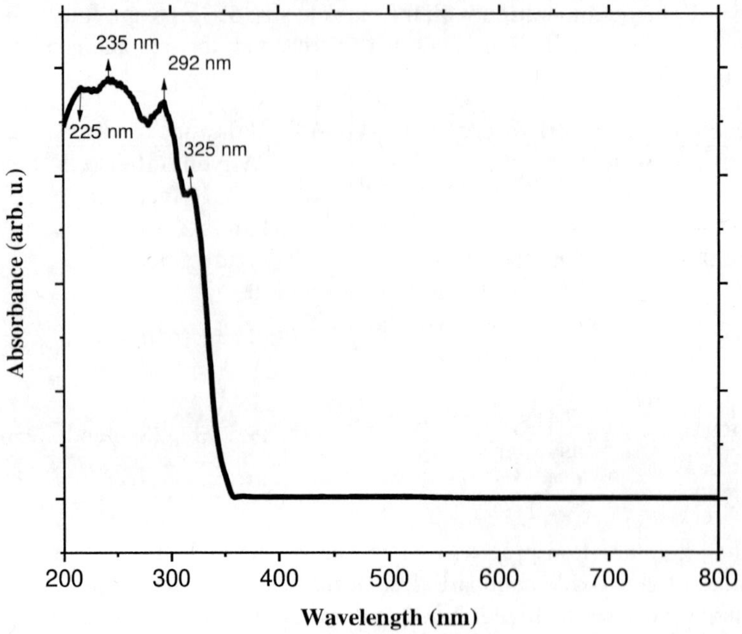

Figure 5. The relation between wavelength and absorbance of the zinc oxide nanoparticles

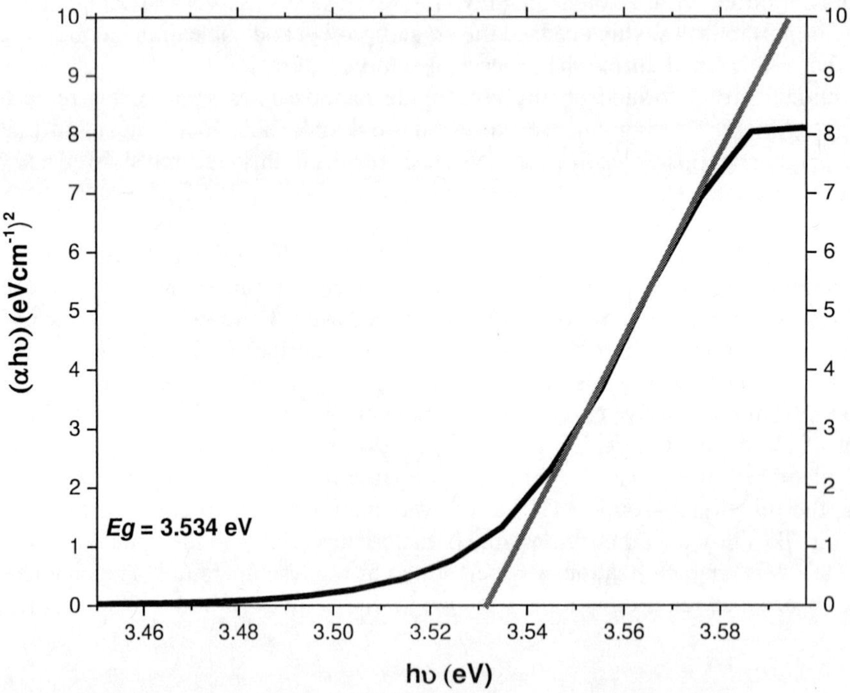

Figure 6. Band gap energy of the zinc oxide nanoparticles

Based on Figure 5, the absorption spectrum of the zinc oxide nanoparticles was identified at the wavelength of 200–380 nm. The maximum absorption peaks of zinc oxide nanoparticles were identified at the wavelength of 225, 235, 292, and 325 nm. Interestingly, the absorption of zinc oxide nanoparticles at the obtained wavelength was far lower compared to that of the absorption of zinc oxide nanoparticles with bulk size that occurred at 388 nm. The emerged absorption at the wavelength of 225 nm can be related to the inter-band transition on the metal electron from the basic level of valence band [25]. On the particle of the sample, a direct transition occurred when the electron transited from the valence band to the conduction band directly. Mathematically, the band gap energy of the zinc oxide nanoparticles was obtained from the linear fitting by using a Davis & Mott's equation as on Equation (2) [26] through the graphic relation of $h\upsilon$ and $(\alpha h\upsilon)^2$ as shown in Figure 6.

$$(\alpha h\upsilon)^{1/n} = A(h\upsilon - Eg) \qquad (2)$$

where $h\upsilon$ is the photon energy (eV) related to the material, Eg is the band gap energy, A is the contradiction constant and, n is the exponential value of ½ for direct band gap energy. Based on the *fit*ting linear results in Figure 7, it was obtained the band gap energy of 3.534 eV. This experimental data corresponds to the previous study conducted by Preethi *et al* (2016) which shows that the value of band gap energy of zinc oxide nanoparticles from the synthesis results with a sol-gel method was approximately 3.5–3.9 eV [27]. In physics, the value of band gap energy identified that the zinc oxide nanoparticles belong to semiconductor material category.

Table 2. The results of antifungal activity test on the zinc oxide nanoparticles

Treatment	**Inhibition Zone Diameter (mm)**		
	Zinc oxide nanoparticles	**Antibiotic (Positive control)**	**H_2O (Negative control)**
1	4.30	26.00	0
2	5.20	25.05	0
3	4.00	25.50	0
4	4.85	24.85	0
Average	**4.59**	**25.35**	**0**

In this research, the antifungal activity test on the zinc oxide nanoparticles was conducted using *Candida albicans* fungus through a well-diffusion method. The antifungal activity test results on zinc oxide nanoparticles (concentration of 54%) were presented in Table 2. The average value of antibiotic inhibition zone diameter (positive control) on *Candida albicans* fungus was 25.35 mm. Meanwhile for aquadest (negative control) did not show any change in the inhibition zone diameter of *Candida albicans* fungus. An adequate significant change occurred on ZnO sample with an antibiotic as the positive control. Based on the results obtained, it showed that the ZnO particles gave a significant effect in inhibiting the growth of *Candida albicans* fungus. The growth inhibition of *Candida albicans* fungus by ZnO particle was likely caused by the ZnO particle has a small size (nanometer scale) which could go through the cell walls due to the direct contact between particles with fungal membrane cells, thus it was able to inhibit the performance of fungal cells which caused the fungal growth inhibition even the fungal cell death. The results of the inhibition zone diameter are almost close to the previous study conducted by Sharma *et al.* which are about 1.5-11.4 mm [28]. Therefore, the powdered zinc oxide nanoparticles prepared in this work becomes imperative materials to be applied as antifungal agent.

4. Conclusion

We conclude that the ZnO sample was successfully fabricated using a sol-gel method. The crystal structure of the ZnO sample was in hexagonal wurtzite and crystallized in nanometric scale of 35.8 nm. Furthermore, the visual morphology if the ZnO sample identified a spherical particle shape with the particle size distribution formed by several crystallites of 38.16 nm. The functional group of the ZnO sample was identified at the wavenumber of 454 and 523 cm^{-1}. The optical properties of ZnO showed that the sample has a gap energy value of 3.534 eV. Meanwhile, antifungal activity of the ZnO sample has an inhibit zone diameter of 4.59 mm.

Acknowledgments
This work is supported by research grant from PNBP Universitas Negeri Malang for AT.

References

[1] Janaki A C, Sailatha E and Gunasekaran S 2015 Synthesis, characteristics and antimicrobial activity of ZnO nanoparticles *Spectrochim. Acta. A. Mol. Biomol. Spectrosc.* **144** 17–22
[2] Zhou Y, Jin C, Li Y and Shen W 2018 Dynamic behavior of metal nanoparticles for catalysis *Nano Today* **20** 101–20
[3] Jayoti D, Malik P and Prasad S K 2018 Effect of ZnO nanoparticles on the morphology, dielectric, electro-optic and photo luminescence properties of a confined ferroelectric liquid crystal material *J. Mol. Liq.* **250** 381–7
[4] Chapel A, Dkhil S B, Therias S, Gardette J-L, Hannani D, Poize G, Gaceur M, Shah S M, Wong-Wah-Chung P, Videlot-Ackermann C, Margeat O, Rivaton A and Ackermann J 2016 Effect of ZnO nanoparticles on the photochemical and electronic stability of P3HT used in polymer solar cells *Sol. Energy Mater. Sol. Cells* **155** 79–87
[5] Gao W, Chen Y, Zhang Y, Zhang Q and Zhang L 2018 Nanoparticle-based local antimicrobial drug delivery *Adv. Drug Deliv. Rev.* **127** 46–57
[6] Barui A K, Kotcherlakota R and Patra C R 2018 Biomedical applications of zinc oxide nanoparticles *Inorganic Frameworks as Smart Nanomedicines* (Elsevier) pp 239–78
[7] Applerot G, Lipovsky A, Dror R, Perkas N, Nitzan Y, Lubart R and Gedanken A 2009 Enhanced Antibacterial Activity of Nanocrystalline ZnO Due to Increased ROS-Mediated Cell Injury *Adv. Funct. Mater.* **19** 842–52
[8] Singh S, Barick K C and Bahadur D 2013 Fe_3O_4 embedded ZnO nanocomposites for the removal of toxic metal ions, organic dyes and bacterial pathogens *J. Mater. Chem. A* **1** 3325
[9] Sharma D, Rajput J, Kaith B S, Kaur M and Sharma S 2010 Synthesis of ZnO nanoparticles and study of their antibacterial and antifungal properties *Thin Solid Films* **519** 1224–9
[10] Romaña D L, Brown K H and Guinard J-X 2002 Sensory Trial to Assess the Acceptability of Zinc Fortificants Added to Iron-fortified Wheat Products *J. Food Sci.* **67** 461–465
[11] Zandi S, Kameli P, Salamati H, Ahmadvand H and Hakimi M 2011 Microstructure and optical properties of ZnO nanoparticles prepared by a simple method *Phys. B Condens. Matter* **406** 3215–8
[12] Jagadish C and Pearton S J 2011 *Zinc oxide bulk, thin films and nanostructures: processing, properties, and applications* (Elsevier)
[13] Taufiq A, Putra E G R, Okazawa A, Watanabe I, Kojima N and Pratapa S 2015 Nanoscale Clustering and Magnetic Properties of $Mn_xFe_{3-x}O_4$ Particles Prepared from Natural Magnetite *J. Supercond. Nov. Magn.* **28** 2855–2863
[14] Hasanpour A, Niyaifar M, Asan M and Amighian J 2013 Synthesis and characterization of Fe_3O_4 and ZnO nanocomposites by the sol–gelmethod *J. Magn. Magn. Mater.* **334** 41–4
[15] Wahab R, Ansari S G, Kim Y S, Seo H K, Kim G S, Khang G and Shin H-S 2007 Low temperature solution synthesis and characterization of ZnO nano-flowers *Mater. Res. Bull.* **42** 1640–8
[16] Kołodziejczak-Radzimska A, Markiewicz E and Jesionowski T 2012 Structural Characterisation of ZnO Particles Obtained by the Emulsion Precipitation Method *J. Nanomater.* **2012** 1–9
[17] Top A and Çetinkaya H 2015 Zinc oxide and zinc hydroxide formation via aqueous precipitation: Effect of the preparation route and lysozyme addition *Mater. Chem. Phys.* **167** 77–87
[18] Zhao L-H, Zhang R, Zhang J and Sun S-Q 2012 Synthesis and characterization of biocompatible ZnO nanoparticles *CrystEngComm* **14** 945–50
[19] Barreto G P, Morales G and Quintanilla M L L 2013 Microwave Assisted Synthesis of ZnO Nanoparticles: Effect of Precursor Reagents, Temperature, Irradiation Time, and Additives on Nano-ZnO Morphology Development *J. Mater.* **2013** 1–11
[20] Xiong G, Pal U, Serrano J G, Ucer K B and Williams R T 2006 Photoluminesence and FTIR study of ZnO nanoparticles: the impurity and defect perspective *Phys. Status Solidi C* **3** 3577–

81

[21] Hlaing Oo W M, McCluskey M D, Lalonde A D and Norton M G 2005 Infrared spectroscopy of ZnO nanoparticles containing CO[sub 2] impurities *Appl. Phys. Lett.* **86** 073111

[22] Vafaee M and Ghamsari M S 2007 Preparation and characterization of ZnO nanoparticles by a novel sol–gel route *Mater. Lett.* **61** 3265–8

[23] Koutu V, Shastri L and Malik M M 2016 Effect of NaOH concentration on optical properties of zinc oxide nanoparticles *Mater. Sci.-Pol.* **34**

[24] Farrokhi M, Hosseini S-C, Yang J-K and Shirzad-Siboni M 2014 Application of ZnO–Fe_3O_4 Nanocomposite on the Removal of Azo Dye from Aqueous Solutions: Kinetics and Equilibrium Studies *Water. Air. Soil Pollut.* **225**

[25] Kumar H and Rani R 2013 Structural characterization of silver nanoparticles synthesized by micro emulsion route *Int J Engg Inno Technol* **3** 344–348

[26] Viezbicke B D, Patel S, Davis B E and Birnie D P 2015 Evaluation of the Tauc method for optical absorption edge determination: ZnO thin films as a model system: Tauc method for optical absorption edge determination *Phys. Status Solidi B* **252** 1700–10

[27] Preethi S, Anitha A and Arulmozhi M 2016 A Comparative Analysis of the Properties of Zinc Oxide (ZnO) Nanoparticles Synthesized by Hydrothermal and Sol-Gel Methods *Indian J. Sci. Technol.* **9**

[28] Sharma R K and Ghose R 2015 Synthesis of zinc oxide nanoparticles by homogeneous precipitation method and its application in antifungal activity against Candida albicans *Ceram. Int.* **41** 967–75

IOP Conf. Series: Journal of Physics: Conf. Series **1093** (2018) 012002 doi:10.1088/1742-6596/1093/1/012002

BCR Sequential Leaching for Geochemical Fractions and Assessment of Fe, Ni, and Mn in the Coastal Sediments Sendang Biru Port, East Java, Indonesia

Anugrah Ricky Wijaya[1,*], Ida Farida[1], Surjani Wonorahardjo[1], Yudhi Utomo[1], Hasan Daupor[2], Md. Sazzad Hossain[3], Tatsuya Kunisue[4]

[1]Department of Chemistry, Faculty of Mathematics and Natural Sciences, Universitas Negeri Malang, Jl. Semarang 5 Malang 65145, Indonesia
[2]Chemistry Major, Faculty of Science Technology and Agriculture, Yala Rajabhat University, Thailand
[3]Department of Aquaculture, Faculty of Fisheries, Bangladesh Agricultural University (BAU), Bangladesh
[4]Environmental Chemistry and Ecotoxicology, Center for Marine Environmental Studies (CMES), Ehime University, Japan

*Corresponding author's email: anugrah.ricky.fmipa@um.ac.id

Abstract. The changes in the aquatic environment by either natural or anthropogenic sources may affect the presence of heavy metals in the sediment of Sendang Biru Port. Heavy metals such as Fe, Ni, and Mn possibly give a negative impact on ecological and biological systems in the sea water. To monitor the status of sediment of the Port Sendang Biru, we investigated the geochemical fraction of Fe, Ni, and Mn in sediment using the BCR (European Community Bureau of Reference) sequential leaching method. The precision test of BCR method showed that Fe, Ni, and Mn has 4.82; 3.73; and 3.16 Relative Standard Deviation (RSD), respectively. The accuracy test of BCR method showed that percentage recovery for Fe, Ni, and Mn were 85.5%; 96.2%; and 63.0%, respectively. The fractionation of Fe, Ni, and Mn in the sediment of Sendang Biru Port showed that Fe was dominant in residual fraction (90.7%), whereas Ni and Mn were dominant in non-residual fraction (78.8% Ni and 77.2% Mn). The dominance of Fe in the residual fraction indicated that Fe was from natural sediment minerals bound to silicate and aluminate, whereas the dominance of Ni and Mn in non-residual fractions was possibly derived from anthropogenic effects. Assessment from index calculation of sediment had the category contaminated at moderate levels by Ni and contaminated at low levels by Fe and Mn.

Keywords: Sediment, metal fractionation, BCR, sequential leaching

1. Introduction

Sendang Biru Port is one of the major ports in East Java. This Port has about 3000 fishermen in Sendang Biru with fish catch reached 11500 tons per year. Some human's activities (anthropogenic) that occurred surrounding Sendang Biru Port are possibly contributed by fishing boat activities, waste disposal from the fish market and domestic waste disposal causing the heavy metal contamination in sediment.

Sediment is the final site of an accumulated heavy metal compound in the aquatic environments [1]. Heavy metals (Fe, Ni, and Mn) in dominant sediment may have a toxic effect on the human being.

Some researchers used the method of partial leaching to assess Fe, Ni, and Mn in sediment and coral [1–4]. This method is only for the first assessment of that metals and cannot explain their concentrations in their sediment fraction. The fraction of metal binding in sediment is very important to understand metals mobility which can affect ecological and biological systems in the aquatic environment surrounding the Port. To understand the geochemical fractions, we used sequential leaching for Fe, Ni, and Mn in sediment to know each of fraction mobility which can influence environmental systems.

One of sequential leaching methods that is commonly used is the BCR (European Community Bureau of Reference) method. The BCR method fractionates metals into four fractions comprising exchangeable and easily soluble fraction (F1), easily reducible fraction (F2), oxidizable fraction (F3) and residual fraction (F4). F1, F2, and F3 easily indicate heavy metal mobility in the sediment or non-residual fractions, whereas F4 represents a difficult fraction mobilized [5]. Here, we report and assess using BCR method to leach Fe, Ni, and Mn in port sediment fractions.

2. Experimental Methods

2.1. Sample Collection and Pretreatment for Analysis

Figure 1 shows the collected samples of sediment in the center of Sendang Biru Port. The samples were collected on November 12, 2016 using a stainless-steel grab, put in the plastic bag, and then transferred to the Analytical Chemistry Laboratory.

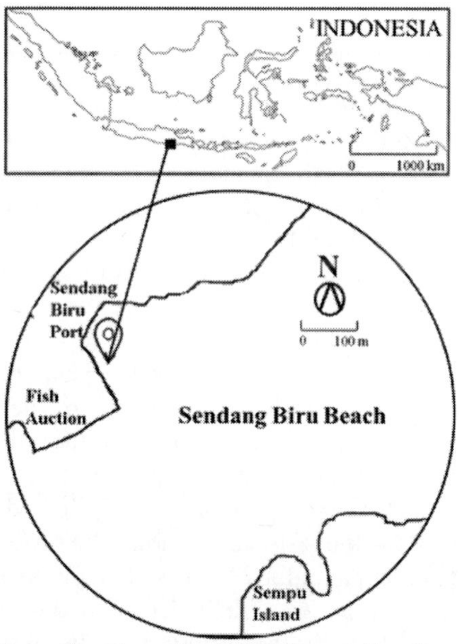

Figure 1. Sampling location of sediment Sendang Biru Port.

The sediment samples are fine grains and composed of mud and silt. The samples were separated from rock, bone, corals, and wood. The sediment samples were homogenized and then dried in an oven at 60 °C [3]. The qualitative metal composition of representative sediment samples was determined using X-ray Fluorescence. The samples were powdered by hand using an agate mortar and pestle. The instrument used was a PANalytical, Minipal 4. The chemical substances in sediment are listed in Table 1.

Table 1. Chemical substances in sediment

Metal	Concentration (%)
Al	6+/-0.6
Si	5.8+/-0.3
K	0.3+/-0.2
S	0.2+/-0.6
Ca	57.9+/-0.003
Ti	1.1+/-0.6
V	0.05+/-0.009
Cr	0.079+/-0.004
Mn	0.17+/-0.004
Fe	18.8+/-0.2
Ni	0.21+/-0.004
Cu	0.15+/-0.005
Zn	0.02+/-0.005
Br	0.21+/-0.01
Sr	4.60+/-0.12
Zr	0.2+/-0.2
Mo	3.28+/-0.94
In	0.8+/-1.0
Ba	0.1+/-0.03
Yb	0.08+/-0.007
Re	0.38+/-0.04

2.2. The Sequential Leaching Experiment

The sample sediments from Sendang Biru Port and reference sediment of Geochemical Survey of Japan (JSD-1) were analyzed for Fe, Ni, and Mn contents in fraction sediment using BCR method. The determination of metal concentration in the reference sediment was used to control the quality of data through precision and accuracy tests, while the determination of metal concentration in Sendang Biru Port's sediment was aimed to find out the concentration of its geochemical fraction and status of Fe, Ni, and Mn.

Analysis of Fe, Ni, and Mn in the sediment consisted of four stages: (1) exchangeable and easily soluble fractions. 0.5 grams of sediment samples were leached with 20 mL of acetic acid solution (0.11 M) for 16 hours; (2) the easily reducible fraction. The residue of the exchangeable and easily soluble fraction was leached with 20 mL of hydroxylamine hydrochloride solution (0.1 M; pH 2) for 16 hours; (3) the oxidizable fraction. The residue from the easily reducible fraction was leached using 5 mL of hot hydrogen peroxide (30%) to nearly dry twice, then leaching followed by 25 mL of ammonium acetate solution (1.0 M; pH 2) for 16 hours; (4) the residual fraction. The residue from the oxidized fraction was leached with 5 mL of concentrated nitric acid for 24 hours, then put in a sonic bath for 30 min and heated to 110 °C. After drying, the residue was added with 20 drops of concentrated nitric acid and 10 drops of fluoride acid, then heated until dry. Finally, the residue was dissolved and diluted with 2 mL of 1% HNO_3 five times [3,6]. The leached sample was then analyzed using an AAS (PerkinElmer A Analyst 700) to determine the concentration of Fe, Ni, and Mn of each fraction in sediment.

3. Results and Discussion

3.1. The BCR Method for Leaching Fe, Ni, and Mn in Sediment

The effectiveness of BCR method in leaching Fe, Ni, and Mn in sediment was tested. This test aimed to ensure that the method is able to produce the correct data which are in accordance with its designation. Testing parameters of the effectiveness of this method are the precision and accuracy.

The results of precision test of BCR method for leaching Fe, Ni, and Mn using RSD are listed in Table 2. The leached Fe, Ni, and Mn has relative standard deviation (RSD) of 4.82, 3.73, and 3.16, respectively. The test of BCR method indicated high precision for leaching Fe, Ni, and Mn which has RSD less than 5%. Therefore, it can be stated that the BCR is a precision method for leaching Fe, Ni, and Mn in Sendang Biru Port sediment.

The accuracy test of BCR method for leaching Fe, Ni, and Mn showed the percentage recovery for Fe (85.5%), Ni (96.2%), and Mn (63.0%) (Table 2). The range of percentage recovery for standard accuracy was 85-110%. This suggests that the applied BCR method is accurate for leaching Fe and Ni, but is less accurate for leaching manganese in sediments.

The low accuracy of Mn obtained using BCR method indicated that Mn was not able to be perfectly leached by this method. According to [7], the low Mn accuracy is due to the inability of the solvent used in the second fraction (F2) of the BCR method for leaching MnO. The low Mn accuracy can also be affected by disturbance of sedimentary matrices such as Cu, Ni, Ca, and Fe [8].

Table 2. The precision and accuracy tests of the BCR method

Metal	RSD	Percentage Recovery (%)
Fe	4.82	85.4
Ni	3.73	96.2
Mn	3.16	62.3

The precision and accuracy tests of BCR method (Table 2) showed that the BCR method is capable of leaching Fe and Ni precisely and accurately, but it is not effective for leaching Mn accurately. This suggests that the BCR method is feasible to be used for leaching Fe and Ni, but less feasible to be used for leaching Mn in Sendang Biru Port sediment.

3.2. Fractionations of Fe, Ni, and Mn Concentrations in Sediment

Metal fractionation in sediments reflects the mechanism of metal binding in sediments. The difference of metal binding in these sediments shows the difference of metal source. Generally, anthropogenic metals are presented in non-residual fractions, whereas metals derived from natural sedimentary materials are in residual fractions [9]. BCR method breaks metals binding in the sediment into four fractions, namely (1) exchangeable and easily soluble fraction (F1), (2) easily reducible fraction (F2), (3) oxidizable fraction (F3), and (4) residual fraction (F4). F1, F2, and F3 are referred to a non-residual fraction.

Three dominant metals are suspected to be anthropogenic contaminants surrounding the port (Table 1). The results of Fe, Ni, and Mn fractionation are illustrated in Figure 2. Based on the Figure, it is known that Fe is dominant in the residual fraction while Ni and Mn are dominant in the non-residual fraction.

As shown in Figure 2 and Table 3, the contents of Fe in ion exchangeable and soluble, easily reducible, and oxidizable fractions are 26.2(104) mg/Kg (1.51%), 25.0(104) mg/Kg (1.44%), and 110(104) mg/Kg (6.33%), respectively. The total contents of Fe in this three fractions indicated the Fe binding in the non-residual fraction in sediment. These fractions may be contributed from anthropogenic source surrounding Sendang Biru Port.

Table 3. Metal contents of Fe, Ni, and Mn in sediment fraction

Metal	Metal Contents (mg/kg)		
	Fe (104)	Ni	Mn
F1	26.2	23.6	69.8
F2	25.0	22.4	58.1
F3	110	22.9	58.6
F4	1572	18.5	55.3

IOP Conf. Series: Journal of Physics: Conf. Series **1093** (2018) 012002 doi:10.1088/1742-6596/1093/1/012002

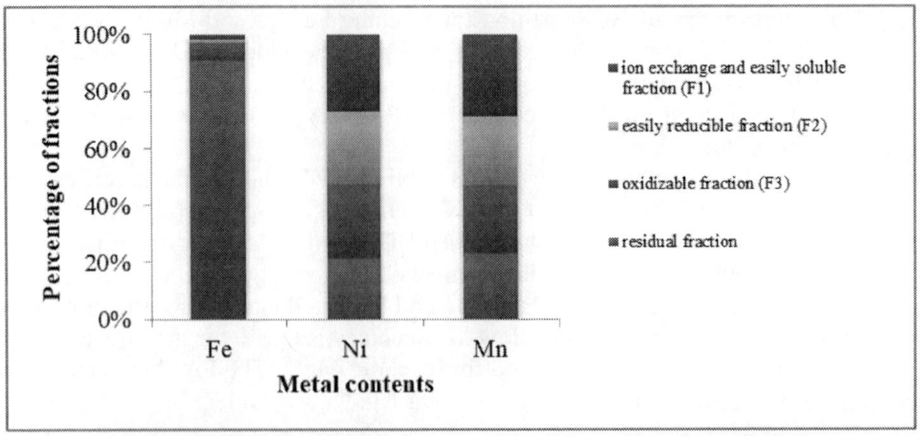

Figure 2. Geochemical fractions of Fe, Ni, and Mn in sediment

The concentration of Fe in residual fraction was 1572(104) mg/Kg (90.7%) (Table 3; Figure 2). The dominance of Fe in residual fractions suggests that Fe in the sediment of Sendang Biru Port is strongly binding with silicate and alumina in natural mineral sediments [10]. The high concentration of Fe in the residual fraction indicated that the sediment was relatively uncontaminated by Fe and it was not harmful to the ecological and biological systems in the waters. The possible chemical reaction in this fraction can be written as follow [3]:

$$\text{Sediment–}SiO_2\text{–}Al_2O_3\text{–Fe(III)}(s) + HNO_3(aq) + HF(aq) \rightarrow Fe(NO_3)_3(aq) + Al(NO_3)_3(aq) + SiF_4(g) + H_2O(l) + H_2(g)$$

$$\text{Sediment–}SiO_2\text{–}Al_2O_3\text{–Fe(II)}(s) + HNO_3(aq) + HF(aq) \rightarrow Fe(NO_3)_2(aq) + Al(NO_3)_3(aq) + SiF_4(g) + H_2O(l) + H_2(g)$$

Fractionation of Ni and Mn in the sediment of Sendang Biru Port shows that both metals are dominant in non-residual fractions (Figure 2). The concentrations of metals in non-residual fraction were 18.53 mg/kg (78.8%) and 186 mg/kg (77.2%), while in residual fraction were 18.5 mg/kg (21.2%) and 55.3 mg/kg (22.9%) (Table 3, Figure 2). These results indicated that Ni and Mn had a little bit similar percentages. The high concentrations of Ni and Mn in non-residual fractions suggested that Ni and Mn were contributed from anthropogenic process surrounding Sendang Biru Port. Ni and Mn in non-residual fractions may be released from sediment when there is a change in the aquatic environment so that it affects the ecological and biological systems in Sendang Biru Port.

The higher concentrations of Ni and Mn in exchangeable and easily soluble fraction were 23.6 and 69.8 mg/kg, respectively (Table 3). The high concentration of Ni and Mn at the most unstable fraction is due to the input of the resident's waste surrounding Sendang Biru Port. Metals in this fraction can impact on ecological and biological systems. These metals are very weakly associated with negative organic or inorganic binding site in the sediment. The presence of a decrease in pH in the aquatic environment can cause Ni and Mn in this fraction dissolved into the surface water so that it can cause the negative impact on the environment.

The high concentrations of Ni and Mn were also found in the oxidizable fraction (22.9 and 58.6 mg/kg) (Table 3). Ni and Mn in this fraction form a stable complex with humic acid (HA) derived from organic matter. The high concentrations of Ni and Mn in this fraction may be contributed from anthropogenic inputs around Sendang Biru Port such as the disposal of fish market waste or domestic waste. This study supports previous similar studies reported by Passos et al. 2010 [9] for suggesting input of human's activity surrounding the site. Ni and Mn can be released into waters when the process degradation of organic matter occurs.

The concentrations of Ni and Mn were found in the easily reducible fraction (22.4 and 58.1 mg/kg) (Table 3). The presence of Ni in this fraction is bound to Fe-Mn oxide, whereas Mn in this fraction is derived from Mn bound to Fe and Mn oxides in its oxidant form [11]. Ni and Mn at a reduced fraction may be due to the activity of fishing vessels and discharges from industry [3]. Ni and Mn contained in this fraction can be released into the waters when the aquatic environment becomes anoxic [12].

The possible chemical reaction of Ni and Mn in non-residual fraction can be written as follows, (which Me = Ni or Mn) [3]:

1. Ion exchangeable and soluble fraction
   Sediment–Me^{2+}*(s)* + CH_3COOH*(aq)* ⇄ Sediment–$(H)_2$*(s)* + $(CH_3COO)_2Me$*(aq)*
   Sediment–$MeCO_3$*(s)* + CH_3COOH*(aq)* ⇄ Sediment*(s)* +$(CH_3COO)_2Me$*(aq)* +H_2CO_3*(aq)*

2. Easily reducible fraction
   Sediment–MnO_2–Me*(s)* + $NH_2OH.HCl$*(aq)* → Sediment*(s)* + $MnCl_2$*(aq)* + $MeCl_2$*(aq)* + NH_3*(g)* + H_2O*(l)* + O_2*(g)*

3. Oxidizable fraction
   Sediment–HA–Me*(s)* + H_2O_2*(aq)* + CH_3COONH_4*(aq)*→ Sediment–HA*(s)* + $(CH_3COO)_3Fe$*(aq)* + NH_3*(g)* + H_2O*(l)*+ O_2*(g)*

The lowest concentration of Ni and Mn were in residual fraction 18.5 mg/kg and 55.3 mg/kg, respectively (Table 3). This is probably related to the abundance of Ni and Mn in the earth's crust, including sediment in Sendang Biru Port. Metals in the residual fraction are strongly bound to natural minerals and resistant components of the solid matrix in sediment so that they are difficult to mobilize into pore-waters through dissociation. The possible chemical reaction in this fraction can be written as follow, (where Me = Ni or Mn) [3]:

Sediment–SiO_2–Al_2O_3–Me*(s)* + HNO_3*(aq)* + HF*(aq)* →$Me(NO_3)_2$*(aq)* + $Al(NO_3)_3$*(aq)* + SiF_4*(g)* + H_2O*(l)* + H_2*(g)*

3.3. Assessment of Fe, Ni, and Mn in Sediment

Assessment of sediment status of Sendang Biru Port is used to know the level of contamination of Fe, Ni, and Mn in the sediment associated with water pollution. The determination of this status is based on the calculation of the contamination factor (CF), Geo-accumulation Index (Igeo), and Pollution Load Index (PLI) [2,13]. The calculations of CF, Igeo, and PLI values are respectively based on the mathematical equations expressed by some researchers. CF and Igeo are used to determine the level of contamination of each heavy metal (Fe, Ni, or Mn) in sediment, whereas PLI is used to determine the level of total heavy metal contamination (Fe, Ni, and Mn) in sediment [13–15].

Table 4. Assessment of Fe and Mn in Sediment

Sampling site	CF		I_{geo}	
	Fe	Mn	Fe	Mn
1	1.94	1.52	0.18	0.02
2	1.74	1.08	0.16	-0.32
3	1.38	1.57	-0.13	0.07
4	2.12	1.36	0.28	-0.10
5	0.36	3.37	-2.06	1.17
6	0.65	1.13	-1.21	-0.17
7	1.00	1.00	-1.69	-0.18
8	1.87	1.42	0.51	-0.06

As listed in Table 4, CF calculations showed that the sediment of Sendang Biru Port has been contaminated by Fe, Ni, and Mn with a moderate category for Ni and low category for Fe and Mn. In case of Igeo calculation, the sediment was contaminated in the low to moderate category by Ni, whereas

Fe and Mn were contaminated in low category low (Table 4). Based on the calculation, the PLI value was 0.95 (Table 4). These results indicated that the sediment of Sendang Biru Port was low contaminated by Fe, Ni, and Mn. The low contaminated sediment can improve the quality of seawater and its productivity such as increasing fish, lobster, and shrimp (Chitosan) [16].

4. Conclusion

The BCR method was effective for leaching Fe and Ni but was less effective for leaching Mn in sediments. The BCR method was capable of leaching Fe and Ni precisely and accurately but was not effective for leaching Mn accurately. Fe dominant was in residual fraction, while Ni and Mn were dominant in a non-residual fraction. The dominance of Fe in the residual fraction indicated Fe was strongly binding with silicates and alumina in the natural minerals of the sediment. Ni and Mn were dominant in the non-residual fraction indicating that they were contributed from anthropogenic around the Sendang Biru Port. Assessment from the index calculation of sediment had a moderate category contaminated by Ni and a low category contaminated by Fe and Mn.

Acknowledgments

This research is supported by KemenristekDikti, Fundamental Grant Scheme (DIPA Number 042.06-0/2016), Indonesian Government.

References

[1] Wijaya A R, Ouchi A K, Tanaka K, Shinjo R and Ohde S 2012 Metal contents and Pb isotopes in road-side dust and sediment of Japan *J. Geochem. Explor.* **118** 68–76

[2] Wijaya A R, Ouchi A K, Tanaka K, Cohen M D, Sirirattanachai S, Shinjo R and Ohde S 2013 Evaluation of heavy metal contents and Pb isotopic compositions in the Chao Phraya River sediments: Implication for anthropogenic inputs from urbanized areas, Bangkok *J. Geochem. Explor.* **126–127** 45–54

[3] Wijaya A R, Ohde S, Shinjo R, Ganmanee M and Cohen M D 2016 Geochemical fractions and modeling adsorption of heavy metals into contaminated river sediments in Japan and Thailand determined by sequential leaching technique using ICP-MS *Arab. J. Chem.*

[4] Sazzad M H, Wijaya A R, Tanaka K and Ohde S 2010 Environmental effects on the stable carbon and oxygen isotopic compositions and skeletal density banding pattern of Porites coral from Khang Khao Island, Thailand *Afr. J. Biotechnol.* **9**

[5] Sahara E, Widihati I A G and Putra I G D 2015 Fraksinasi dan bioavailabilitas logam berat Fe dan Mn pada sedimen di Pelabuhan Benoa *J. Kim.* **9** 124–31

[6] Mester Z, Cremisini C, Ghiara E and Morabito R 1998 Comparison of two sequential extraction procedures for metal fractionation in sediment samples *Anal. Chim. Acta* **359** 133–42

[7] Martínez-Fernández M, Barciela-Alonso M C, Moreda-Piñeiro A and Bermejo-Barrera P 2011 Matrix solid phase dispersion-assisted BCR sequential extraction method for metal partitioning in surface estuarine sediments *Talanta* **83** 840–9

[8] Davidson C M, Thomas R P, McVey S E, Perala R, Littlejohn D and Ure A M 1994 Evaluation of a sequential extraction procedure for the speciation of heavy metals in sediments *Anal. Chim. Acta* **291** 277–86

[9] Passos E de A, Alves J C, dos Santos I S, Alves J do P H, Garcia C A B and Spinola Costa A C 2010 Assessment of trace metals contamination in estuarine sediments using a sequential extraction technique and principal component analysis *Microchem. J.* **96** 50–7

[10] Kartal Ş, Aydın Z and Tokalıoğlu Ş 2006 Fractionation of metals in street sediment samples by using the BCR sequential extraction procedure and multivariate statistical elucidation of the data *J. Hazard. Mater.* **132** 80–9

[11] Tokalioğlu Ş, Kartal Ş and Elçi L 2000 Determination of heavy metals and their speciation in lake sediments by flame atomic absorption spectrometry after a four-stage sequential extraction procedure *Anal. Chim. Acta* **413** 33–40

[12] Li X and Thornton I 2001 Chemical partitioning of trace and major elements in soils contaminated by mining and smelting activities *Appl. Geochem.* **16** 1693–706

[13] Mulyaningsih T R, Alfian A and Sutisna S 2015 Distribusi Logam Berat Dalam Sedimen Daerah Aliran Sungai Ciujung Banten *J. Teknol. Reakt. Nukl. Tri Dasa Mega* **14**

[14] Muller G 1969 Index of geoaccumulation in sediments of the Rhine River *Geojournal* **2** 108–18

[15] Tomlinson D L, Wilson J G, Harris C R and Jeffrey D W 1980 Problems in the assessment of heavy-metal levels in estuaries and the formation of a pollution index *Helgoländer Meeresunters.* **33** 566–75

[16] Ariadi Lusiana R, Putri Protoningtyas W, Ricky Wijaya A, Siswanta D, Mudasir M and Juari Santosa S 2017 Chitosan-Tripoly Phosphate (CS-TPP) Synthesis Through Cross-linking Process: the Effect of Concentration Towards Membrane Mechanical Characteristic and Urea Permeation *Orient. J. Chem.* **33** 2913–9

Effect of ZnO and Annealing on the Hydrophobic Performance of x(ZnO)-CA-PLA

Suci E I Suryani[1], Ulwiyatus Sa'adah[1], Wahida N L Amini[1], and Thathit Suprayogi[1], Ayu A Mustikasari[1], Ahmad Taufiq[1,2], Sunaryono[1,2], Markus Diantoro[1,2,*], Hadi Nur[3]

[1]Department of Physics, Universitas Negeri Malang, Faculty of Mathematics and Natural Sciences, Jl. Semarang 5 Malang 65145, Indonesia
[2]Central Laboratory of Mineral and Advanced Materials, Universitas Negeri Malang, Jl. Semarang 5 Malang 65145, Indonesia
[3]Centre for Sustainable Nanomaterials, Ibnu Sina Institute for Scientifc and Industrial Research, Universiti Teknologi Malaysia, Skudai, Johor, Malaysia

*Corresponding author's email: markus.diantoro.fmipa@um.ac.id

Abstract. The advantages of hydrophobic surface are keeping the surface always dry, clean, and reducing fluid friction. The maintenance cost could be reduced significantly due to its naturally self-cleaning properties. Although many studies have been performed, the use of metal oxide combined cellulose acetate PLA has rarely informed. The purpose of this research was to fabricate a hydrophobic coating on a glass surface of ZnO-CA-PLA with various compositions of ZnO nanoparticles (ZnO NPs), annealing temperature, the temperature of measuring condition, and time exposure on its hydrophobicity which represented by the contact angle. The film was prepared using the spin-coating method. The crystallinity and morphology of ZnO-CA-PLA films were characterized using XRD, SEM-EDAX, and FTIR. The contact angle of hydrophobicity was measured using a digital microscope. It was revealed that the ZnO-CA-PLA composite film shows an excellent hydrophobic character. The increase of ZnO NPs gives rise to increase contact angle. A similar trend also achieved by increasing of annealing time. In contrary, the contact angle of hydrophobicity may reduce by prolonged exposure time and the increase of measuring temperature.

Keywords: ZnO-CA-PLA, dopant, annealing, time exposure, hydrophobic

1. Introduction

Biodegradable polymers are an area of interest for scientists to develop. PLA has attracted the considerable attention of many research groups due to produce from renewable resources, which make them biodegradable and compostable [1]. Poly Lactic acid (PLA) as a biodegradable material with excellent mechanical and biocompatibility properties, but it shows weak hydrophobic, low thermal stability, cannot withstand high temperatures, and sensitive to moisture [2].

Many efforts have been reported on the work of hydrophobic materials. Some previous researchers also use one or two components of ZnO-CA-PLA. The research studies of PLA biocomposites coupled with cellulose nanowhiskers show an increase in temperature stability [3,4]. CA has a good potential for electronic application with specific modifications such as inducing metal oxide [5]. Some elements in the transition group such as Mn, Fe, Cu, Zn, Pt, Au, Ag has been used to modify the properties of the material [6]. The research studies of metal oxide additions such as ZnO show a change of morphology, structures, increased hydrophobic properties, antibacterial used as applications in medical [7,8]. ZnO as a semiconductor material that has unique electrical properties, optical and biocompatible can be used as a dopant in preparing a composite material. It is found that the biocomposite films are characterized by a good dispersion of the ZnO particles in the PLA matrix. It was also rarely reported that no pre-treatment was performed on ZnO particles, such as silanization, to decrease its incompatibility with the polymer, excellent mechanical properties, decrease of permeability to CO_2 and O_2 [2], [9].

We propose to optimize the use of three materials consist of ZnO-CA-PLA with various composition, annealing temperature and time exposure. It is expected to achieve superior properties and excellent performance on the films for the application. This research was conducted to obtain the information of the effect of ZnO dopant and annealing on the performance of ZnO-CA- PLA hydrophobic.

A surface with hydrophobic properties can keep the surface dry, clean, reducing fluid friction with the surface. When it applied to the glass, then it can clean by itself naturally (self-cleaning), so the cost of maintenance is cheaper. The potential of hydrophobic films produced by this study can be applied as waterproof coating paint for marine shipyard paint, and ponds paint.

2. Materials and Methods

The raw materials used in the synthesizing process were Zinc Acetate Dihydrate (p.a.) Cellulose Acetate (p.a.), Poly Lactic Acid, DMF (dimethylformamide) 99,8%, Aceton (p. a.) Chloroform, Deionized water, and Methanol. The materials were provided by Merck and Sigma-Aldrich. The instruments used were, ultrasonic cleaner power sonic 405 model LUC, SCILOGEX MS-H280-Pro hotplate stirrer, furnace 48000 Thermolyne, and Spincoater TC100 precision.

The fabricating of ZnO-CA-PLA/glass film is briefly described as follows. Firstly, a blend of 0.03 g CA and 0.25 g PLA were dissolved in chloroform and various amount of ZnO of 0, 0.1, 0.2, 0.3, and 0.4 g to a solvent of DMF: Aceton. Every solution was mixed for 2 hours at 80 °C with the speed of 600 rpm. The yielded solution was deposited on a glass substrate by a using a spin coating method using 2000 rpm for 30 seconds. Subsequently, ZnO-CA-PLA/glass film was annealed at 50°C for 2 hours. The ZnO-CA-PLA/glass film properties were characterized by means of XRD, FTIR, SEM-EDX, and Digital Microscope.

3. Results and Discussion

3.1. Functional Group of CA-PLA-ZnO by FTIR

IR Spectra is one of the ways to obtain information about a functional group of organic material existing in the system. By using IR spectra, one could obtain the information what was the mechanism in the process of synthesis. One may result in a new compound, a doped compound, or just composite systems. The effect of ZnO NPs of CA-ZnO NPs-PLA IR spectra is shown in Figure 1.

CA has a specific functional group which is shown by chain structure. The transmittance peaks on wavenumber of 1732, 1239 and 1367 cm^{-1} indicate the appearance of carbonyl bonding of C=O *stretching*, acetyl –C-O *stretching* from acetate [9], and a bond of methyl group which stating of acetate group. IR spectra show the adding of ZnO in CA-ZnO-PLA at 1141 cm^{-1}. It means that introducing ZnO on the CA-PLA showing a composite system of the film.

3.2. Crystal Structure of CA-PLA-ZnO Analysis.

XRD analysis informs that there is a higher PLA composition than ZnO. The portion of PLA to ZnO is supported by EDAX analysis.

From the patterns of Figure 2, we calculate crystal size using Scherrer formula and also calculate the associated crystallinity. The crystallinity was calculated by area based on the X-RD patterns. The results are presented in Table 1.

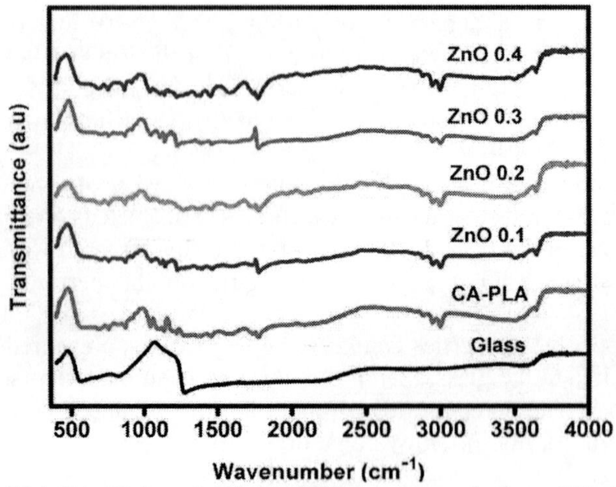

Figure 1. IR Spectra of (a) CA-PLA and CA-ZnO-PLA with variation of ZnO (b) 0.1, (c) 0.2, (d) 0.3, and (e) 0.4.

Based on Table 1, it is obtained that the introducing of ZnO make the ZnO-CA-PLA has the crystal size between 12.97- 19.01 nm. The crystal size may be insignificantly affected by the amount of ZnO NPs in the films. The size of crystal at the nanometer scale, it indicated that the film shows a large surface area. The grain size or surface area can be associated with the performance of hydrophobic properties of water on the surface. The crystallinity is in the range of 2.11% - 8.57%. The crystallinity increases with increasing ZnO concentration. The increase of crystallinity in the ZnO/CA/PLA film indicates that ZnO contributes to crystallinity in the ZnO-CA-PLA film.

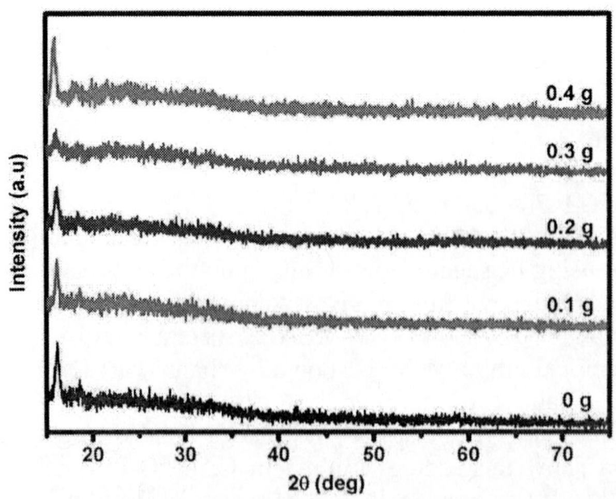

Figure 2. Diffraction pattern of ZnO-CA-PLA/glass composite film for various ZnO.

Table 1. Crystal Size of the variation ZnO mass on ZnO-CA-PLA.

Sample	**Crystal size (nm)**	**Crystallinity (%)**
CA-PLA	15.89	2.11
0,1 ZnO-CA-PLA	19.01	3.49
0,2 ZnO-CA-PLA	14.93	2.92
0,3 ZnO-CA-PLA	12.97	6.90
0,4 ZnO-CA-PLA	15.04	8.57

3.3 The Morphology of ZnO-CA-PLA/glass film

The images of SEM show that nano ZnO-CA-PLA/glass is the porous film. Based on Figure 3, it is shown that the morphology of cellulose acetate (CA) slightly smooth on its surface. It means that the film has less porous. The introducing PLA is on CA, provide the surface more porous as indicated by Fig. 3b. The introducing of ZnO made the films more porous and agglomerate between ZnO and CA-PLA. More ZnO added on CA- PLA gives more agglomeration of its surface which is shown in Fig. 3c to 3f. The average size of the porous is in the range micrometer.

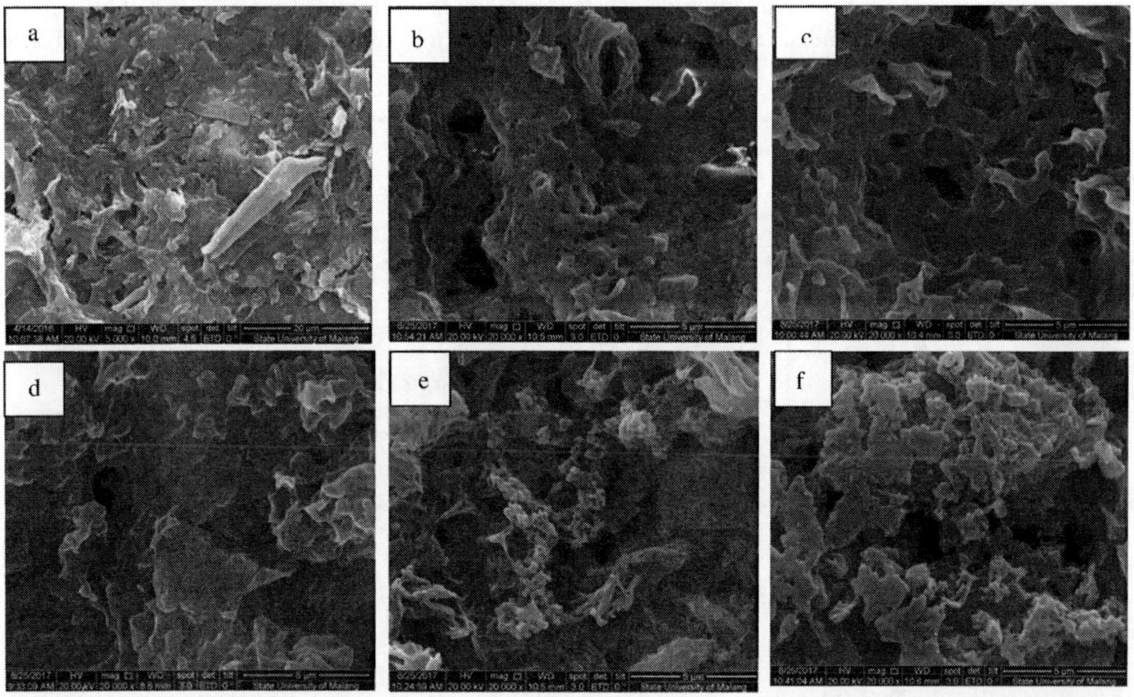

Figure 3. The morphology of (a) CA; (b) CA-PLA; and ZnO-CA-PLA with variation of ZnO for 0.1 (c); 0.2 (d); 0.3 (e); and 0.4 (f).

3.4 Hydrophobicity of CA-ZnO-PLA

Contact angle measurement can be used to show the hydrophobic properties [10]. Figure 4, shows that the addition of ZnO significantly increases its contact angle. The highest contact angle was obtained by the sample with 0.3 grams of ZnO.

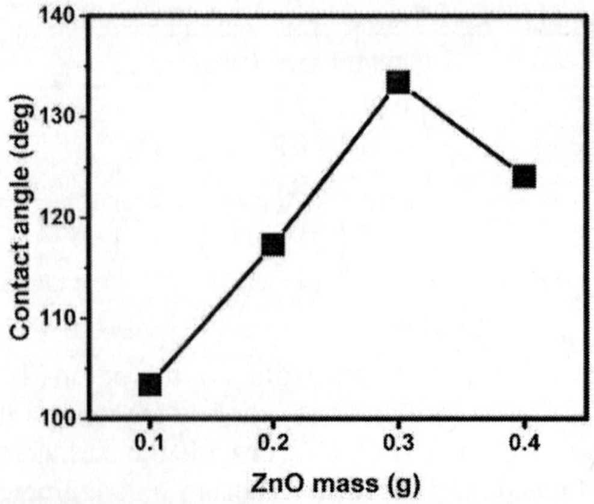

Figure 4. The effect of ZnO addition at annealing 1 hours on contact angle.

We also obtained a significant improvement of the annealing effect on contact angle. The longer annealing times led to the more hydrophobic surface as depicted in Figure 5 and Table 2. The increase of hydrophobicity is depended on the annealing time. This trend is occurring at all ranges of ZnO dopant. The optimum value does not change by the annealing time.

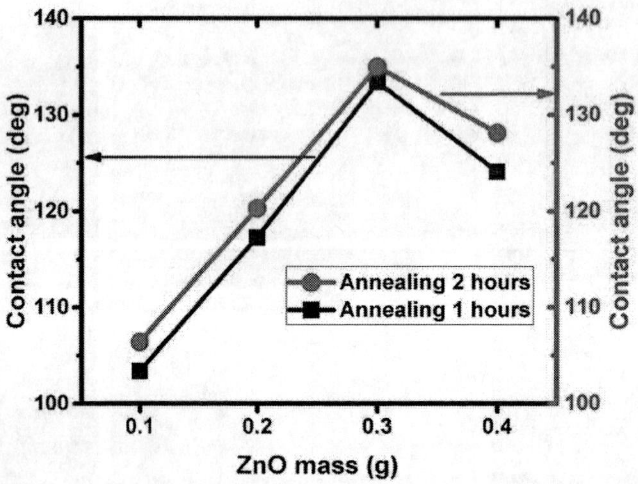

Figure 5. The effect of annealing times to the contact angle

Table 2. The contact angle of water as a function of ZnO annealing time.

ZnO (g)	1 (h)	2 (h)
0.1	103.4	106.4
0.2	117.3	120.3
0.3	133.4	135.0
0.4	124.1	128.1

Other impressive results show by the contact angle of water droplet on the film as a variation of temperature measuring condition. As indicated in Figure 7, the room temperature contact angle is slightly declined while exposing up to 1 minute. The contact angle still exhibits a superhydrophobic performance

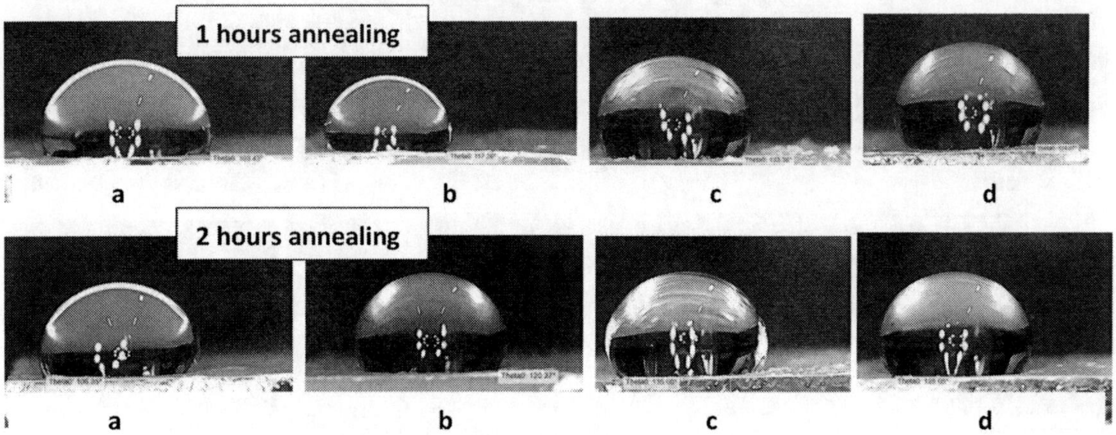

Figure 6. The visual contact angle of water drop on ZnO-CA-PLA film with various ZnO of (a) 0.1, (b) 0.2, (c) 0.3, and (d) 0.4 grams.

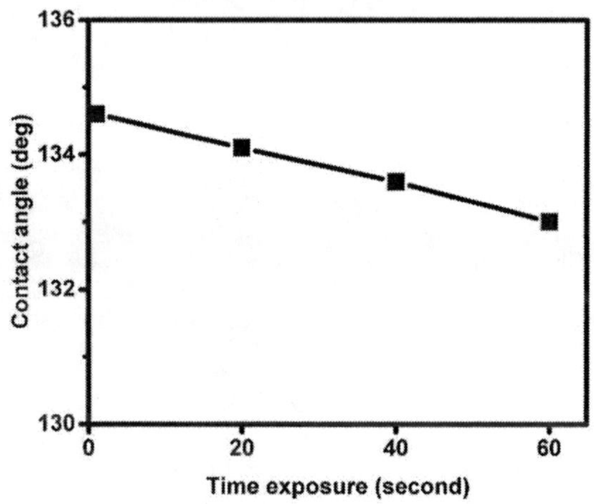

Figure 7. The effect of exposure time on the contact angle

A further experiment for the same films, we measured the influence of temperature measuring condition while exposed in different duration. As expected, we found that the higher temperature tends to decrease its contact angle. On the other hand, the time exposure also slightly reduce the contact angle. The performance of hydrophobic water on the films under the influence of time exposure and temperature condition is shown in Figure 8.

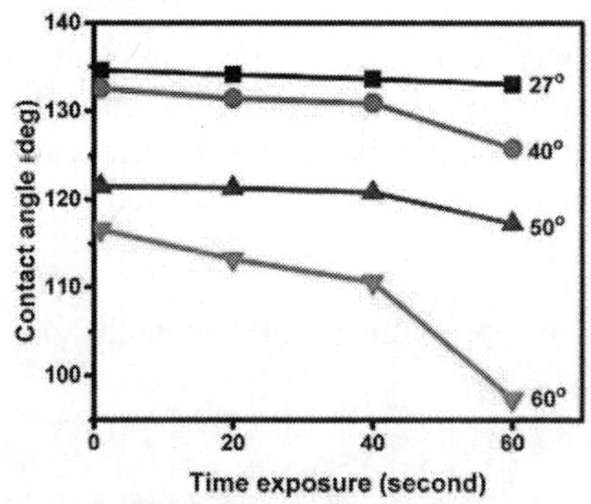

Figure 8. The effect of temperature on the contact angle of water

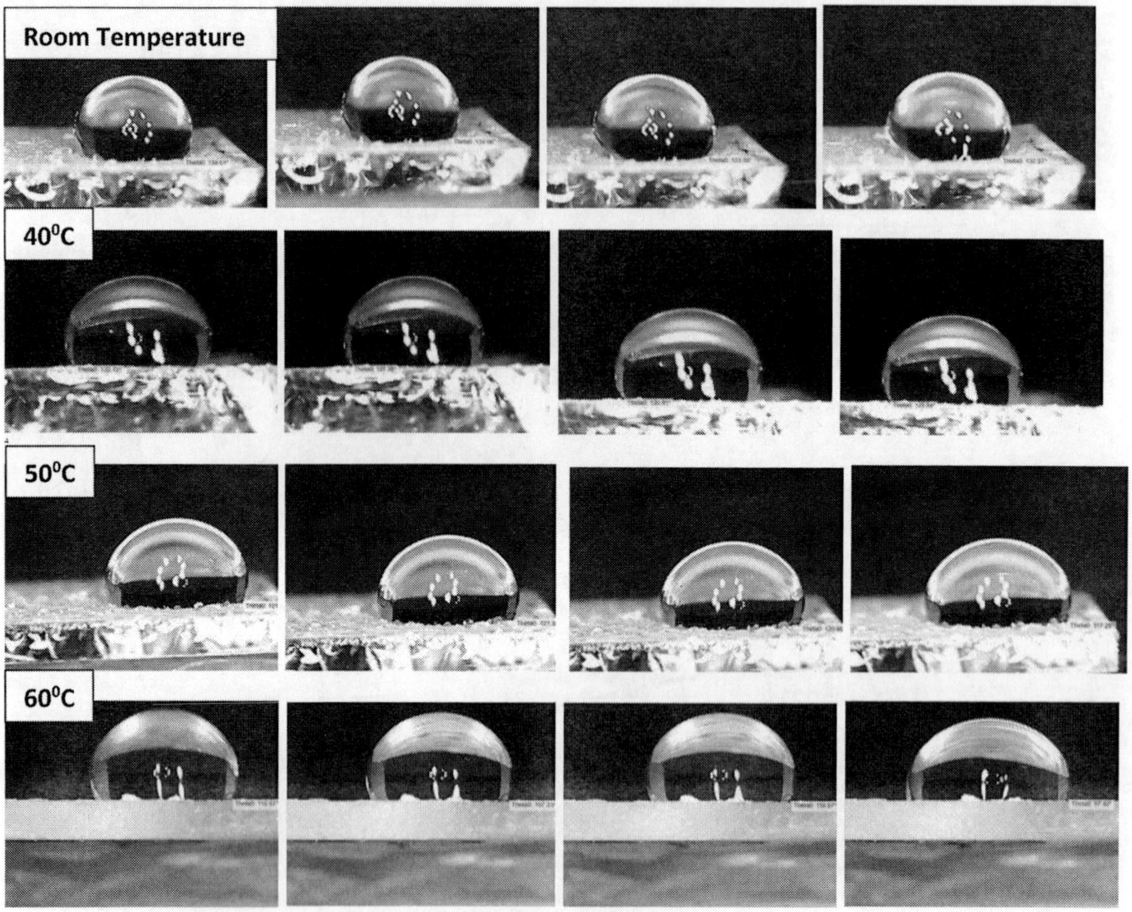

Figure 9. The contact angle of water droplet on ZnO-CA-PLA film at various temperatures.

A visual of the contact angle of water droplet on the films as a function of temperature condition and time exposure is depicted in Figure 9. Several aspects which influencing the hydrophobic performance of ZnONPs-CA-PLA has been studied. There two categories of the factors. One type may improve the hydrophobic performance, and the other one is a group of factors which could reduce the contact angle. Although these results are relatively similar to other report [11], this work shows more comprehensive results than other published works. It is known that ZnO-CA is an excellent potential material for electrical properties [12]. It would be good when the electronic device to fabricate also showing a good of hydrophobic performance.

4. Conclusion

ZnO-CA-PLA hydrophobic film has been successfully prepared through the use of spin coating methods. Contact angle represents the performance of hydrophobicity of the films. The introducing of ZnO NPs and increase of annealing time give rise to increase the contact angle. On the other hand, the rise in time exposure and the rise in the temperature measurement tend to decrease the contact angle.

Acknowledgment

Part of this work was supported by the Ministry of Research and Higher Education of Indonesia through the PNBP UM of MD.

References

[1] Castro-Aguirre E, Iñiguez-Franco F, Samsudin H, Fang X and Auras R 2016 Poly(lactic acid)—Mass production, processing, industrial applications, and end of life *Adv. Drug Deliv. Rev.* **107** 333–66

[2] Zhen W and Zheng Y 2016 Synthesis, characterization, and thermal stability of poly (lactic acid)/zinc oxide pillared organic saponite nanocomposites via ring-opening polymerization of D , L -lactide: PLA/Zinc Oxide Pillared Organic Saponite Nanocomposites *Polym. Adv. Technol.* **27** 606–14

[3] Hossain K M Z, Ahmed I, Parsons A J, Scotchford C A, Walker G S, Thielemans W and Rudd C D 2012 Physico-chemical and mechanical properties of nanocomposites prepared using cellulose nanowhiskers and poly(lactic acid) *J. Mater. Sci.* **47** 2675–86

[4] Khoo R Z, Ismail H and Chow W S 2016 Thermal and Morphological Properties of Poly (Lactic Acid)/Nanocellulose Nanocomposites *Procedia Chem.* **19** 788–94

[5] Diantoro M, Mustikasari A A, Wijayanti N, Yogihati C and Taufiq A 2017 Microstructure and dielectric properties of cellulose acetate-ZnO/ITO composite films based on water hyacinth *J. Phys. Conf. Ser.* **853** 012047

[6] Diantoro M, Hidayati N N S, Latifah R, Fuad A, Nasikhudin, Sujito and Hidayat A 2016 Electrical conductivity modification using silver nano particles of Jatropha Multifida L. and Pterocarpus Indicus w. extracts films p 030034

[7] Anitha S, Brabu B, John Thiruvadigal D, Gopalakrishnan C and Natarajan T S 2013 Optical, bactericidal and water repellent properties of electrospun nano-composite membranes of cellulose acetate and ZnO *Carbohydr. Polym.* **97** 856–63

[8] Myint M T Z, Kumar N S, Hornyak G L and Dutta J 2013 Hydrophobic/hydrophilic switching on zinc oxide micro-textured surface *Appl. Surf. Sci.* **264** 344–8

[9] Zhang C, Zhong S and Yang Z 2008 Cellulose acetate-based molecularly imprinted polymeric membrane for separation of vanillin and o-vanillin *Braz. J. Chem. Eng.* **25** 365–373

[10] Shinde V R, Lokhande C D, Mane R S and Han S-H 2005 Hydrophobic and textured ZnO films deposited by chemical bath deposition: annealing effect *Appl. Surf. Sci.* **245** 407–13

[11] A. Pedna, L. Pinho, P. Frediani, M. J. Mosquera, "Obtaining SiO2–fluorinated PLA bio-nanocomposites with applications reversible and highly-hydrophobic coatings of buildings", Progress in Organic Coatings 90 (2016) 91–100.

[12] M. Diantoro, T. Suprayogi, U. Sa'adah, N. Mufti, A. Fuad, A. Hidayat, and H. Nur (July 18th 2018). Modification of Electrical Properties of Silver Nanoparticle, Silver Nanoparticles Khan Maaz, IntechOpen, DOI: 10.5772/intechopen.75682.

The Growth of ZnO Nanorods on Stainless-steel foils and Its Application for Piezoelectric Nanogenerator

Nandang Mufti[1,3,*], Anggreta Damayanti[1], Ariprihartha[2], Arramel[4], Ahmad Taufiq[1,3], Sunaryono[1,3]

[1]Department of Physics, Faculty of Mathematics and Natural Sciences, Universitas Negeri Malang, Jl. Semarang 5, Malang 65145, Indonesia
[2]Department of Electrical Engineering, Faculty of Engineering, Universitas Negeri Malang, Jl. Semarang 5, Malang 65145, Indonesia
[3]Centre of Advanced Materials for Renewable Energy, Universitas Negeri Malang, Jl. Semarang 5, Malang 65145, Indonesia
[4]Department of Physics, National University of Singapore, 2 Science Drive 3, 117542, Singapore

*Corresponding author's email: nandang.mufti.fmipa@um.ac.id

Abstract. ZnO nanorods have attracted much attention because of their potential for harvesting energy device. In this study, we investigate the effect of hexamethylenamin concentration to a characteristic of ZnO nanorods on a stainless-steel foil. ZnO nanorods were synthesized by simple hydrothermal methods on Stainless-steel (SS) by varying the precursor ratio of hexamethylenediamine and Zinc nitrate tetrahydrate. The ZnO nanorods samples were characterized by X-ray diffraction for crystal structure properties, and Scanning Electron Microscope (SEM) for their morphology. The piezoelectric nanogenerator performance was characterized by measuring current and voltage with electrometer and oscilloscope, respectively. The lattice parameters of ZnO nanorods are *a, b* = 3.2495 Å, and *c* = 5.2069 Å with the ratio c/a is *1.6* indicating that ZnO nanorods have hexagonal wurtzite structure. The length and of ZnO nanorods increases by increasing the precursor ratio from 0.347 to 1.229 μm. However, the diameter can be varied from 63 to 102 nm. The small diameter and longest of ZnO nanorod are the best for nanogenerator applications.

Keywords: Growth, ZnO Nanorod, stainless-steel, piezoelectric, nanogenerator

1. Introduction

The desire to produce self-powered electronic led to increasing the research in energy harvesting device in the last few years [1]. Along with the progress in nanotechnology, the demand for low power consumption in electronic applications has attracted much attention. One way to solve this problem is by using an energy harvesting device that can collect energy from environmental vibration by using piezoelectric materials [2,3]. Piezoelectric materials have a unique ability to convert mechanical vibrations into electrical energy, and this material becomes a very promising for the renewable energy source. Many studies have been interested in the use of one-dimensional (1D) piezoelectric nanostructures such as PZT, $BaTiO_3$, CdS, ZnO, and TiO_2 for mechanical energy harvesting [4]. ZnO is one of an n-type semiconductor that has a wide band gap of 3.37 eV with an excitonic binding energy of about 60 meV at room temperature [5]. Moreover, ZnO is also relatively cheap material, non-toxic, environmentally friendly and has been widely used in applications such as light-emitting diodes [6], UV sensors [7], Gas Sensors [8], and Solar Cells [9]. ZnO in the form of nanorods can improve physical properties such as light absorption, due to its high surface area [10]

ZnO nanorods (ZnO NRs) have high piezoelectric coefficient values and can be grown at low temperatures on almost any kind of substrate. Piezoelectric properties of ZnO can be utilized to convert the environmental vibration into electricity called piezoelectric nanogenerators. However, the current and voltage generated by ZnO nanorods are still relatively small and grown on the ITO/FTO substrate. In order to improve the efficiency of nanogenerator, it important to optimize the ZnO NRs growth to increase the total surface area by tuning growth parameters such as the concentration of precursor, growth temperature, time, [11]. Therefore, we investigate the effect of precursor concentration between zinc nitrate tetrahydrate (ZNT) and Hexamethylenetetramine (HMT) to characteristics of ZnO NRs on stainless-steel foil due to cheaper and good conductivity.

2. Materials and Methods

The ZnO NRs were grown by the hydrothermal method on a stainless-steel foil. The stainless-steel foil was cleaned by dipping into acetone on the ultrasonic cleaner for 10 minutes, then washed with deionized water. Then, the stainless-steel is stored and dried on a clean tissue. ZnO solution was made by dissolving the Zn-acetate powder ($Zn(CH_3COO)_2.2H_2O$) in ethanol solution. The mixture was heated and stirred at temperature of 70 °C for 45 minutes by magnetic stirrer hotplate. Then, Monoethanolamine (MEA) was added dropwise to the solution at temperature of 700 ^{0}C for 2 hours. The concentration ratio between Zn-acetate and MEA is equal. The solution was cooled at room temperature for 24 hours.

ZnO films were synthesized by a spin coating method. The ZnO solution was dropped on the clean stainless-steel substrate that was placed on the spin coater plate, then spinning at 300 rpm for 25 seconds. The preheating substrate was done at 150 ^{0}C for 10 minutes, whereas the annealing of ZnO film was done at 550 °C for 2 hours

ZnO NRs were grown by the hydrothermal method. The growth solution was prepared using zinc nitrate tetrahydrate (ZNT) and Hexamethylenetetramine (HMT) in deionized water. The concentration of zinc nitrate tetrahydrate was 45 mM, and the concentration of HMT was varied at 45 mM, 56 mM, and 68 mM. Therefore, the molar ratio between ZNT and HMT was 1: 1.1; 1.25; and 1: 1.5. The solution was stirred using a magnetic stirrer for 45 minutes at room temperature. Then, the ZnO film was immersed in the prepared solution for 6 hours with temperature of 90 °C. The sample was washed with DI-Water to remove the salt residue from the sample surface and was dried with a dry blower. Finally, The ZnO NRs on the Stainless-steel foil were annealed at 550 °C for 2 hours in the furnace. Beside ZnO nanorods, The PVA films on stainless-steel foil was prepared by using spin coating method with speed of 3000 rpm for 20 seconds.

All ZnO nanorods samples were characterized by X-ray diffraction (XRD) for structural properties investigation, Scanning Electron Microscope (SEM) to determine the length and the diameter of ZnO nanorods. The oscilloscope was used for voltage measurements.

3. Results and Discussion

Figure 1 shows the X-ray diffraction patterns of ZnO NRs with variations of precursors ratio between ZNT and HMT. The XRD patterns of ZnO seed layers and ZnO NRs are similar to high-intensity peak at the Bragg plane of (100), (002), and (101). Other peaks are also observed as stainless-steel peaks. The crystal structure of the ZnO NRs was analyzed using Rietica software and summarized in Table 1. The lattice parameters of ZnO NRs and ZnO bulk from JPCDS standard (No. 36-1451) were alike, that is a,b = 3.2495 Å, and c = 5.2069 Å with the ratio c/a is *1.6* indicating that ZnO nanorods have hexagonal wurtzite structure. This result is similar with lattice parameters of ZnO NRs that previously reported [12,13]

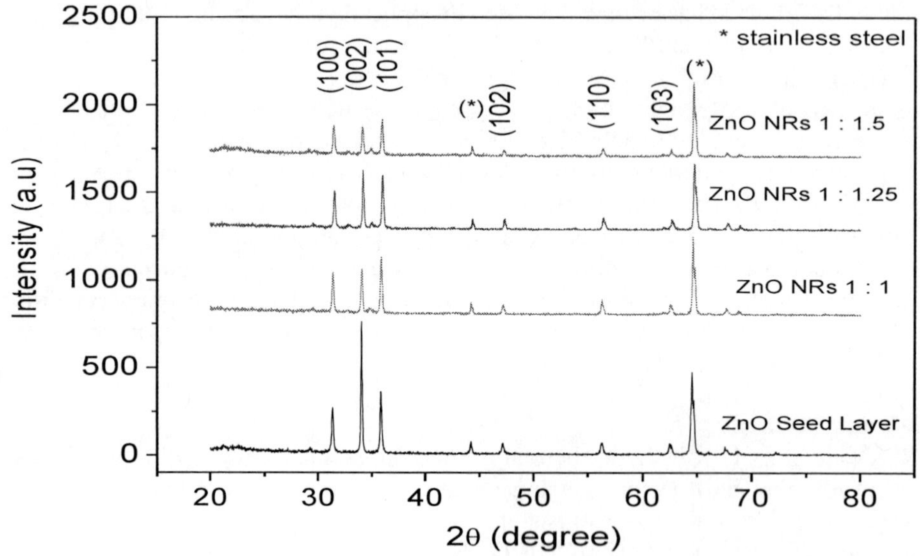

Figure 1. The XRD patterns of ZnO nanorods with different precursor concentration ratio between ZNT and HMT

The lattice parameter indicated that the ZnO NRs had a crystalline form and a single phase. The peak (101) was seen to have a higher intensity compared with other peaks, except for the ZnO NRs with precursor concentration ratio of 1 : 1.25 that has the highest peak (002). It indicated that the ZnO NRs grew along the –*c* direction, while the peak (101) was caused by the growth of some nanorods that was out from the c-direction. The growth orientation of ZnO NRs on the plane (002) was similar to results that reported by other researchers [9,14].

Table 1. The Crystal Structure of ZnO nanorods with different precursor concentration between ZNT and HMT

Lattice Parameters	Model ZnO	ZnO NRs 1:1	ZnO NRs 1:1.25	ZnO NRs 1:1.5
$a = b$(A)	3.249	3.2495	3.2495	3.2495
c = (A)	5.2038	5.2069	5.2069	5.2069
c/a	1.6	1.6	1.6	1.6

Figure 2 shows SEM photos of ZnO NRs at different precursor concentration ratio between ZNT and HMT. The morphology of ZnO nanorods leads to a hexagonal form. The highest diameter shows at ZnO nanorods with precursor concentration ratio between ZNT and HMT at 1: 1.25. The insert in Figure 2 shows a cross-section of ZnO NRs. In general, the thickness of ZnO NRs Increases by increasing precursor concentration. The longest rod obtained at precursor concentration ratio 1:1.5. The diameters and rod lengths summarized in Tabel 2. However, the diameter of ZnO NRs varies between 63.4 to 102.5 nm.

Table 2. The lengths and diamcters of ZnO NRs with different precursor concentration between ZNT and HMT

precursor concentration ratio between ZNT and HMT	Length of ZnO nanorods (μm)	Diameter (nm)
1:1	0.347	63.4
1:1.25	0.437	102.5
1:1.5	1.229	73.4

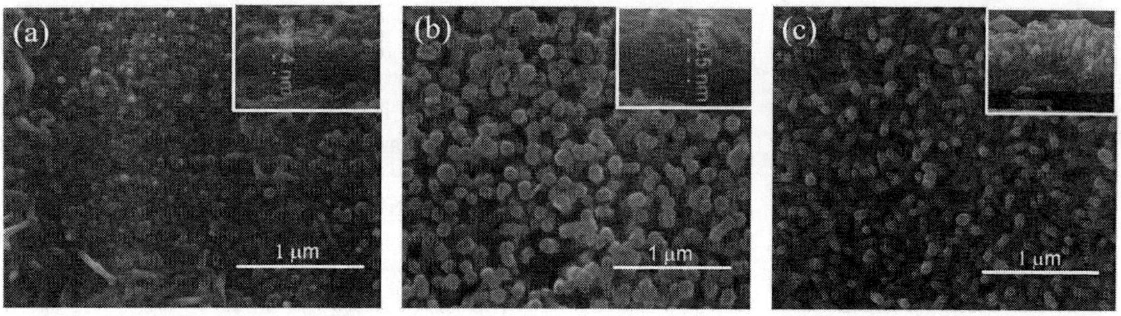

Figure 2. SEM Photos of ZnO NRs with different precursor concentration ratio between ZNT and HMT (a) 1:1, (b) 1:1.25, and (c) 1: 1.5. The Inserts are SEM photos for the cross-section of ZnO NRs.

In order to measure the nanogenerator performance of ZnO NRs, the voltage of ZnO NRs that have been combined with PVA measured by oscilloscope as shown in Figure 3. The results show that ZnO nanorods at precursor concentration ratio between ZNT and HMT at 1:1 and 1:1.5 has the highest voltage of around 2.6 Volt. This output Voltage is higher compared with nanogenerator from ZnO/PVA nanofiber previously reported [15]. However, the voltage fluctuations at 1:1.5 is more stable rather than 1 :1. The voltage measurement was done by pressing ZnO NRss/PVA around 6 N/m^2. This result indicates that small diameter and longest of ZnO NRs are the best for nanogenerator applications due to the large surface area.

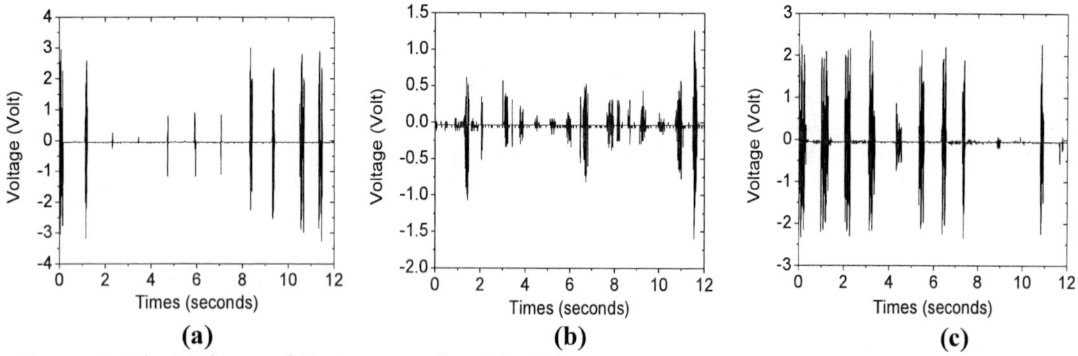

Figure 3. The Voltage of ZnO nanorods with different precursor concentration ratio between ZNT and HMT of (a) 1:1, (b) 1:1.25 (c) 1:1.5, in combination with PVA coated on stainless steel.

4. Conclusion

The ZnO nanorods on Stainless-steel foil have been successfully synthesized by simple hydrothermal methods. The crystal structure of ZnO nanorods is hexagonal wurtzite with *a, b* parameters of 3.2495 Å, *c*-parameter of 5.2069 Å and c/a ratio of 1.6. The increasing precursor concentration ratio between ZNT and HMT increases the length of ZnO nanorods from 0.347 μm for 1:1, to 0.437 μm for 1:1.25 and 1.229 μm for 1:1.5. While the diameters of ZnO nanorods are varying between 63.44 nm to 102.5 nm. The best and stable output voltage of 2.6 volt obtained at precursor concentration ratio 1:1.5 indicating that small diameter and longest of ZnO nanorod is the best for nanogenerator applications.

Acknowledgments

The authors would like to thank DRPM, Ministry of Research, Technology and Higher Education (Ristekdikti) Republic of Indonesia for supporting a research PBK grant for NM.

References

[1] Xu S, Qin Y, Xu C, Wei Y, Yang R and Wang Z L 2010 Self-powered nanowire devices *Nat. Nanotechnol.* **5** 366–73

[2] Du S, Jia Y, Chen S-T, Zhao C, Sun B, Arroyo E and Seshia A A 2017 A new electrode design method in piezoelectric vibration energy harvesters to maximize output power *Sens. Actuators Phys.* **263** 693–701

[3] Fan F R, Tang W and Wang Z L 2016 Flexible Nanogenerators for Energy Harvesting and Self-Powered Electronics *Adv. Mater.* **28** 4283–305

[4] Nechibvute A, Chawanda A and Luhanga P 2012 Piezoelectric Energy Harvesting Devices: An Alternative Energy Source for Wireless Sensors *Smart Mater. Res.* **2012** 1–13

[5] Özgür ü., Alivov Y I, Liu C, Teke A, Reshchikov M A, Doğan S, Avrutin V, Cho S-J and Morkoç H 2005 A comprehensive review of ZnO materials and devices *J. Appl. Phys.* **98** 041301

[6] Tsai J, Shih J, Wu T and Meen T 2012 n-ZnO nanorods/p+-Si (111) heterojunction light emitting diodes *Nanoscale Res. Lett.* **7** 664

[7] Lupan O, Postica V, Gröttrup J, Mishra A K, de Leeuw N H and Adelung R 2017 Enhanced UV and ethanol vapour sensing of a single 3-D ZnO tetrapod alloyed with Fe 2 O 3 nanoparticles *Sens. Actuators B Chem.* **245** 448–61

[8] Sayago I, Hontanon E, Aleixandre M, Fernandez M J, Santos J P and Gracia I 2017 ZnO and ZnO/SnO2 nanofibers as resistive gas sensors for NO2 detection (IEEE) pp 1–4

[9] Meng Y, Lin Y and Yang J 2013 Synthesis of rod-cluster ZnO nanostructures and their application to dye-sensitized solar cells *Appl. Surf. Sci.* **268** 561–5

[10] Mehrabian M, Azimirad R, Mirabbaszadeh K, Afarideh H and Davoudian M 2011 UV detecting properties of hydrothermal synthesized ZnO nanorods *Phys. E Low-Dimens. Syst. Nanostructures* **43** 1141–5

[11] Briscoe J, Jalali N, Woolliams P, Stewart M, Weaver P M, Cain M and Dunn S 2013 Measurement techniques for piezoelectric nanogenerators *Energy Environ. Sci.* **6** 3035

[12] Fuad A, Fibriyanti A A, Subakti, Mufti N and Taufiq A 2017 Effect of Precursor Concentration Ratio on the Crystal Structure, Morphology, and Band Gap of ZnO Nanorods IOP Conference Series: Materials Science and Engineering vol 202

[13] Idiawati R, Mufti N, Taufiq A, Wisodo H, Laila I K R, Fuad A and Sunaryono 2017 Effect of Growth Time on the Characteristics of ZnO Nanorods IOP Conference Series: Materials Science and Engineering vol 202

[14] Foo K L, Hashim U, Muhammad K and Voon C H 2014 Sol–gel synthesized zinc oxide nanorods and their structural and optical investigation for optoelectronic application *Nanoscale Res. Lett.* **9** 429

[15] Subagiyo D and Suyitno 2017 Enhancing Output Voltage of Piezoelectric-Based Nanogenerators Using Zinc Oxide Material Codoped with Aluminum and Cobalt *IOP Conf. Ser. Mater. Sci. Eng.* **267** 012037

Effect of Polyethylene Glycol (PEG) on Particle Distribution of $Mn_{0.25}Fe_{2.75}O_4$-PEG 6000 Nanoparticles

Sunaryono[1,2,*], Kormil Saputra[1], Rosabiela Irfa Andina[1], Nurul Hidayat[1], Ahmad Taufiq[1,2], Hendra Susanto[3], Nandang Mufti[1,2], Arif Hidayat[1], Chusnana Insjaf Yogihati[1], Sigit Triwicaksono[4], Siriwat Soontaranon[5]

[1]Department of Physics, Faculty of Mathematics and Natural Sciences, Universitas Negeri Malang, Jl. Semarang 5 Malang 65145, Indonesia
[2]Research Center of Minerals and Advanced Materials, Faculty of Mathematics and Natural Sciences, Universitas Negeri Malang, Jl. Semarang 5 Malang 65145, Indonesia
[3]Department of Biology, Faculty of Mathematics and Natural Sciences, Universitas Negeri Malang, Jl. Semarang 5 Malang 65145, Indonesia
[4]Department of Metalurgy, FTI, ITS Surabaya, Jl Arif Rcahman Hakim, Surabaya, Indonesia
[5]Synchrotron Light Research Institute, 111 University Venue, Muang District, Nakhon Ratchasima 30000, Thailand.

*Corresponding author: sunaryono.fmipa@um.ac.id

Abstract. $Mn_{0.25}Fe_{2.75}O_4$ nanoparticles have been successfully synthesized using a coprecipitation method. PEG 6000 polymer was added on the synthesis of $Mn_{0.25}Fe_{2.75}O_4$ which aimed at decreasing and preventing the emergence of aggregation between particles. The characterization of the phase structure, absorption of functional groups, and distribution of $Mn_{0.25}Fe_{2.75}O_4$-PEG 6000 nanoparticles have been successfully performed using the instruments of XRD, FTIR, and Small Angel X-Ray Scatterring (SAXS) respectively. XRD characterization results showed that the phase structure formed was in the form of magnetite and corresponded to the Inorganic Crystal Structure Database (ICSD) number 9013529. The bond of Fe-O and Mn-O consecutively on the wave numbers of 430 and 482 cm^{-1} which were the representation of $Mn_{0.25}Fe_{2.75}O_4$ material have been successfully confirmed through the characterization using FTIR, meanwhile the stretching vibrating on the wave numbers of 1251, 1352, 1472, and 1629 cm^{-1} for functional groups C-H is the representation of PEG template which has the chemical structure of -OH[-CH_2-CH_2-]n OH. SAXS characterization showed that the primary and secondary particle sizes from the $Mn_{0,25}Fe_{2,75}O_4$–PEG 6000 nanoparticle were 3.08 and 11.4 nm respectively. The value is the optimum results of the PEG 6000 template influence on the synthesize of $Mn_{0.25}Fe_{2.75}O_4$-PEG 6000 nanoparticles which are potential to be a raw material of magnetic fluid and gel on hyperthermia therapy.

Keywords: $Mn_{0,25}Fe_{2,75}O_4$, PEG 6000, Template, SAXS

1. Introduction

Nowadays, a study on magnetic nanomaterial has been an interest of many researchers because the material ingredient has a wide potential to be applied on various fields such as in the fields of medical engineering, industry, and magnetic engineering. Among other magnetic materials, iron oxide (Fe_3O_4) is one of the interesting materials to be studied due to its unique magnetic property, chemical stability, and biocompatibility [1]. Magnetic materials based on Fe_3O_4 have been utilized in several applications including Magnetic Resonance Imaging (MRI) [2,3], anti-radar material [4], cancer therapy [5,6], drug delivery system [7] and et cetera.

In obtaining a magnetite nanoparticle, various ways and methods have been reported by researchers. Shahid *et al.* 2018 [8] synthesizing Fe_3O_4 particles with the materials of steel industry waste. Other researchers reported that Fe_3O_4 nanoparticles could be synthesized using a simple solvothermal method at 200 °C for 8 hours [9]. The method was using poly (vinyl) pyrrolidone (PVP) with a different concentration to gain unique nanoparticles with the average particle size of ~ 50–90 nm [9]. In addition, in the previous research [10], the researcher has successfully fabricated Fe_3O_4 nanoparticles using a coprecipitation method and obtained an average particle size of 15 nm.

To gain an optimum particle size and magnetization value from Fe_3O_4 nanoparticles, several ways and methods have been performed by the researchers, one of them is by performing a certain ion doping treatment such as Zn^{2+} [11], Ni^{2+} [12], and Mn^{2+} [13]. Specifically, on the ion doping on Mn^{2+}, the distribution of Cation on metal ion can give an effect on the magnetic properties of Fe_3O_4 nanoparticles [14]. Taufiq *et al.* 2015 also reported that the addition of Mn^{2+} ions to Fe_3O_4 which form $Mn_xFe_{3-x}O_4$ can increase the saturation magnetization value of Fe_3O_4. This increase in magnetization is caused by Mn^{2+} ions which have a magnetic moment of 25% higher than Fe^{2+} ions, where the value of Mn^{2+} magnetic moment is recorded at 5μB while Fe^2 + is 4μB [13]. This result showed that there is a significant effect of Fe_3O_4 nanoparticle doping process by Mn^{2+} ion, however, the synthesis and the characterization of PEG 6000 template effect on the synthesis process of $Mn_{0.25}Fe_{2.75}O_4$-PEG 6000 nanoparticles is rarely reported. This research report also reviewed the PEG 6000 template effect on the structure and distribution of $Mn_{0.25}Fe_{2.75}O_4$-PEG 6000 nanoparticles.

2. Materials and Methods

2.1. Synthesis of $Mn_{0,25}Fe_{2,75}O_4$-PEG 6000 nanoparticles

The materials used in this research were iron sand from Tulungagung city Beach, hydrochloric acid (HCl), manganese (II) chloride tetrahydrate ($MnCl_2.4H_2O$), ammonium hydroxide (NH_4OH) and aquadest. The synthesis of Fe_3O_4 nanoparticles was begun by the separation process of iron sand using a permanent magnet to generate magnetic powder with a low impurity. Through a coprecipitation method, the magnetic powder obtained from the separation process was mixed with HCl on a certain concentration to generate $FeCl_2$ and $FeCl_3$ solutions. The solution was then added to $MnCl_2.4H_2O$ with a concentration of $x = 0.25$. The next step is titration with an NH_4OH solution and processed using a hot plate magnetic stirrer. The titration result was washed with distilled water until it reached normal pH (pH 7) which was then filtered using filter paper with a certain mess scale. The obtained $Mn_{0.25}Fe_{2.75}O_4$ nanoparticles were then mixed with 20 ml of PEG 6000 solution and processed on a hot plate magnetic stirrer for 1 hour at room temperature. The results of the mixture of nanoparticles $Mn_{0.25}Fe_{2.75}O_4$ and PEG 6000 are then filtered again and in the oven for 1 hour. Finally, the formed $Mn_{0.25}Fe_{2.75}O_4$-PEG 6000 nanoparticles were able to be XRD, FTIR, dan SAXS characterized.

2.2. Characterization

In order to find out the effect of PEG 6000 template on particle distribution of $Mn_{0.25}Fe_{2.75}O_4$-PEG 6000 nanoparticles, the samples from this work were characterized using XRD, FTIR, and SAXS. the XRD instrument was used to find out the crystal structure of $Mn_{0.25}Fe_{2.75}O_4$-PEG 6000 nanoparticles as phase composition, crystal size, and lattice parameter. FTIR characterization was conducted to know the functional groups from $Mn_{0.25}Fe_{2.75}O_4$-PEG 6000 nanoparticles. While the SAXS was used to find out the particle distribution of primary and secondary particle sizes.

3. Results and Discussion

3.1. XRD Test Results

The XRD characterization of $Mn_{0.25}Fe_{2.75}O_4$-PEG 6000 nanoparticles was performed at the range of 10^0 to 80^0. Figure 1 shows the diffraction pattern of $Mn_{0.25}Fe_{2.75}O_4$-PEG 6000 nanoparticles in which the refinement was performed using Rietica software. From the analysis results, it was obtained that the phase of $Mn_{0.25}Fe_{2.75}O_4$-PEG 6000 nanoparticles was magnetite phase which was identifies based on the database of Inorganic Crystal Structure Database (ICSD) number 9013529. The noted magnetite phase has the highest crystal peak at 35.4° in the field hkl (311) while the distribution of other crystal peaks is identified at the angles (2θ) of 30.15°, 43.05°, 53.5°, 56.97°, 62.59°, dan 74.28° and each corresponded to the Miller index position (111), (220), (400), (422), (511), dan (440). The lattice parameter and the comparison results of crystal grain size using a Rietica analysis and Debye Scherrer presented on Tabel 1. The lattice parameter results obtained between the data model with the refinement results showed the corresponding results. It can be concluded that the formed $Mn_{0.25}Fe_{2.75}O_4$-PEG 6000 is a magnetite phase representation.

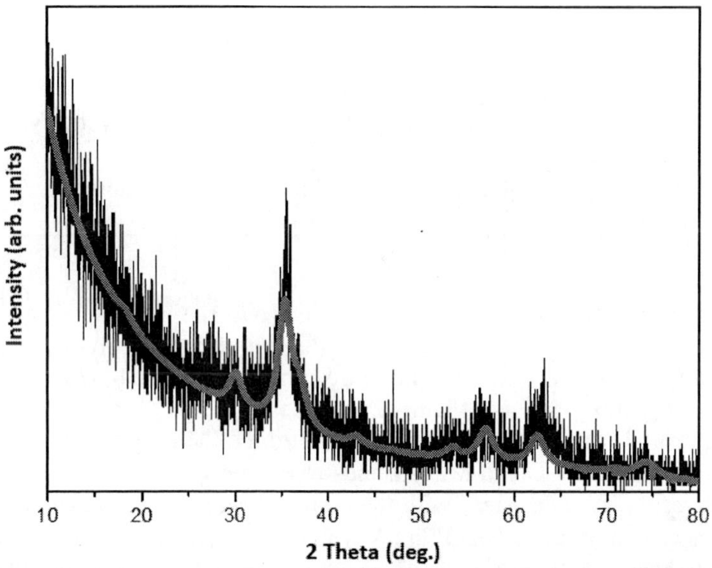

Figure 1. Refinement results of $Mn_{0.25}Fe_{2.75}O_4$-PEG 6000 nanoparticles diffraction patterns

Through a Rietica analysis, the crystal grain size of $Mn_{0.25}Fe_{2.75}O_4$-PEG 6000 nanoparticles of 4.97 nm. This is in line with the average size of Mn-ferrites nanoparticles which is 5.1 nm as reported by Bahtiar *et al.* 2017 [15]. In addition to using Reitica analysis, $Mn_{0.25}Fe_{2.75}O_4$-PEG 6000 nanoparticles were also carried out particle size analysis using the Debye Scherrer method. In this method, the value of Full Width Half Maximum (FWHM) which is the reference is FWHM from the highest peak which is at the angle (2θ) of 35.46° as shown in Figure 2. From the results of refinement, crystal grain size of $Mn_{0.25}Fe_{2.75}O_4$-PEG 6000 nanoparticles is 6.29 nm. This result is slightly greater than the value obtained using Reitica analysis.

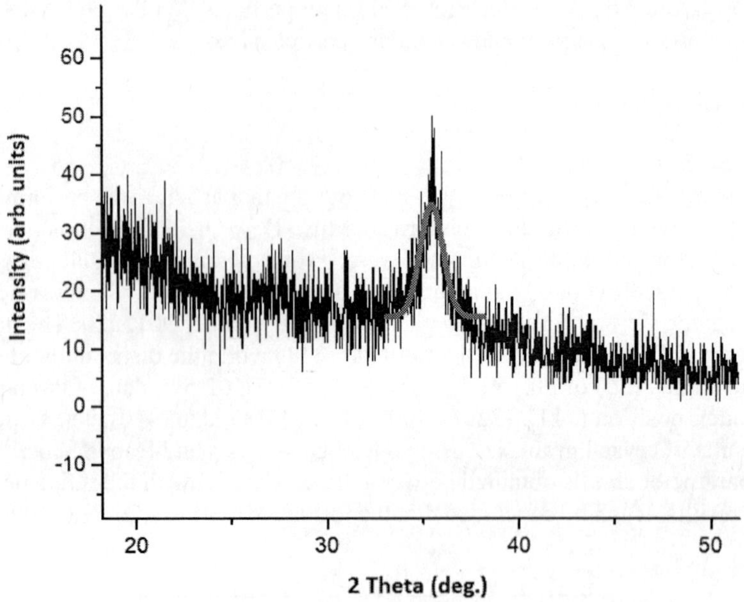

Figure 2. Fitting results of $Mn_{0.25}Fe_{2.75}O_4$-PEG 6000 nanoparticles diffraction peak

Table 1. The value of lattice parameter and crystal grain size $Mn_{0.25}Fe_{2.75}O_4$-PEG 6000

Sample	**Lattice Parameter** $a=b=c$ **(Å)**		**Crystal Grain Size (nm)**	
Analysis	Rietica	Model	Rietica	Debye-Scherrer
$Mn_{0.25}Fe_{2.75}O_4$-PEG 6000	8.38	8.40	4.97	6.29

3.2. FTIR Characterization Results

The FTIR characterization was conducted to find out the functional groups of $Mn_{0.25}Fe_{2.75}O_4$ nanoparticles after given the PEG 6000 template. PEG 6000 is the polymer that is easy to dissolve in water and non-toxic with the chemical structure of -OH[-CH_2-CH_2-]$_n$OH. The FTIR test result of $Mn_{0.25}Fe_{2.75}O_4$-PEG 6000 is shown in Figure 3.

Based on the results of the data obtained, the spectrum of $Mn_{0.25}Fe_{2.75}O_4$ nanoparticles was recorded on the functional groups of Fe-O and Mn-O bonds were detected respectively at wave numbers 430 and 482 cm^{-1}. These results are consistent with the results of research conducted by previous researchers [16,17] that is there is a bond of Fe-O dan Mn-O on the wavenumber below 700 cm^{-1}. There are also other studies reported that the functional groups of Fe-O dan Mn-O bond are at the wavenumber above 685 cm^{-1} [18,19]. Furthermore, the analysis result of functional groups on nanoparticles was also supported by other research results, where the functional groups of Fe-O and Mn-O bound will face a stretching vibrating at the wavenumber of 430 – 580 cm^{-1} [20–22].

Besides that, PEG polymer with the chemical structure of -OH[-CH_2-CH_2-]nOH which becomes the template of $Mn_{0.25}Fe_{2.75}O_4$ experiences stretching vibrating at the wavenumbers of 1251, 1352, 1472, 1629, and 2879 cm^{-1}. The wavenumber is the representations of functional groups of C-H. The absorption of wave numbers also occurs at position 1300 cm^{-1} which shows the representation of the functional group C=O. These results are consistent with previous research reports [23], functional group C=O vibrating is in the wavenumber of 1250 cm^{-1} while the C-H functional group will experience stretching vibrating on the wavenumbers of 1298, 1350, 1456-1471, dan 2879-2791 cm^{-1} [23]. The spectrum of nanoparticles was also observed in wave numbers 3346 cm^{-1}. The wavelength is a stretching of O-H functional groups from the sample. This is in line with research conducted by Vamvakidis *et al.*

[22,24,25]. They also report that the functional group of O-H Stretching is at a wavenumber of ~ 3439 nm^{-1} or in the range of ~3500 cm^{-1} [20].

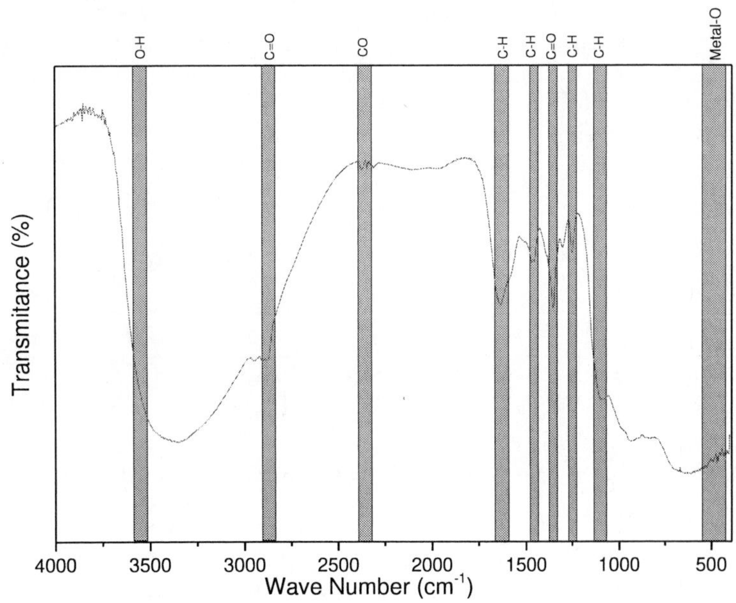

Figure 3. FTIR characterization results for $Mn_{0.25}Fe_{2.75}O_4$-PEG 6000 nanoparticles

3.3. The SAXS Test Results

SAXS Instrument is one of the sophisticated tools to be used to find out the structure distribution pattern of $Mn_{0.25}Fe_{2.75}O_4$-PEG 6000 nanoparticles. Figure 4 is the fitting result of SAXS characterization of $Mn_{0.25}Fe_{2.75}O_4$–PEG 6000 nanoparticles using global fitting performed by using a two-lognormal distribution pattern via SASfit software.

The global fitting results will obtain a fractal dimension (D), primary particle (r_1), and secondary particle (r_2). The global fitting pattern can be determined by matching the exponentials between the fractal area and the scattering intensity between the experimental data and the theoretical calculation as written in Equation 1 [26].

$$I(q) \propto \int_0^\infty N_1(r_1)F_N^2(q,r_1)S(q,\varsigma,D,r_1)dr_1 + \int_0^\infty N_2(r_2)F_N^2(q,r_2)S(q,\varsigma,D,r_2)dr_2 \quad (1)$$

Equation 1 is a theoretical calculation for global fitting using the two-lognormal size distribution with $I(q)$ is intensity, $[F(q,r)]^2$ is the scattering amplitude from the information of electron density and variables $S(q,\varsigma,D,r)$ are structure factor.

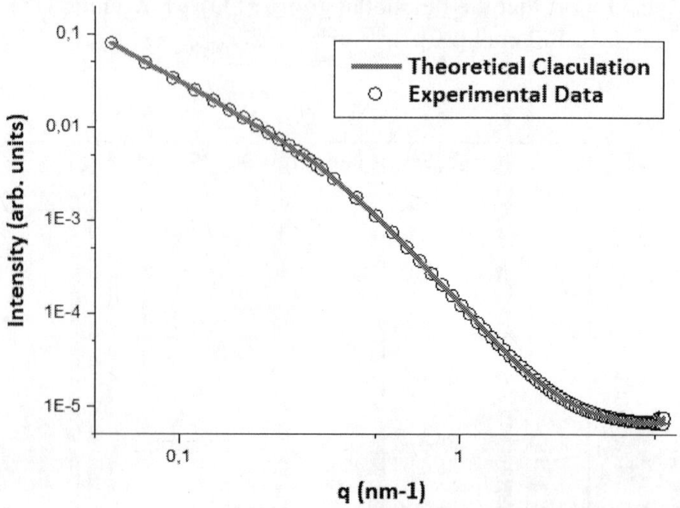

Figure 4. Global fitting results of SAXS test on $Mn_{0.25}Fe_{2.75}O_4$-PEG 6000 nanoparticles

A two-lognormal distribution that was used in SAXS analysis would generate a size distribution as a representation from structure factor and sphere as a representation from form factor. The global fitting analysis results generated red and black-colored curves. The red-colored curve is the result of global fitting based on theoretical calculation and the black-colored curve is the curve of experimental data results. From the global fitting analysis, it is found that the primary particle size from $Mn_{0.25}Fe_{2.75}O_4$–PEG 6000 nanoparticles is 3.08 nm and the secondary particle size of $Mn_{0.25}Fe_{2.75}O_4$–PEG 6000 nanoparticles is 11.4 nm. This result is not significantly different from the previous studies which stated that the primary and secondary particles of $Mn_{0.25}Fe_{2.75}O_4$ nanoparticles are 3.8 and 9.6 nm respectively [27]. However, the secondary particle from $Mn_{0.25}Fe_{2.75}O_4$–PEG 6000 showed a bigger value than nano $Mn_{0.25}Fe_{2.75}O_4$ nanoparticles without PEG 6000 [27]. The difference in particle sizes between the $Mn_{0.25}Fe_{2.75}O_4$ nanoparticles with and without PEG 6000 is caused by the PEG 6000 polymer template. The existence of PEG 6000 polymer template will cause $Mn_{0.25}Fe_{2.75}O_4$ nanoparticles to be trapped inside the long PEG polymer network eventually raises an aggregation.

4. Conclusion

The synthesis of $Mn_{0.25}Fe_{2.75}O_4$-PEG 6000 nanoparticles has been successfully performed using a coprecipitation method. $Mn_{0.25}Fe_{2.75}O_4$-PEG 6000 nanoparticles have magnetite phase with the database of Inorganic Crystal Structure Database (ICSD) number 9013529 which can be observed from the XRD data analysis and has the average particle size of 4.97 nm. Through a global fitting analysis via two-lognormal distribution, SAXS characterization results showed that the primary and secondary particle sizes of $Mn_{0.25}Fe_{2.75}O_4$–PEG 6000 are 3.08 and 11.4 nm respectively. The aggregation of $Mn_{0.25}Fe_{2.75}O_4$–PEG 6000 nanoparticles occurred as an effect of the PEG 6000 polymer template existence. $Mn_{0.25}Fe_{2.75}O_4$ nanoparticles were trapped in a long network of PEG 6000 polymer chains

Acknowledgments

The authors wish to express their gratitude for the research funding support which is fully funded by the PNBP UM on behalf of SN.

IOP Conf. Series: Journal of Physics: Conf. Series **1093** (2018) 012005 doi:10.1088/1742-6596/1093/1/012005

References

[1] Tzitzios V, Basina G, Bakandritsos A, Hadjipanayis C G, Mao H, Niarchos D, Hadjipanayis G C, Tucek J and Zboril R 2010 Immobilization of magnetic iron oxide nanoparticles on laponite discs – an easy way to biocompatible ferrofluids and ferrogels *Journal of Materials Chemistry* **20** 5418

[2] Gitsioudis G, Stuber M, Arend I, Thomas M, Yu J, Hilbel T, Giannitsis E, Katus H A and Korosoglou G 2011 IRON magnetic resonance: a robust technique for angiography providing high blood-to-tissue contrast within the clinically approved dosage of superparamagnetic nanoparticles *Journal of Cardiovascular Magnetic Resonance* **13** P377

[3] Jin R, Lin B, Li D and Ai H 2014 Superparamagnetic iron oxide nanoparticles for MR imaging and therapy: design considerations and clinical applications *Current Opinion in Pharmacology* **18** 18–27

[4] Taufiq A, Bahtiar S, Sunaryono, Hidayat N, Hidayat A, Mufti N, Diantoro M, Fuad A, Munasir, Rahmawati R, Adi W A, Pratapa S and Darminto 2017 Preparation of Superparamagnetic $Zn_{0.5}Mn_{0.5}Fe_2O_4$ Particle by Coprecipitation-Sonochemical Method for Radar Absorbing Material *IOP Conference Series: Materials Science and Engineering* **202** 012024

[5] Kandasamy G and Maity D 2015 Recent advances in superparamagnetic iron oxide nanoparticles (SPIONs) for in vitro and in vivo cancer nanotheranostics *International Journal of Pharmaceutics* **496** 191–218

[6] Kafrouni L and Savadogo O 2016 Recent progress on magnetic nanoparticles for magnetic hyperthermia *Progress in Biomaterials* **5** 147–60

[7] Nizam El-Din H M and El-Naggar A W M 2014 Radiation synthesis of magnetic sensitive ferrogels from poly(ethylene glycol) and methacrylic acid as drug delivery matrices for vitamin B12 *Designed Monomers and Polymers* **17** 322–33

[8] Shahid M K, Phearom S and Choi Y-G 2018 Synthesis of magnetite from raw mill scale and its application for arsenate adsorption from contaminated water *Chemosphere* **203** 90–5

[9] Tipsawat P, Wongpratat U, Phumying S, Chanlek N, Chokprasombat K and Maensiri S 2018 Magnetite (Fe3O4) nanoparticles: Synthesis, characterization and electrochemical properties *Applied Surface Science* **446** 287–92

[10] Sunaryono, Taufiq A, Munaji, Indarto B, Triwikantoro, Zainuri M and Darminto 2013 Magneto-elasticity in hydrogels containing Fe[sub 3]O[sub 4] nanoparticles and their potential applications pp 53–6

[11] Ferrari S, Bilovol V, Pampillo L G, Grinblat F and Errandonea D 2018 High pressure in-situ X-ray diffraction study on Zn-doped magnetite nanoparticles *Solid State Sciences* **77** 1–4

[12] Mohapatra M, Pandey B, Upadhyay C, Anand S, Das R P and Verma H C 2005 Effect of Ni doping on the properties of fine magnetite particles *Journal of Magnetism and Magnetic Materials* **295** 44–50

[13] Taufiq A, Sunaryono, Rachman Putra E G, Okazawa A, Watanabe I, Kojima N, Pratapa S and Darminto 2015 Nanoscale Clustering and Magnetic Properties of Mn x Fe3−x O4 Particles Prepared from Natural Magnetite *Journal of Superconductivity and Novel Magnetism* **28** 2855–63

[14] Güner S, Amir M, Geleri M, Sertkol M and Baykal A 2015 Magneto-optical properties of Mn3+ substituted Fe3O4 nanoparticles *Ceramics International* **41** 10915–22

[15] Bahtiar S, Taufiq A, Sunaryono, Hidayat A, Hidayat N, Diantoro M, Mufti N and Mujamilah 2017 Synthesis, Investigation on Structural and Magnetic Behaviors of Spinel M-Ferrite [M = Fe; Zn; Mn] Nanoparticles from Iron Sand *IOP Conference Series: Materials Science and Engineering* **202** 012052

[16] Veisi H, Gholami J, Ueda H, Mohammadi P and Noroozi M 2015 Magnetically palladium catalyst stabilized by diaminoglyoxime-functionalized magnetic Fe3O4 nanoparticles as active and reusable catalyst for Suzuki coupling reactions *Journal of Molecular Catalysis A: Chemical* **396** 216–23

[17] Zhang B, Tang G, Yan Z, Wang Z, Yang Q and Cui J 2007 Synthesis of magnetic manganese ferrite *Journal of Wuhan University of Technology-Mater. Sci. Ed.* **22** 514–7

[18] Meng X and Zhang Z 2017 Synthesis and characterization of plasmonic and magnetically separable Ag/AgCl-Bi 2 WO 6 @ Fe 3 O 4 @SiO 2 core-shell composites for visible light-induced water detoxification *Journal of Colloid and Interface Science* **485** 296–307

[19] Sundari R, Hua T I, Aziz M and Nizar D 2014 The characterization study of ferrites (magnesium and manganese) using sol gel method *Malay. J. Analyt. Sci* **18** 485–490

[20] Singh N H, Bera A, Debnath A and Saha B 2018 Mixed Phase Crystalline Hausmannite and Manganese Ferrite Nanoparticles with Magnetic Properties for Environmental Application *Materials Today: Proceedings* **5** 2300–2305

[21] Wang G, Zhao D, Ma Y, Zhang Z, Che H, Mu J, Zhang X and Zhang Z 2018 Synthesis and characterization of polymer-coated manganese ferrite nanoparticles as controlled drug delivery *Applied Surface Science* **428** 258–63

[22] Vamvakidis K, Katsikini M, Vourlias G, Angelakeris M, Paloura E C and Dendrinou-Samara C 2015 Composition and hydrophilicity control of Mn-doped ferrite ($Mn_x Fe_{3-x} O_4$) nanoparticles induced by polyol differentiation *Dalton Transactions* **44** 5396–406

[23] Catauro M, Renella R A, Papale F and Vecchio Ciprioti S 2016 Investigation of bioactivity, biocompatibility and thermal behavior of sol–gel silica glass containing a high PEG percentage *Materials Science and Engineering: C* **61** 51–5

[24] Ghosh R, Pradhan L, Devi Y P, Meena S S, Tewari R, Kumar A, Sharma S, Gajbhiye N S, Vatsa R K, Pandey B N and Ningthoujam R S 2011 Induction heating studies of Fe3O4 magnetic nanoparticles capped with oleic acid and polyethylene glycol for hyperthermia *Journal of Materials Chemistry* **21** 13388

[25] Sun J, Zhou S, Hou P, Yang Y, Weng J, Li X and Li M 2007 Synthesis and characterization of biocompatible Fe3O4 nanoparticles *Journal of Biomedical Materials Research Part A* **80A** 333–41

[26] Yusuf S M, Mukadam M D, De Teresa J M, Ibarra M R, Kohlbrecher J, Heinemann A and Wiedenmann A 2010 Structural and magnetic properties of amorphous iron oxide *Physica B: Condensed Matter* **405** 1202–6

[27] Taufiq A, Sunaryono, Rachman Putra E G, Okazawa A, Watanabe I, Kojima N, Pratapa S and Darminto 2015 Nanoscale Clustering and Magnetic Properties of Mn x Fe3−x O4 Particles Prepared from Natural Magnetite *Journal of Superconductivity and Novel Magnetism* **28** 2855–63

The Effects of Catalyst Weight and Mole Ratio of Oil-Methanol on Crude Palm Oil (CPO) Esterification using H_2SO_4 (3M)/ Clay Catalyst

Nurhayati*, **Sofia Anita, Tengku Ariful Amri, Muhdarina**

Department of Chemistry FMIPA University of Riau Kampus Bina Widya Km. 12.5 Simpang Baru, Pekanbaru, Indonesia 28293.

*Corresponding author's email: nurhayati@lecturer.unri.ac.id

Abstract. This study aimed to synthesize biodiesel from CPO using heterogeneous clay catalyst activated with H_2SO_4 (3M) in the esterification step. The catalyst was synthesized by the method of reflux at a temperature of 85 °C. The catalyst was characterized using XRD and XRF. The synthesized catalyst was used to make biodiesel from CPO by esterification process with the variations of catalyst weight and the mole ratio of oil to methanol. The purity of biodiesel was analyzed using GC. The XRD results showed that sulfuric acid activated clay consisted of sharp intensities of silica and this was supported by XRF results in which the silica content was 87.05% while Al_2O_3 was 11.90%. Heterogeneous acid catalyst of H_2SO_4 / clay with 3M H_2SO_4 concentration could reduce a free fatty acid content of the oil as much as 85.2%, ie from 6.61% to 0.98%. The maximum Biodiesel yield after the process of transesterification using CaO catalyst from shells of blood cockle (*anadara granosa*) was 71.90%. The maximum Biodiesel yield was obtained at the optimum esterification reaction conditions that were the weight of the catalyst of 2%, the mole ratio (oil: methanol) 1:18 with a reaction temperature of 65 ± 2 °C for 3 hours.

Keywords: Clay catalyst, *anadara granosa*, oil-methanol, crude palm oil, esterification

1. Introduction

Biodiesel is an alternative diesel oil that is generally defined as a Fatty Acid Methyl Ester (FAME) of plant oil, or animal fat. Because of the increasingly low availability of petroleum fuels, biodiesel has attracted attention lately as it comes from renewable sources. Biodiesel may be obtained from the reaction of esterification and transesterification between oils with monohydric alcohols in an acidic and alkaline catalyst [1-4]. But the cost of production is higher than the production cost derived from petroleum fuel. To lower prices and make biodiesel compete with petroleum, the selection of cheaper materials and catalysts for biodiesel production needs to be taken into account.

Crude palm oil (CPO) has the best potential in the process of biodiesel synthesis because the available amount is abundant and its properties so it is worth more economically. According to the Minister of Industry, in 2013 the need for Indonesia Oil Palm (MGS) in 2013 was estimated to reach 5.22 million tons or an average of 435 thousand tons per month. CPO or unrefined palm oil is also much cheaper than refined cooking oil, however, it also contains high free fatty acids [1,3]. Free fatty acids contained in oil when reacting with an alkaline catalyst will produce soap, resulting in reduced biodiesel production. Thus, to increase the biodiesel yield processed from the oil it is necessary to do two synthesis steps, the first step is esterification of free fatty acid with an acid catalyst and followed by transesterification with base catalyst.

The synthesis of biodiesel by esterification process of CPO using homogeneous acid catalyst and continued with transesterification using CaO catalyst has been reported [1]. The use of the homogeneous acid catalyst (H_2SO_4) in esterification has high biodiesel yield, but it has some disadvantages, such as less economical because once used, it requires a lot of solvents to wash the biodiesel product. Promising alternatives are the use of heterogeneous acid catalysts.

The use of clay as a heterogeneous catalyst has several advantages, such as environmentally friendly, has a large surface and has a high cation exchange capability. Several studies have been conducted to activate the natural clay and used as a heterogeneous catalyst in the transesterification process [4,5]. Activation natural clay with a strong acid such as sulfuric acid (H_2SO_4) is expected to increase the acidity and surface area of the clay, and it can be used as a heterogeneous acid catalyst for the esterification process in biodiesel synthesis. According to Michelle and Angelo, the acid treatment of a Brazilian natural clay gave excellent catalytic properties and provided good conversion in the esterification of fatty acids [4]. In addition, H-Mordenite (Al/Si ratio = 19) treated with phosphoric acid is an efficient alternative to a homogeneous acid catalyst for esterification of free fatty acid in neem oil [5]. Therefore, in this study biodiesel synthesis was done by two stages, esterification using heterogeneous acid catalyst activated with sulfuric acid (H_2SO_4 / Clay) and transesterification using CaO catalyst from the shell of blood cockle (*anadara granosa).*

2. Materials and Methods

2.1. Instrumentation and materials

The instrumentation used in this research was analytical balance (Mettler Toledo AL204), oven (Heraeus Instrument D-63456), desiccator, furnace (Naberthem type L31R), crucible, mortar, hotplate stirrer (RSH-IDR), magnetic stirrer (spin bar), three-neck flask, condenser, thermometer, separating funnel, Ostwald viscometer, pycnometer, Cleveland flash point tester and other glassware. The materials used wee Crude palm oil (CPO), natural clay from Maredan Village, isopropyl alcohol (IPA), phenolphthalein indicator, KOH 0.1 N, potassium hydrogen phthalate, acetone, HCl 0.5 N, CCl_4, reagent wijs, KI, $Na_2S_2O_3$, and starch solution.

2.2. Procedure

2.2.1. Activation and characterization of clay heterogeneous acid catalyst

The Maredan natural clays were washed several times to remove the impurities, then they were dried overnight in an oven at 105 °C. Chemical activation of natural clay was carried out by adding 300 ml of H_2SO_4 (3M) to 30 grams of clay, then the mixture was refluxed for 1 hour at a temperature of 85 °C. The resulting catalyst was heated to 105 °C, washed several times, dried and calcined at 500 °C for 3 hours.

2.2.2. Biodiesel synthesis

The esterification process was carried out by mixing the catalyst of H_2SO_4 (3M)/clay and methanol refluxed while stirring at 70 °C for 3 hours. After reacting, the mixture was inserted into the separating funnel and washed with warm water. The bottom wash water was removed and the top was inserted into a three-neck flask for the transesterification process [1]. The transesterification process was done using CaO heterogeneous catalyst derived from *anadara granosa* [1, 9] calcined at 900 °C for 10 h.

3. Results and Discussion

3.1. Catalyst Characterization

Characterization of mineral content and chemical composition on catalyst clay activated with sulfuric acid (3M) analyzed using X-ray diffraction (XRD) and XRF respectively. The XRD results were compared to JCPDS (Join Committee Powder Diffraction Structure) standard by Markgraf and Reeder [6]. The XRD data of the H_2SO_4/clay heterogeneous catalyst are shown in figure 1.

The XRD pattern on the activated Maredan clay of 3 M sulfuric acid (H_2SO_4) shows some peaks kaolinite loss. The main peak appear was non-mineral (Quartz) which was found at 2 thetas of 26.61°, 20.84° and 50.09°. The peak characterization of muscovite can be seen from the diffraction peak at 2θ of 36.52, 68.3 and 42.43, and this is in accordance with the research that has been done by Bhattacharyya and Gupta [7]. Furthermore, from the result of chemical composition analysis using XRF (Table 1), it can be seen that the main content of catalyst H_2SO_4 (3M)/ clay catalyst was silica (SiO_2) of 87.05%, followed by alumina (Al_2O_3) of 11.91%.

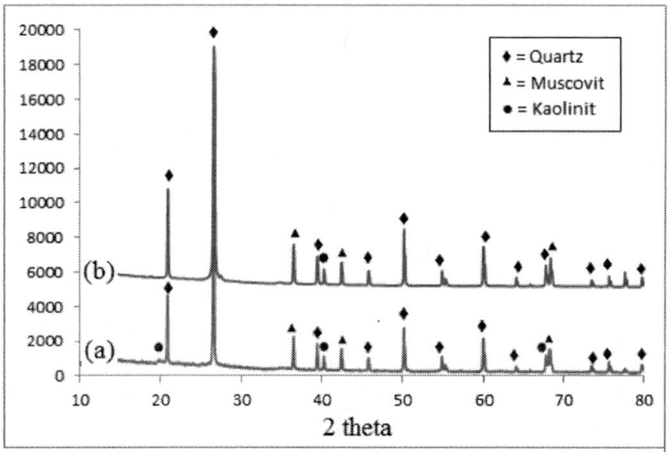

Figure 1. XRD patterns of (a) natural clay and (b) H_2SO_4 (3M)/ natural clay

Table 1. XRF results of clay activated H_2SO_4 (3M)

Elements	Concentration (%)
Al	10.229
Si	87.502
P	0.422
K	0.607
Ca	0.119
Ti	0.581
Cr	0.008
Fe	0.374

3.2. Catalyst effect on biodiesel analysis

In biodiesel synthesis by esterification process, free fatty acid content was always controlled (determined). This was done to find out how much free fatty acid (FFA) was reduced in the process. This is due to the transesterification process (continued from the esterification process) the FFA content in oil should not exceed 2%. If a sample has free fatty acid content and water content > 1% then when reacted with the catalyst will form an emulsion that can cause the formation of soap. The soap formation will affect the biodiesel separation process. Therefore, before the transesterification process, esterification process was firstly done to convert fatty acid into methyl ester while the oil having water content of <1% can be directly carried out by transesterification reaction [8].

Table 2 shows the results of biodiesel synthesis with variations in catalyst weight (1, 2, and 3%) and the mole ratio of oil to methanol (1: 6, 1:12, and 1:18). From the table, it shows that the content of FFA decreased from 6.61% to 0.98 %. The optimum biodiesel yield was obtained at 71.90% found when the catalyst weight of 3M H_2SO_4/clay was 2%, the mole ratio of oil to methanol 1:18 and the reaction temperature of 65 ± 2 °C for 3 hours. When compared to the biodiesel yield using catalyst natural Maredan clay (64.04%), biodiesel yield using H_2SO_4(3M)/clay catalyst showed greater results.

Table 2. The Effect of catalyst weight and the mole ratio of oil to methanol on biodiesel yield

Catalyst weight	Molar ratio oil: Methanol	FFA after esterification (%)	Water content (%)	Biodiesel yield (%)
1%	1:6	1.83	0.07	55.37
	1:12	1.70	0.06	50.82
	1:18	1.96	0.06	31.95
2%	1:6	1.40	0.05	43.75
	1:12	1.40	0.08	49.35
	1:18	**1.83**	**0.05**	**71.90**
3%	1:6	1.40	0.07	42.67
	1:12	0.98	0.06	45.92
	1:18	1.28	0.06	31.95

From the Table 2, it can be seen that when the number of catalysts increased the number of molecules formed would increase and the rate of reaction would also increase until it reached the maximum state (2%). The addition of the catalyst further does not increase the yield of biodiesel but otherwise, the biodiesel yield decreases. This is because a certain amount of catalyst can react with triglycerides and methanol. When the use of a catalyst with excess amounts then the mass transfer between the catalyst and the reactants will decrease so that the interaction decreases and the biodiesel yield is reduced. In addition. By adding an excess amount of catalyst will cause the catalyst to accumulate in the glass reactor wall and will affect the equilibrium of the reaction. so that at the end of the reaction will produce more by products such as soap formation [10].

3.3. Characterization biodiesel by GC analysis

GC analysis was used to determine the compound of methyl ester of biodiesel produced. The chromatogram of GC results can be seen in Figure 2 which shows the presence of 12 peaks detected as methyl esters at the first peak with an area of 26.95% with a retention time of 33.848 minutes (pentadecanoate acid/methyl myristate). The second largest area followed by pale eleven with area of 21.00% with a retention time of 39.951 min (hexadecanoic acid / methyl palmitate). and the third largest area at the fourth peak of 18.90% with retention time of 35.78 min (9 octadecanoic acid/methyl oleate).

IOP Conf. Series: Journal of Physics: Conf. Series **1093** (2018) 012006 doi:10.1088/1742-6596/1093/1/012006

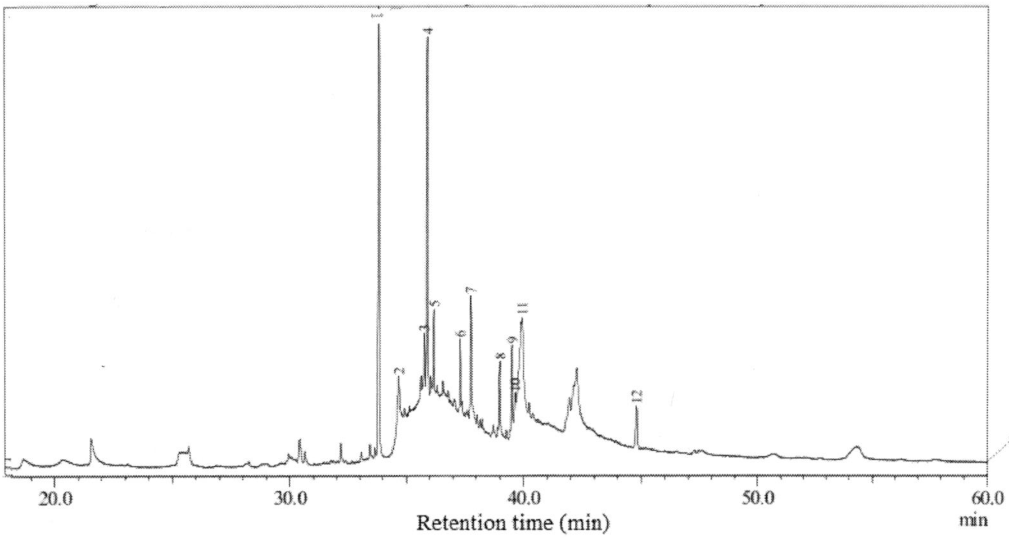

Figure 2. The chromatogram GC of biodiesel

4. Conclusion

The synthesis of biodiesel was done by esterification process using H_2SO_4 (3M)/clay of heterogeneous catalyst and transesterification process using CaO catalyst from blood cockle shell calcined at 900 °C for 10 hours. The mineral types in H_2SO_4 (3M)/clay heterogeneous catalysts were kaolinite and muscovite while non-mineral type was quartz. with silica (SiO_2) content was 87.05% and alumina (Al_2O_3) was 10.229. The heterogeneous acid catalyst could reduce free fatty acid CPO by 85% from 6.61% to 0.98%. The maximum yield of biodiesel after going through two stages (esterification and transesterification) was 71.90% obtained under optimum esterification conditions with a heterogeneous catalyst H_2SO_4 (3M)/clay 2% w/w and a mole ratio of oil to methanol was 1:18.

Acknowledgments

The authors acknowledge the support of the Ministry of Research and Technology. Indonesian Directorate General of Higher Education through LPPM UR for the financial support given in this research under PTUPT Grant 2017.

References

[1] Nurhayati. Anita S. Amri TA and Linggawati A 2017 Indones. J. Chem. 17 (2) 309–315
[2] Michelle. J.C.R and Angelo C.P.. Renewable Energy 92 (2016) 171-177
[3] Devitria. Nurhayati. Sofia A 2013 Indonesian Chemia Acta 3 (2) 39-45
[4] Birla A. Singh B. Upadhyay SN. and Sharma YC 2012 Bioresource Technology 106 95–100
[5] Sathya SV. Varathachary TK. Selvaraj DK. Ponnusamy V. Subramanian S. 2010 Bioresource Technol. 101 5897–902.
[6] Markgraf S A and Reeder R J 1985 American Mineralogist 70 590-600
[7] Bhattacharyya KG and Gupta SS 2008 Advances in Colloid and interface Sci.. 140 (2). 114-131
[8] Kouzu M. Kasuno T. Tajika M. Sugimoto Y. Yamanaka and Hidaka. 2008 J. *Fuel* 87 2798-2806
[9] Buasri A. Chaiyut N. Loryuenyong V. Worawanitchaphong P and Trongyong V 2013 *The Scientific World Journal* 2013. Article ID 460923
[10] Hayyan A. Alam Md Z. Mirghani MES. Kabbashi NA. Hakimi NINM Siran YM and Tahiruddin S 2010 *Bioresourse Technology* 101 7804-11

Moringa oleifera Leaf Powder Madura Variety: Characterization and Biomaterials Property for Biomedical and Nanotechnology Application

H Susanto[1,*], A Taufiq[2], S Sunaryono[2], S Soontaranon[3], Y A Hariyanto[2], A I Mawardi[4], N G Adreyanto[1], D T Yunisa[1], F Rufiandita[1], F Nizarghazi[1], G Alifi[1], L N Putri[1], S D M Kurnia[1], Sumardi[1]

[1]Department of Biology, Faculty of Mathematics and Natural Sciences, Universitas Negeri Malang, Jl. Semarang 5 Malang 65145, Indonesia
[2]Department of Physics, Faculty of Mathematics and Natural Sciences, Universitas Negeri Malang, Jl. Semarang 5 Malang 65145, Indonesia
[3]Synchrotron Light Research Institute, 111 University Venue, Nakhon Ratchasima 30000, Thailand
[4]Department of Post Graduates Studies, Universitas Islam Negeri Sunan Ampel, Surabaya 60237, Indonesia

*Corresponding author's email: hendrabio@um.ac.id

Abstract. Biomaterials have a potential property and become the future target in the advanced bioengineering and its application. One of the local biodiversities is *Moringa oleifera* widely spread in the East region of Indonesian archipelago especially Madura islands. *Moringa oleifera* leaf ingredients are a promising local source for a future target in biomaterials development, biomedical science, bionanotechnology, and engineering. Importantly, *Moringa oleifera* leaf powder consists of a higher concentration of organics materials and some essential minerals. However, it is still lack of information of the specific organic compounds and minerals within the Madura variety samples and how the structural characteristics of this biomaterials. The goal of this study was to characterize the essential ingredients and its structural properties within the *Moringa oleifera* leaf powder from Madura variety. The samples of *Moringa oleifera* leaf powder were characterized by X-ray fluorescence spectroscopy (XRF), small angle X-ray scattering (SAXS), and organic compound analysis by the proximate test. Interestingly, the significant content of protein, carbohydrate, Vitamin A, and vitamin C was observed in these samples while the lower level of lipid was confirmed. The higher level of Calcium, Phosphor, Ferrum, Potassium, Sodium, Sulfur, Molybdenum, Barium, and Manganese was found in the samples. The primary and secondary particle size and fractal dimension were obtained from SAXS data analysis proving that the synthesized green nanoparticles from *Moringa oleifera* leaf powder Madura variety have a significant feature as a potential candidate for nanomaterials in biomedical administration. Thus, the underlying data of organic ingredients, minerals concentration, and structural feature provide a hallmark for the *Moringa oleifera* leaf powder Madura variety as the potential future candidate in biomaterials and biomedical application linked to bioengineering development.

Keywords: *Moringa oleifera*, leaf powder, biomaterial, biomedical, nanotechnology.

1. Introduction

Nowadays, the advanced development of materials based on biocompatibility is a long-term target for bioengineering, medical application, and biomaterials-based study. The characterization of materials property and chemical activity of this biomaterials has been attracting more attention to exploring as the potential future candidate for natural and harmless materials. Natural materials sources have been widely used in the several fields including the miracle flora from Asia, *Moringa oleifera*. This plant has been established as the novel biomaterials resources and was explored to be a potential natural material for environmental safety and bioengineering [1–4], bioenergy [5–7], biomedical sciences [8–11], poultry science [12], and others. The greener synthesis of nanoparticles from natural resources suggested the harmless waste materials and environmental friendly [13].

Importantly, the recent report showed that malnutrition and metabolic syndrome played a valuable contribution to the global health problem [14]. Increasing the mortality incidence in younger age or infant is significantly associated with short-term breastfeeding period, metabolic perturbation, and caloric imbalance [15]. The lack of access to nutritional food in the lower income community especially from developing countries results in severe undernutrition status in the long-term period [14,16]. As a consequence, the susceptibility of this population to several metabolic and infectious diseases gradually increased in the last decade [17,18]. According to the local wisdom and previous data, *Moringa oleifera* is the promising material combating nutritional deficiency and metabolic imbalance in the population [19]. The Moringa is well established as the traditional medicinal plant with higher nutrition due to the presence of some essential nutrients in their leaf [20,21].

Despite serving as the future candidate for undernutrition therapy in the clinical or biomedical application, in advanced bioengineering development, *Moringa oleifera* is essential to produce biomaterials or green materials [22,23]. Several previous works reported that *Moringa oleifera* has been applied as a novel nano-hydroxyapatite for bone repair [24], electrochemical application [25], ZnO green nanomaterials [26], NiO nanoparticles and cancer cell proliferation inhibition [27,28], TiO nanoparticles synthesis for wound healing model [29], and antibacterial agent [30]. Importantly, the moringa leaf easily found abundantly around the Indonesian archipelago, especially in Madura islands opens new opportunity to be used for producing the bionanomaterials. Therefore, due to the simplicity, stability, and biocompatibility of this biomaterials, in this work, the leaf powder from the local bioresources was selected to analyze the chemical property and its potential application. Even though some studies have been done with their preliminary laboratory work related to green synthesis of moringa nanoparticles, however, there is limited information on how the fundamental molecular and structural characteristics of moringa leaf powder originated from Madura variety. Moreover, there is no report on biomaterials synthesis by using moringa leaf powder of Madura variety.

Here, in this study, the preparation of moringa leaf powder using a simple method provided an opportunity to produce the organic leaf powder with higher nutrients concentration and potential rough biomaterials for applied nanotechnology. The employment of organic and an organics protocol in this work is related to its advantages in producing high-quality bionanomaterials. The characterization of green nanomaterials from local biodiversity in particular moringa leaf powder Madura variety by employing co-precipitation technique will be explored in the preliminary laboratory work. Furthermore, the selected properties such as organic compounds, minerals, structure, and particles size of the samples are also discussed.

2. Materials and Methods

The samples preparation were conducted by collecting Moringa leaf from a local variety of Madura islands. The Moringa leaf powder was processed from the fresh leaf harvested in the local farming area. The leaf of *Moringa oleifera* was washed by clean water and dried at room temperature (25-27 °C) for 4-5 days. The room humidity was maintained to avoid ingredient degradation during the drying step. Next, the dried leaf was collected and drilled followed by filtering step with specific mesh. The Moringa leaf powder samples were then stored at sterilized aluminum foil bag for the green nanomaterials characterization.

Green nanomaterials synthesis of *Moringa oleifera* Madura variety was fabricated using the co-precipitation method. *Moringa oleifera* leaf powder was dissolved in 45 mL HCl at room temperature, and the solution then was filtered. In the following step, about 15 mL solution was titrated by using 28 mL NH_4OH and stirred for 30 min. Furthermore, the resulted solution was washed in the serial process and filtered immediately. The measurement of nutritional ingredient (proximate analysis) and bio-nanomaterials characterization (X-ray fluorescence/XRF spectroscopy) were conducted at *Sentral Ilmu Hayati* Laboratory, Brawijaya University, and Sentral Laboratory, Universitas Negeri Malang, Indonesia. In the final laboratory work, to investigate the particle size and structural properties, the samples were also characterized by using synchrotron-based small-angle angle X-ray scattering (SAXS) owing by SLRI, Thailand.

3. Results and Discussion

The primary step in our preliminary study was to access the essential macro and micronutrient ingredients within Moringa leaf powder Madura variety. Organic analysis showed that Moringa leaf powder Madura sample has a higher concentration of protein (> 32% /100 g dried leaf powder), lower lipid level (<6% /100 g dried leaf powder), and significantly higher for carbohydrate (> 37% /100 g dried leaf powder) (Figure 1A). Furthermore, very importantly, the significance of vitamin A and vitamin C concentration was also observed in this sample while no significant difference for Vitamin B1, B2, and B3 in the same sample (Figure 1B). Next, the primary characteristics of the basic compound within moringa leaf powder are shown in Table 1.

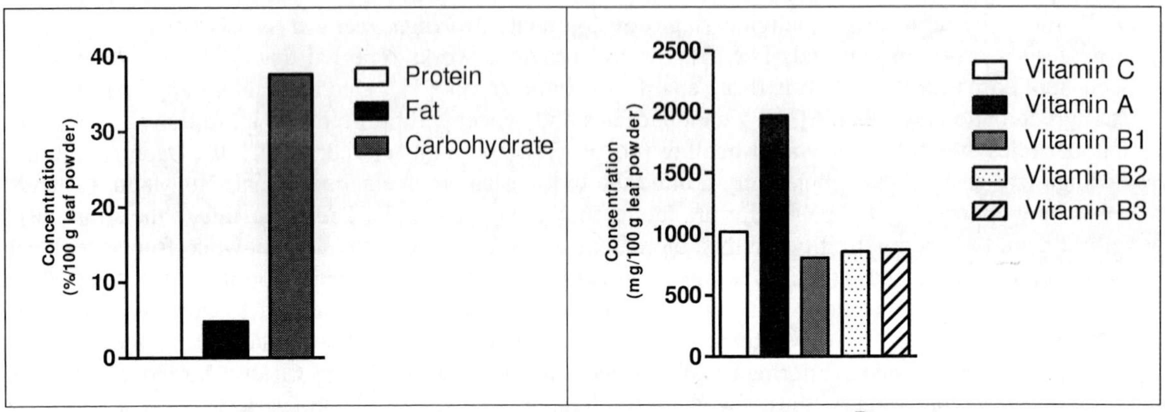

A B

Figure 1. The basic bioproperties of Moringa leaf powder Madura variety. (A). Macronutrients concentration within the dried leaf of *Moringa oleifera*, (B). Vitamins concentration within the dried leaf of *Moringa oleifera*

The raw and synthesized *Moringa oleifera* was characterized using synchrotron SAXS. The scattering profiles of the samples and their fitting analysis results are shown in Figure 2 and Table 2, respectively. The particles distribution equation of the primary particles was employed to the log-normal model [31]. In this case, we combined bilog normal spherical and mass fractal models to investigate the size distribution, structural and form factors as well as primary and secondary particles size distributions of the samples [31, 32].

Table 1. Elemental content of *Moringa oleifera* leaf powder Madura variety

No	Elemental content	Concentration (%)
1.	Ca	68.5
2.	Mg	0.5
3.	P	1.6
4.	Na	3.8
5.	K	14.7
6.	Fe	2.1
7.	S	3.0
8.	Cu	0.3
9.	Mn	10
10.	Ni	0.1
11.	Mo	8.3
12.	Ba	3.4
13.	Ti	0.2
14.	Si	0.5
15.	Cr	1.1
16.	Yb	0.5

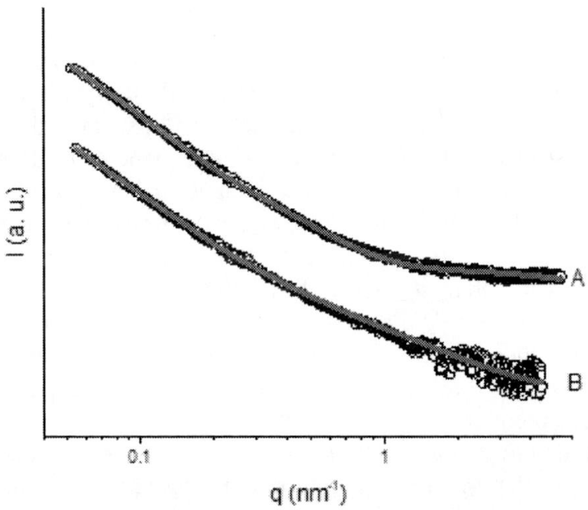

Figure 2. Synchrotron SAXS profiles of the (A) raw and (B) synthesized *Moringa oleifera*

Based on Figure 2, the experimental data (represented by circles) were fitted by the mathematical model (represented by red line) through SASfit program [33]. Based on Table 2, it is shown that the primary particle size of the Moringa leaf powder decreased significantly after the synthesizing process. This data implied that the co-precipitation method at room temperature can be applied to produce the smaller particle size of green nanomaterials from Moringa leaf powder. This result is in line with the previous work presenting that the natural nano-sized magnetite could be prepared effectively by employing such a method [31]. The decrease in the particle size confirmed that the green nanoparticles are able to be economically and easily synthesized by this method. Therefore, the prepared samples in this work are potentially able to be used in reducing the toxicity effect and increasing biocompatibility in biomedical applications.

Table 2. SAXS analysis of *Moringa oleifera* leaf powder Madura Variety

	A	**B**
Primary particle size (nm)	3.45	1.75
Secondary particle size (nm)	8.90	5.10
Fractal Dimension	2.90	2.88

The major problem of biomaterial application in medical administration is strongly associated with toxicity and how the biomaterial absorption related to the particle size. Thus, the primary concern for advanced green nanomaterial application should be focused on size and harmless activity in the living materials, particularly in human physiology. According to our previous work, it has been proven that the co-precipitation method at an average temperature increased particles solubility and decreased toxicity effect [34]. Also, this study showed that the green nanoparticles have a compact and smooth surface property even synthesized by a simple method. In line with another laboratory data, the material synthesis by using co-precipitation method results in the homogeny nanoparticles related to the surface property and its basic structure [35]. This data was confirmed by the fractal dimension value within a range from 2.88 to 2.90. The fractal dimension illustrated that green nanomaterials from *Moringa oleifera* leaf powder Madura variety performed a higher solubility within the solution because this value indicated that *Moringa Oleifera* has a roughness surface. Importantly, the solubility level of specific materials linearly correlated to the particles size which decreases the particle size will improve the particle solubility in the solution [36].

In the advanced nanotechnology and materials science, the critical process was addressed on nanoparticles synthesis process. Prominently, the synthesis method will determine the physicochemical properties and thermodynamics activity within living materials. The solubility of nanomaterials related to particle size in the living organism fluid is essential for bio-nanomaterials. It will implicate the biological feature of these nanoparticles results in the different activity among those nanoparticles. The fast expansion of biotechnology mixed nanotechnology in the biomedical field provides a central hallmark to explore the activity of specific nanoparticles in biological processes. Also, the higher biocompatibility of nanoparticles was synthesized from green materials promising candidates as a safety agent in the living organism. Importantly, there is limited information of the essential chemical and structural characteristics of moringa leaf powder Madura islands variety. Based on the above explanation, *Moringa oleifera* leaf powder produced by co-precipitation method can be proposed as the novel green materials candidate in medical application. Hence, bionanomaterials from *Moringa oleifera* leaf powder Madura variety can be proposed as an additional and potential biomaterials in medical treatment and others application.

Nowadays, the essential role of *Moringa oleifera* as green biomaterials has been proposed in the biomedical area. *Moringa oleifera* leaf has a potent activity to reduce obesity and diabetes. Isothiocyanate within moringa leaf significantly reduced body weight, gluconeogenesis, alleviate insulin resistance, and post-prandial plasma glucose in T2DM patients [37–39]. Moringa leaf extract was also potential agent against hyperlipidemia [40], reduced hyperglycemia through enhancing mitochondrial respiratory chain [41], children with undernutrition [42,43], propagated neural differentiation [44], and became vital source of vitamin A for younger age [45]. Moreover, the anti-cancer performance of green nanomaterials has received global attention since they showed a significant activity to inhibit the cell proliferation rate and proposed as nanomedicine. Pre-treatment with *Moringa oleifera* showed anti-proliferation effect that tended to reduce the cancer cell growth and induced the cell death as entirely [28]. Based on our preliminary investigation, it was suggested that the *Moringa oleifera* leaf powder Madura variety might have a similar potential role in combating hyperglycemia, hyperlipidemia, protein undernutrition, vitamin A deficiency in a child, and anti-proliferation natural agent against cancer. Also, we hypothesized that the co-interaction between green nanoparticles and cancer cell membrane corroborates the disruption of the cancer cell membrane.

4. Conclusion
Based on the data analysis, we conclude that the green nanomaterials of Moringa leaf powder show better characteristics as a potential candidate for biomedical and nanotechnology than that of other similar materials. Therefore, the Moringa leaf powder Madura variety becomes a novel and essential new alternative biomaterial to be applied in medical treatment and nano-engineering.

Acknowledgments
We thank Prof. Rasjad Indra for supporting proximate analysis and additional data from minerals analysis, and CV. Nurul Jannah Sumenep-Madura for providing primary sample preparation.

References
[1] K.S. George, K.B. Revathi, N. Deepa, C.P. Sheregar, T.S. Ashwini, S. Das, A Study on the Potential of Moringa Leaf and Bark Extract in Bioremediation of Heavy Metals from Water Collected from Various Lakes in Bangalore, Procedia Environ. Sci. 35 (2016) 869–880. doi:10.1016/j.proenv.2016.07.104.
[2] B. Xiong, B. Piechowicz, Z. Wang, R. Marinaro, E. Clement, T. Carlin, A. Uliana, M. Kumar, S.B. Velegol, Moringa oleifera f-sand Filters for Sustainable Water Purification, Environ. Sci. Technol. Lett. 5 (2018) 38–42. doi:10.1021/acs.estlett.7b00490.
[3] L. Muruganandam, M.P. Saravana Kumar, A. Jena, S. Gulla, B. Godhwani, Treatment of waste water by coagulation and flocculation using biomaterials, IOP Conf. Ser. Mater. Sci. Eng. 263 (2017) 032006. doi:10.1088/1757-899X/263/3/032006.
[4] G.A.P. Mateus, M.P. Paludo, T.R.T. dos Santos, M.F. Silva, L. Nishi, M.R. Fagundes-Klen, R.G. Gomes, R. Bergamasco, Obtaining drinking water using a magnetic coagulant composed of magnetite nanoparticles functionalized with Moringa oleifera seed extract, J. Environ. Chem. Eng. 6 (2018) 4084–4092. doi:10.1016/j.jece.2018.05.050.
[5] D.M. Fernandes, R.M.F. Sousa, A. de Oliveira, S.A.L. Morais, E.M. Richter, R.A.A. Muñoz, Moringa oleifera: A potential source for production of biodiesel and antioxidant additives, Fuel. 146 (2015) 75–80. doi:10.1016/j.fuel.2014.12.081.
[6] A.K. Azad, M.G. Rasul, M.M.K. Khan, S.C. Sharma, R. Islam, Prospect of Moringa Seed Oil as a Sustainable Biodiesel Fuel in Australia: A Review, Procedia Eng. 105 (2015) 601–606. doi:10.1016/j.proeng.2015.05.037.
[7] B. Mendieta-Araica, E. Spörndly, N. Reyes-Sánchez, F. Salmerón-Miranda, M. Halling, Biomass production and chemical composition of Moringa oleifera under different planting densities and levels of nitrogen fertilization, Agrofor. Syst. 87 (2013) 81–92. doi:10.1007/s10457-012-9525-5.
[8] M. López, M. Ríos-Silva, M. Huerta, Y. Cárdenas, J.A. Bricio-Barrios, M.I. Díaz-Reval, Z. Urzúa, M. Huerta-Trujillo, K. López-Quezada, X. Trujillo, Effects of Moringa oleifera leaf powder on metabolic syndrome induced in male Wistar rats: a preliminary study ,Effects of Moringa oleifera leaf powder on metabolic syndrome induced in male Wistar rats: a preliminary study, J. Int. Med. Res. (2018) 0300060518781726. doi:10.1177/0300060518781726.
[9] M. El-Khadragy, E.M. Alolayan, D.M. Metwally, M.F.S. El-Din, S.S. Alobud, N.I. Alsultan, S.S. Alsaif, M.A. Awad, A.E. Abdel Moneim, Clinical Efficacy Associated with Enhanced Antioxidant Enzyme Activities of Silver Nanoparticles Biosynthesized Using Moringa oleifera Leaf Extract, Against Cutaneous Leishmaniasis in a Murine Model of Leishmania major, Int. J. Environ. Res. Public. Health. 15 (2018). doi:10.3390/ijerph15051037.
[10] Y.A.R. Mahaman, F. Huang, M. Wu, Y. Wang, Z. Wei, J. Bao, M.T.M. Salissou, D. Ke, Q. Wang, R. Liu, J.-Z. Wang, B. Zhang, D. Chen, X. Wang, Moringa Oleifera Alleviates Homocysteine-Induced Alzheimer's Disease-Like Pathology and Cognitive Impairments, J. Alzheimers Dis. JAD. 63 (2018) 1141–1159. doi:10.3233/JAD-180091.
[11] M. Alejandra Sánchez-Muñoz, M.A. Valdez-Solana, M.I. Campos-Almazán, Ó. Flores-Herrera,

M. Esparza-Perusquía, S. Olvera-Sánchez, G. García-Arenas, C. Avitia-Domínguez, A. Téllez-Valencia, E. Sierra-Campos, Streptozotocin-Induced Adaptive Modification of Mitochondrial Supercomplexes in Liver of Wistar Rats and the Protective Effect of Moringa oleifera Lam, Biochem. Res. Int. 2018 (2018) 5681081. doi:10.1155/2018/5681081.

[12] Y. Cui, J. Wang, W. Lu, H. Zhang, S. Wu, G. Qi, Effect of dietary supplementation with Moringa Oleifera leaf on performance, meat quality, and oxidative stability of meat in broilers, Poult. Sci. (n.d.). doi:10.3382/ps/pey122.

[13] O.V. Kharissova, H.V.R. Dias, B.I. Kharisov, B.O. Pérez, V.M.J. Pérez, The greener synthesis of nanoparticles, Trends Biotechnol. 31 (2013) 240–248. doi:10.1016/j.tibtech.2013.01.003.

[14] I. Kolčić, Double burden of malnutrition: A silent driver of double burden of disease in low– and middle–income countries, J. Glob. Health. 2 (2012). doi:10.7189/jogh.02.020303.

[15] D. Walters, S. Horton, A.Y.M. Siregar, P. Pitriyan, N. Hajeebhoy, R. Mathisen, L.T.H. Phan, C. Rudert, The cost of not breastfeeding in Southeast Asia, Health Policy Plan. 31 (2016) 1107–1116. doi:10.1093/heapol/czw044.

[16] H. Delisle, M. Batal, The double burden of malnutrition associated with poverty, The Lancet. 387 (2016) 2504–2505. doi:10.1016/S0140-6736(16)30795-4.

[17] SEANUTS-Proceedings_KL.pdf, (n.d.). https://www.frieslandcampinainstitute.com/app/uploads/2012/11/SEANUTS-Proceedings_KL.pdf (accessed July 5, 2018).

[18] A.R. El-Ghannam, The global problems of child malnutrition and mortality in different world regions, J. Health Soc. Policy. 16 (2003) 1–26. doi:10.1300/J045v16n04_01.

[19] M.D. Thurber, J.W. Fahey, Adoption of Moringa oleifera to combat under-nutrition viewed through the lens of the "Diffusion of Innovations" theory, Ecol. Food Nutr. 48 (2009) 212–225. doi:10.1080/03670240902794598.

[20] maria isabel grijalva, I.V. Fernandez, J.A. Ponce, E. Artalejo, amparo nieblas, L.E. Gonzalez, Supplement Moringa oleifera leaves and their impact on the nutritional status of vitamin A, iron and zinc in preschools: A pilot study, FASEB J. 25 (2011) lb175–lb175. doi:10.1096/fasebj.25.1_supplement.lb175.

[21] R.-Y. Yang, S.C.S. Tsou, T.-C. Lee, L.-C. Chang, G. Kuo, P.-Y. Lai, Moringa, a Novel Plant Rich in Antioxidants, Bioavailable Iron, and Nutrients, in: Herbs Chall. Chem. Biol., American Chemical Society, 2006: pp. 224–239. doi:10.1021/bk-2006-0925.ch017.

[22] Kharissova, O. V., The greener synthesis of nanoparticles - ScienceDirect, (n.d.). https://www.sciencedirect.com/science/article/pii/S0167779913000152 (accessed July 4, 2018).

[23] R. Sood, D.S. Chopra, Metal –plant frameworks in nanotechnology: an overview, Phytomedicine. (2017). doi:10.1016/j.phymed.2017.08.025.

[24] D. Govindaraj, M. Rajan, Synthesis and Spectral Characterization of Novel nano-Hydroxyapatite from Moringaoleifera Leaves, Mater. Today Proc. 3 (2016) 2394–2398. doi:10.1016/j.matpr.2016.04.153.

[25] N. Matinise, K. Kaviyarasu, N. Mongwaketsi, S. Khamlich, L. Kotsedi, N. Mayedwa, M. Maaza, Green synthesis of novel zinc iron oxide (ZnFe2O4) nanocomposite via Moringa Oleifera natural extract for electrochemical applications, Appl. Surf. Sci. 446 (2018) 66–73. doi:10.1016/j.apsusc.2018.02.187.

[26] N. Matinise, X.G. Fuku, K. Kaviyarasu, N. Mayedwa, M. Maaza, ZnO nanoparticles via Moringa oleifera green synthesis: Physical properties & mechanism of formation, Appl. Surf. Sci. 406 (2017) 339–347. doi:10.1016/j.apsusc.2017.01.219.

[27] A.A. Ezhilarasi, J.J. Vijaya, K. Kaviyarasu, M. Maaza, A. Ayeshamariam, L.J. Kennedy, Green synthesis of NiO nanoparticles using Moringa oleifera extract and their biomedical applications: Cytotoxicity effect of nanoparticles against HT-29 cancer cells, J. Photochem. Photobiol. B. 164 (2016) 352–360. doi:10.1016/j.jphotobiol.2016.10.003.

[28] K. Vasanth, K. Ilango, R. MohanKumar, A. Agrawal, G.P. Dubey, Anticancer activity of

Moringa oleifera mediated silver nanoparticles on human cervical carcinoma cells by apoptosis induction, Colloids Surf. B Biointerfaces. 117 (2014) 354–359. doi:10.1016/j.colsurfb.2014.02.052.

[29] V. Sivaranjani, P. Philominathan, Synthesize of Titanium dioxide nanoparticles using Moringa oleifera leaves and evaluation of wound healing activity, Wound Med. Complete (2016) 1–5. doi:10.1016/j.wndm.2015.11.002.

[30] L. Katata-Seru, T. Moremedi, O.S. Aremu, I. Bahadur, Green synthesis of iron nanoparticles using Moringa oleifera extracts and their applications: Removal of nitrate from water and antibacterial activity against Escherichia coli, J. Mol. Liq. 256 (2018) 296–304. doi:10.1016/j.molliq.2017.11.093.

[31] A. Taufiq, Sunaryono, N. Hidayat, A. Hidayat, E.G.R. Putra, A. Okazawa, I. Watanabe, N. Kojima, S. Pratapa, Darminto, Studies on Nanostructure and Magnetic Behaviors of Mn-Doped Black Iron Oxide Magnetic Fluids Synthesized from Iron Sand, Nano. 12 (2017) 1750110.

[32] A. Taufiq, E.G.R. Putra, A. Okazawa, I. Watanabe, N. Kojima, S. Pratapa, Nanoscale Clustering and Magnetic Properties of MnxFe3- xO4 Particles Prepared from Natural Magnetite, J. Supercond. Nov. Magn. 28 (2015) 2855–2863.

[33] J. Kohlbrecher, I. Bressler, Software Package SASfit for Fitting Small-Angle Scattering Curves, (2011). https://kur.web.psi.ch/sans1/SANSSoft/sasfit.html.

[34] R.H. Müller, K. Peters, Nanosuspensions for the formulation of poorly soluble drugs: I. Preparation by a size-reduction technique, Int. J. Pharm. 160 (1998) 229–237.

[35] D.W. Schaefer, Fractal models and the structure of materials, MRS Bull. 13 (1988) 22–27.

[36] R.H. Müller, K. Peters, Nanosuspensions for the formulation of poorly soluble drugs, Int. J. Pharm. 160 (1998) 229–237. doi:10.1016/S0378-5173(97)00311-6.

[37] R. Gupta, M. Mathur, V.K. Bajaj, P. Katariya, S. Yadav, R. Kamal, R.S. Gupta, Evaluation of antidiabetic and antioxidant activity of Moringa oleifera in experimental diabetes: Antidiabetic activity of M. oleifera, J. Diabetes. 4 (2012) 164–171. doi:10.1111/j.1753-0407.2011.00173.x.

[38] C. Waterman, P. Rojas-Silva, T.B. Tumer, P. Kuhn, A.J. Richard, S. Wicks, J.M. Stephens, Z. Wang, R. Mynatt, W. Cefalu, I. Raskin, Isothiocyanate-rich Moringa oleifera extract reduces weight gain, insulin resistance, and hepatic gluconeogenesis in mice, Mol. Nutr. Food Res. 59 (2015) 1013–1024. doi:10.1002/mnfr.201400679.

[39] M. Mbikay, Therapeutic Potential of Moringa oleifera Leaves in Chronic Hyperglycemia and Dyslipidemia: A Review, Front. Pharmacol. 3 (2012). doi:10.3389/fphar.2012.00024.

[40] S. Adisakwattana, B. Chanathong, Alpha-glucosidase inhibitory activity and lipid-lowering mechanisms of Moringa oleifera leaf extract, Eur. Rev. Med. Pharmacol. Sci. 15 (2011) 803–808.

[41] J. Sosa-Gutiérrez, M. Valdéz-Solana, T. Forbes-Hernández, C. Avitia-Domínguez, G. Garcia-Vargas, J. Salas-Pacheco, O. Flores-Herrera, A. Téllez-Valencia, M. Battino, E. Sierra-Campos, Effects of Moringa oleifera Leaves Extract on High Glucose-Induced Metabolic Changes in HepG2 Cells, Biology. 7 (2018) 37. doi:10.3390/biology7030037.

[42] M. Barichella, G. Pezzoli, S.A. Faierman, B. Raspini, M. Rimoldi, E. Cassani, S. Bertoli, A. Battezzati, A. Leone, L. Iorio, V. Ferri, G. Pinelli, C. Pusani, C. Bolliri, R. Cilia, S. Caronni, P. De Marco, E. Cereda, Nutritional characterisation of Zambian Moringa oleifera: acceptability and safety of short-term daily supplementation in a group of malnourished girls, Int. J. Food Sci. Nutr. (2018) 1–9. doi:10.1080/09637486.2018.1475550.

[43] M. Glover-Amengor, R. Aryeetey, E. Afari, A. Nyarko, Micronutrient composition and acceptability of Moringa oleifera leaf-fortified dishes by children in Ada-East district, Ghana, Food Sci. Nutr. 5 (2017) 317–323. doi:10.1002/fsn3.395.

[44] L. Romeo, F. Diomede, A. Gugliandolo, D. Scionti, F. Lo Giudice, V. Lanza Cariccio, R. Iori, P. Bramanti, O. Trubiani, E. Mazzon, Moringin Induces Neural Differentiation in the Stem

Cell of the Human Periodontal Ligament, Sci. Rep. 8 (2018). doi:10.1038/s41598-018-27492-0.

[45] V. Lopez-Teros, J.L. Ford, M.H. Green, G. Tang, M.A. Grusak, L. Quihui-Cota, T. Muzhingi, M. Paz-Cassini, H. Astiazaran-Garcia, Use of a "Super-child" Approach to Assess the Vitamin A Equivalence of Moringa oleifera Leaves, Develop a Compartmental Model for Vitamin A Kinetics, and Estimate Vitamin A Total Body Stores in Young Mexican Children, J. Nutr. 147 (2017) 2356–2363. doi:10.3945/jn.117.256974.

IOP Conf. Series: Journal of Physics: Conf. Series **1093** (2018) 012008 doi:10.1088/1742-6596/1093/1/012008

Indonesian Black Tea Classification Using Fourier-Transform Near-Infrared Spectroscopy and a Principal Component Analysis

R O Anindya[1], J Muninggar[2], and F S Rondonuwu[2,*]

[1] Department of Science and Mathematics, Universitas Kristen Satya Wacana, Jl. Diponegoro 52-60, Salatiga 50711, Indonesia
[2] Research Center of Near-Infrared Applications, Universitas Kristen Satya Wacana, Jl. Diponegoro 52-60, Salatiga 50711, Indonesia

*Corresponding author's email: ferdy@staff.uksw.edu

Abstract. Indonesian black tea (*Camellia sinensis*) is processed based on several standard methods to achieve a certain grade. There are more than 25 grades produced including Dust one (D1), Fanning's (FANN) and Pekoe Fanning's (PF). In general, the physical appearances are similar to each other, which makes it difficult to be differentiated from visual observations by non-experts. During the packaging process, a tea grade can easily be packed using a bag with a different label of the grade. In the global market, such a mistake should be avoided before shipping. Since shipping usually involves a large number of bags, then rapid, accurate, and nondestructive methods are critical. This paper discusses a method for classifying tea grades rapidly, accurately and nondestructively using Fourier-transform near-infrared spectroscopy (FT-NIRS) and a principal component analysis (PCA). Sixty-nine tea samples from grade Dust one (D1), Fanning's (FANN) and Pekoe Fanning's (PF) were separated into two groups; 46 samples were used for the training sets and 30 samples for the validation sets. The analysis successfully differentiated that of 3 different grades, indicating that the method is a potential alternative for tea classification in quality control processes.

Keywords: Indonesian black tea, FTIR, principal, quality control

1. Introduction

By 2015 Indonesia was- the seventh largest exporter of tea in the world after Sri Lanka, China, India, Malawi, and Vietnam. It was a rank lower than that in 2012 where Indonesia occupied the sixth position. The decline in the export quantity is due to the decline in the quality of black tea (*Camellia sinensis*) [1]. The quality of black tea is determined by several characteristics: particle size, appearance, color, liquor taste and appearance in the infusion. The two main factors affecting the grading of black tea are the processing methods (orthodox, and CTC) and the tea particle sizes [2]. Tea processing is done through several stages, namely: the acceptance of raw materials, sifting, grinding, fermentation, drying, and sorting [3-5].

The tea processing in the mass production process with long stages allows the occurrence of errors, such as a mismatch of the content with the grade labels on the packaging, resulting in an inconsistent product quality that decreases the level of market appreciation, which further reduces sales prices.

Each type of tea has a different quality and different selling price; therefore to reduce errors in the production line, a fast, as possible, accurate, and non-destructive examination is absolutely required. Thus, a method using Fourier-transform near-infrared (FT-NIR) spectroscopy and PCA (principal component analysis) to a set of black tea samples were developed.

FT-NIR spectroscopy is well known and widely applied in specific areas of analysis, such as the analysis and identification of components in pharmaceuticals, food, and beverages [6-8]. Those applications are possible because of the FT-NIR ability to penetrate the solid and liquid samples with an electromagnetic spectrum that reaches around 1000 - 2.500 nm or an energy equivalent of about 10.000 - 4.000 cm^{-1}. Research using FT-NIR was also applied in the food study that examined the ratio of sugar content in mango Gedong Gincu [9]. In health, FT-NIR was used to measure uric acid levels in urine using enzymatic methods [10]. With its wide capabilities and applications, FT-NIR was utilized to examine dry and solid tea samples.

This paper reports study of a rapid and non-destructive technique for classifying black tea using FT-NIR spectroscopy and PCA using three different black tea grades from Dust one (D1), Fanning's (FANN), and Pekoe Fanning's one (PFI).

2. Experimental Methods

The method used in this study was basically similar to the one developed by Shou-He Yan [11] with modifications. Three types of tea from various estates in Indonesia namely D1, FANN, and PFI were used as a set of samples, consisting of 96 data sets divided into two groups of 66 training sets in which each grade in training sets consists of 22 samples and 30 validation sets in which each grade in validation sets consists of 10 samples, *vide infra*. Each of the tea samples was stored in an aluminum package to avoid light radiations. Before performing the measurements, each tea sample was placed into a small container and mixed for sample homogeneity. After the mixing process, the tea samples (2-3 grams each) were put into a measuring tube (vial).

Spectral acquisitions were performed using a near infrared spectrometer Buchi Nirfelx 500 solid® that operated on a 4.000-10.000 cm^{-1} energy span. The samples were loaded into vials that were ready for spectral acquisition with the utilization of NIRware spectrometer control software. Prior to analysis using the PCA method, the spectra were initially smoothened by the Savitszky-Golay method that was applied to the 1st and the 2nd derivative signals. The aim of the 1st derivation of the spectrum was to correct the signal baseline by scattering and multiplication, while the 2nd derivation aimed to increase the separation of each absorption peak. After the smoothing of the spectrum, the spectral data were analyzed using PCA. The method of PCA has been clearly explained elsewhere [12]. The results of the PCA analysis were then displayed in three-dimensional space by PC1, PC2, and PC3 axis.

3. Results and Discussion

A total of 96 tea samples, as explained in the previous section, consisted of grade D1, FANN, and PFI. From each grade, data were retrieved from 22 training sets (TS) and 10 validation sets (VS), as listed in Table 1. The TS was served as a reference for classification, whereas the VS was used to confirm the results of the classification based on the analysis used in accordance with the actual grade.

Table 1. Black tea samples along with the quantity of sample used as the training sets (TS) and the validation sets (VS).

No	Grade	Sample Quantity		
		TS	VS	Total
1	D1	22	10	32
2	FANN	22	10	32
3	PFI	22	10	32

The results of the NIR reflection spectra shown in Figure 1 are the results of 96 tea sample measurements. The spectral patterns are similar to one another because the tea varieties used have common similarities. However, the baseline and the intensity of each spectrum is varied. This is due to the influence of the used sample particles which have different textures that generate varying light scattering [13]. The scattering by different sizes of particles in the diffused-reflection mode of NIR spectral acquisition cannot be avoided. However, 1st derivative spectra are generally applied to bring all spectra into zero baselines [14].

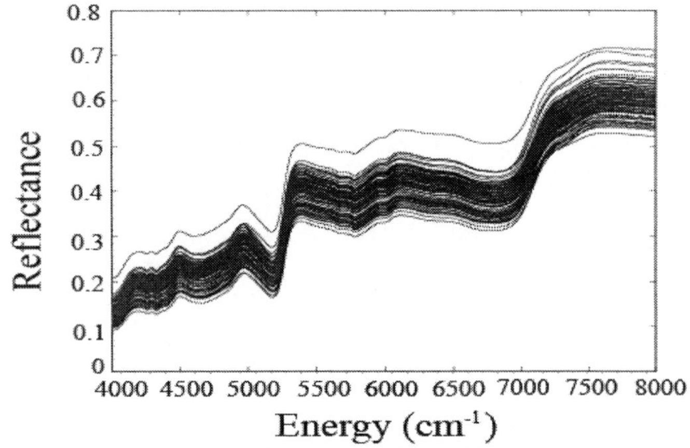

Figure 1. Reflection spectra of 96 black tea samples.

Figure 1 shows the NIR reflectance spectra of 96 black tea samples. This reflectance spectrum contains not only signal from samples but also background and noise. To separate the signal from noise and background for PCA analysis, a pre-process treatment was applied including smoothing and derivations. Smoothing avoids local maximum and minimum artifacts and suppresses noise during spectral derivations, while 2nd derivative spectra improve peaks positions and spectral resolutions [15]. The smoothing was performed using Savitzky-Golay's method before the 1st and 2nd derivative processes. The results of 2nd derivative spectra are shown in Figure 2.

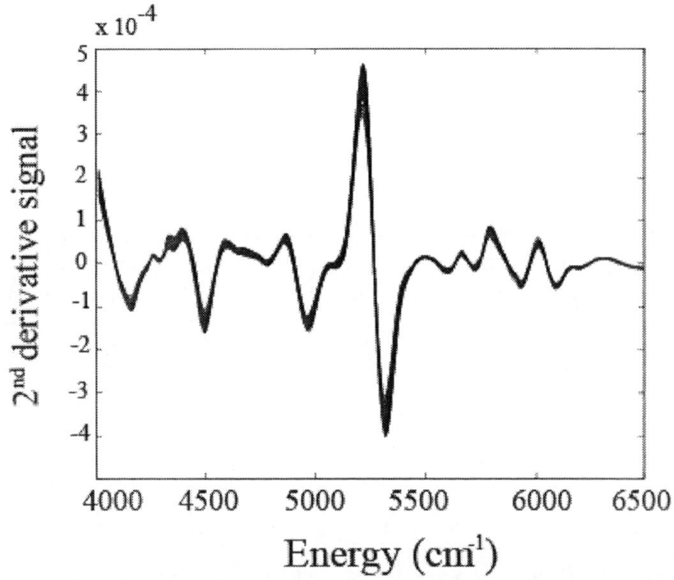

Figure 2. The NIR 2nd derivative spectra of 96 black tea samples.

Figure 2 shows that the spectra are not completely overlapped both in positions and relative intensity. Close examinations showed several peak positions differed up to several tens of wavenumber, indicating slightly deference chemical dynamics are present from sample-to-sample. Several interactions between the chemical components present in the analysed material were well recognized such as the combination of the vibration of the CH molecular bonds in the energy span of 4545-4065 cm^{-1}, the combination of NH and OH in the energy span of 5000 - 4545 cm^{-1}, and the CH overtone in the energy span of 6173 - 5556 cm^{-1}.

Tea grade classifications were performed using the NIR 2nd derivative spectra as it successfully displayed different spectral characteristics among samples. PCA provides the principal component (PC) scores, which was the representations of the analyzed data matrix in PC space. PCA analysis on NIR 2nd derivative spectra of 96 black tea samples showed the first three vectors of PC score in PC space, PC1, PC2, and PC3 successfully classified the training sets and validation sets of samples into three main clusters in which each cluster was practically occupied by the same grade of tea.

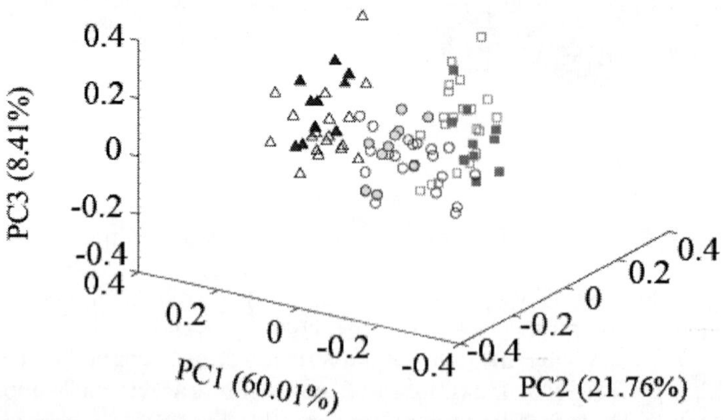

Figure 3. PCA plot for 3 grades of tea: D1 (red square), FANN (blue circle), PFI (blue triangle). The open symbol indicates the training set, and the filled symbol indicates the validation set. The percentage shown in each axis represents total variance explained.

Figure 3 shows three different clusters namely red squared (D1), blue circle (FANN), and the blue triangle (PFI). The open symbol is the training set, and the filled symbol is the validation set. Based on the PCA calculation for 66 training sets data, 60.01% of the variance can be explained by PC1, 21.76% by PC2 and 8.41% by PC3, resulted 90.18% variance can be explained using the above 3 PC's. To validate the PCA classification method, 30 samples consisting of 10 samples for each grade were used as a blind sample test. The validation procedure was done by incorporated only one validation sample in every PCA analysis of validation to ensure that changes of the training sets parameters were minimum. The validation procedures were then repeated for all 22 validation samples. The overall results of the PCA analysis could differentiate 3 grades into PC1, PC2, and PC3, although a very few samples were misclassified. One D1 sample set falls in the FANN cluster, five FANN fall in the dust cluster, and one FANN data falls in the PFI cluster. Complete PCA analysis both for training sets and validation sets are summarized in Table 2.

Table 2. Results of the analysis using PCA on the 3 grades of tested samples.

Data	Grade	Number of Samples	No of Samples Classified by PCA: DUST	FANN	PF I
Training	DUST	22	21	1	0
	FANN	22	5	16	1
	PF I	22	0	0	22
Validation	DUST	10	10		
	FANN	10		10	
	PF I	10			10

Table 2 shows the percentages of the successful prediction by the PCA analysis reaching the accuracy of more than 91% (16 out of 22 samples were correctly classified) for the training sets and 100% (30 out of 30 samples were correctly classified) for the validation data. These results indicated that the method used in this study can be applied to classify the grade of black tea rapidly, accurately, and non-destructively.

4. Conclusion

A combination of Fourier-Transform Near-Infrared Diffusion-Reflection (FT-NIR) spectroscopy and a principal component analysis (PCA) can be used as an alternative method to examine the grade classifications of black tea with a rapid and non-destructive treatment. Tea samples of the same grade will clump together in the same classification, as in this study the samples were classified as PC1, PC2, and PC3. Although it is still necessary to add variations in training sets with different grades, this study showed that the general accuracy level of this classification method is over 90%.

Acknowledgments

The FSR was funded by an internal grant from Universitas Kristen Satya Wacana and Kemenristekdikti (grand 2016). The authors would like to thank Tafip Haryanto for preparing the NIR spectrometer prior measurements.

References

[1] International Tea Committee, 2015, *World Tea Production & Consumption,* The United Kingdom.

[2] Wibowo Z. S. and Sugiat, 1998, *Indonesian tea between east and west,* The 5th Int'l Symposium on tea culture pp 103-117.

[3] Abass T., Kustaniyati B. and Fa S., 2008, *Rancang Bangun Pengendali Sistem Pengolahan Teh Hitam.*

[4] D Setyamidjaja, 2000 ,*Teh: Budi Daya dan Pengolahan Pascapanen* (Yogyakarta: Kanisius).

[5] Hartoyo A., 2003, *Teh dan khasiatnya bagi kesehatan* (Yogyakarta: Kanisius).

[6] Khorasani M., Amigo J. M., Sonnergaard J., Olsen P., Bertelsen P. and Rantanen J., 2015, Visualization and prediction of porosity in roller compacted ribbons with near-infrared chemical imaging (NIR-CI). *J Pharm Biomed Anal.* 109 11-17.

[7] Meng Y., Wan S., Cai R., Jiang B. and Zhao W., 2015, Discrimination and content analysis of fritillaria using near-infrared spectroscopy *J. Anal. Methods Chem.* 2015 1-8.

[8] Chang C. W., Laird D. A., Mausbach M. J. and Hurburgh C. R., 2001, Near-infrared reflectance spectroscopy–principal components regression analyses of soil properties. *Soil Sci. Soc. Am. J* 65(2) 480-90.

[9] Fahri N., Purwanto Y. A. and Budiastra I. W., 2017, Penggolongan Mangga Gedong Gincu

Berdasarkan Rasio Kandungan Gula Asam Menggunakan Prediksi Near Infrared Spectroscopy. *Jurnal Keteknikan Pertanian* 4(1) 31-36.

[10] Johnson R. A. and Wichern D. W., 2002, *Applied multivariate statistical analysis*, New Jersey: Prentice Hall 5(8).

[11] Yan S. -H., 2007, NIR evaluation of the quality tea and its market price, *Spectroscopy Europe*, 16(2) 16-19.

[12] Abdi H., 2010, Partial least squares regression and projection on latent structure regression (PLS Regression), *Wires Computational Statistics*, John Wiley & Sons Inc. 1-10.

[13] Tan S. M., Luo R. M., Zhou Y. P., Gong H. and Tan Z., 2012, Rapid and non-destructive discrimination of tea varieties by near-infrared diffuse reflection spectroscopy coupled with classification and regression trees *Afr. J. Biotechnol.* 11(9) 2303-12.

[14] Dryden G. M., 2003, *Near Infrared Reflectance Spectroscopy: Applications in Deer Nutrition* (Canberra: RIRDC Publication).

[15] Cen H. and He Y. 2007 Theory and application of Near Infrared Reflectance Spectroscopy in Determination of Food Quality *J. Trends in Food Sci. & Technol.* 18 72-83.

IOP Conf. Series: Journal of Physics: Conf. Series **1093** (2018) 012009 doi:10.1088/1742-6596/1093/1/012009

Identification of Peptides Compounds from *Mirabilis jalapa* L. (Caryophyllales: Nyctaginaceae) Potentially as a Biopesticide

Dina Maulina[1,2,*], Sutiman Bambang Sumitro[3], Mohamad Amin[1], Sri Rahayu Lestari[1]

[1]Graduate Program for Biology Education, Universitas Negeri Malang, Jl. Semarang 5 Malang 651465, Indonesia
[2]Faculty of Teacher Training and Education, University of Lampung, Jalan Prof. Dr. Soemantri Brojonegoro No.1, Lampung 35145, Indonesia
[3]Department of Biology, Faculty of Mathematics and Natural Sciences, University of Brawijaya, Jalan Veteran No. 1, Malang, East Java 65145, Indonesia

*Corresponding author's email: dina.maulina@fkip.unila.ac.id.

Abstract. Mirabilis Antiviral Protein (MAP) is a specific protein found in the *Mirabilis jalapa*. The MAP has many promising benefits, one of them is that it can be used as biopesticides. However, the structure and visualization of protein from MAP have not been identified. This study investigated the modelling structures of MAP compounds which potentially become bioinsecticide for insect and the interaction between bonding ligand-receptor in the surface cell. This research used In silico models that refers to the BIOPEP database with the bioinformatics tools methods. The screening of MAP compounds was conducted through hydrolysis steps by 30 enzymes to become new peptides. The new Antimicrobial Peptides (AMPs) were analyzed by a multidimensional statistical analysis using four predicted algorithms, Support Vector Machines (SVM), Random Forest (RF), a Neural Network Artificial (ANN) and Discriminant Analysis (DA) that were available in the Anti-Microbial Peptides Collection (CAMP Database). The results were identified six new types of AMPs from 449 AMPs during the in-silico proteolysis process toward *M. jalapa*. This peptide predicted the function become antimicrobials by inactivating the ribosomes. There were obtained three of the best peptide structures from *M. jalapa* that have the potential as toxin compounds (biopesticides). Through the visualization of the three AMPs, models of biopesticide structures are WIFKTVEEIKLVMGLLKSS, IKLVMGLLKSS, and ITNIRTKVA with the residual category 6-22, and the characterization of molecular weight was 561,7 to 2971,4 g/mol. The AMPs fulfill the Boman Index which the measurement of protein-peptide affinity to build the biological interactions, ranging from -0,51 to 2,98. Further, these renewable peptides can be used to determine the pathways that occur in targeted insect pests as biopesticides.

Keywords: Antimicrobial peptides (AMPs), biopesticide, mirabilis antiviral protein (MAP), *Mirabilis jalapa*.

1. Introduction

The various research about the benefits of secondary metabolite compounds from plants has been studied recently [1-3]. There have been many studies regarding their benefits in pharmacology, medicine and including applications for botanical insecticides [4, 5]. The use of botanical insecticides which derived from natural substances from plant becomes one of the safest alternatives to reduce the number of pest populations. Their applications have not caused chemical residues that would be safe for the environment [6, 7].

One of the botanical insecticides which intended as biopesticides for insect pests is *Mirabilis jalapa* L. (Caryophyllales: Nyctaginaceae). *M. jalapa* has many natural chemical compounds containing triterpenes, proteins, flavonoids, alkaloids, and steroids, alanine, alpha-amyrins, arabinose, beta-amyrins, campesterol, daucosterol, and dopamine [8]. The results of the cDNA sequences of *M. jalapa* plants contain the Mirabilis Antiviral Protein (MAP) compounds [9]. It was potentially as repellent and antifeedant. The MAP has an antiviral and antiviroid substance that functions to eliminating the pathogens that damage the body [10]. The MAP has the ability as an insecticide substance and inactivating the ribosomes' target cells. The specific protein affects to weaken the defense mechanism function of the pest body both physiological and anatomical [11, 12].

The previous study showed that *M. jalapa* able to decreasing the body defense. Target pest by the change number and activity of hemocyte cells from the Spodoptera pest [13]. The exposed larva with *M. jalapa* extract in 0.8% (w/v) concentration for 24 hours experienced the decreased body function signed by decreased movements. The avoiding movement behavior from food sources, vomiting and diarrhea. It shows that the weakening of the insect's immune system caused by inactivating ribosomes which also causes the changes in the mechanism and metabolism of *Spodoptera litura* cells and lead to the changes in overall body activity [14]. The antifeedant compounds which contained in the extract of *M. jalapa* also cause metabolic changes in the function of the gastrointestinal organs.

A number of previous studies on many kinds proteins in *M. jalapa* plant have been widely conducted. However, the special antimicrobial peptides (AMPs) as potentially as a biopesticide and the mechanism in the insect body is not known. The interaction between AMPs with cell membranes in the insect body becomes essential, to further detections the pathway ligand AMPs and receptor binding.

The information related to the structure, visualization and interaction MAP is very critical to understand. It is fundamental to determine the application of *M. jalapa* as a biopesticide. Unfortunately, there is no previous research which revealed the AMPs of *M. jalapa* that potentially as a natural pest control agent. The purpose of this study was to investigate the modeling structures and determine the optimization of the peptides from *M. jalapa*. Moreover, the modeling of the structure of the MAP protein will be known how the cellular interaction occurs inside the insect's body.

2. Materials and Methods

This study used In silico method utilizing bioinformatics tools. The identification of AMPs analyzed the revealed type of protein data sequences. This research is referring to the method developed by Pangastuti *et al.* [15, 16] and Ilmawati *et al.* [17].

2.1. Retrieval of sample:

The peptides compounds in *M. jalapa* was obtained from a Clavilia database [18] of protein structures in the Raintree Nutritions. The cDNA for Mirabilis antiviral protein (MAP) peptide chains were saved in the BIOPEP and protein data bank (PDB).

2.2. Analysis of AMPs from Mirabilis jalapa:

The virtual screening is used to observe the new AMPs from *M. jalapa* as potential as a biopesticide. The analysis was identified by (a) searching for new proteins through FASTA sequences from MAP by the UniProt database [19], (b) predicting the proteins by cutting the structure using enzymes in the BIOPEP database (c) discovering antimicrobial peptides in the Anti-Microbial Peptide collection

(CAMP) database (d) using physicochemical properties in the Innovagen database [20] and antimicrobial database (APD2) [21] e) making prediction of new AMPs, and (f) determining the best protein.

Initially, the collection of the MAP FASTA sequence from the UniProt database to search the new of AMPs. Moreover, the database of sensory peptides and amino acids has an identical structure to the BIOPEP database of biologically active peptides. The information about sensory peptides and amino acids was inserted into the database using standard BIOPEP layouts for bioactive peptides. Information about the biological activity of sensory peptides. The sequence in the BIOPEP database to obtain a new peptide with peptide cutting process using 30 enzymes available in the database.

In addition, the new peptide analyzed by multidimensional statistical analysis as a support vector machine (SVM), random forest (RF), artificial neural network (JST) and discriminant analysis (DA) that available in the CAMP database. The Acquisition of AMPs was selected and compared to the data source through the new AMPs obtained.

The prediction of new AMP was analyzed with the following parameters, such as molecular weight, residual amount, extinction coefficient, iso-electric point, net cost at pH 7, and the predicted solubility using Innovagen and Boman Index using APD2 database. The last stage, the determination of the best protein structure which declared as a biopesticide compound from *M. jalapa* was visualized in 3D.

3. Results and Discussion

3.1. Results

The acquisition of information on the *M. jalapa* potentially as biopesticide was obtained the MAP sequence in the UniProt database. The BIOPEP data source would identify the MAP proteins contained in *M. jalapa*. Further, the peptide chains were cut by using the enzymes that produce polypeptide or peptide. Further, the pieces of polypeptide and peptide were analyzed to find out which polypeptide or peptide which has the function as Anti-Microbial Peptide by selecting to the CAMP database.

The results of the CAMP database by multidimensional statistical analysis supported vector machine (SVM), random forest (RF), artificial neural network (JST) and discriminant analysis (DA). The analysis results showed that there are 6 of 449 potential peptides as antimicrobial peptides. These AMPs compared with manual tools in the BIOPEP database to confirm the recency of the acquired AMP peptide. The six new AMP peptides from *M. jalapa* are summarized in Table 1, and the individual physicochemical AMP properties of *M. jalapa* was presented in Table 2.

There were three amino acids from the analysis according to the best characteristics (Seq. Id. 102, 113, and 395) with the residual category 6-22. This characterization is based on the molecular weight of 561.7 to 2971.4 g/mol. The positively charged peptides are the majority of antimicrobial peptides known as cationic antimicrobial peptides (CAP) [22], which have an isoelectric point close to 10, meaning that the AMP tends to precipitate and lose its biological function [23]. All those AMPs fulfill the Boman Index which the measurement of protein-peptide affinity and its ability to build the biological interactions, ranging from -0.51 to 2.98 for the AMP protein from *M. jalapa* [24]. According to the physicochemical nature of AMP, it was found that AMP with the sequence number ID 102, 113, and 395 is the best AMP (Figure 1).

IOP Conf. Series: Journal of Physics: Conf. Series **1093** (2018) 012009 doi:10.1088/1742-6596/1093/1/012009

Table 1. The new AMPs of *Mirabilis jalapa* with physicochemical properties support

seq. id	Sequence	SVM	RF	ANN	DA
98	LGIFRLENSIVNIYGKAGDV KKQAKFFLLAIQMVSEAARFKYISDKIP	0.527	0.5285	AMP	0.732
102	WIFKTVEEIKLVMGLLKSS	0.858	0.7505	AMP	0.681
113	IKLVMGLLKSS	0.982	0.5875	AMP	0.806
336	SKPSTTTATKCQLATSPVTISPWIFKTVE	0.982	0.5875	AMP	0.806
381	YGDLEKNGGLRKDNPLGIFRLEN	0.820	0.598	AMP	0.656
395	ITNIRTKVA	0.987	0.530	AMP	0.810

Super vector machine (SVM), random forest (RF), artificial neural network (ANN) and discriminant analysis (DA) multidimensional statistical analysis.

The 3D visualization peptides result from the MAP individual above becomes a bright spot to understand the bounding among membranes that may occur inside the target insect body. The visual formation of specific proteins in *M. jalapa* plants is a reference for knowing the cellular pathway in weakening and controlling insect populations. The cellular pathway determination will be an important reference to knowing the further study and the synergism of these compounds in the insect body. Therefore, these findings indicate that *M. jalapa* plant as biopesticide that can be used in controlling insect pest populations which can be assessed through cellular levels.

Table 2. The physicochemical characterization of the AMPs from *Mirabilis jalapa*

Seq. ID	Sequence	Number of residues	Mass weight (g/mol)	Gravy	Boman index (kcal/mol)	Extinction coefficient ($M^{-1}cm^{-1}$)	Isoelectric point (pH)	Net charge at pH 7	Estimated solubility	Aliphatic index	Instability index
98	LGIFRLENSIVNIY GKAGDV KKQAKFFLLAIQM VSEAARFKYISDKI P	48	5426.45	0.18	0.86	2560	10.08	2	Good	109.79	12.95
102	WIFKTVEEIKLVM GLLKSS	19	2221.73	0.59	-0.08	5690	9.71	1	Good	133.16	25.59
113	IKLVMGLLKSS	11	1188.54	1.11	-0.82	0	10.73	2	Poor	168.18	3.45
336	SKPSTTTATKCQL ATSPVTISPWIFKT VE	29	3105.61	0.08	0.94	5690	9.58	1	Poor	67.24	38.94
381	YGDLEKNGGLRK DNPLGIFRLEN	23	2612.93	-1.01	2.64	1280	0	-1	Good	84.78	8.13
395	ITNIRTKVA	9	1015.22	0.19	1.84	0	11.41	2	Good	130	31.58

The grey column shows the AMPs from the analysis according to the best characteristics.

3.2. Discussion

This study observed the MAP compound peptide *M. jalapa* which is an antimicrobial compound potentially as a biopesticide. There are three types of peptide sequences were found WIFKTVEEIKLVMGLLKSS, IKLVMGLLKSS, and ITNIRTKVA. AMPs of these findings are specific compounds of *M. jalapa* containing tannin, triterpene, flavonoids, isoflavones, alkaloids, coumarins, and proteins. This component was secondary metabolites with the specific function as antifeedant and repellent for insects.

Flavonoids are antibacterial compounds that have complex proteins extracellular structure and dissolved. This structure caused damage to the cells followed by the discharge of intracellular component [25]. Also, to their role in DNA-RNA inhibition and synthesis by intercalation or hydrogen bonding with a nucleic acid base accumulation. Flavonoids played a role in inhibitor energy metabolism according to absorb various active metabolites and macromolecular biosynthesis when the cells require sufficient energy.

Tannin is a compound that has a destructive mechanism in insect cells as an inhibitor with activates cell adhesion (the binding molecule on the host). The located of binding was on the surface of cells and enzymes to disrupt the transport of proteins inner layer of cells. Tannins have the receptor targets in the cell membranes polypeptides and capable of damaging it. Since tannins are parts of phenol compounds, it causes to readily be a soluble in the water.

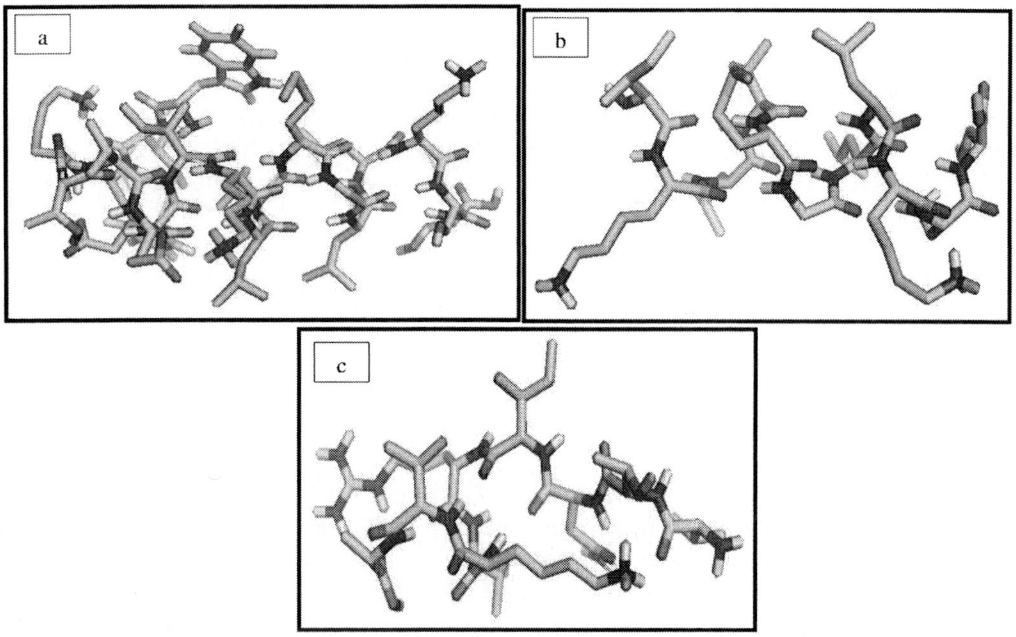

Figure 1. The AMPs sequence of *Mirabilis jalapa* visualized using PyMOL$^+$ software (a) Sequence 102 WIFKTVEEIKLVMGLLKSS (b) Sequence 113 IKLVMGLLKSS (c) Sequence 395 ITNIRTKVA.

The phenol compounds will attack a polar group (phosphate group). Additionally, the phospholipid molecule will break down into glycerol, carboxylic acids, and phosphoric acid. The phospholipid components were not able to recover the damaging cell membrane and impact to the cell membrane would be detrimental and experiencing growth barrier. The next stage, when the phospholipid layer was a very thin condition, phenol could penetrate easily and damage cell contents.

The mechanism of antimicrobial works in inhibiting the organism is varied and complex. Antibacterial compounds can work bacteriostatic ally, bacteriocidal, and bacteriologically [26]. The previous studies reveal that MAP is a compound that has the ability as a protein, also capable of disabling ribosomal activity in the target cells [10]. These compounds are allegedly damaged to target insects cell. Based on the previous review, the information obtained that MAP is antiviral and antiviroid substances for *M. jalapa* plants which eliminate pathogens that damage the body. The identification toward the active compounds contained shows that there are four main components of the MAP: 3,3methylenebis (4-hydroxycoumarin); N-D-alpha-phenyl glycine; Laminaribiitol; and 3-(4-(dimethylamino) cinnamoyl) -4hydroxycoumarin [27]. The potential of MAP compounds as biopesticides are proved by the results of studies showed a decreasing in physiological function indicated the cellular and humoral immune responses as parameters the decrease in hemocytic amount, phagocytosis activity, and lectin activity when exposed with *M. jalapa* [11, 14].

The function of *M. jalapa* as a biopesticide principally intended to regulate the number of pests with active MAP compounds contained therein which do not harm other non-target organisms. Based on the characteristics of insect mortality caused by the application of *M. jalapa*, it can be used as a reference in natural pest control. The larvacidal feature that causes death on insects, the content of flavonoids, saponins, and tannins from *M. jalapa* leaf extract is a repellent compound for insects. The saponins compounds, flavonoids, triterpenoids, alkaloids and fatty oils are larvicidal (toxic) for insects [28]. The content of these compounds can disrupt the function of nerves by inhibiting the acetylcholinesterase enzymes in breaking the neurotransmitters into acetic acid and choline, which affect in the coordination disorders of the larval muscle system [29, 30].

4. Conclusion

M. jalapa has six peptides with the potential as antimicrobials based on the CAMP database. According to the physicochemical characteristic of AMP, the three new peptides were found with the best characterization of *M. jalapa* plants that have the potential as a biopesticide; there were sequence ID 102 WIFKTVEEIKLVMGLLKSS, sequence ID 113 IKLVMGLLKSS, and sequence ID 395 ITNIRTKVA. These renewable peptides used to determine the pathways that occur in targeted insects as biopesticides.

Acknowledgments

The authors express their sincere thanks to Ms. Ardini Pangastuti, S. Pd., M.Si. for her assistance in In-silico methods. We would like to thank Doctoral Dissertation Research Grant, Directorate of Research and Community Service, Directorate General of Research and Development Strengthening of Research Ministry, Technology and Higher Education of Indonesia which has provided a funding for this research.

References

[1] Wink M. (2016). Secondary metabolites, the role in planta diversification of. In: Reference Module in Life Sciences: Encyclopedia of Evolutionary Biology. pp. 1-9. Elsevier, Amsterdam. doi: 10.1016/B987-0-12-800049-6.00263-8.

[2] Zhong JJ. (2011). Plant secondary metabolite. In: Reference Module in Life Science: Comprehensive Biotechnology (second ed.). pp. 299-308. Academic Press. 299-308. doi: 10.1016/B978-0-08-088504-9.00190-2.

[3] Rattan RS. (2010). Mechanism of action of secondary insecticidal metabolites of plant origin. Crop Protect. 299: 913-920.

[4] Delgoda R., and Murray JE. (2008). An evolutionary perspective on the role of secondary plant metabolites. In: Pharmacognosy. Nirali Prakashan, Pune, India (Eds). Kokae, C.K., Purohit, A.P., Gokhale, A.B. 93-100. doi: 10.1016/B978-0-12-802104-0.00007-X.

[5] Pavarini DP., Pavarini SP., Nieheus M., Lopes NP. (2012). Exogenous influence on plant secondary metabolites levels. Journal of Animal Feed Science and Technology 176: 5-16. doi: 10.1016/j.anifeedsci.2012.07.002.

[6] Leng, P.; Zhang, Z.; Pan, G.; Zhao, M. (2011). Applications and development trends in biopesticide. Afr. J. Biotech., 10, 19864-19873. doi:10.5897/AJBX11.009.

[7] Kandagal, A.S.; Khetagoudar, M.C. (2011). Study on larvicidal activity on weed extract against Spodoptera litura. J. Environ. Biol., 34, 253-257. doi:10.1007/s00436-008-1142-x.

[8] Gogoi J., and Veer V. (2016). Isolation and characterization of bioactive components from Mirabilis jalapa L. radix. Journal of Traditional and Complementary Medicine 6: 41-47. doi: 10.1016/j.jtcme.2014.11.028

[9] Kataoka, J., Miyano, M., Habuka, N., Masuta, C. & Koiwai, A. (1993). A Genomic gene for the MAP, A ribosome-inactivating protein from Mirabilis jalapa, contains an intron, Nucleic Acid Research, 21, 108 – 114.

[10] Vivanco, M.J. (1999). The antiviral and antiviroid activity of MAP-containing extracts from

Mirabilis jalapa roots. Plant Dis., 83, 1116-1121. doi:10.1094/PDIS.1999.83.12.1116.

[11] Suryani, I.; Anggraeni, T. (2014). The effect of leaf biopesticide Mirabilis jalapa and fungi Metarhizium anisopliae to immune response and mortality of Spodoptera exigua Instar IV. AIP Conference Proceedings 2014, 1589, 312. doi:10.1063/1.4868808.

[12] Sirajuddin, N. T. & Anggraeni, T. (2014). The effect of leaf biopesticide Mirabilis jalapa and entomopathogenic fungi beauveria basiana combinations to some physiological characters and histology of Crocidolomia pavonana (Lepidoptera: Pyralidae) larva. AIP Conference Proceedings. 1589: 308 (2014). Doi: https://doi.org/10.1063/1.4868807

[13] Pramita, M., & Anggraeni, T. (2015). The effect of Mirabilis jalapa leaves biopesticide treatment on the mycelium growth of entomopathogenic fungi Beauveria basiiana inside the larva body Crocidolomia binotalis. AIP Conference Proceedings 2015. 1677, 090010, doi:10.1063/1.4930755.

[14] Maulina D., and Anggraeni, T. (2014). The effect of the combination of two biological control agents, Mirabilis jalapa and Bacillus thuringiensis, to Spodoptera exigua's immune responds and their mortality. AIP Conference Proceeding 1589: 316. doi: 10.1063/1.4868809.

[15] Pangastuti A., Amin IF., Amin AZ., and Amin M. (2016). The natural bioactive compound from Moringa oleifera against cancer-based on in silico screening. Jurnal Teknologi (Sciences & Engineering).78: 315–318. Doi: 10.11113/jt.v78.8328

[16] Pangastuti, A, Indriwati, S.I. and Amin, M. (2018). Investigation of the anti-aging properties of allicin from Allium sativum L bulb extracts by a reverse docking approach. Tropical Journal of Pharmaceutical Research 17 (4): 635-639. Retrieved from https://www.ajol.info/index.php/tjpr/article/view/170514.

[17] Ilmawati, R.R.; Amin, A.Z. and Amin, M. (2017). Alliin as a natural bioactive from single bulb garlic (Allium sativum) for Nitric Oxide (NO) is increasing in the atherosclerotic process based on Insilico Screening. Biology, Medicine, & Natural Product Chemistry, 6, 59-62. doi: 10.14421/biomedich.2017.62.59-62.

[18] Clavillia Plant Database. (2004). Raintree Nutrition. www.rain-tree.com. (accessed on February 2017).

[19] UniProt. http://www.uniprot.org/ (accessed on February 2017).

[20] Innovagen. http://www.innovagen.com/ (accessed on February 2017).

[21] Antimicrobial database (ADP2). http://aps.unmc.edu/AP/ (accessed on February 2017)

[22] Dziuba, B. & Dziuba, M. (2014). New milk protein-derived peptides with potential antimicrobial activity: An approach based on bioinformatic studies. International Journal of Molecular Science. Vol 15, 14531-14545. Doi: https://doi.org/10.3390/ijms150814531

[23] Osorio, D. (2015). Rondon-Villareal, P. & Torres, R. Peptides: A package of data mining of antimicrobial peptides. The R Journal. Vol 7, 1, 4-14. ISSN: 2073-4859.

[24] Almaali, A. (2014). Antimicrobial peptides potential of some microbial histones: in silico modeling. International Journal of Recent Scientific Research. Vol. 5, 3, 649-655. ISSN: 0976-3031.

[25] Nuria, M.C., A. Faizatun., dan Sumantri. (2009). Antibacterial test of ethanol extract of Jatropha curcas L. fibers against Staphylococcus aureus ATCC 25923 Bacteria, Escherichia coli ATCC 25922, and Salmonella typhi ATCC 1408. Journal of Agricultural Sciences. 5: 26 – 37. [In Indonesia].

[26] Hadioetomo, R.S. (2005). The basic of microbiology. Jakarta: UI Press. [Indonesian version: Pelczar, M. Jr. dan E.C.S.Chan].

[27] Mahalingam, R., Bharathidasan, R., Ambikapathy, V., Panneerselvam, A. (2012). GC-MS determination of bioactive compounds of Mirabilis jalapa. Asian Journal of Plant Science and Research, 2 (3):224-227.

[28] Sutami, I. (2007) Effectiveness of pare leaf extract (Momordica charantia) as larvacide against Aedes Aegypti In Indonesia. [Thesis: Magister]. Maranatha University, Bandung

[29] Mishbach C., Dweck H., Vogel H., Vilsincaskas., Stensmyr M., Hansson B., and Grosse-Wilde

E. (2014). Evolution of insect olfactory receptor. eLife, 3; e02115. doi: 10.755/eLife.02115.

[30] Kolodziejczyk A., Sun X., Meinrzhagen IA., and Nassel DRR. (2008). Glutamate, GABA, and acetylcholine signaling components in the lamina of the Drosophila visual system. Plos One, 3, 2110. Doi: 10.1371/journal.pone.0002110.

Functional Group and Magnetic Properties of Fe_3O_4 Ferrofluids: The Impact of Dispersion Agent Composition

Ahmad Taufiq[1,*], Rosy Eko Saputro[1], Sunaryono[1], Yuanita Amalia Hariyanto[1], Nurul Hidayat[1], Arif Hidayat[1], Nandang Mufti[1], Hendra Susanto[2], Mujamilah[3]

[1]Department of Physics, Faculty of Mathematics and Natural Sciences, Universitas Negeri Malang, Jl. Semarang 5 Malang 65145, Indonesia
[2]Department of Biology, Faculty of Mathematics and Natural Sciences, Universitas Negeri Malang, Jl. Semarang 5 Malang 65145, Indonesia
[3]Centre for Science and Technology of Advanced Materials, National Nuclear Energy Agency of Indonesia, Tangerang 15314, Indonesia

*Corresponding author's email: ahmad.taufiq.fmipa@um.ac.id

Abstract. This work was aimed to identify the functional group and magnetic properties of the Fe_3O_4 ferrofluids coated by oleic acid regarding different compositions of olive oil as a dispersing agent. The results of the x-ray diffraction (XRD) data presented that the phase of the samples was a Fe_3O_4 phase, which was identified based on the diffraction peaks with highest peak at 35.6° with the particle size of 8.8 nm. The results of the Fourier transform infrared (FTIR) characterization on the ferrofluids showed those functional groups representing the existence of magnetic and dispersing agent particles. Based on the data analysis of the magnetization data, the composition of the dispersing agent influenced the saturation magnetization with the highest value of approximately 18 emu/gram. Therefore, the prepared ferrofluids in this work open a potency to be applied in various fields, such as for an antibacterial agent.

Keywords: Ferrofluid, Fe_3O_4, iron sand, dispersing agent, magnetization.

1. Introduction

In general, ferrofluid can be defined as a colloidal suspension consisting of magnetic nanoparticles which dispersed in a carrier liquid [1]. Ferrofluid is classified as a unique engineering material since it could maintain its liquid characteristics in a magnetic field and its particles would not be separated from the dispersant [2]. Therefore, in developing specific applications, ferrofluid has several advantages and opens new potential applications in various fields.

In the previous work, the development of ferrofluid using surfactant oleic acid was done by using different synthesis methods with the purpose to achieve the stable magnetic particle in nanometric size, one of them is the co-precipitation method [3]. However, some drawbacks were still found in some research above, namely the large size of the ferrofluid magnetic particle of 20 nm [4], and the agglomeration phenomenon that hampers the monodispersed particles, as well as its low durability of around one month [5]. Such disadvantages influence the magnetic properties of the ferrofluid since the superparamagnetic properties are not sustainable. Therefore, it is imperative to find out an alternative method for preparing the stable magnetic ferrofluids-based.

In order to control the stability of the ferrofluids is by selecting and controlling surfactants and dispersing agents of the fluids. In our previous work, the ferrofluids were successfully synthesized by selecting TMAH as a surfactant and controlling H_2O composition as a dispersing agent [6]. However, the stability of the ferrofluids was not appropriately achieved due to the presence of the aggregations and clusters. Therefore, selecting a suitable surfactant and controlling dispersing agent composition become an essential thing to be developed. Based on the above explanation, we developed the preparation of magnetic nanoparticles–based ferrofluids coated by a surfactant using oleic acid. Furthermore, in order to control the stability of the ferrofluid, we varied the composition of olive oil as a dispersing agent. To complete our study, the crystal structure, morphology, magnetic properties, and their correlation were also investigated in this work.

2. Materials and Methods

The natural sand from Banyuwangi, Indonesia was rinsed using distilled water and then sifted to obtain completely smooth sand. The natural sand was then separated using a strong permanent magnet to obtain the Fe_3O_4 powder. Subsequently, the separated Fe_3O_4 powder was reacted with HCl at room temperature while stirred. The results were then filtered to be then titrated with NH_4OH while being stirred continuously using a magnetic stirrer. Black sediment was then formed following the titration process. Next, the sediment was washed using distilled water with up to pH = 7 and then separated from the distilled water using a permanent magnet. To produce ferrofluids, the first step taken was preparing the Fe_3O_4 as a filler with 20% of the ferrofluid mass. It was performed based on the ratio of olive oil as a dispersion agent to oleic acid with a constant volume. Three variations of olive oil volumes were used to produce the samples coded by C, B, and A. C refers to the sample with a 2:1 ratio of oleic acid-olive oil content. B refers to the sample wih a 2:1.4 ratio of oleic acid-olive oil content, and A refers to the sample with a 2:0.6 ratio of oleic acid-olive oil content. After weighing, the Fe_3O_4 was heated on a hotplate magnetic stirrer while still being stirred. Such heating aimed to produce highly homogeneous particles.

Characterizations were performed using the first instrument, i.e. XRD, to identify the composition of phase, crystal structure, and lattice parameter of the sample. The characterization was employed by an X'Pert Pro, XRD was utilized. Meanwhile, we applied an FTIR (Shimadzu) to investigate the functional group of the ferrofluids. Magnetic characterization was done to identify the magnetic properties of the samples using a vibrating sample magnetometer (VSM).

3. Results and Discussion

The XRD data and refinement analysis for the Fe_3O_4 powder performed through Rietica software are presented in Figure 1. The FTIR spectra of the three Fe_3O_4 ferrofluid samples are shown in Figure 2. Meanwhile, the results of the VSM characterization of the three Fe_3O_4 ferrofluid samples are shown in Figure 3.

The process of refinement utilized data models from the database with a code of AMCSD 0002763. The results of refinement showed the values of Rp = 18.78; Rwp = 25.83, and chi-square of 1.11. The particle size was then known to be 8.8 nm. Also, the highest peak was identified to be at the angle of 35.65° with an *hkl* of (311) and a lattice parameter of 8.37 Å. The previous reports confirm such results; even the obtained particle size was smaller than those reported previously [7].

Based on Figure 2, two magnetite tapes representing the existence of magnetite (Fe-O) were found at 590.22 and 603.72 cm^{-1}. Such results are in accordance with the studies reported previously [8]. Also, four tapes that related to the vibration from carboxyl group (COO^-) were found at 1417.68, 1435.04, 1523.76, and 1591.27 cm^{-1}. Besides, the vibrations of C-H bond also emerged at 1118.71 cm^{-1} [9], CH_2 bond at 2854.65 and 2929.87 cm^{-1} [10], CH_3 bond at 3464.15 cm^{-1}, and C=C bond at 1377.17 cm^{-1} [9]. The overall vibrations of carboxyl group and C-H bond in the sample showed the existence of oleic acid that successfully bonded with the magnetic nanoparticles.

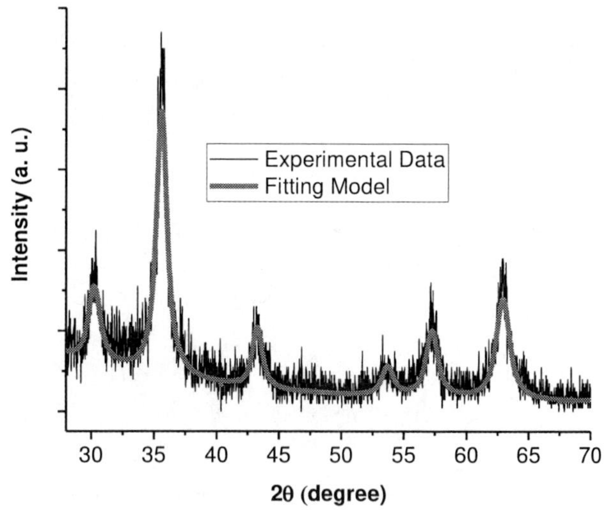

Figure 1. XRD and refinement patterns of the Fe_3O_4 powder

Subsequently, the presence of olive oil functional group was also essential as a dispersion medium. Two tapes represented the vibrations of the olive oil conjugation bond were found at 2729.27, 2752.42, 3645.15, 3672.47, and 3834.49 cm^{-1}. Furthermore, there was also a vibration on a tape at 1743.65 cm^{-1}, which resulted from the ester carbonyl bond (C=O). The ester bond (C=O) was represented by a bond showing the olive oil saturation at 3005.10 and 1170.79 cm^{-1}. Such results are in agreement with the previous research [11]. The mentioned vibrations revealed that olive oil had been successfully bond with the magnetic-oleic acid nanoparticle system.

The magnetization data (Figure 3) showed that the sample C with a 2:1 ratio of oleic acid-olive oil compositions had the highest saturation magnetization value of approximately 18 emu/gram, followed by the sample B with a 2:1.4 ratio of oleic acid-olive oil compositions having the saturation magnetization value of approximately 15 emu/gram. Meanwhile, the sample A with a 2:0.6 ratio of oleic acid-olive oil compositions had a relatively low magnetization value of approximately 10 emu/gram. The results obtained from the three samples that were already coated by oleic acid already had superparamagnetic properties, but still far from the required value of magnetization for bulk, namely 90 emu/gram [12]. Such a decline could occur when the size of the ferrofluid magnetic particle is below 25 nm [13]. Besides, the uniformity of magnetic nanoparticle size is determined by its chelation with the support of the oleic acid coating that could stabilize the magnetic particle. The oleic acid that covered the magnetic particle surface could reduce the surface activity and contact of the particle [14].

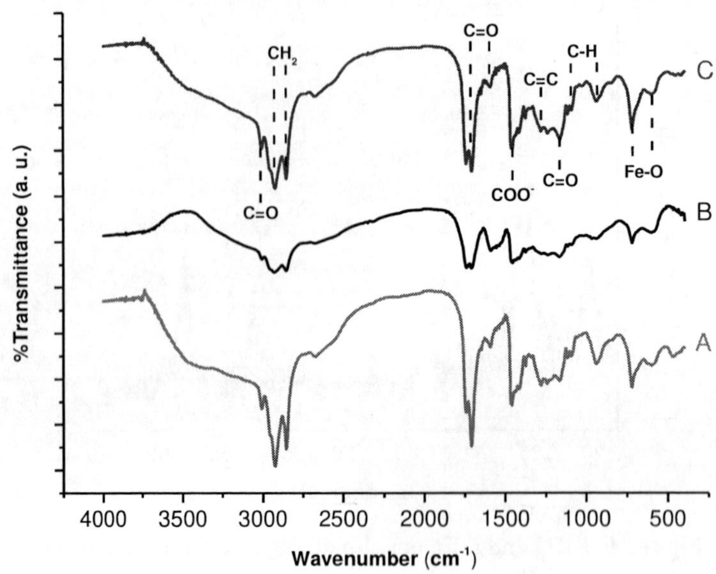

Figure 2. FTIR spectra of the three Fe_3O_4 ferrofluid samples

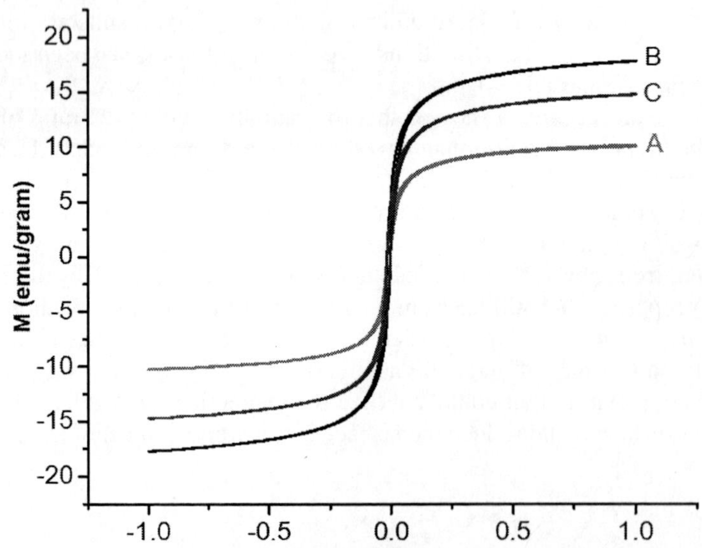

Figure 3. Hysteresis curve of the three Fe_3O_4 ferrofluid samples

Based on the figure, it also can be seen that the B sample had an optimum composition. An aggregation was found on the A sample that influenced its magnetic response. Therefore, the magnetization data of such two samples seemed low. In the meantime, a reduction from the maximum condition was found in the C sample. It was due to the transition from the optimum condition to unstable condition because the C composition exceeded the effective composition, namely the B sample composition [15]. Furthermore, the coercivity field (*Hc*), remanent magnetization (*Mr),* and saturation magnetization (*Ms*) of all the obtained data are presented as follow.

Table 1. *Hc*, *Mr*, and *Ms* values of the three samples through magnetization data analysis

Sample	H_c (T)	M_r (emu/gram)	M_s (emu/gram)
A	0.0157	0.0073	10.140
B	0.0163	0.0038	17.805
C	0.0166	0.0039	14.811

According to the data in Table 1, the average value of the external magnetic field, in which the remanent magnetization occurred, was 0.016 T. The remanent magnetization of the A until C samples declined. Such decline was possible since the magnetic nanoparticles that acted as filler in A sample experienced an aggregation so that its remanent magnetization was relatively high, i.e. 0.006 emu/gram [16]. The remanent magnetization was reduced on the B and C samples as an impact of the oleic acid that was able to coat the magnetic nanoparticle so that the aggregation could be minimized. Theoretically, an increasing remanent magnetization is caused by an aggregation of magnetic nanoparticles. Based on the data above, it could be confirmed that the use of oleic acid could increase a magnetic response and also minimize the aggregation causing the declining value of remanent magnetization [17].

Table 2. Particle sizes and magnetic moment of the samples through magnetization data analysis

Sample	Particle Size (nm)	Magnetic Moment (J/T)	Magnetic Moment (μB)
A	8.37	7.35×10^{-19}	79,235
B	8.43	9.17×10^{-19}	98,878
C	8.41	9.08×10^{-19}	97,908

In this study, the particle size was also obtained from the magnetization data as presented in Table 2. Based on the table, the particle sizes for the two samples are small and suitable to be called as superparamagnetic [18]. The magnetic characterization was only performed on the sample that is in a liquid form and not purely Fe_3O_4 particles. The TEM characterization was done by using a bright-field mode in order to recognize the surface morphology of the sample since it allows a larger observable diameter compared with the dark-field mode on VSM characterization [19].

Another factor that is still relevant to the declining particle size is the nanoparticle of which surface and volume ratio is higher than its surface and bulk ratio. Such high comparison caused the decrease of the crystal transitional symmetry on the ferrofluid surface layer so that a magnetization was not found in this area [20]. Besides, the strong interaction between the Fe_3O_4 filler and surfactant oleic acid cause an irregular spin that resulted in a magnetically dead layer [21]. However, theoretically, the three samples above already had suitable results calculated from the Equation (1) [22].

$$D = \left(\frac{18 k_B T}{\pi} \frac{\chi_i}{\rho M_S} \right)^{1/3} \quad (1)$$

where ρ is the density of Fe_3O_4, χ_i is initial susceptibility, M_s was explained above, T is the room temperature, and k_B is Stefann Boltzman's constant which is equal 1.38×10^{-23} J/K. The higher the saturation magnetization value of the sample, the larger the particle size, to the extent that the sample is in a single domain condition [23]. The magnetic nanoparticle that is categorized as having a single domain should fulfill some criteria, namely having a size below the critical size (25 nm) [24]. Such result is in line with a previous study [25]. Besides the particle size, the magnetic moment of a sample is also important to be studied. Based on the calculation of the obtained data, the values of the magnetic moment are shown in Table 2.

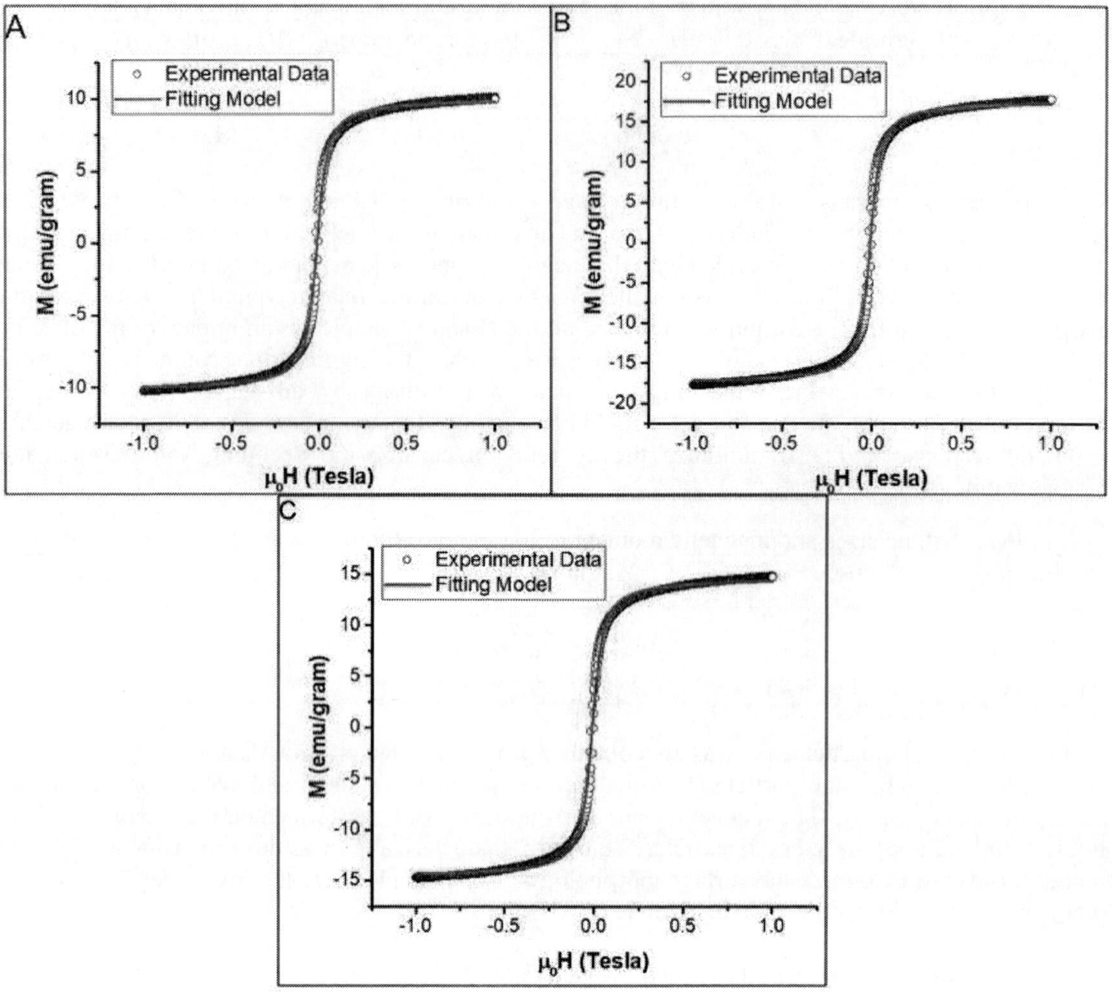

Figure 4. Hysteresis curve of the three Fe_3O_4 ferrofluid samples fitted with Langevin equation without susceptibility

Table 2 shows that the magnetic moment of the two samples are in line accordance with magnetic moment value of the Fe_3O_4 nanoparticle shown in the previous research [26]. Theoretically, when characterized using Equation 2, the value of the magnetic moment was in the same direction as the cubic equation of the particle size and saturation magnetization. Hence, a large particle size and saturation magnetization of a sample shall increase the magnetic moment that could be obtained. In relation to the magnetic moment, the previous research results showed that the magnetic moment of ferrofluid Fe_3O_4 was ranged between 30,000 to 70,000 μB [26]. When compared to this research data, the magnetic moment value in this research is relatively large. This indicates that the sample has been made entirely superparamagnetic with a good magnetic response. The influence of the dispersion medium on sample magnetic moment value also need to be calculated. It is because the medium influences the filler stability inside it. The utilization of olive oil was proven to stabilize filler better than the other dispersion medium such as kerosene and chloroform. It is based on the review of ferrofluid magnetic moment value with kerosene dispersion medium and chloroform whose magnetic moment value ranged between 10,000 to 20,000 μB [27].

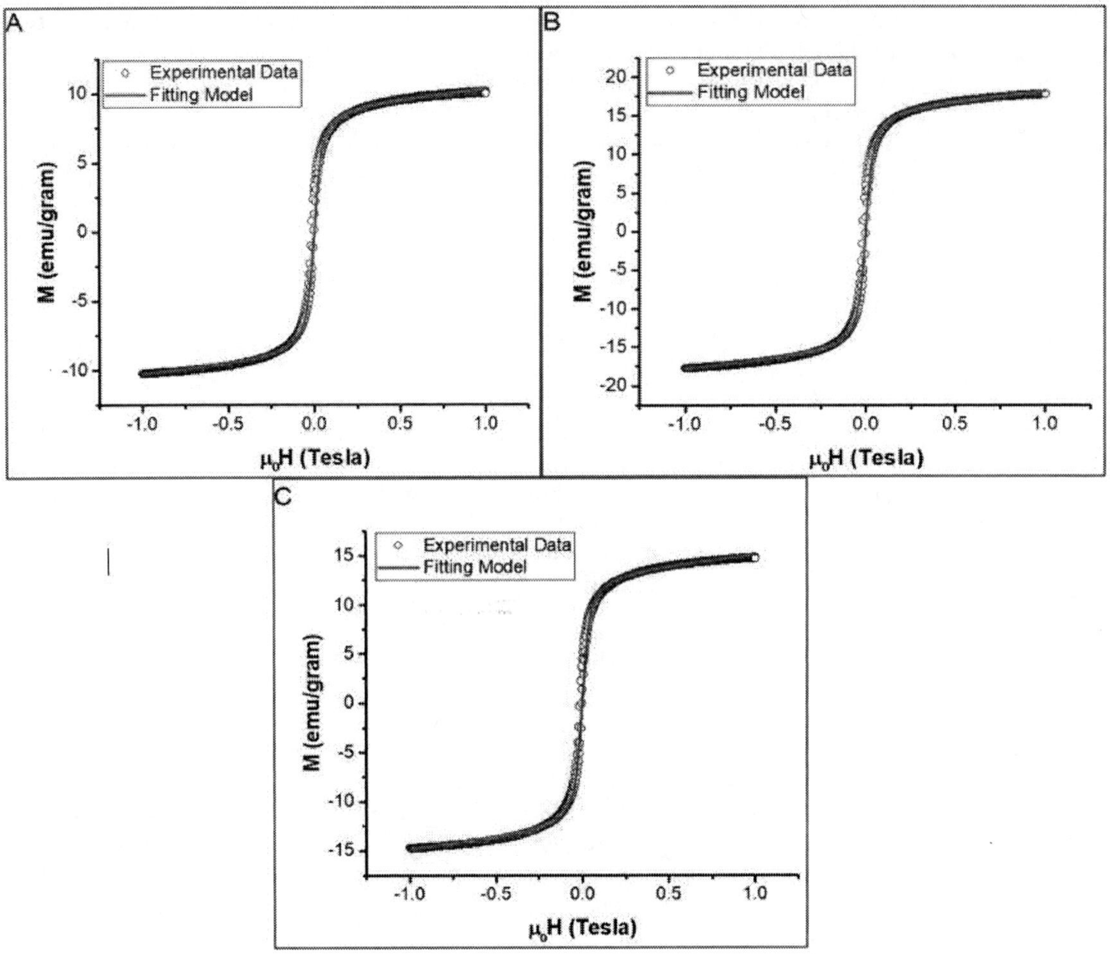

Figure 5. Hysteresis curve of the three Fe_3O_4 ferrofluid samples fitted with Langevin equation with susceptibility

Based on the hysteresis curves as shown in Figures 4 and 5, the magnetization data was fitted using two different Langevin equations. In Figure 4, the hysteresis curve was fitted using a general Langevin equation on Equation (2). Meanwhile, the data for Figure 5 was fitted with a modified Langevin equation with susceptibility on Equation (3) [28].

$$M = M_S\left(\coth\left(\frac{\mu H}{k_B T}\right) - \frac{k_B T}{\mu H}\right) \quad (2)$$

$$M = M_S L\left(\frac{\mu_p H}{k_B T}\right) + \chi H \quad (3)$$

where M is the magnetization, H is the magnetic field and the other quantities have been described above. When noted in Figures 4 and 5, the generated fitting curve looks different, even though they are derived from the same equation. Susceptibility was the factor influencing those differences. In physics, susceptibility in the magnetic context indicates the magnetism level of an object for magnetization, in which it is highly affected by the mineral and iron oxide content of a material [29]. Mathematically, the material susceptibility (χ) in general fulfills the Equation 4 [30].

$$M = \chi H \quad (4)$$

Since the magnetization and magnetic field in international system of units has the same unit that is ampere per meter (A/m) then the magnetic susceptibility is gradient on Equation 4. In this research, the main material used was Fe_3O_4. However, every sample has different iron oxide contents, which were caused by the filler variation performed on the sample preparation stage. In relation to theory, for the entire Figure 4 the susceptibility on the modified Langevin was ignored, while for the whole Figure 5 the susceptibility on each sample was included in refinement parameter. Thus, theoretically, the obtained fitting results also have results closer to the Fe_3O_4 susceptibility. Based on the fitting analysis, Equation 3 with is more satisfied in fitting analysis for the magnetization data.

The susceptibility values of the samples are 55.43, 61.32, and 59.32 for the respective A, B, and C. Based on the data, it suggests that the sample having the highest susceptibility is sample B, this fulfills the theoretical requirement declared on Equation 4 when the second sample has the highest magnetisation when compared to others, which is 17.805 emu/gram. Furthermore, the third and smallest sample is the first sample. A logical linkage can be made to sample B which has the best stability and a good magnetic response as well, however generally the three samples that have been made have fulfilled the theoretical aspects and showed appropriate results with the theoretical review aforementioned. Previous studies concern the magnetic susceptibility quantitatively, however some discuss it qualitatively. When compared to the previous studies, this research has a good appropriateness, that the greater the magnetization of material saturation will result in increased susceptibility [31]. Therefore, filler variation has an important role in generating a stable ferrofluid with controllable susceptibility. Based on the above discussions, the prepared magnetic nanoparticles in the ferrofluids exposed a potency to be functionalized such as for an antibacterial agent and other aspects.

4. Conclusion

A synthesis of the Fe_3O_4 particle as a ferrofluid filler with surfactant oleic acid dispersed in olive oil has successfully been performed. The magnetic nanoparticle has a particle size of 8.8 nm sized particles. The XRD data analysis confirmed such results well. Based on the XRD pattern, it can be identified that the sample has a Fe_3O_4 phase and is already representative to be applied as ferrofluid filler. Furthermore, the prepared ferrofluids had the functional groups representing the existence of magnetic and dispersing agent particles. According to the magnetics characterization, the C sample with a ratio of oleic acid-olive oil composition of 2:1 presented the largest magnetization value, and the use of oleic acid could improve the magnetic response of the samples.

Acknowledgment

We would like to thank KEMENRISTEKDIKTI RI for providing "PDUPT" for AT.

References

[1] R. Olaru, A. Arcire, C. Petrescu, and M.-M. Mihai, "Study of the magnetic force delivered by an actuator with nonlinear ferrofluid and permanent magnets," *IEEJ Trans. Electr. Electron. Eng.*, vol. 12, no. 1, pp. 24–30, 2017.

[2] D. Ramimoghadam, S. Bagheri, and S. B. A. Hamid, "In-situ precipitation of ultra-stable nano-magnetite slurry," *J. Magn. Magn. Mater.*, vol. 379, pp. 74–79, 2015.

[3] A. Taufiq, S. Pratapa, and M. Zainuri, "Various Magnetic Properties of Magnetite Nanoparticles Synthesized from Iron-Sands by Coprecipitation Method at Room Temperature," in *Materials Science Forum*, 2015, vol. 827, pp. 229–234.

[4] X. Chen, M. G. Organ, and W. J. Pietro, "One-pot Synthesis of Size Controllable Amine-Functionalized Core-Shell Magnetic Nanoparticles for Use in Microfluidic Flow Separators," *J Nanosci Adv Tech*, vol. 1, no. 3, pp. 25–31, 2016.

[5] V. Patsula, M. Moskvin, S. Dutz, and D. Horák, "Size-dependent magnetic properties of iron oxide nanoparticles," *J. Phys. Chem. Solids*, vol. 88, pp. 24–30, 2016.

[6] A. Taufiq *et al.*, "Studies on Nanostructure and Magnetic Behaviors of Mn-Doped Black Iron Oxide Magnetic Fluids Synthesized from Iron Sand," *Nano*, vol. 12, no. 09, p. 1750110, Aug.

2017.

[7] R. Govindaraj, R. Mythili, and G. Amarendra, "Stability and magnetic interactions between magnetite nanoparticles dispersed in zeolite as studied using Mössbauer spectroscopy," *J. Magn. Magn. Mater.*, vol. 418, pp. 248–252, 2016.

[8] S. Bhattacharya, K. Jenamoni, and S. Nayar, "Stability of biomimetic ferrofluids established by a systematic study using microwave irradiation at defined wattages," *J. Magn. Magn. Mater.*, vol. 324, no. 20, pp. 3261–3266, 2012.

[9] J. Dhumal, S. Bandgar, K. Zipare, and G. Shahane, "Fe 3 O 4 ferrofluid nanoparticles: synthesis and rheological behavior," *Int J Mater Chem Phys*, vol. 1, pp. 141–145, 2015.

[10] M. S. Darwish, "Effect of carriers on heating efficiency of oleic acid-stabilized magnetite nanoparticles," *J. Mol. Liq.*, vol. 231, pp. 80–85, 2017.

[11] F. Laachari, H. Maâtaoui, F. El Bergadi, A. Boukir, and S. K. Ibnsouda, "Improvement of a lipolytic activity assay protocol to increase the efficiency of the lipases activity measurement and validation by infrared spectroscopy," *Afr. J. Biotechnol.*, vol. 14, 2015.

[12] O. V. Minaeva *et al.*, "Comparative study of cytotoxicity of ferromagnetic nanoparticles and magnetitecontaining polyelectrolyte microcapsules," in *Journal of Physics: Conference Series*, 2017, vol. 784, p. 012038.

[13] A. A. Galhoum, M. G. Mahfouz, N. M. Gomaa, T. Vincent, and E. Guibal, "Chemical modifications of chitosan nano-based magnetic particles for enhanced uranyl sorption," *Hydrometallurgy*, 2016.

[14] M. A. Fontecha-Cámara, C. Moreno-Castilla, M. V. López-Ramón, and M. A. Álvarez, "Mixed iron oxides as Fenton catalysts for gallic acid removal from aqueous solutions," *Appl. Catal. B Environ.*, vol. 196, pp. 207–215, 2016.

[15] N. Di Maggio *et al.*, "Rapid and efficient magnetization of mesenchymal stem cells by dendrimer-functionalized magnetic nanoparticles," *Nanomed.*, vol. 11, no. 12, pp. 1519–1534, 2016.

[16] S. Qin and Q. Xu, "Room temperature ferromagnetism in N_2 plasma treated graphene oxide," *J. Alloys Compd.*, vol. 692, pp. 332–338, 2017.

[17] S. Jovanović, M. Spreitzer, M. Tramšek, Z. Trontelj, and D. Suvorov, "Effect of oleic acid concentration on the physicochemical properties of cobalt ferrite nanoparticles," *J. Phys. Chem. C*, vol. 118, no. 25, pp. 13844–13856, 2014.

[18] H. Shokrollahi, "Structure, synthetic methods, magnetic properties and biomedical applications of ferrofluids," *Mater. Sci. Eng. C*, vol. 33, no. 5, pp. 2476–2487, 2013.

[19] K. Y. Yoon *et al.*, "Control of magnetite primary particle size in aqueous dispersions of nanoclusters for high magnetic susceptibilities," *J. Colloid Interface Sci.*, vol. 462, pp. 359–367, 2016.

[20] B. Aslibeiki, G. Varvaro, D. Peddis, and P. Kameli, "Particle size, spin wave and surface effects on magnetic properties of $MgFe_2O_4$ nanoparticles," *J. Magn. Magn. Mater.*, vol. 422, pp. 7–12, 2017.

[21] J. Wu, L. Pei, S. Xuan, Q. Yan, and X. Gong, "Particle size dependent rheological property in magnetic fluid," *J. Magn. Magn. Mater.*, vol. 408, pp. 18–25, 2016.

[22] Z. L. Liu, Y. J. Liu, K. L. Yao, Z. H. Ding, J. Tao, and X. Wang, "Synthesis and magnetic properties of Fe_3O_4 nanoparticles," *J. Mater. Synth. Process.*, vol. 10, no. 2, pp. 83–87, 2002.

[23] M. Houshiar, F. Zebhi, Z. J. Razi, A. Alidoust, and Z. Askari, "Synthesis of cobalt ferrite (CoFe 2 O 4) nanoparticles using combustion, coprecipitation, and precipitation methods: a comparison study of size, structural, and magnetic properties," *J. Magn. Magn. Mater.*, vol. 371, pp. 43–48, 2014.

[24] J. Carrey, B. Mehdaoui, and M. Respaud, "Simple models for dynamic hysteresis loop calculations of magnetic single-domain nanoparticles: Application to magnetic hyperthermia optimization," *J. Appl. Phys.*, vol. 109, no. 8, p. 083921, 2011.

[25] S. A. Maboudi, S. A. Shojaosadati, and A. Arpanaei, "Synthesis and characterization of multilayered nanobiohybrid magnetic particles for biomedical applications," *Mater. Des.*, vol.

115, pp. 317–324, 2017.
[26] S. V. Dyachenko, M. A. Vaseshenkova, K. D. Martinson, I. A. Cherepkova, and A. I. Zhernovoi, "Synthesis and properties of magnetic fluids produced on the basis of magnetite particles," *Russ. J. Appl. Chem.*, vol. 89, no. 5, pp. 690–696, 2016.
[27] M. T. López-López, J. D. G. Durán, A. V. Delgado, and F. González-Caballero, "Stability and magnetic characterization of oleate-covered magnetite ferrofluids in different nonpolar carriers," *J. Colloid Interface Sci.*, vol. 291, no. 1, pp. 144–151, 2005.
[28] N. Rezaei, M. H. Ehsani, M. Aghazadeh, and I. Karimzadeh, "An Investigation on Magnetic-Interacting Fe_3O_4 Nanoparticles Prepared by Electrochemical Synthesis Method," *J. Supercond. Nov. Magn.*, vol. 31, no. 7, pp. 2139–2147, 2018.
[29] B. D. Cullity and C. D. Graham, *Introduction to magnetic materials*. John Wiley & Sons, 2011.
[30] M. Sheikholeslami and H. B. Rokni, "Simulation of nanofluid heat transfer in presence of magnetic field: a review," *Int. J. Heat Mass Transf.*, vol. 115, pp. 1203–1233, 2017.
[31] C. Fei *et al.*, "Synthesis and magnetic properties of hard magnetic (CoFe2O4)–soft magnetic (Fe3O4) nano-composite ceramics by SPS technology," *J. Magn. Magn. Mater.*, vol. 323, no. 13, pp. 1811–1816, 2011.

Effectiveness of Modified Natural Zeolite through Acid Activation as A Catalyst on Cellulose Conversion into Glucose from Cotton Assisted by Ultrasonic

Sumari*, Ni'matus Sholihah, Mahrullina Mahirotul Aisiyah, Intan Oktaviani, Natasha Khilmi, Yana Fajar Prakasa

Department of Chemistry, Faculty of Mathematics and Natural Sciences, Universitas Negeri Malang, Jl. Semarang 5 Malang 65145, Indonesia

*Corresponding author's email: sumari.fmipa@um.ac.id

Abstract. This study aimed to convert cotton cellulose into glucose using the catalyst of activated natural zeolite assisted by ultrasonic. The catalyst of hydrolysis of cellulose into glucose used acid-activated natural zeolite. Preparation and activation of natural zeolite used HCl solution 1M, 2M, and 3M. This activated natural zeolite was characterized using Freundlich adsorption isotherms, XRD, and XRF. Hydrolysis of cotton cellulose assisted by ultrasonic-waves was conducted for 1, 2, and 4 hours. Identification of the degradation product used Fehling test and Nelson-Somogyi test. Based on the Freundlich adsorption isotherm analysis, the natural zeolite activated by 1M HCl solution had the largest surface area of 120 m^2/g. The XRD diffractogram showed that zeolite before and after the activation process had a similar type structure that is mordenite. Meanwhile, the XRF test showed a decrease in the relative percentage of impurities followed by the increase in Si and Al percentage on activated natural zeolite with Si/Al ratio of 7.13. Based on the Fehling and Nelson-Somogyi test, the highest yield of glucose of 4.51% was obtained when using a catalyst natural zeolite activated with HCl 1 M for 4 hours sonication time.

Keywords: Cellulose, activated natural zeolite, glucose, catalyst, ultrasonic

1. Introduction

Cellulose is an organic compound of a polysaccharide as the main content of plant cell wall. Based on the statistical data, the biomass waste production in Indonesia reaches 147.6 million ton every year. The content of cellulose within the biomass waste is approximately 30 – 50%. Therefore, cellulose obtained from the biomass waste cultivation in Indonesia reaches about 59 million ton every year [1]. One of the biomass wastes is cotton that also causes environmental water pollution. The abundant cellulose has a significant potential to be used as a renewable energy source.

Cellulose can be hydrolyzed to be glucose by the enzyme of an acid catalyst [2]. Glucose is needed by human as the main energy source of their body [3]. Besides, glucose is a raw material of vitamin C synthesis through the Reichstein process, sorbitol, and bioethanol as a renewable energy fuel [4]. Nevertheless, this poly-glucose compound has a complex and strong structure so that it is difficult to be hydrolyzed into glucose. One of the research methods that have been developed for cellulose hydrolysis is hydrolysis method by using 2% of acid catalyst yielding the glucose of 5.0675% [5].

However, the direct use of acid is not environmentally friendly since acid is corrosive and toxic. Another research was conducted that doing cellulose hydrolysis become glucose using high temperature-ultra and the explosion of vapor pressure and it resulted in glucose of 67.7% (60 atm, 1 minute) [6]. Cellulose hydrolysis becoming glucose by using hydrothermal at the high pressure and temperature (250 atm, 300 °C) was also done, and it yielded 63.8% of glucose [7]. Another research by applying base pre-treatment method at the temperature of 121 °C continued with fermentation by cellulose enzyme yielded the high glucose of 45.69% [8]. However, those methods mentioned before were done at the high temperature and pressure so that they were less practical and safe. Moreover, the cellulose enzyme is relatively high-cost. Cellulose hydrolysis using the catalyst of activated natural zeolite (calcination) that was impregnated by Ni. This process resulted in glucose with a percent yield of 0.967 %wt [9]. This process was basically environmentally friendly, but the percent yield was still low, and thus the catalyst modification is needed to make the process more efficient.

In this research, the cellulose hydrolysis process was by using natural zeolite activated by acid (HCl) helped by ultrasonic wave. Zeolite activated by acid was expected would be filled by H^+ ion of an acid. The existence of H^+ ion and ultrasonic wave helped to break the glycosidic bond easily in the cellulose so that it resulted in glucose and the other much more reducer sugar.

The study of cellulose hydrolysis by using this H-zeolite catalyst was fascinating since it fulfilled the principle of green chemistry; energy is more efficient since it is done at the room temperature and pressure. The glucose production that can be from biomass waste is beneficial since it does not compete with the food need. Besides, the glucose product can be fermented easily into bioethanol as the renewable energy source as the solution to face the crisis of energy with the raw material of fossil that is getting lower.

2. Materials and Methods

2.1. *Equipment*
Beaker glass of 200 mL and 50 mL, crucible, 100 mL Erlenmeyer, filter paper of Whatmann 42, stirrer, Buchner funnel, mortar and pestle, oven, 100 mesh strainer, wrap plastic, spatula, drop pipette, 100 mL measuring cylinder, furnace, burette, autoclave, fume hood, ultrasonic, spectronic-20, digital analytical balance, conical funnel, magnetic stirrer, magnetic stirrer, 200 mL and 50 mL volumetric flash, watch glass, evaporation bowl, three legs, stative and clamp, filler, 10 mL graduated pipette, 10 mL volume pipette, rubbing alcohol burner, netting wire, shaker, XRF PANalytical with the type of MiniPal 4, and XRD PANalytical with the type of type E'xpert Pro.

2.2. *Materials*
Cellulose of cotton, distilled water, natural zeolite from Malang regency, the solution of 3 M HCl, glucose, Nelson-Somogyi reagent, Fehling reagent, NaOH (s), $H_2C_2O_4.2H_2O$ (s), pp indicator.

2.3. *Procedures*

2.3.1. Natural Zeolite Activation
Zeolite was pounded became powder, and it was sieved with a measurement of 100 mesh. After that, it was soaked in the solution of 1 M, 2 M, and 3 M of HCl for 24 hours. The result then was heated at the temperature of 500 °C for four hours. The active zeolite was then characterized by using Freundlich adsorption isotherm, XRF, and XRD.

2.3.2. Freundlich Adsorption Isotherm Analysis
The solution of oxalate acid was prepared with the concentration of 0.05 M, 0.1 M, 0.2 M, 0.3 M, 0.4M, 0.7 M, and 1.1M. One gram of active zeolite was soaked in the oxalate acid solution. It then was shaken for 30 minutes, and it was untreated for 24 hours. After that, the filtering was done to get a

filtrate. The filtrate was titrated with 1 M NaOH solution. The following is the equation of Freundlich adsorption isotherm.

$$\frac{x}{m} = kc^{\frac{1}{n}} \quad (1)$$

can be used to calculate the area of catalyst surface like zeolite. The data obtained from this research was the number of oxalate acid mole adsorbed, and then the surface area was calculated with the following formulation.

$$surface\ area = adsorbed\ mole \times \left(6023 \times 10^{23} \frac{molecule}{mole}\right) \times 20\text{Å}^2 \quad (2)$$

where x/m is the mass adsorbed per unit mass of adsorbent, c is the concentration, and k and n are constants [10].

2.3.3. *Analysis of zeolite using XRD and XRF*

The zeolite before and after activation was characterized by using XRD and XRF. XRD test aimed to determine the type of mineral in the zeolite. XRF test was intended to know the metals contained in the zeolite.

2.3.4. *The Process of Cellulose Conversion using Zeolite Catalyst*

The cellulose of cotton and water (1:8) was put into a reactor added by active zeolite catalyst which was 3% of cellulose mass. The tools were set to reach the set point and sonication was done for one, two, and four hours. The filtrate obtained was tested by Fehling and Nelson-Somogyi tests.

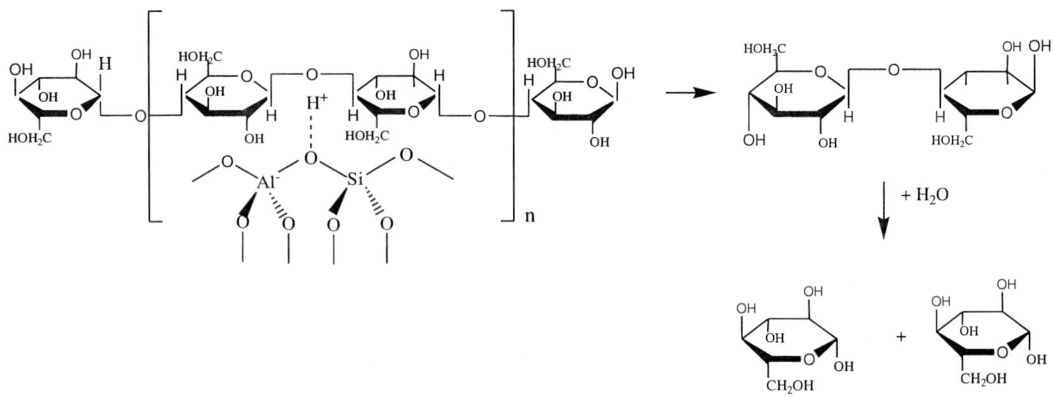

Figure 1. Hydrolysis reaction of cellulose into glucose using acid activated zeolite catalyst

2.3.5. *Fehling Test*

Fehling test is a qualitative test to know the existence of reducer sugar like glucose. This test was done by taking a little of filtrate as the result of sonication. After that, it was added with Fehling reagent, heated and cooled. The positive test was indicated by the change of blue into sorrel.

2.3.6. *Nelson-Somogyi Test*

Nelson-Somogyi test is a quantitative test to know the existence of reducer sugar like glucose. This test was done by taking a little of filtrate as the result of sonication. After that, it was added with Nelson-Somogyi regen, heated and cooled. The next step was measuring the absorbance with spectronic-20 on the wavelength of 540 nm. The measurement in this test was based on the intensity of the blue color (showing the amount of glucose in the sample). In this test, the curve of the calibration of the concentration glucose standard solution versus absorbance was obtained. Such calibration curve was used to calculate the concentration of sample glucose from the absorbance data. The percentage of glucose can be calculated by using the following equation:

$$\text{yield percentage of glucose} = \frac{mass\ of\ glucose\ (experiment)}{\text{mass of glucose (theory)}} \times 100\%$$

3. Results and Discussion

3.1. *Natural Zeolite Activation*

The natural zeolite sample used as the initial material in this research was natural zeolite from Sumbermanjing, South Malang, Indonesia. The analysis results showed that HCl could reduce the polluter metal of the zeolite and increase the percentage of the main components of zeolite which were Si and Al through ion-exchange with H^+ ion. The existence of H^+ ion can make the breaking of the glycosidic bond (1.4'-β glycosides) among the glucose units easy. The obtained results were subsequently heated on a furnace to delete the ions as the rest of acid like CO_2 and water molecule absorbed by the zeolite pores so that its surface area increased. Physically, the appearance was the color change of natural zeolite from light green into brown.

(a) **(b)**

Figure 2. The sample of non-activated natural zeolite (a) Activated natural zeolite (b)

3.2. *Freundlich Isotherm Test of Activated Natural Zeolite*

The followings are the surface area data as the results of the Freundlich isotherm test.

Table 1. Data for Freundlich isotherm test

Sample	Specific Surface Area (m^2/g)
Non-activated natural zeolite	85.52
Natural zeolite activated by1 M HCl (ZATA-1M)	120.00
Natural zeolite activated by 2 M HCl (ZATA-2M)	60.20
Natural zeolite activated by 3 M HCl (ZATA-3M)	45.17

The research results showed the surface area of active zeolite increased by the activation but it decreased again. The largest surface area was natural zeolite activated by 1 M HCl. The surface area decreased by the increase in acid concentration. This case was possibly caused by the high acid concentration that can damage the structure and pores of the zeolite. The surface area of all these non-activated and activated natural zeolites was smaller than that of synthetic zeolite made by Selim et al. [11] that was 374 m^2/g. The more concentrated of HCl, the surface of the activated natural zeolite was smaller. The surface area of the material can be affected by temperature activation, impregnation, and concentration of acid used in the activation [12].

3.3. *Characterization of Activated Natural Zeolite Using XRD and XRF*

3.3.1. XRD Analysis

The spectra results of X-ray diffraction of natural zeolite before and after activation using acid solution had peaks pattern at 2θ angles like to those of zeolite mordenite (MOR) according to the diffractogram of zeolite MOR listed at the data of JCPDS (Joint Committee on Powder Diffraction Standards) [13].

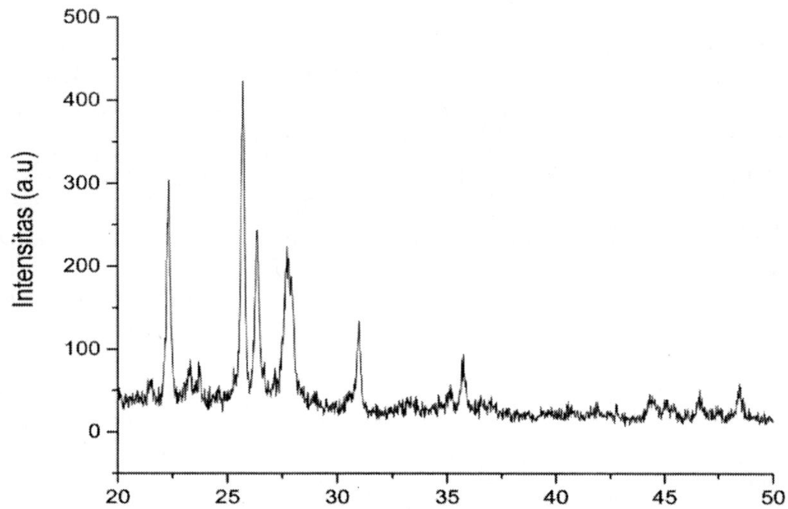

(a)

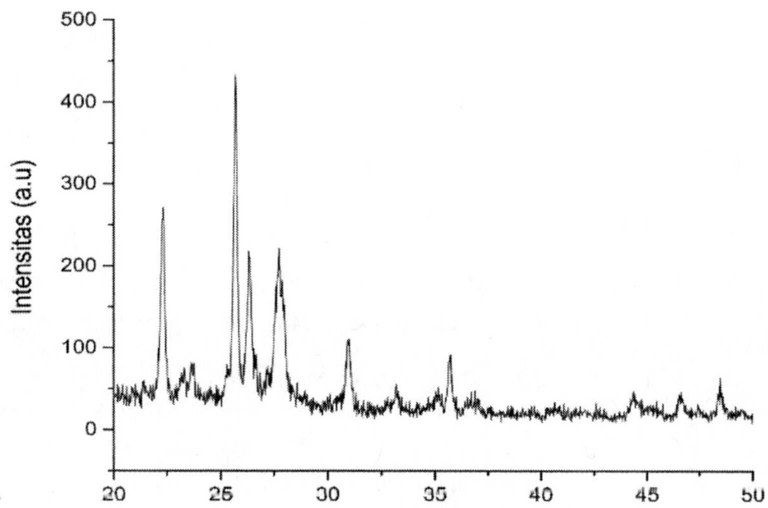

(b)

Figure 3. (a) Diffractogram of non-activated natural zeolite (b) Diffractogram of natural zeolite activated by acid

Such an assumption was strengthened by the similarity of the value of 2θ and d-spacing of the sample with MOR-zeolite. The comparison of the position of 2θ angles at each peak of the sample non-activated natural zeolite and activated natural zeolite with zeolite MOR standard (JCPDS) diffractogram is shown in Table 2.

Based on the data at Table 2, the number of 2θ angels at each peak of the natural zeolite before and after activation was relatively the same. Therefore, it can be said that the natural zeolite is mordenite (MOR). The similarity of the diffractogram pattern of the main peaks in the same 2θ angles show that the structure of the materials is similar [13, 14].

Table 2. (a) Position of 2θ angles of non-activation natural zeolite and MOR (standard) (b) Position of 2θ angles of activated natural zeolite and the type of MOR (standard)

Non-activated Natural Zeolite		**MOR-Zeolite**		**Natural Zeolite Activated by Acid**		**MOR-Zeolite**	
2θ (deg)	***d*-spacing [Å]**	**2θ (deg)**	***d*-spacing [Å]**	**2θ (deg)**	***d*-spacing [Å]**	**2θ (deg)**	***d*-spacing [Å]**
22.3268	3.98195	22.20	4.004	22.3008	3.98653	22.20	4.004
23.2382	3.82780	23.64	3.764	23.6717	3.75868	23.64	3.764
25.6998	3.46648	25.63	3.476	25.6851	3.46843	25.63	3.476
26.3431	3.38327	26.25	3.395	26.3323	3.38464	26.25	3.395
27.7154	3.21879	27.67	3.223	27.6996	3.22059	27.67	3.223
30.9649	2.88802	30.89	2.894	30.9365	2.89061	30.89	2.894
35.1136	2.55572	35.61	2.521	35.6965	2.51532	35.61	2.521
44.3947	2.04061	44.13	2.053	44.3943	2.04062	4 .13	2.053
46.6322	1.94778	46.58	1.950	46.5795	1.94986	46.58	1.950
48.4647	1.87677	48.45	1.879	48.4729	1.87648	48.45	1.879

(a) **(b)**

3.3.2. XRF Analysis

Based on the data obtained from the XRF analysis above, we can know that there was a decrease in the relative percentage of polluter metal except for Si and Al as the main components of zeolite. The increase in Si and Al was from 43.9%, and it had not been detected in 48.8% and 6.6%. It was caused by HCl with a dilute or concentrated concentration that could dissolve the metal such as Fe, Zn, Mn, and the others (Svehla,1990). The followings are the reactions of polluter metal and acid.

Table 3. The result of XRF analysis of natural zeolite before and after activation

Element	Si (%)	Al (%)	K (%)	Ca (%)	Ti (%)	V (%)	Cr (%)	Mn (%)	Fe (%)	Ni (%)	Cu (%)	Zn (%)	Sr (%)	Eu (%)	Re (%)
Before activation	43.9	-	3.32	13.5	2.23	0.06	0.088	0.42	30.2	3.54	0.53	0.23	1.4	0.2	0.3
After activation	48.8	6.6	3.51	12.3	2.34	0.07	0.081	0.26	21.0	2.91	0.45	0.17	1.2	0.10	0.3

$$Fe + 2HCl \rightarrow Fe^{2+} + 2Cl^{-} + H_2$$
$$Zn + 2H^{+} \rightarrow Zn^{2+} + H_2$$
$$MnO_2 + 6HCl \rightarrow 2Mn^{2+} + Cl_2 + 4Cl^{-} + 3H_2O$$

Besides, based on the data above, we can know that the ratio of Si/Al was 7.13. The higher the ratio of Si/Al, the acidity, and crystallinity of the zeolite catalyst, the catalytic character increased as well. The increase in acid strength of this natural zeolite was seen from the theory of Bronsted-Lowry [15]. The ratio of Si/Al influences the catalytic character of the catalysts like the active site and its acidity [16].

IOP Conf. Series: Journal of Physics: Conf. Series **1093** (2018) 012011 doi:10.1088/1742-6596/1093/1/012011

3.4. *Qualitative Test of Glucose by Using Fehling Reagent*

Qualitative test as the result of hydrolysis presented in Table 4.

Table 4. Data of Fehling test of each active zeolite sample

The converted cellulose by using activated natural zeolite (acid)	Fehling test of the Result of Sonication of Cotton Cellulose		
	One Hour	Two Hours	Four hours
1	++	+	++++
2	+	++	+++
3	+	+++	++

Notes :

+ : the glucose formed was very little (greenish-blue)
++ : the glucose formed was little (green)
+++ : the glucose formed was much (yellow)
++++ : the glucose formed was very much (red)

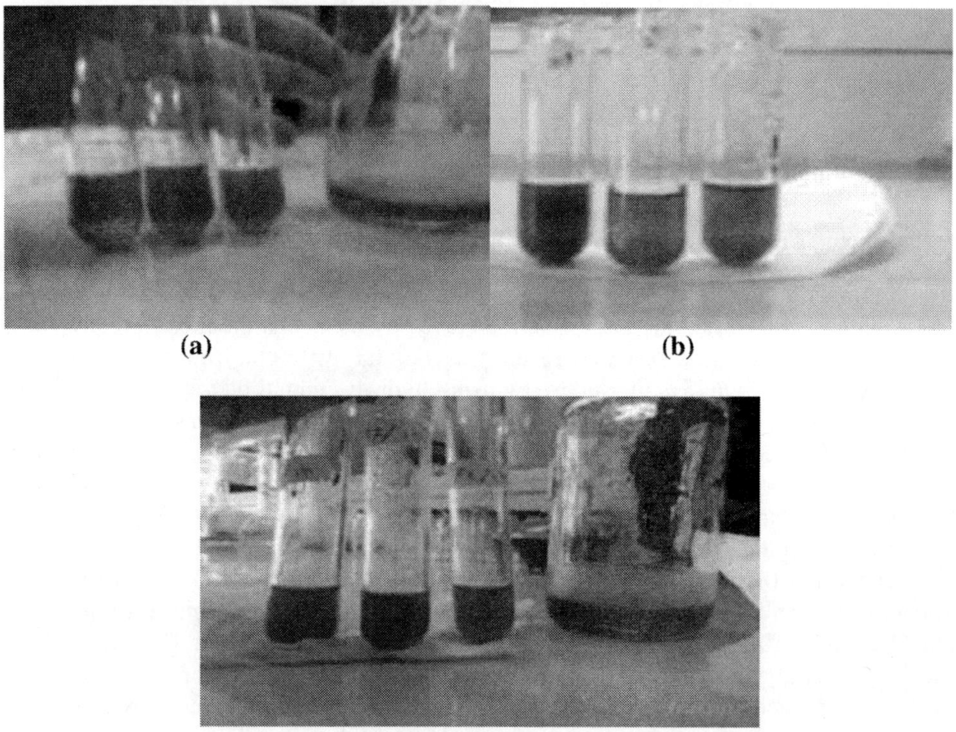

(a) (b) (c)

Figure 4. (a) One-hour sonication (b) Two hours sonication (c) Four hours sonication

The highest amount of glucose was apparent in four hours sonication. This case means the sonication time of four hours was the best time in cellulose hydrolysis process becoming glucose. The longer the sonication time, the glucose concentration formed will be higher as well [17]. Such a statement is suitable for the data of experiment results shown by the color formed.

3.5. *Quantitative Test of Glucose Product Using Nelson-Somogyi Reagent*

Nelson-Somogy test is a quantitative test to know the existence of reducer sugar (like glucose). By this test, the glucose concentration was obtained in ppm unit by using the equation of standard solution line. The yield percentage of glucose as the result of cellulose conversion of cotton is presented in Figure 5.

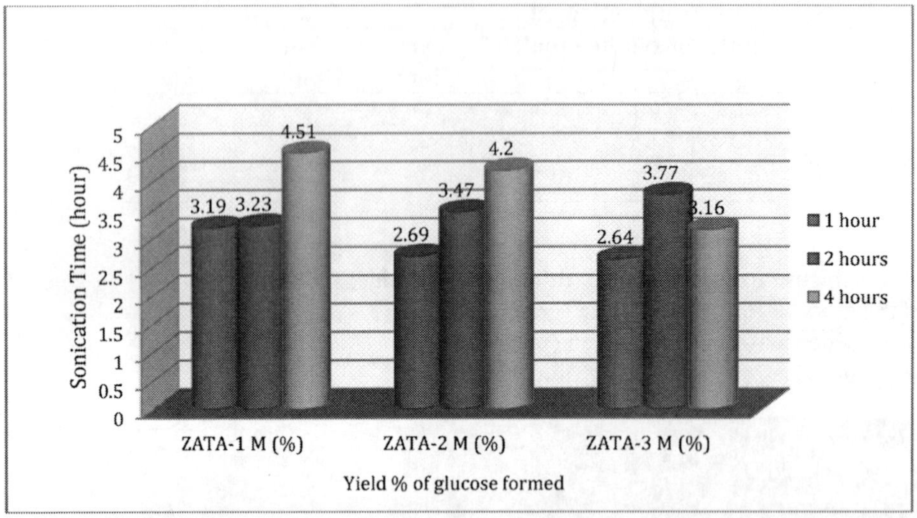

Figure 5: Influence of acid concentration on activation of natural zeolite (ZATA) catalyst and sonication time to the yield % of glucose in the degradation process of cotton cellulose.

Based on the data above, the highest percentage of glucose was 4.51% from the result of cellulose hydrolysis of cotton. The products were obtained when the active zeolite catalyst (ZATA-1M) was used with the sonication time of four hours. The obtained results were in line with the surface area in which ZATA with 1 M HCl had the largest surface area so that it was the most effective to hydrolyze the cellulose into glucose. This research result was not far different from the research results of Osvaldo et al. [5] getting 5.0675% of glucose by using hydrolysis method with 2% of the acid catalyst [18]. Although the yield percentage of glucose resulted from this research was lower than the result of Osvaldo et al.[5], the method that we offer is more environmentally friendly. This research result also showed that the yield percentage of glucose was five times higher than the yield percentage of glucose resulted from the research conducted by Fungky [9] that was 0.967%, where he did cellulose hydrolysis assisted by ultrasonic with natural zeolite catalyst activated physically (calcination) with or without impregnation of Ni metal. The benefit of this method compared to the other previous research is it is more environmentally friendly and the energy used is more efficient since it does not need high temperature like what had been done by Fajriutami [8] and Sumari et al.[7]. Besides, it is more environmentally friendly than the method that used acid reagent [19]. Moreover, the excellence of using natural zeolite activated by acid (ZATA) is this zeolite can be recovered easily and used repeatedly.

4. Conclusion

Natural zeolite before and after activated by HCl was confirmed in the form of mordenite type. HCl used in the activation process could reduce the impurity through ion exchange with H^+ ion. Moreover, the natural zeolite catalyst activated by 1 M of HCl was the most effective one to hydrolyze the cellulose from the cotton into glucose with the yield percentage of 4.71%. The ultrasonic used in the hydrolysis process gave influence in converting the cellulose into glucose with the highest yield through four hours sonication to the natural zeolite activated by 1 M of HCl.

IOP Conf. Series: Journal of Physics: Conf. Series **1093** (2018) 012011 doi:10.1088/1742-6596/1093/1/012011

References

[1] ZREU (Zentrum fur rationell Energieanwendung und Umwelt GmbH). 2000. Biomass in Indonesia. (German Energy Saving Project)
[2] Rispiandi. 2011..Jurnal Fluida: Bandung. 7 (1) :1-11
[3] Irawan MA. 2007. Metabolisme energi. Polton Sports Science & Performance Lab.1, 06 1-5
[4] Schenck F. W. 2006, Glucose and Glucose-Containing Syrups in Ullmann's Encyclopedia of Industrial Chemistry, (Weinheim Wiley-VCH)
[5] Osvaldo Z,S., Panca Putra S., M. Faizal. 2012. Jurnal Teknik Kimia.18 (2) 52-62
[6] Sasaki, C., Keisuke S., Chikako A., Yoshitoshi N..2012..Carbohydrate Polymers Journal. 89 298-301
[7] Sumari S., Roesyadi A, Sumarno. 2013. Scientific Study & Research Chemistry & Chemical Engineering, Biotechnology, Food Industry. ISSN 1582-540X 14(4) 229-239
[8] Fajriutami, T., Widya F, Hermiati E. 2016. Jurnal Riset Industri. Bogor. 10 (3): 147-161
[9] Fungky.2016. Penggunaan Katalis Nikel-Zeolit Alam Aktif pada Hidrolisis Selulosa menjadi Glukosa dengan Bantuan Ultrasonik.(Malang: State University of Malang)
[10] Castellan. 1983. Physical Chemistry 3rd edition. University of Maryland.
[11] Selim, M.M., El-Mekkawi, D.M., Aboelenin, R.M.M., Ahmed, S.A.S,. Mohamed, G.M., 2017. Journal of the Association of Arab Universities for Basic and Applied Sciences 24, 19–25
[12] Kumar A. & Jena H.M. 2016.. Results in Physics, 6, 651-658
[13] Treacy, M.M.J & Higgins, J.B. 2001. Collection of Simulated XRD Powder Patterns for Zeolites. Elseiver. Published on behalf of the StuctureCommision of the International Zeolite Association Fourth Revised Edition.
[14] Kalla, R., Sumarno, S., Mahfud, M., 2015. Bulletin of Chemical Reaction Engineering & Catalysis, 10 (3), 304-312.
[15] Budianto A., Sumari S. and Udyani K. 2015. ARPN Journal of Engineering and Applied Sciences, 10 (22):10317-10323.
[16] Bhadauria,J., Singh,B.K., Tomar,A.,& Tomar,R. 2011.. J.Chem.Pharm.Res, 3, 245-257
[17] Vityuk, A., Hristiyan A. A., Georgi N. V., Shuguo Ma, Oleg S. A., Michael D.A..2014..Physical Chemistry Journal:A-Q
[18] Anggraini, P. A., Andri R. R., Sumarno. 2014. Jurnal Teknik. 2 (1): 9 – 20.
[19] Sun, B., Gege P., Lian D., Aihua Xu, Xiaoxia Li.2015. Bioresource Technology Journal. 196, 454-458

Synthesis and Characterization of ZnO Nanorods by Hydrothermal Methods and Its Application on Perovskite Solar Cells

Ifa K R Laila[1], Nandang Mufti[1,2,*], Siti Maryam[1], Abdullah Fuad[1,2], Ahmad Taufiq[1,2], and Sunaryono[1,2]

[1]Department of Physics, Faculty of Mathematics and Natural Sciences, Universitas Negeri Malang, Jl. Semarang 5 Malang 65145, Indonesia
[2]Centre of Advanced Materials for Renewable Energy, Universitas Negeri Malang, Jl. Semarang 5 Malang 65145, Indonesia

*Corresponding author's email: nandang.mufti.fmipa@um.ac.id

Abstract. The aim of this research is to investigate the effect of ZnO Nanorods (ZnO NRs) morphology synthesized by the hydrothermal method to electrical properties of $CH_3NH_3PbI_3$/ZnO NRs perovskite solar cell (PSCs). ZnO nanorods were synthesized on the ITO substrate by a hydrothermal method. ZnO NRs were synthesized using hexamethylenetetramine (HMT) and zinc nitrate with a 1:1 molar ratio for 6 h. The growth temperature varied at 90°C and 100°C. The zinc nitrate concentration also varied at 25mM and 50 mM. The perovskite was made through a two-step deposition by spin coating with PbI_2 and CH_3NH_3I as the main ingredients. The effects of the synthesis conditions on ZnO NRs and Perovskite films were systematically investigated by X-Ray Diffraction (XRD), Scanning Electron Microscopy (SEM), Fourier transform infrared spectroscopy (FTIR), LCR AC meters and multimeters. The SEM results showed that as the temperature and concentration of zinc nitrate increase the size of the diameter and length of the rods are increasing. ZnO NRs synthesized with 50 mM concentrations of zinc nitrate at a growth temperature of 90 °C temperature showed the best results in terms of ZnO NRs morphology. The XRD characterization results showed that the formed $CH_3NH_3PbI_3$ film contained PbI_2 impurities. The existence of PbI_2 was suspected in opening the gap of recombination causing in low current and high dielectric constant.

Keywords: ZnO nanorods, perovskite solar cell, $CH_3NH_3PbI_3$ and hydrothermal method.

1. Introduction

The increasing energy demand, environmental issues, and limited availability of fossil fuels are demanding the research on sustainable and renewable energy resources [1]. The sunlight is an ultimate source to accomplish clean energy demand. Solar cell technology provides an eco-friendly and renewable energy route to convert photon energy into electricity directly [2,3]. The organometallic halide of $CH_3NH_3PbI_3$ has attracted interest due to high efficiency, transparent and easy to synthesis. $CH_3NH_3PbI_3$ has a direct band gap of 1.55 eV determining an offset absorption of up to 800 nm [4]. $CH_3NH_3PbI_3$ has met 19 % efficiency in numerous studies, and it is expected to achieve the target of 20% efficiency [3]. However, the research on $CH_3NH_3PbI_3$ PSCs stability to UV lighting and efficiency

IOP Conf. Series: Journal of Physics: Conf. Series **1093** (2018) 012012 doi:10.1088/1742-6596/1093/1/012012

is still developed. The efficiency and stability of PCS can be improved by using an electron transport material (ETM) and hole transported material (HTM) to prevent the electron-hole pair recombination [3].

The ZnO is one of ETM that widely for PSCs [6]. ZnO in nanoscale have better electronic and optical properties compared to bulk materials [7]. It has been considered that vertical 1D structures with high crystallinity are important keys in the manufacture of good solar cells [8,9]. One of the advantages of the 1D structure is that it can provide a ballistic transport effect, in which electrons can move in a medium regardless of the electrical resistance caused by scattering. Thus, the possibility of energy loss in the electron transport process can be minimized. Moreover, scattering in the transport process can occur due to impurities, defects in the crystal and structures that allow the oscillation of electron motion in the transport process [10]. Among the various forms of the structure modifications, the formation of ZnO with rods has been studied intensively [11].

ZnO is a compound that does not need a high temperature to show a better performance; therefore, it could function as ETM instead of other metal oxides which are processed at high temperatures. In a low-temperature PSCs, ZnO has been coated on both rigid and flexible substrates, by different techniques such as Electrodeposition [12], chemical bath deposition [13], sonochemical [14], sol-gel [15], hydrothermal [16] and solvothermal [17]. Meanwhile, the steam phase methods such as chemical vapor deposition [18] and vapor-liquid-solid (VLS) [19] have been successfully employed for the growth of ZnO NRs on a glass substrate. However, the steam phase methods have some weaknesses that include the need for a high temperature, high operational cost, and a relatively long time.

The ZnO NRs can be improved by optimizing the length and diameter of the rods formed. The size of the diameter and length of the rods formed will have an effect on the surface area [20]. Therefore, it is important to investigate growth parameters like growth temperature and concentration of zinc nitrate to optimize diameter and length of the ZnO NRs. Moreover, it is also important to study the effect of perovskite impurities to output current and dielectric of PSCs

2. Materials and Methods

2.1. *ITO Preparation*

The ITO substrate with 2.5 cm x 2.5 cm dimension was washed for 10 minutes in an ultrasonic cleaner with acetone and then continued washed by Deionized Water for 10 minutes. The clean ITO was dried at room temperature. The ITO surface was checked by a multimeter to identify the conductive surface.

2.2. *Preparation of ZnO seed layer*

The ZnO seed layer was synthesized by employing Zinc acetate dihydrate ($Zn(CH_3COO)_2 \cdot 2H_2O$, Aldrich, 98 %) as a precursor, ethanol as a solvent, and monoethanolamine (MEA) as a stabilizer. The concentration ratio ($Zn(CH_3COO)_2.2H_2O$): MEA was maintained at 1:1 molar. The ZnO seed layer deposition coated by a spin coating method at the rate of 3000 rpm for 20 seconds. A preheating process at 150 °C was performed for 10 minutes, and then continued by an annealing process at 550 °C for 2 hours in order to eliminate the organic and solvent components.

2.3. *The Growth of ZnO NRs*

The ZnO NRs were grown by using a hydrothermal method on ZnO seed layer substrate. The concentration of precursors used was 25 mM and 50mM and the molar ratio between Zn $(NO_3)_2.4H_2O$ and HMT was 1:1. During the growing process, the solution was heated in a variety of growth temperatures of 90 °C and 100 °C for 6 hours. The sample was cleaned by DI-Water to remove the remaining salt from the sample surface and dried at room temperature. Lastly, the ZnO nanorods on the ITO substrate were annealed for 2 hours at 550 °C. When the hexamethylenetetramine (HMT) and $Zn(NH_3)_4{}^{2+}$ was used as a precursor, the chemical reaction could be stated in the following mechanism [14]:

$$(CH)_6N_4 + 6H_2O \rightarrow 6HCHO + NH_3 \quad (1)$$

$$NH_3 + H_2O \rightarrow NH_4^+ + OH \quad (2)$$

$$Zn^{2+} + NH_3 \rightarrow Zn\,(NH_3)_4^{2+} \quad (3)$$

$$Zn^{2+} + 4OH^- \rightarrow Zn(OH)_4^{2-} \quad (4)$$

$$Zn(NH_3)_4^{2+} + 2OH^- \rightarrow ZnO + 4NH_3 + H_2O \quad (5)$$

$$Zn(OH)_4^{2-} \rightarrow ZnO + H_2O + 2OH^- \quad (6)$$

On the initial growth of the ZnO nanorods, the HMT started to decompose into ammonia (Equation (1)) and produce hydroxide ion, namely OH^- (Equation (2)). The Zn^{2+} Cation further reacted to NH_3 and OH^- anion forming ZnO core (Equation (3) and Equation (4)). The crystal core would develop, and further degraded into the ZnO core nanorods under the influence of more OH ion at a specific temperature, according to Equation (5) and (6) [30]. Hence, as the time passed by, the ZnO core would keep growing. Subsequently, the Zn on the surface of the Zn substrate would oxidize and form ZnO nanorods on the substrate. In the meantime, the oxidation process resulted in the lateral growth of the ZnO core during the attachment of Zn atom on the ZnO nuclei core [20].

2.4. $CH_3NH_3PbI_3$ Synthesis

The next deposition following the ZnO nanorod formation was the fabrication of the $CH_3NH_3PbI_3$ perovskite device by a two-step deposition. First, a Methyl-ammonium iodide (MAI) solution was generated from a mixture of 30 ml hydroiodic acid of 0.227 mol (57 wt% in water) and 27.8 ml methylamine of 0.273 mol (40 wt% in methanol). Second, a solution was made from 2.3-gram PbI_2 powder and dissolved in 5 ml dimethylformamide (DMF) stirred for 30 minutes. The two solutions fabricated in the first and second methods were prepared for the deposition process via a spin coating method on the substrate of ITO/ ZnO NRs. The first deposition layer was the PbI_2 solution followed by the CH_3NH_3I solution. Then, a preheating step was carried at 100 °C for 30 minutes on each stage of deposition. Next, the layered substrate was kept in a tightly sealed and put in a dry place.

2.5. Material Characterization

The ZnO NRs surface morphology analyzed by Scanning Electron Microscopy (SEM, FEI INSPECT-S50 type). The crystallization and structural properties of ZnO NRs characterized by X-Ray Diffraction (XRD, Philips X'Pert PRO) with a Cu Kα radiation (λ = 1.54 Å). Fourier-Transform Infrared Spectroscopy (FTIR) was used to investigate the optical property of the ZnO NRs. The dielectric constant and current of the $CH_3NH_3PbI_3$/ZnO NRs PSCs were measures by LCR meter and multimeter.

3. Results and Discussion

3.1. XRD Diffraction

The structural properties of ZnO NRs characterized by XRD in 2θ range between 10 – 90 degrees with a Cu Kα wavelength (λ) of 1.54 Å. Figure 1 shows the diffraction patterns of ZnO NRs at different concentration and growth temperature. The diffraction peak at the 2θ of 31.08° (d_{100}); 34.56° (d_{002}); and 35.54° (d_{101}) is identified as hexagonal crystal system (*wurtzite*). The x-ray diffraction pattern of this ZnO NRs is in agreement with JCPDS standard (No.36-1451). The three main observed peaks observed related with Bragg plane of (100), (002) and (101).

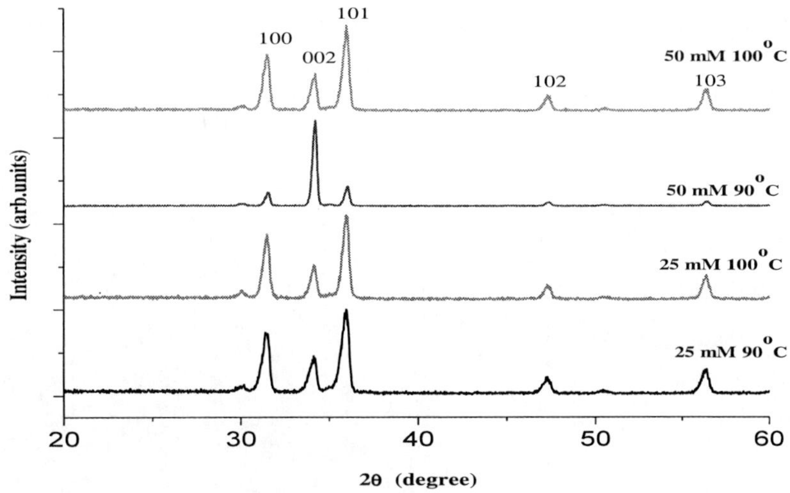

Figure 1. XRD patterns of the ZnO Nanorods at different zinc nitrate concentration and growth temperatures

The indication of a tetragonal $CH_3NH_3PbI_3$ crystal structure formation could be seen from the XRD peaks indicated at 2θ angles, namely 14.05°, 24.9 °, 28.45°, 31.87°, 40.45°, and 43.13° that were generated at XRD fields of (001), (202), (220), (310), (224), and (314), respectively [21]. Meanwhile, the highest peak of the perovskite crystal structure was found at (001) crystal field, and the PbI_2 peak was at 12.4° (001), 25.5° (002), 38.7° (003), and 52.5° (004). It is in excellent agreement with the research conducted by Li showing PbI_2 peaks at the (001), (002), (003), (004) fields [22]. Figure 2 shows the diffraction patterns of perovskite films synthesized by a two-step solution deposition. The perovskite $CH_3NH_3PbI_3$ diffraction peaks were observed at the (001), (202), and (220) planes. the PbI_2 diffraction peaks observed as impurities.

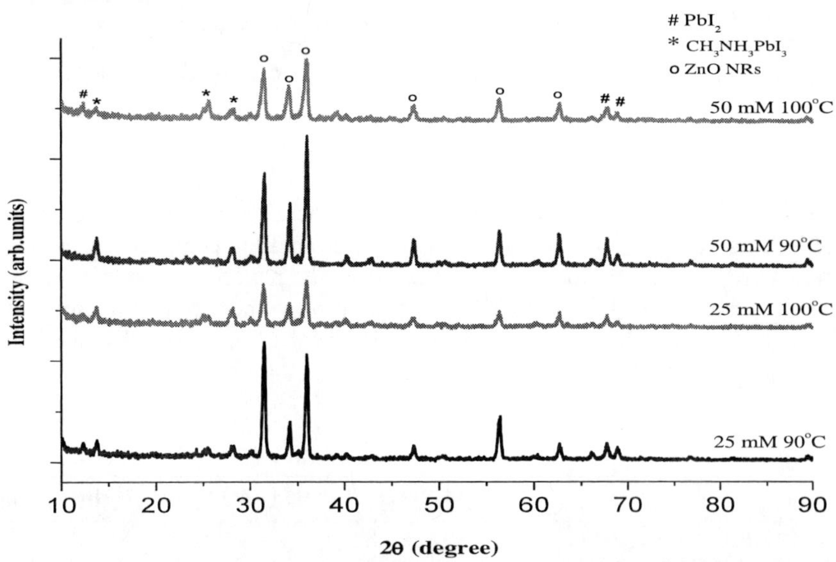

Figure 2. Diffraction patterns of $CH_3NH_3PbI_3$/ZnO NRs/ITO perovskite crystal with a variation of growth temperature and zinc nitrate concentration.

3.2. SEM Characterization

The surface morphologies of ZnO NRs were analyzed by using a scanning electron microscope (SEM) and presented in Figure 3. While, the cross-section of ZnO NRs shows in the insert to measure the length of the rods. Figure 3 shows that all ZnO NRs that have been synthesized have a hexagonal form, growth vertically and well-distributed. The diameter and length of ZnO NRs increase by increasing growth temperature and zinc nitrate concentration. The diameter and length of ZnO NRs summarize in Table 1.

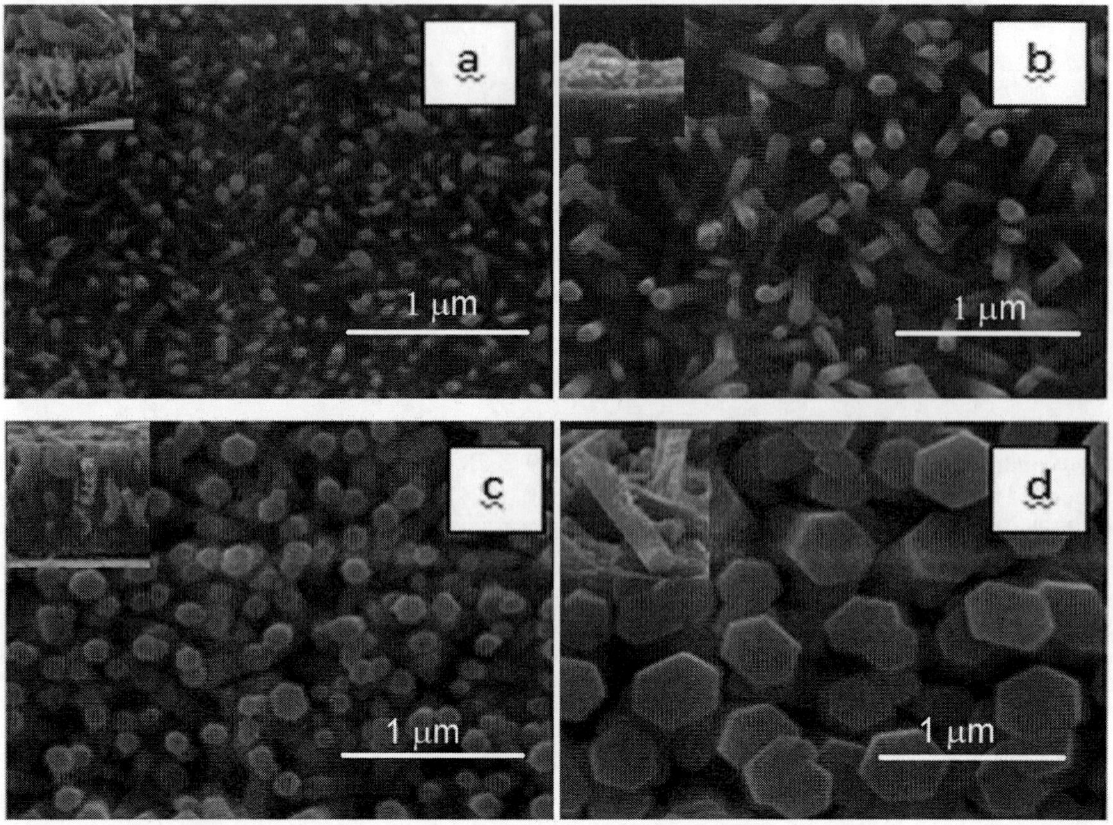

Figure 3. ZnO NRs morphology growth at (a) 25 mM 90 ^{O}C (b) 25 mM 100 ^{O}C (c) 50 mM 90 ^{O}C (d) 50 mM 100 ^{O}C for 6 hours. The insert is the cross section of ZnO NRs.

The increasing diameter and length of ZnO NRS by increasing precursor concentration and growth temperature is in good agreement with a study performed by Rani et al. (2018) [23]. Table 1 shows that growth temperature from 90 °C to 100 °C can improve the nanorods diameter from 81 nm to 365 nm and improve the length from 1.2 *µm* to 5.1 *µm*. This improvement due to a temperature growth slows down ZnO nucleation rate and increases the development of $Zn(OH)_2$ that acts as the intermediate compound of the ZnO NRs[19].

Table 1. The average diameter and length of ZnO NRs with variation of growth temperature and zinc nitrate concentration.

	Diameter of rods (nm)	**Lengths of rods (µm)**
ZnO NRs 25 mM 90 ^{O}C	81	1.2
ZnO NRs 25 mM 100 ^{O}C	92	1.5
ZnO NRs 50 mM 90 ^{O}C	148	4.8
ZnO NRs 50 mM 100 ^{O}C	365	5.1

3.3. FTIR Characterization
Figure 4 illustrates the FITR spectra of $CH_3NH_3PbI_3$ perovskite film with a 25 and 50 mM zinc nitrate concentration and some variation of the growth temperature of 90 °C and 100 °C. It could be seen that the absorption area of the material ranged from 500 to 4000 cm^{-1}. In general, the references present that the FTIR spectra of ZnO NRs are approximately 406 cm^{-1} and around 540 cm^{-1}–560 cm^{-1} in the blue shift range [34]. In this study, the FTIR spectra of ZnO NRs are between 485 cm^{-1} and 562 cm^{-1}.

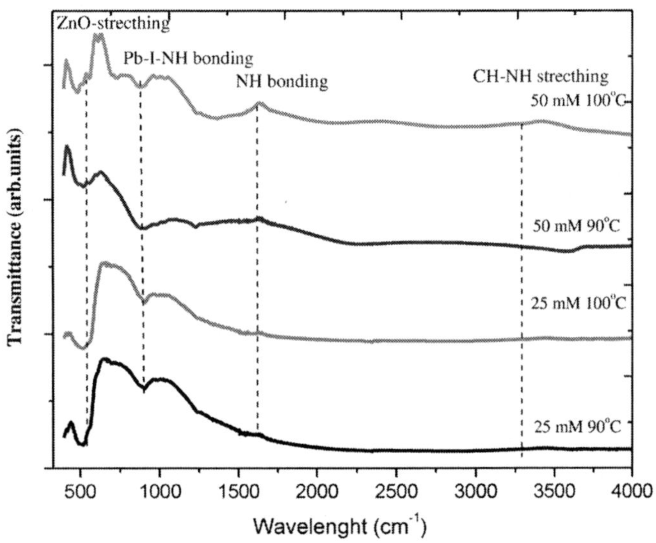

Figure 4. FTIR spectra of $CH_3NH_3PbI_3$/ZnO NRs/ITO perovskite film with variation of growth temperature and zinc nitrate concentration.

The chemical structure of $CH_3NH_3PbI_3$ particles is shown by FTIR (Figure 4). The spectrum displays the main characteristics of the peak areas of 1500 cm^{-1} and 956 cm^{-1}, where an N-H bond and Pb-I-NH stretching were found, respectively. Subsequently, a CH-NH stretching was found at the peak area of around 3215 cm^{-1}. Such results are in good agreement with the conversion results of the research done by Kesari (2015), which compared three spectra of pure PbI_2, MAI, and perovskite showing that the perovskite peak area could be found after a 20-minute sonication process [34].

3.4. Current Measurement
The output current and resistivity of PCs device are summarized in Table 2. Even though the perovskite solar cell film has a small size, it also produced a current. The higher the lux and the light distance, the smaller the current production. The results showed that sample 50 mM 100 °C had the most significant current among the others, namely of 2.35 x 10^2 μA. The relatively high current is also supported by the nanorod morphology that is relatively uniform and homogeneous. Such homogeneity would result in higher light absorption. Generally, a high current shows a declining resistivity value. the output current is lower compared to perovskite PCSs that reported previously [21,21] due to the $CH_3NH_3PbI_3$ perovskite phase is not well synthesized as shown in XRD results. The PSCs device still contains impurities of PbI_2 phase.

Table 2. Currents and resistivity of the Perovskite solar cells with variation of growth temperature and zinc nitrate concentration.

	Type of *Film*			
	25 mM 90 °C	**25 mM 100 °C**	**50 mM 90 °C**	**50 mM 100 °C**
Output Current (μA)	6.42	17.7	2.35	235
Resistivity (Ohm.cm)	2.94 x 10^4	1.48 x 10^4	1.81 x 10^4	5.66 x 10^2

3.5. Dielectric Constant Measurement

The measurement of dielectric constant of the ITO/ZnONRs/$CH_3NH_3PbI_3$ perovskite solar cell device was performed by using an LCR meter AC. The measurement was done via a two-probe method under the influence of the variation of frequencies towards the dielectric constant. Some variation of frequencies includes 100 Hz, 1 kHz, 10 kHz, 100 kHz, and 200 kHz. The measurement would result in the values of capacity that could be directly observed from the LCR monitor screen. The dielectric constant could be measured by using the following equation:

$$\varepsilon_r = \frac{C}{\varepsilon_0} \cdot \frac{d}{A} \tag{1}$$

Where ε_r is dielectric constant, ε_0 is vacuum permittivity (8.854 x 10^{-12} F/m), C is capacitance (F), *d* is the thickness of the perovskite film (m), and *A* is the surface of the perovskite film active area (m^2). The measurement results of the impact of frequency on the dielectric constant of ITO/ZnONRs/$CH_3NH_3PbI_3$ solar cells are illustrated in Figure 5.

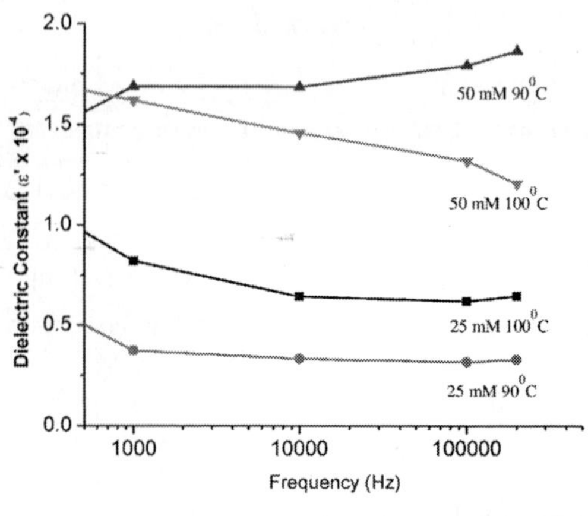

Figure 5. Dielectric constant of perovskite solar cell as a function of frequency with a variation of growth temperature and zinc nitrate concentration.

A high frequency had a low dielectric constant since the dipoles could not maintain its delivery on the alternating current. A steep decrease of dielectric constant was found at low frequencies, namely from 100 Hz – 1 kHz. It could be observed at the sample 25 mM, 50 mM 90 °C and 25 mM, 50 mM 100 °C film showing a steeply decreasing trend with a frequency of 200 kHz and dielectric constant 6.4x10^{-5}, 1.8x10^{-5} and 3.2x10^{-4},1.2x10^{-4}, respectively. A small dielectric constant of material would result in high conductivity.

4. Conclusion

In this study, ZnO NRs were synthesized by hydrothermal method with variation of growth temperature and zinc nitrate concentration. The SEM showed that the ZnO NRs have r diameters from 81 nm to 365 nm with the average rod length between 1.2 μm and 5.1 μm. The XRD diffraction showed that the ZnO NRs have a good crystal quality and orientations at (101) field. The ZnO NRs that had been synthesized by using 50 mM 90 °C presented the best result about structural, optical properties and dielectric constant. The existence of PbI_2 in $CH_3NH_3PbI_3$ PSCs might cause low output current with the maximum current of 235 μA. The structure and synthesis condition of $CH_3NH_3PbI_3$ Perovskite is very important for PCSs efficiency.

Acknowledgments

This research was support by the Ministry of Research, Technology, and Higher Education of the Republic of Indonesia for the research funding through PDUPT grant for NM.

References

[1] Abdmouleh Z, Gastli A, Ben-Brahim L, Haouari M and Al-Emadi N A 2017 Review of optimization techniques applied for the integration of distributed generation from renewable energy sources *Renew. Energy* **113** 266–80

[2] Bi D, Xu B, Gao P, Sun L, Grätzel M and Hagfeldt A 2016 Facile synthesized organic hole transporting material for perovskite solar cell with efficiency of 19.8% *Nano Energy* **23** 138–44

[3] Green M A, Ho-Baillie A and Snaith H J 2014 The emergence of perovskite solar cells *Nat. Photonics* **8** 506–14

[4] Nagarjuna P, Narayanaswamy K, Swetha T, Rao G H, Singh S P and Sharma G D 2015 CH3NH3PbI3 Perovskite Sensitized Solar Cells Using a D-A Copolymer as Hole Transport Material *Electrochimica Acta* **151** 21–6

[5] He X, Wu J, Tu Y, Xie Y, Dong J, Jia J, Wei Y and Lan Z 2016 TiO2 single crystalline nanorod compact layer for high-performance CH3NH3PbI3 perovskite solar cells with an efficiency exceeding 17% *J. Power Sources* **332** 366–71

[6] Ebadi M, Zarghami Z and Motevalli K 2017 40% Efficiency enhancement in solar cells using ZnO nanorods as shell prepared via novel hydrothermal synthesis *Phys. E Low-Dimens. Syst. Nanostructures* **87** 199–204

[7] Wu S, Li J, Lo S-C, Tai Q and Yan F 2012 Enhanced performance of hybrid solar cells based on ordered electrospun ZnO nanofibers modified with CdS on the surface *Org. Electron.* **13** 1569–75

[8] Feng T, Wen L, Wang D, Xie D and Wu M 2016 Structure transformation of hydrothermally treated ZnO nanorod array *Mater. Lett.* **163** 102–5

[9] Liu Y and Gao W 2015 Growth process, crystal size and alignment of ZnO nanorods synthesized under neutral and acid conditions *J. Alloys Compd.* **629** 84–91

[10] Xu Y, Liu Tian, Li Z, Feng B, Li S, Duan J, Ye C, Zhang J and Wang H 2016 Preparation and photovoltaic properties of perovskite solar cell based on ZnO nanorod arrays *Appl. Surf. Sci.* **388** 89–96

[11] Zafar M, Yun J-Y and Kim D-H 2017 Performance of inverted polymer solar cells with randomly oriented ZnO nanorods coupled with atomic layer deposited ZnO *Appl. Surf. Sci.* **398** 9–14

[12] Hou Q, Zhu L, Chen H, Liu H and Li W 2013 Highly regular and ultra-thin porous ZnO nanosheets: An indirect electrodeposition method using acetate-containing precursor and their application in quantum dots-sensitized solar cells *Electrochimica Acta* **94** 72–9

[13] Taunk P B, Das R, Bisen D P, Tamrakar R K and Rathor N 2015 Synthesis and optical properties of chemical bath deposited ZnO thin film *Karbala Int. J. Mod. Sci.* **1** 159–65

[14] Chen S-W and Wu J-M 2011 Nucleation mechanisms and their influences on characteristics of ZnO nanorod arrays prepared by a hydrothermal method *Acta Mater.* **59** 841–7

[15] Yoon Y-C, Park K-S and Kim S-D 2015 Effects of low preheating temperature for ZnO seed layer deposited by sol–gel spin coating on the structural properties of hydrothermal ZnO nanorods *Thin Solid Films* **597** 125–30

[16] Hassanpour A, Bogdan N, Capobianco J A and Bianucci P 2017 Hydrothermal selective growth of low aspect ratio isolated ZnO nanorods *Mater. Des.* **119** 464–9

[17] Ullah S, Ahmed F, Badshah A, Ali Altaf A, Raza R, Lal B and Hussain R 2013 Solvothermal Preparation of ZnO Nanorods as Anode Material for Improved Cycle Life Zn/AgO Batteries ed B Xu *PLoS ONE* **8** e75999

[18] Ye Z, Wang T, Wu S, Ji X and Zhang Q 2017 Na-doped ZnO nanorods fabricated by chemical vapor deposition and their optoelectrical properties *J. Alloys Compd.* **690** 189–94

[19] Alvi N H, Hassan W ul, Farooq B, Nur O and Willander M 2013 Influence of different growth environments on the luminescence properties of ZnO nanorods grown by the vapor–liquid–solid (VLS) method *Mater. Lett.* **106** 158–63

[20] Zhou W D, Wu X, Zhang Y C and Zhang M 2007 Solvothermal synthesis of hexagonal ZnO nanorods and their photoluminescence properties *Mater. Lett.* **61** 2054–7

[21] Phillips L J, Rashed A M, Treharne R E, Kay J, Yates P, Mitrovic I Z, Weerakkody A, Hall S and Durose K 2016 Maximizing the optical performance of planar CH 3 NH 3 PbI 3 hybrid perovskite heterojunction stacks *Sol. Energy Mater. Sol. Cells* **147** 327–33

[22] Li S, Yang B, Wu R, Zhang C, Zhang C, Tang X-F, Liu G, Liu P, Zhou C, Gao Y, Meng J-Q and Yang J 2016 High-quality CH 3 NH 3 PbI 3 thin film fabricated via intramolecular exchange for efficient planar heterojunction perovskite solar cells *Org. Electron.* **39** 304–10

[23] Rani R A, Ghafar S A, Zoolfakar A S and Rusop M 2018 Growth of ZnO nanorods on glass substrate deposited using dip coating method p 020072

Distribution and Assessment of Fe and Mn in the Coastal Sediments of Sendang Biru, East Java, Indonesia

Anugrah Ricky Wijaya[1,*], Ida Farida[1], Atik Sakbaniah[1], Anisa Mai Rahmawati[1], Endang Budiasih[1], Hasan Daupor[2], Md. Sazzad Hossain[3], Tatsuya Kunisue[4]

[1]Department of Chemistry, Faculty of Mathematics and Natural Sciences, Universitas Negeri Malang, Jl. Semarang 5 Malang 65145, Indonesia
[2]Chemistry Major, Faculty of Science Technology and Agriculture, Yala Rajabhat University, Yala 95000, Thailand
[3]Department of Aquaculture, Faculty of Fisheries, Bangladesh Agricultural University (BAU), Mymensingh-2202, Bangladesh
[4]Environmental Chemistry and Ecotoxicology, Center for Marine Environmental Studies (CMES), Ehime University, Bunkyo-cho 2-5, Matsuyama 790-8577, Japan

*Corresponding author's email: anugrah.ricky.fmipa@um.ac.id

Abstract. The high presence of heavy metals in sediment can affect the ecological and biological systems in East Java Indonesia. To monitor the environmental status of Fe and Mn, we investigated geochemical fraction and distribution of these metals at 8 sites in the sediment of Sendang Biru. BCR and Tessier sequential leaching methods were applied to leach Fe and Mn, respectively, due to high precision and accuracy of these methods. The pattern of geochemical fractions in sediment samples showed the maximum leached levels of Fe and Mn (>50%) in residual fractions in Sendang Biru Beach, indicating the natural effects surrounding sites. The portions of Fe and Mn in non-residual fractions located at Sendang Biru Port were higher compared with the site in the adjacent sea, indicating these metals were derived possibly from anthropogenic effects. The assessment from CF and I_{geo} calculations of Fe and Mn in sediment samples showed that the sediment near the Sendang Biru Port was moderately contaminated by Fe and Mn.

Keywords: Sediment, fraction, BCR, Tessier, leaching, Sendang Biru Beach

1. Introduction

The port in Sendang Biru Beach is one of the famous and big ports in East Java, Indonesia. The anthropogenic and natural activities can contribute heavy metal in seawater surrounding Sendang Biru Port. Sediment and coral can be used as media which reflect heavy metal contents including Fe and Mn in seawater [1]. To understand the Fe and Mn mobilities in each of fraction in sediment associated with the healthy environment, we used sequential leaching method to evaluate concentration and assessment of these elements. The dominant presence of Fe and Mn in Sendang Biru sediment suspected influences ecological and biological systems in the aquatic environment. To understand metals mobility which can influence environmental systems in Sendang Biru Beach, we used sequential leaching method for Fe and Mn contents in sediment. These metals have the main pathways residing in their fractions through chemical and physical processes [2–4]. The adsorption, complexation, and precipitation of Fe and Mn in sediment are the main processes of these elements cooperated with iron and manganese hydroxides, particulate organic matter and clay minerals due to their strong affinities.

Commonly, sequential leaching of Fe and Mn from different sediment fractions were investigated using Tessier and BCR (Bureau Community of Reference) methods [5,6]. Some researchers recorded Fe in the fraction of sediment and applied Tessier method releasing with the range 80-100% recovery [7,8] and 0.74-4.40 %RSD [9,10]. BCR method of leached Mn releasing with the range 95-104 %recovery and 0.80-1.61 %RSD [5,6,11].

Now, the coastal area of Sendang Biru is prepared to improve towards international fishing and others marine products [12]. It is very important to monitor heavy metals such as Fe and Mn in seawater by sediment fraction. Here we applied these methods for determining Fe and Mn concentrations in the geochemical fractions of sediment of Sendang Biru Beach. The major objective of this study was to elevate the distribution and assessment of Fe and Mn in sediment surrounding Sendang Biru Beach.

2. Experimental Methods

The samples were collected on 12th of November 2016. The sediment samples were taken surrounding Sendang Biru Port with the geographical location between latitude -8°25'57.9" S and longitude 112°41'02.5" E (Figure 1). The sediments were taken from the surface of Sendang Biru Port with sediment grab. After the sediment grab was carefully opened, the sediment was transferred into the plastic bag. The samples were then leached using BCR method to leach Fe and Tessier method to leach Mn (Table 1).

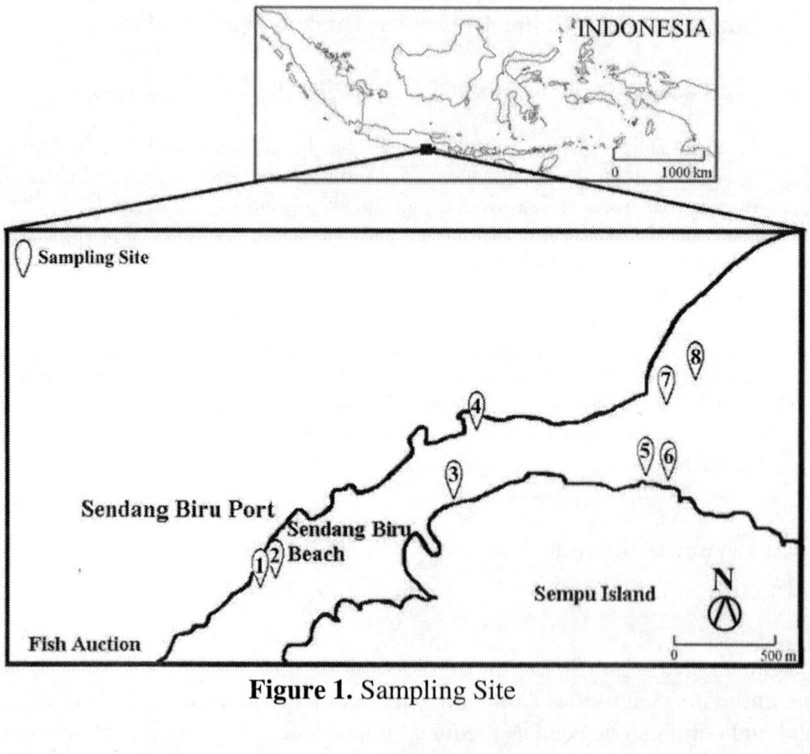

Figure 1. Sampling Site

Table 1. The procedure of sequential leaching of BCR and Tessier methods

Fe-BCR	Metal Mobility	Mn-Tessier	Metal mobility
20 mL of CH_3COOH 0.11 M solution, room temperature, shake in 16 hours	Acid Soluble (F1)	4mL of $MgCl_2$ 1 M solution, pH = 7, room temperature, shake in 1 hour	Exchangeable (F1)
20 mL of $NH_2OH.HCl$ 0.5 M solution, pH 2, room temperature, shake in 16 hours	Reducible (F2)	4 mL of CH_3COONa 1 M, pH = 5 with CH_3COOH, room temperature, shake in 5 hours	Associated to carbonate (F2)
5 mL of H_2O_2 8.8 M solution, pH 2, room temperature, shake in 1 hour, repeat in 85 °C, shake in 1 hour 5 mL of H_2O_2 8.8 M solution, pH 2, 85 °C, shake in 1 hour 25 mL of CH_3COONH_4 1M, pH 2, room temperature, shake in 16 hours	Oxidizable (F3)	10 mL of $NH_2OH.HCl$ 0.04 M solution in 25% CH_3COOH solution, 96 °C, shake in 6 hours	Reducible (F3)
1 mL of 65% HNO_3 solution and 3 mL of 36% HCl solution, room temperature, shake in 2 hours, add 3 drops of HF solution	Residual (F4)	1.5 mL of HNO_3 0.02 M solution and 2.5 mL of 30% H_2O_2 solution, pH = 2, 85 °C, shake in 2 hours, add 1.5 mL of 30% H_2O_2 solution, 85 °C, shake in 3 hours 2.5 mL of CH_3COONH_4 3.2 M solution in 20% HNO_3 solution, room temperature, and shake in a half hour	Oxidizable (F4)
		1 mL of 65% HNO_3 solution and 3 mL of 36% HCl solution, room temperature, shake in 2 hours and add 3 drops of HF solution	Residual (F5)

3. Results and Discussion

3.1. *Distribution of Fe and Mn contents in sediment*

Fe and Mn concentrations, for the Sendang Biru samples, are reported in Table 2. The content of Fe was divided into four fractions (F1, F2, F3, and F4) followed by BCR method. The explanation of F1, F2, F3, and F4 is listed in Table 1. The range of Fe contents in F1, F2, F3, and F4 were 369-5621; 461-5853; 994-10019; and 5480-25965 mg/Kg and total fraction of Fe contents varied from 7304 to 42884 mg/Kg. The distribution of Fe contents in sediment has the same pattern values from F1 to F3. The lowest concentrations of Fe were detected at site 5 and the highest values at site 1. This result indicated Fe content was contributed by anthropogenic activities in Sendang Biru Port and then continued by the process of Fe dilution in the open sea.

The chemical reactions in the F1, F2, and F3 can be explained as follows [2]:

F1: Sediment–Fe^{3+}*(s)* + CH_3COOH*(aq)* ⇄ Sediment–$(H)_3$*(s)* + $(CH_3COO)_3Fe$*(aq)*

Sediment–$Fe_2(CO_3)_3$*(s)* + CH_3COOH*(aq)* ⇄ Sediment*(s)* + $(CH_3COO)_3Fe$*(aq)* + H_2CO_3*(aq)*

Sediment–Fe^{2+}*(s)* + CH_3COOH*(aq)* ⇄ Sediment–$(H)_2$*(s)* + $(CH_3COO)_2Fe$*(aq)*

Sediment–$Fe_2(CO_3)_2$*(s)* + CH_3COOH*(aq)* ⇄ Sediment*(s)* + $(CH_3COO)_2Fe$*(aq)* + H_2CO_3*(aq)*

F2: Sediment–MnO_2–Fe*(s)* + $NH_2OH.HCl$*(aq)* → Sediment*(s)* + $MnCl_2$*(aq)* + $FeCl_3$*(aq)* + NH_3*(g)* + H_2O*(l)* + O_2*(g)*

Sediment–MnO_2–Fe*(s)* + $NH_2OH.HCl$*(aq)* → Sediment*(s)* + $MnCl_2$*(aq)* + $FeCl_2$*(aq)* + NH_3*(g)* + H_2O*(l)*+ O_2*(g)*

F3: Sediment–HA–Fe*(s)* + H_2O_2*(aq)* + CH_3COONH_4*(aq)*→ Sediment–HA*(s)* +$(CH_3COO)_3Fe$*(aq)* + NH_3*(g)* + H_2O*(l)*+ O_2*(g)*

Sediment–HA–Fe*(s)* + H_2O_2*(aq)* + CH_3COONH_4*(aq)*→ Sediment–HA*(s)* +$(CH_3COO)_2Fe$*(aq)* + NH_3*(g)* + H_2O*(l)*+ O_2*(g)*

The concentration of Fe in F4 ranged from 5480-25965 mg/Kg (Table 2). The highest portion of Fe in F4 compared those in F1-F3 suggested the effect of lithogenic and terrestrial associated with natural source. Fe in F4 refers to the content of Fe in residual fraction. Possible of chemical reaction is followed [2]:

F4: Sediment–SiO_2–Al_2O_3–Fe*(s)* + HNO_3*(aq)* + HF*(aq)* →$Fe(NO_3)_3$*(aq)* + $Al(NO_3)_3$*(aq)* + SiF_4*(g)* + H_2O*(l)* + H_2*(g)*

Sediment–SiO_2–Al_2O_3–Fe*(s)* + HNO_3*(aq)* + HF*(aq)* →$Fe(NO_3)_2$*(aq)* + $Al(NO_3)_3$*(aq)* + SiF_4*(g)* + H_2O*(l)* + H_2*(g)*

The contents of Mn in each fraction from site 1 to site 8 are listed in Table 2. The leached and determined of Mn concentrations followed the method by Tessier which divided as five fractions (F1, F2, F3, F4, and F5) (Table 1). The possibilities of chemical reaction are explained as followed [2]:

F1: Sediment-Mn^{2+}(*s*) + $MgCl_2$(*aq*) ⇄ Sediment-Mg^{2+}(*s*) + $2MnCl_2$(*aq*)

F2: Sediment-$2MnCO_3$(*s*) + $4CH_3COONa$(*aq*) ⇄ $2Na_2CO_3$(*aq*) + $2Mn(CH_3COO)_2$(*aq*)

F3: Sediment-Fe_2O_3-Mn(*s*) + $2NH_2OH.HCl$(*aq*) → $2Fe^{2+}$(*aq*) + Cl_2(*g*) + Mn^{2+}(*aq*) + $2H_2O$(*l*) + $2NH_3$(*g*)

Sediment-$Fe(OH)_3$-Mn(s) + $2NH_2OH.HCl$(*aq*) → 2 Fe^{2+}(*aq*) + Cl_2(*g*) + Mn^{2+}(*aq*) + H_2O(*l*) + 2 NH_3(*g*)

F4: Sediment–HA–Mn*(s)* + H_2O_2*(aq)* + CH_3COONH_4*(aq)*→ Sediment–HA*(s)* + NH_3(g) +H_2O(l) + O_2(g) + $(CH_3COO)_2Mn$*(aq)* +

F5: Sediment–SiO_2–Al_2O_3–Fe*(s)* + HNO_3*(aq)* + HF*(aq)* → $Fe(NO_3)_2$*(aq)* + $Al(NO_3)_3$*(aq)* + SiF_4*(g)* + H_2O*(l)* + H_2*(g)*

The patterns of Mn contents in F1 and F2 were the same at site 1 and site 4 and the patterns in F3-F5 were different at site 2, 4, 5, 6, and 7 (Table 2; Figure 2). The portion of Mn contents in F1 and F2 tended to be stable due to the balance of cation and anion (carbonate) exchange from the open sea. The proportion of Mn contents ranged from 7.96-16.0 mg/Kg in F1 and 7.26-22.0 mg/Kg in F2 (Table 2).

Table 2. Distribution of Fe and Mn contents in sediment's fraction of Sendang Biru Beach

Sampling	Fe content (mg/Kg)					Mn content (mg/Kg)					
Site	F1	F2	F3	F4	Total	F1	F2	F3	F4	F5	Total
1	5621	5853	10019	17803	39297	16.0	22.0	25.2	8.80	52.4	124
2	5110	5406	9102	15581	35200	8.53	7.94	13.7	5.97	52.0	88.1
3	3806	4049	7108	12921	27884	12.5	13.5	31.0	18.6	53.0	129
4	4541	4592	8156	25595	42884	13.4	21.9	23.4	8.24	44.4	111
5	369	461	994	5480	7304	10.9	10.6	38.5	15.1	201	276
6	1149	1657	2613	7711	13129	11.1	12.6	20.1	7.96	57.6	109
7	2337	2382	4179	11338	20236	9.15	15.2	10.1	5.70	41.6	81.8
8	2969	3207	5628	25965	37769	7.96	7.26	14.0	9.79	77.6	117

The range of total Fe concentrations (7304-42884 mg/Kg) in the coastal area of Sendang Biru exceeded the chronic criterion for protection of aquatic organism for Fe (~1000 mg/kg) in natural marine sediment, USA [13]. Conversely, most sediment samples had a range of Mn concentrations (818-276 mg/Kg) that were below the 452 mg/Kg for standard marine sediment ocean [14].

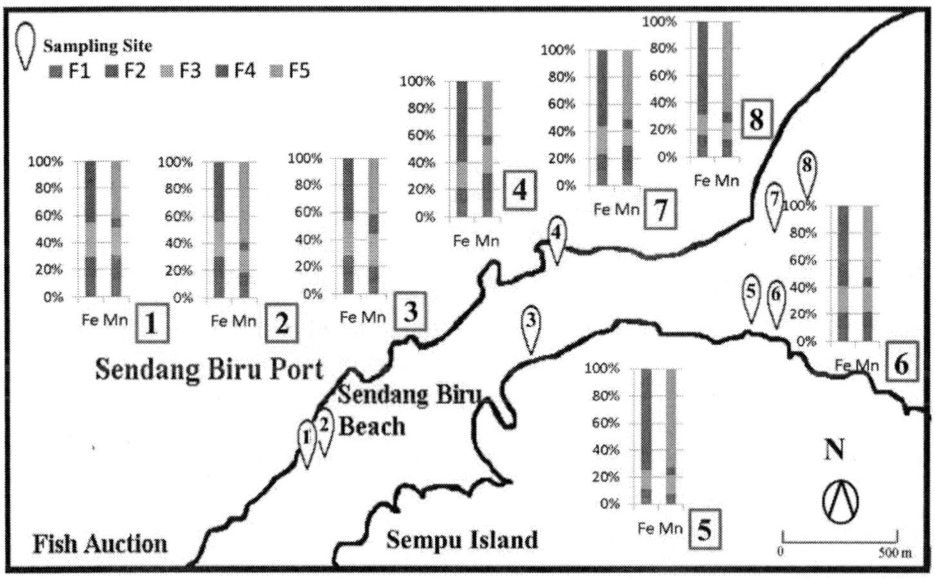

Figure 2. The portion of Fe and Mn Contents in Sediment's Fraction of Sendang Biru Beach

As shown in Figure 2, the pattern of geochemical fractions in sediment samples showed the maxima leached levels of Fe and Mn in the residual fraction from site 1 to site 8. The Fe and Mn contents in sediment possibly were contaminated by natural sources such as soil, agricultural, minerals surrounding Sendang Biru Beach bounded in silica and alumina. The portions of Fe and Mn in non-residual fractions were higher in Sendang Biru Port compared with the site in the adjacent sea indicating these metals possibly from anthropogenic effects.

3.2. *Assessment of Fe and Mn in Sediment*

To understand the level of contamination of Fe and Mn contents in sediment associated with water pollution, we performed an assessment using the calculation index of status pollution in sediments such as contamination factor (CF) and geoaccumulation index (Igeo) [3,15]. CF was calculated using $C_{metal}/C_{background}$. In this study, the content of Fe and Mn at site 7 (the lowest concentration) was considered as background.

The classification of CF is explained as follows: CF<1: low contamination factor, $1 \leq CF < 3$: moderate contamination factor; $3 \leq CF < 6$: considerable contamination factor; $CF \geq 6$: very high contamination factor. As listed in Table 3, the value of CF from site 5 to site 7 in the calculation of Fe were recorded from 0.36 to 1.00. This indicates that Fe contents at site 5-7 reflected low risk, whereas the values of Fe contents in sediment from site 1 to site 1 were moderate risk. In the case of CF in Mn sediment, site 1 to site 8 was categorized as moderate contamination.

The second assessment of Fe and Mn contents using the geo-accumulation index (I_{geo}). I_{geo} can be calculated using $I_{geo} = \log_2 (C_n/1.5\ B_n)$, whereas C_n is the concentration of Fe or Mn. The value of 1.5 is the factor of lithogenic effect and B_n is the background as explained C background in CF index. I_{geo} value is classified as follows: $I_{geo} \leq 0$, class 0, unpolluted; $0 < I_{geo} \leq 1$, class 1, from unpolluted to moderately polluted; and $1 < I_{geo} \leq 2$, class 2, moderately polluted. As listed in Table 3, the I_{geo} of Fe and Mn contents ranged from -2.06 to 1.17 indicating the status of quality in sediment categorized from class 0 to class 2. These values reflect the status of sediment in the range from unpolluted to moderately pollution.

Table 3. Assessment of Fe and Mn in Sediment

Sampling site	CF		I_{geo}	
	Fe	Mn	Fe	Mn
1	1.94	1.52	0.18	0.02
2	1.74	1.08	0.16	-0.32
3	1.38	1.57	-0.13	0.07
4	2.12	1.36	0.28	-0.10
5	0.36	3.37	-2.06	1.17
6	0.65	1.13	-1.21	-0.17
7	1.00	1.00	-1.69	-0.18
8	1.87	1.42	0.51	-0.06

4. Conclusion

In this Sendang Biru area, the majority of geochemical fractions of Fe and Mn contents in sediment were found within the high portions of Fe and Mn in residual fractions. This can be attributed substantially to the natural sources such as weathering or leaching of the rocks containing Fe and Mn minerals in the sediment. The distribution of Fe and Mn contents in non-fraction of sediment surrounding Sendang Biru Port showed highly compared with those in adjacent of the sea possibly associated with the dilution process of that metals through the sea current. The assessment of Fe and Mn contents in Sendang Biru sediment using CF and I_{geo} revealed moderately health risk.

Acknowledgments
The present work is funded by KEMENRISTEKDIKTI, Fundamental Grant Scheme, 2015-2016, DIPA Number 042.06-0/2016.

References

[1] Sazzad M H, Wijaya A R, Tanaka K and Ohde S 2010 Environmental effects on the stable carbon and oxygen isotopic compositions and skeletal density banding pattern of Porites coral from Khang Khao Island, Thailand *Afr. J. Biotechnol.* **9**

[2] Wijaya A R, Ohde S, Shinjo R, Ganmanee M and Cohen M D 2016 Geochemical fractions and modeling adsorption of heavy metals into contaminated river sediments in Japan and Thailand determined by sequential leaching technique using ICP-MS *Arab. J. Chem.*

[3] Wijaya A R, Ouchi A K, Tanaka K, Cohen M D, Sirirattanachai S, Shinjo R and Ohde S 2013 Evaluation of heavy metal contents and Pb isotopic compositions in the Chao Phraya River sediments: Implication for anthropogenic inputs from urbanized areas, Bangkok *J. Geochem. Explor.* **126–127** 45–54

[4] Wijaya A R, Ouchi A K, Tanaka K, Shinjo R and Ohde S 2012 Metal contents and Pb isotopes in road-side dust and sediment of Japan *J. Geochem. Explor.* **118** 68–76
[5] Kartal Ş, Aydın Z and Tokalıoğlu Ş 2006 Fractionation of metals in street sediment samples by using the BCR sequential extraction procedure and multivariate statistical elucidation of the data *J. Hazard. Mater.* **132** 80–9
[6] Ben Salem Z and Ayadi H 2016 Heavy metal accumulation in Diplodus annularis, Liza aurata, and Solea vulgaris relevant to their concentration in water and sediment from the southwestern Mediterranean (coast of Sfax) *Environ. Sci. Pollut. Res.* **23** 13895–906
[7] Martínez-Fernández M, Barciela-Alonso M C, Moreda-Piñeiro A and Bermejo-Barrera P 2011 Matrix solid phase dispersion-assisted BCR sequential extraction method for metal partitioning in surface estuarine sediments *Talanta* **83** 840–9
[8] Sun Y Z, Zhang B, Gao L R, Liu Z T and Zheng M H 2005 Polychlorinated Dibenzo-p-Dioxins and Dibenzofurans in Surface Sediments from the Estuary Area of Yangtze River, People's Republic of China *Bull. Environ. Contam. Toxicol.* **75** 910–4
[9] Gulmini M, Ostacoli G, Zelano V and Torazzo A 1994 Comparison between microwave and conventional heating procedures in Tessier's extractions of calcium, copper, iron and manganese in a Lagoon sediment *The Analyst* **119** 2075
[10] Tessier A, Carignan R and Belzile N 1994 Processes occurring at the sediment–water interface: emphasis on trace elements *Chem. Biol. Regul. Aquat. Syst. Lewis Publ. Boca Raton* 137–73
[11] Davidson C M, Thomas R P, McVey S E, Perala R, Littlejohn D and Ure A M 1994 Evaluation of a sequential extraction procedure for the speciation of heavy metals in sediments *Anal. Chim. Acta* **291** 277–86
[12] Ariadi Lusiana R, Putri Protoningtyas W, Ricky Wijaya A, Siswanta D, Mudasir M and Juari Santosa S 2017 Chitosan-Tripoly Phosphate (CS-TPP) Synthesis Through Cross-linking Process: the Effect of Concentration Towards Membrane Mechanical Characteristic and Urea Permeation *Orient. J. Chem.* **33** 2913–9
[13] Gray J E and Eppinger R G 2012 Distribution of Cu, Co, As, and Fe in mine waste, sediment, soil, and water in and around mineral deposits and mines of the Idaho Cobalt Belt, USA *Appl. Geochem.* **27** 1053–62
[14] Garrett C 2006 Turbulent dispersion in the ocean *Prog. Oceanogr.* **70** 113–25
[15] Mulyaningsih T R, Alfian A and Sutisna S 2015 Distribusi Logam Berat Dalam Sedimen Daerah Aliran Sungai Ciujung Banten *J. Teknol. Reakt. Nukl. Tri Dasa Mega* **14**

Potential Induction Heating using MOSFET H-Bridge Circuit for Material Synthesis

Subakti[1,3], IWayan Dasna[2,3,*], Markus Diantoro[1,3], Samsul Hidayat[1,3]

[1]Department of Physics, Faculty of Mathematics and Natural Sciences, Universitas Negeri Malang, Jl. Semarang 5 Malang 65145, Indonesia
[2]Department of Chemistry, Faculty of Mathematics and Natural Sciences, Universitas Negeri Malang, Jl. Semarang 5 Malang 65145, Indonesia
[3]Central Laboratory of Mineral and Advanced Materials, Universitas Negeri Malang, Jl. Semarang 5 Malang 65145, Indonesia

*Corresponding author's email: idasna@um.ac.id

Abstract. Here we report our effort to integrate microcontroller to control AC current-generating MOSFET H-bridge circuit-based device. We have carried out the tests of the potential application of the induction heating generated by the device for material synthesis by controlling the variation of AC current at constant frequency value. The results showed that at the constant frequency value, the heating process depended on the amount of AC current. The different responses of different kind of materials to the different frequencies at constant AC current value have also been observed. The results provide the fundamental base for further development of induction heating furnace using microcontroller based AC current generation device.

Keywords: Induction Heating, MOSFET H-bridge, material analysis, current generation device.

1. Introduction

In material science, synthesis is the backbone of any efforts to develop advanced materials for both fundamental research and technological applications. Among many effective synthesis routes, solid state reaction is still the first option for many material scientists to look at every time they need to prepare a compound or further heating treatment to study. However, solid-state reaction and annealing routes require a heating process that sometimes consumes electrical energy in a huge amount. We are considering induction heating as an alternative method to be applied in the laboratory to carry out the solid-state reaction process.

To get into the principle of induction heating, we start from a stationary electric current (DC). According to the Biot-Savart law, the magnitude of the magnetic field generated by the electric current is steady at any points around the path (conductor) of DC current. In other words, the magnetic flux does not change. On the other hand, the AC electric current will produce magnetic flux changing around the conductor. If we place a conducting material (material with intrinsic electric resistant) around the pathway of this AC current, it will experience the magnetic flux fluctuation. As the magnetic flux generated by AC current is also changing in direction, the so-called eddy current is then generated inside the material [1-3]. This phenomenon has a wide potential technological application such as in medical physics treatment of cancer [4]. The schematic diagram to give a picturesque explanation of the principle of induction heating by AC electric current is given in Figure 1. In such configuration, if one places a metal such as iron in the inductor, it will be heated rapidly by generating Eddy current within the metal due to the changing of AC magnetic flux [5-7].

The challenge is how to control the AC current phase for induction heating. The AC current fluctuations can be controlled by electronic circuit, both its frequency and the current strength. Therefore, we are applying MOSFET with H-bridge circuit configuration to control the current strength flowing into the load copper coil. In this way, the period of the current flow and the direction of the current flow are fully controlled by the microcontroller. Some parameters came up during the induction heating performance testing are frequency variation, the control of current strength by the duty cycle of the MOSFET modification, load coil geometry which has also been reported [6].

This contribution deals only with controlling the frequencies and the amount of AC current as the first step of many of our research sequences to design and construct AC current generating H-Bridge configuration based induction heating furnace for Universitas Negeri Malang (UM), Indonesia, with the effectiveuse of electrical current. Further design and optimizations of the device as well as its evaluation related to the maximum value of applied current are still ongoing.

2. Experimental Methods

2.1. MOSFET H-bridge

Four N-channels MOSFETs were employed to control the current flow in the coil. To prevent short currently in the MOSFET, the microcontroller was programmed to manipulate its delay time. When the delay is in progress, all MOSFETs are not transmitting the current. The schematic diagram of how current flows in the H-bridge MOSFETs circuit can be seen in Figure 1. The MOSFETs driver is constructed using the optocoupler and power supply with a different ground. The bipolar transistors are used for driving 4 gates MOSFET. It is used to get the low-frequency work, DC to several hundred hertz AC electric current. The driver modification was done to provide a high-frequency working range. This is different from commercial MOSFET driver which is only capable of working below one hundred Hz.

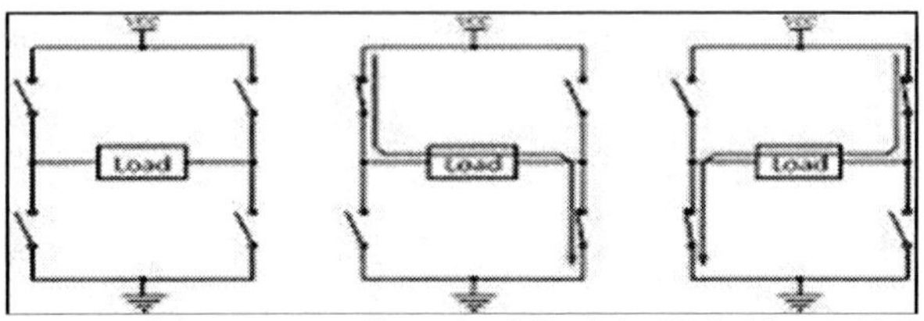

Figure 1. Schematic diagram of the current flow (red line) in H-bridge MOSFET circuit: (a) off, (b) the current flows to the right, and (c) the current flows to the left

We employed an ATMega8 microcontroller to generate controlled square-signals. The period of these signals can be controlled by a microcontroller with a variation value starting from 4.6 μs. The duration of this 'OFF' signal can also be controlled by a microcontroller with variations ranging from 4.96 μs. The signal 'ON' allows the current to flow into the load, the 'OFF' signal makes the current does not flow on the load.

Acting as the load here is a 50 mm diameter copper enameled wire coil, the coil diameter is 18 mm, 50 μH inductance value. By using microcontroller, the 'UP' signal will allow the current moving from one end to another one and the 'DOWN' signal will reverse the direction. These 'UP' and 'DOWN' signals control the MOSFET accordingly, so the MOSFET has the lowest possible RDS value. This low RDS value is variable and can be used to suppress the heat dissipation in the MOSFET and maximize the current flow on the coil. Previous research says that using the coil as an inductor can generate the heating rate up to 8.0 °C/s. From an initial temperature of 40 °C to the surface temperature of 159.9 °C took 15 seconds of the heating process [7].

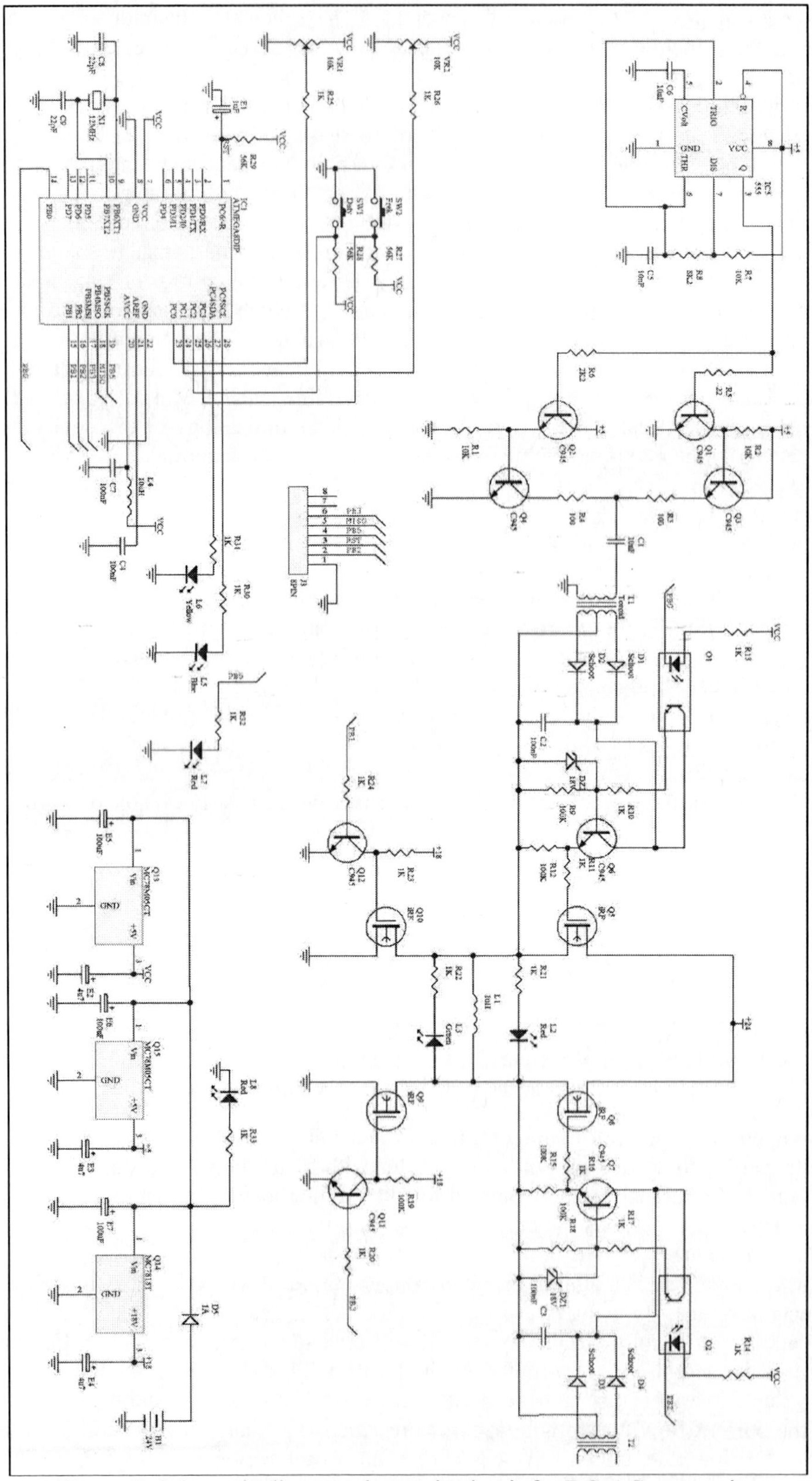

Figure 2. Schematic diagram electronic circuit for DC-AC conversion

2.2. Heating Performance

To attest to the performance of H-bridge based device shown in Figure 2, we applied the generated AC current into the coiled copper-enameled wire. The diameter of the wire is 1.5 mm and the diameter of the coil was 18 mm. The explanation of possible heating effect is given by schematic diagram in Figure 3.

We used 3.13 grams iron ball for the material. Two strategies were carried out to study the effect of AC current value (duty-cycle) and frequencies. We measured the heating temperature of the iron ball by using thermocouple thermometer. The 337 Hz frequency was chosen to keep the MOSFET in the safe working range to anticipate any unexpected heating process. At constant frequency number, duty-cycle was varied between 149-248 μs. We varied the frequencies between 773-864 Hz at constant duty cycle value 244 μs to study the effect of frequency on induction heating effect.

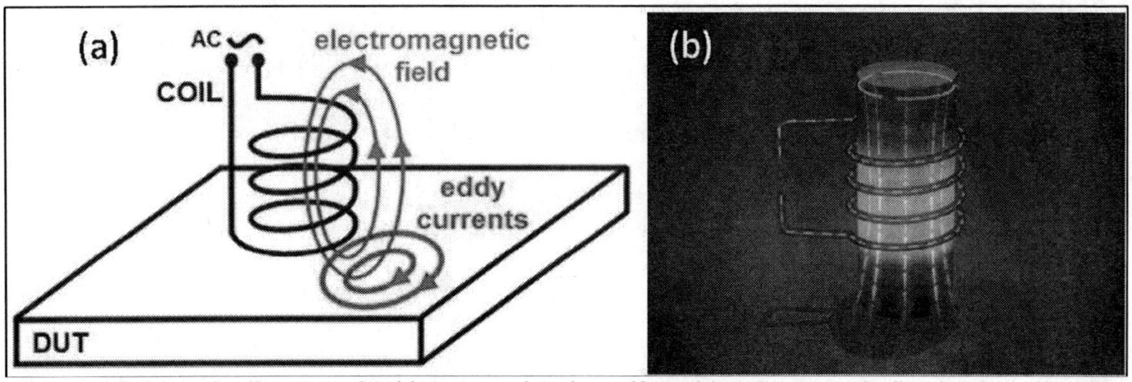

Figure 3. Schematic diagram of Eddy current heating effect, (a) AC current is flowing in coil results in electromagnetic with directions represented by blue lines. Any material placed in the coil will be experiencing Eddy current represented by red lines and accordingly will be heated up by this phenomenon, (b) visual effect of heating process induced by electromagnetic effect from (a).

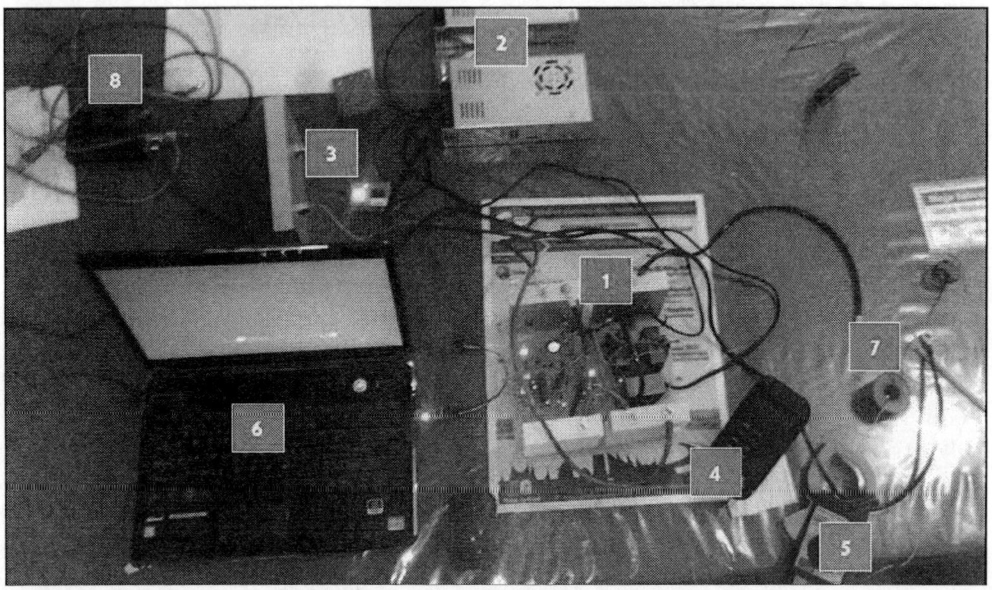

Figure 4. Overall microcontrolled induction heating device: 1) DC-AC convertion, 2) power adaptor, 3) electric current safety control, 4) thermocouple thermometer, 5) generated AC current amperemeter, 6) windows operating system to control the microcontroler, 7) copper enameled coil system, can be considered as a very small laboratory scale, for safety reason.

3. Results and Discussion

From the experimental results given in figure 5 and 6, we can see that the longer duty cycle thebigger the effect of the increase in heating temperature. This is possible as a result of the current that causes the vortex of the heated material to be longer. The smaller duty cycle also causes the current vortex to be shorter anyway. In the 337Hz frequency range, for 149μs current flow, causing the iron ball possesses Eddy current approximately 5% of the cycle and was able to heat the iron ballup to 37.5°C. Under these conditions, the effective current consumption of 2.3A, the voltage applied to the system is 12 V. Therefore, theeffective power is 2.3A × 12V =27.6Watt.

For AC current frequency 864Hz and duty cycle 244 μs, current flow is causing the material to experience Eddy current about 21% of the cycle. It can raise the temperature up to 112.3°C for 30 minutes heating time. At that condition, effective current consumption equal to 6.3A, the voltage used in system of 12V. So, the effective power is 6.3 A × 12V = 75.6Watt.

During the device performance tests, we maintained the duty cycle under 300 μs for safety reason. The duty cycle value is proportionally affecting the heating temperature at a constant frequency, 149μs provides a maximum temperature of 37.5 °C (60 minutes heating), 174 μs duty cycle providing maximum temperature of 42.7 °C (60 minutes heating), 198 μs provides a maximum temperature of 43.1 °C (60 minutes heating), 223 μs provides a maximum temperature of 47.1 °C (60 minutes heating), 248 μs provides a maximum temperature of 58.7 °C (60 minutes heating). At this stage, it is important to elaborate the results with details calculation as reported by Aung [10] for further optimization and power output analysis.

Interestingly, the variation of AC current frequency at the constant duty cycle value 244μs (frequency range 773Hz-864Hz) give maximum temperature as high as 112.3 °C in only 30 minutes heating period, for AC current frequency 864Hz. This heating temperature achievement is enough to be applied for micro-furnace for annealing process of small material samples as reported by Lu [11]. From these results, we have conceived that for the device scale up, we need to optimize the MOSFETs configuration first to obtain optimum highest duty cycle and then to optimize the frequency generation.

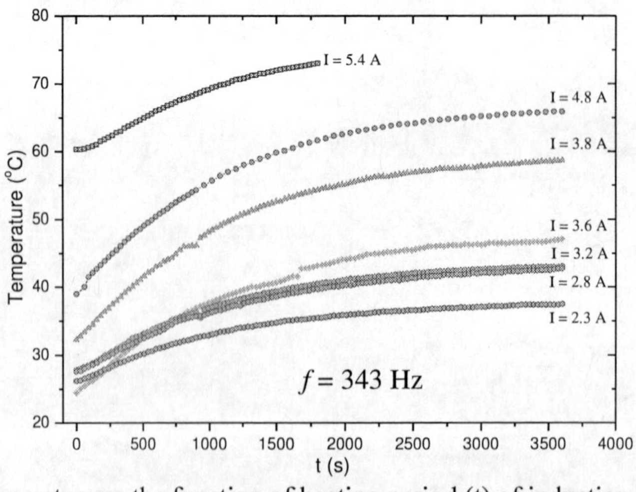

Figure 5. Heating temperature as the function of heating period (t) of induction heater at constant AC current frequency 343 Hz.

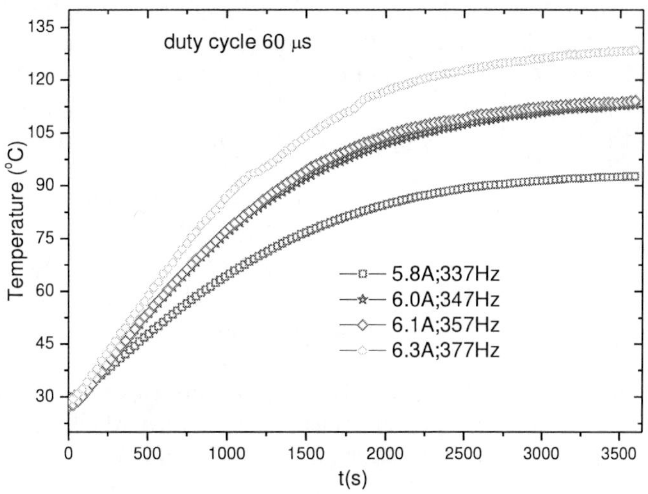

Figure 6. Heating temperature as the function of heating period (t) of induction heater at the constant duty cycle. Note that the AC current is degrading from 6.3 to 5.8 A despite the duty cycle is constant, another question for further development.

4. Conclusion

We have developed a MOSFET H-bridge configuration based induction heating micro-furnace as a fundamental step for further development of induction heating furnace in Universitas Negeri Malang (UM), Indonesia. The microcontroller has been integrated into the device system to manipulate the AC current frequency and duty cycle. There is saturation heating temperature for every duty cycle and frequency values. Up to this report, we have reached a maximum heating temperature of 112.3 °C at the duty cycle of 244μs andAC current frequency 864Hz provides for a heating period only in 30 minutes. These small-scale induction heating tests clearly showed the potential of induction heating to be applied for clean solid-state reaction process as well as providing important basis for its further scale up development.

Acknowledgments

The authors acknowledge the financial support provided by the Ministry of Research, Technology and Higher Education Republic of Indonesia through PUPT project funding scheme.

References

[1] Griffiths, David J. (1998), Introduction to Electrodynamics (3rd ed.), Prentice Hall. pp. 222–224, 435–440. ISBN 0-13-805326-X

[2] Jackson, John David (1999), Classical Electrodynamics (3rd ed.), New York: Wiley, Chapter 5. ISBN 0-471-30932-X

[3] I.S. Grant, W.R. Phillips, Electromagnetism (2nd Edition), Manchester Physics, John Wiley & Sons, 2008, ISBN 978-0-471-92712-9

[4] Robert P. Stigliano, Fridon Shubitidze, James D. Petryk, Levan Shoshiashvili, Alicia A. Petryk & P. Jack Hoopes, (2016), Mitigation of eddy current heating during magnetic nanoparticle hyperthermia therapy, International Journal of Hyperthermia, Volume 32, Issue 7, DOI:10.1080/02656736.2016.1195018

[5] Bae, J., Hwang, J., Park, J., & Kwag, D, (2009), Modeling And Experiments On Eddy Current Damping Caused By A Permanent Magnet In A Conductive Tube, 23, 3024–3035. DOI:10.1007/s12206-009-0819-0

[6] Du, H., Li, J., & Qu, Y, (2014), Mathematical Modeling of Eddy-Current Loss for a New Induction Heating Device,

[7] Lin, H., Chen, S., Jeng, M., Son, P., Chang, J., & Hwang, J, (2012), Induction Heating With The Ring Effect For Injection Molding Plates, International Communications In Heat And Mass Transfer, DOI:10.1016/j.icheatmasstransfer.2012.02.009

[8] Bowmaker, G., A., Effendy, Fariati, Rahajoe, S., I., Skelton, B., W., & White, A., H., 2011, Structural and Infrared Spectroscopic Studies of Some Adducts of Divalent Metal Dihalides (MX2, M = Zn, Cd; X = CI, Br, I) with Variously Hindered Monodentate Nitrogen (Pyridine) Base Ligands (L = Pyridine, 2-Methylpyridine, and Quinoline) of 1:2 Stoichiometry, Z. Anorg. Allg. Chem, 637 : 1361–1370.

[9] Qing, W., A., Kun, Z., F., Zhen, C., L., Cong, G., G., & Shun, H., J., 2004, Synthesis and Crystal Structures of Two Novel 1-D Metal Complexes with Dicyanamide: [Zn(pheen)(dca)2]n and [Cu(quin)2(dca)2]n, Chinese Journal Structure Chemistry, 23(10) : 1148

[10] Soe Sandar Aung, Han Phyo Wai, and Nyein Nyein Soe, Design Calculation and Performance Testing of Heating Coil in Induction Surface Hardening Machine, International Journal of Electrical, Computer, Energetic, Electronic and Communication Engineering Vol:2, No:6, 2008

[11] Zhiqiu Lu, Kirill Poletkin, Ulrike Wallrabe and Vlad Badilita, Performance Characterization of Micromachined Inductive Suspensions Based on 3D Wire-Bonded Microcoils, Micromachines 2014, 5, 1469-1484; doi:10.3390/mi5041469

Synthesis and Characterization of γ-Al_2O_3/SiO_2 Composite Materials

Munasir[1,2,*], Diah Hari Kusumawati[1,2], Sunaryono[3,4], Nugrahani Primary Putri[1,2], Nurul Hidayat[3], Ahmad Taufiq[3,4], Zainul Arifin Imam Supardi[1,2]

[1]Department of Physics, Faculty of Mathematics and Natural Sciences, Universitas Negeri Surabaya, Kampus Ketintang, Jl. Ketintang, Surabaya, 60231.
[2]Research Center of Advanced Materials, Universitas Negeri Surabaya, Kampus Ketintang, Jl. Ketintang, Surabaya 60231.
[3]Department of Physics, Faculty of Mathematics and Natural Sciences, Universitas Negeri Malang, Jl. Semarang 5 Malang 64145, Indonesia.
[4]Research Center of Minerals and Advanced Materials, Faculty of Mathematics and Natural Science, Universitas Negeri Malang, Jl. Semarang 5 Malang 64145, Indonesia.

*Corresponding author's email: munasir_physics@unesa.ac.id

Abstract. This article reports the synthesis process of γ-alumina/SiO_2 composite from aluminium powders and amorphous nanosilica through the means of tetra-methyl-ammonium hydroxide (TMAH) as the mixing medium which produced aluminium hydroxide $Al(OH)_3$ in the form of white powders. The crystal structures, microstructures, and mapping of the atomic constituent of γ-Al_2O_3 and γ-Al_2O_3/SiO_2 samples were respectively analyzed via X-ray diffraction (XRD), and scanning electron microscopy-energy dispersive X-ray SEM-EDX. At the calcination temperature to 300 °C, the samples still had boehmite phase, and the boehmite phase transformation to γ-alumina phase occurred at temperature of 700-900 °C. The γ-alumina phase began to form at the calcination temperature of 500-700 °C with the crystal structure of (440), (511), (400), (222), (311), and (111). The similar diffraction pattern was also shown by the γ-Al_2O_3/SiO_2 composite for the calcination temperature of 700 °C and 900 °C. The microstructure analysis of the γ-Al_2O_3/SiO_2 showed that SiO_2 particles were smaller and round while the γ-Al_2O_3 particles were bigger and elongated. Additionally, the mapping results showed SiO_2 and γ-Al_2O_3 particles were homogenously distributed.

Keywords: Composite, γ-Al_2O_3, SiO_2, composite, calcination.

1. Introduction

Nowadays, knowledge and technology have rapidly developed. This case is inseparable from the dominant role of research on materials which supports the production process in the industrial field. Almost 75% of chemical manufacturing processes require the catalyst assistance. Such fact has encouraged a high number of researchers to synthesize both inorganic and organic catalysts for further development. The purpose of such synthesis is to create a wide utilization of catalysts with minimum side effects on the environment [1]. One of the frequently used catalysts in industrial fields is γ-Al_2O_3. The use of γ-Al_2O_3 as a catalyst and catalyst support [2-4] is based on several advantages. The first advantage is that γ-Al_2O_3 has a broad surface area and ability to capture undesirable dirt. The second advantage is that γ-Al_2O_3 possesses a relatively high thermal stability it has a high melting point (2.318°C). Third, the structure of γ-Al_2O_3 simplifies the combination process of its molecules with active phase molecules, so that a compact catalytic structure can be formed [5]. Mesoporous γ-Al_2O_3 is one of the widely-used supporting catalysts for commercial application. The catalytic performance highly depends on the supporting combination of texture, chemical, and physical characteristics [4].

Tetragonal γ-Al_2O_3 can be formed from boehmite at calcination temperature between 450 and 750 °C [2, 11]. The γ-Al_2O_3 is a solid nanoporous material with low bulk density, low thermal conductivity, and has a tendency to have an amorphous structure, high porosity, and high specific surface area [12]. Nano-sized SiO_2, as a porous, non-conductive, and multi-compatible solid material, tends to have an amorphous structure [13]. SiO_2 nanoporous material has an extremely broad application, e.g. as absorbent [9], packaging [15], separator and electrode in Li-ion battery system [16, 17], drug delivery agent [18], and so on. There are numerous researchers conducting research on alumina-SiO_2 composite material with the focus on its low bulk density characteristic, and SiO_2 doping effect on high heat resistance and the surface area at temperature changing [7]. This study focused on the γ-Al_2O_3 synthesis and the formation of nano SiO_2 composite (amorphous) as the filler.

2. Materials and Methods

Aluminum powder (Merck product) is a matrix with a high purity (98.1%), 2.70 gram/cm^3 density, and a melting point at 600 °C. The second material was a nanometer-sized SiO_2 powder with a high amorphous phase and high purity (>95%) which was prepared from natural quartz through a hydrothermal-coprecipitation process [8]. Lastly, a *Merck* pro-analysis TMAH (*tetramethylammonium hydroxide*) solution was utilized as the mixing medium. The sample preparation consisted of two stages. The first step was preparing the γ-Al_2O_3 sample by adding aluminum powder (Al) into the TMAH solution to produce a white colored solid. The second step was a combustion process in a furnace at 300 °C for 2 hours until it produced boehmite $Al(OH)_3$. The next step was creating the γ-Al_2O_3/SiO_2 sample by adding the prepared SiO_2 into the TMAH solution and then stirred until it became homogeneous (~1-2 hours). The γ-alumina/ SiO_2 was formed through a calcination process at 700 °C temperature in the oven. Characterization was done through XRD to identify the phase and structure, FTIR to analyze the functional radical-group, and SEM-EDX to analyze the microstructure profile of γ-Al_2O_3/SiO_2.

3. Results and Discussion

Figure 1 presents the diffraction pattern of $Al(OH)_3/SiO_2$ (aluminum hydroxide) pattern with the nanosilica particle as the filler. There were ten diffraction peaks, which include the $Al(OH)_3$ and SiO_2 peaks. The existence of the $Al(OH)_3$ peak showed that the material produced at this stage had a boehmite phase [5,6,14–20]. It was due to the Al as a matrix reaction to an oxidation of TMAH as well as the role of Al as a dispersion medium of SiO_2 in the first process, as shown in Figure 2. TMAH is a quaternary ammonium salt, with the chemical formula of $N(CH_3)_4^+ OH^-$, which relatively oxidized the Al. Such process would initiate an exothermic process, namely a process of releasing C, N, and H_2 gasses, characterized by a production of steam. The Al and SiO_2 powder in TMAH formed a new compound of $Al(OH)_3$, SiO_2, and $C_3H_{12}N$ group released during the process.

However, such peak of aluminum hydroxide $Al(OH)_3$ could not be identified anymore after the calcination process had reached 700 °C, yet, there is a new emerging peak which is identified as alumina phase. There was a transformation from the boehmite phase into γ-alumina phase [20]. The tetra-methyl-ammonium cation $[C_4H_{12}N]^+$ which was bound to an oxygen atom would be released after the calcination process had reached 300 °C. Based on the diffraction patterns after the calcination process had reached 300°C in 2 hours, the synthesized white powdered material is dominated by Boehmite phase (Figure 1) [20]. The diffraction patterns identify many less sharp peaks originated from impurities (not alumina), such as nitrate, carbonate, and other impurities which can be found in

Boehmite/SiO_2 composite, refer to the chemical equation in Figure 2 (there are C, N, and H elements) [21].

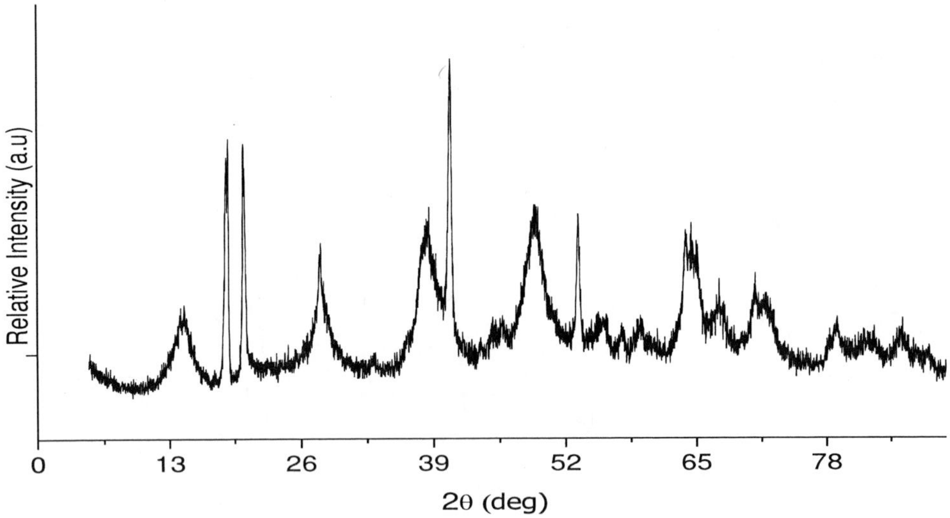

Figure 1. Diffraction patterns of Al[OH]$_3$/SiO_2 (amorphous) sample

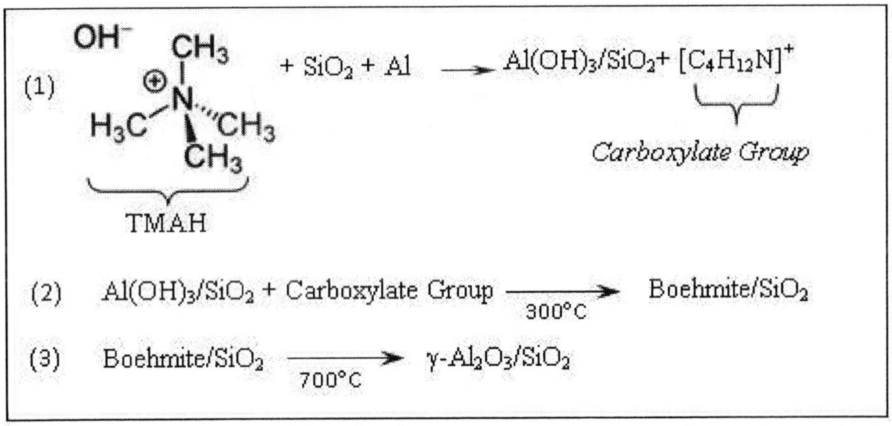

Figure 2. Reaction of amorphous Al and SiO_2 via TMAH: (1-2) Boehmite/SiO_2 and (3) gamma-alumina/SiO_2

Figure 3 identifies the X-ray diffraction patterns of γ-Al_2O_3 and γ-Al_2O_3/SiO_2. There was a significant match [18, 19] in the comparison between the data of the X-ray diffraction test and the data of standard XRD on *gamma* alumina (reference). The relatively similar peaks also have fairly similar intensity values (see Table 1). From the intensity values between gamma alumina/SiO_2 which exceeded 50%, it can be seen that the synthesized gamma alumina and the standard gamma alumina were relatively similar (39.40°). However, there were several peaks, which were visible in the γ-Al_2O_3 /SiO_2. The peaks could not be identified in the conventional gamma alumina, but they could be identified in the synthesized γ-Al_2O_3 sample, namely at 2θ 34.83 angle with 21% relative intensity value, 37.38 angle with 64.09% intensity value, and at 2θ 41.26 angle with 4.55% relative intensity value. It is interesting since, at the angles of 2θ 19.58°, 37.79°, and 60.85°, the relative intensity values of 6.96%, 59.19%, and 10.31% shifted on the standard γ-Al_2O_3 (namely 2θ = 19.45°, 37.60°, and 60.90°). The different diffraction patterns between γ-Al_2O_3 and γ-Al_2O_3/SiO_2 and the standard data (standard γ-Al_2O_3) are likely to be influenced by the process and the core material of the γ-Al_2O_3 phase

IOP Conf. Series: Journal of Physics: Conf. Series **1093** (2018) 012015 doi:10.1088/1742-6596/1093/1/012015

formation [7, 22]. One important factor in the preparation method is aluminum hydroxide decomposition temperature. The phase formation from boehmite into γ-Al_2O_3 requires a temperature higher than 300 °C. It is in accordance with the boehmite phase at the temperature of 300 °C in this study (Fig.1). The γ-Al_2O_3 phase requires the temperature of 500-700 °C [23] which is also in line with the γ-Al_2O_3 phase formed in this research at 700 °C and a new phase of δ-Al_2O_3 formation would emerge at the temperature of 900°C (Fig. 3(a-b)) [19,20].

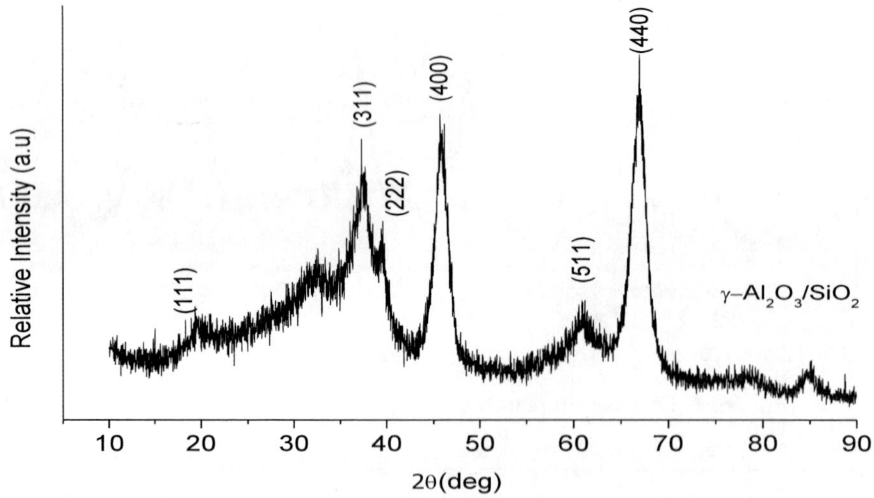

Figure 3. Diffraction patterns of γAl_2O_3/SiO_2 sample

Table 1. XRD data of γ-Al_2O_3 and γ-Al_2O_3/SiO_2

2*θ* (deg)	**γ-Al_2O_3/SiO_2 [b]**		**hkl, ICDD: 29-1486, 29-0063 10-0425**
	***d*-spacing (A°)**	**I/*Io* (%)**	
19.58	4.529	6.96	[111]
34.83	2.573	21.00	[220]
37.38	2.403	64.09	[311]
39.40	2.284	36.76	[222]
41.26	2.186	4.55	
45.73	1.980	77.00	[400]
60.85	1.521	10.31	[511]
66.85	1.398	100	[440]
84.87	1.142	6.97	[444]

Figure 3 (a-b) shows the XRD diffraction patterns of synthesized γ-Al_2O_3 and γ-Al_2O_3/SiO_2 samples for 20% SiO_2 with a calcination treatment at the temperatures of 700 °C. It can be seen that the gamma alumina phase was dominantly formed in the crystal structures of (440), (511), (400), (222), (311) and (111) (6, 18, 20, 24, 25). The SiO_2 content in γ-Al_2O_3/SiO_2 sample was not identified in the diffraction pattern due to the small relative intensity possessed by the SiO_2 particle. The SiO_2 diffraction pattern (nanometer order) is usually cannot be seen since it has an amorphous phase. In this case, the calcination still has the amorphous phase up to the temperature of 700°C, such as at the temperature of 300 °C in which the boehmite Al(OH)$_3$ phase was still dominant [15,21,26].

Figure 4. SEM image of (a) SiO_2 and (b) γ-Al_2O_3 /SiO_2 samples

The morphology and microstructures of SiO_2 and γ-Al_2O_3/SiO_2 particles are shown in Figure 4. (a) The SiO_2 nanoparticle profile was smaller than the nanoparticle profile of γ-Al_2O_3 and had the tendency to be in round shapes, while (b) the profiles of γ-Al_2O_3 and SiO_2 particles substituted inside them forming γ-Al_2O_3/SiO_2. It can be seen that the γ-Al_2O_3 form elongated and is bigger than the form of SiO_2, while the SiO_2 particle was filling the gaps among the Al_2O_3. The distribution of SiO_2 and γ-Al_2O_3 particles can be seen in Figure 5. Meanwhile, the SEM-EDX test results of the γ-Al_2O_3/SiO_2 sample (20%) were characterized with their respective *At%* (atomic percent) of Al, Si and O of 38.81; 3.83 and 57.36%; the mapping results identified the homogeneity of SiO_2 and γ-Al_2O_3 distributions [5, 14, 19].

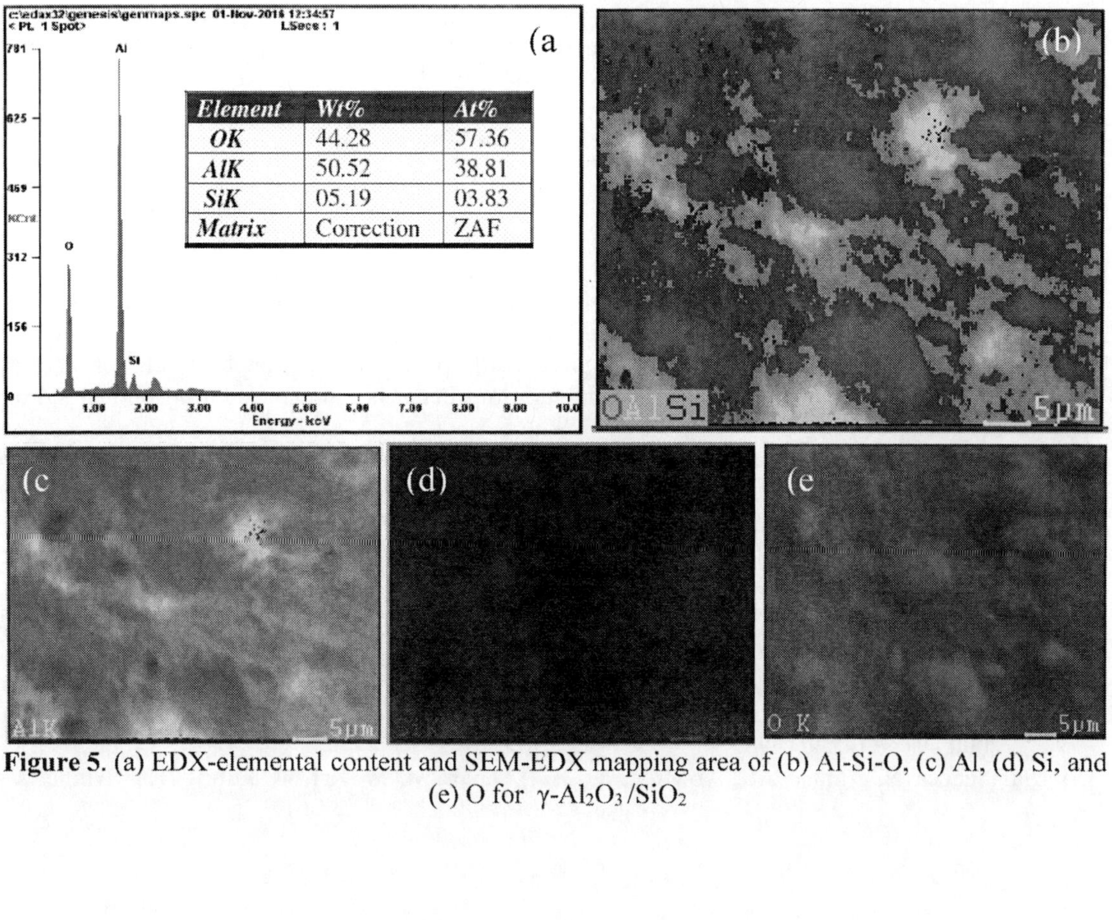

Figure 5. (a) EDX-elemental content and SEM-EDX mapping area of (b) Al-Si-O, (c) Al, (d) Si, and (e) O for γ-Al_2O_3 /SiO_2

4. Conclusion

The γ-Al_2O_3/SiO_2 composite material has been synthesized from the aluminum base material using TMAH as a medium. The mixing of SiO_2 nanoparticles (amorphous) and γ-alumina particles was by a wet-mixing method, assisting in the formation of a homogeneous γ-Al_2O_3/SiO_2. The γ-Al_2O_3 particle shape in γ-Al_2O_3/SiO_2 appeared larger and elongated; the composite structure appeared to be more dominant g-alumina, amorphous SiO_2 particles. The active medium of TMAH can be as a solvent for dispersing SiO_2 nanoparticles and in the mixing process of al and SiO_2 are oxidizing, activating al powder and transforming into γ -alumina.

Acknowledgments

This research was supported by the Ministry of Research and Technology and Higher Education of Indonesia.

References

[1] Kobayashi K, Nagai M. Active sites of sulfided $NiMo/Al_2O_3$ catalysts for 4,6-dimethyldibenzothiophene hydrodesulfurization-effects of Ni and Mo components, sulfidation, citric acid and phosphate addition. Catal Today. 2017 Feb (*Available online 16 February 2017)*

[2] Vangestel J. Catalytic Properties of a $CoMo/Al_2O_3$ Catalyst Presulfided with Alkyl Polysulfides: Comparison with Conventional Sulfiding. J Catal. 1994 Feb;145(2):429–36.

[3] Choi K-H, Korai Y, Mochida I. Preparation and characterization of nano-sized $CoMo/Al_2O_3$ catalyst for hydrodesulfurization. Appl Catal Gen. 2004 Apr;260(2):229–36.

[4] Kuchonthara P, Puttasawat B, Piumsomboon P, Mekasut L, Vitidsant T. Catalytic steam reforming of biomass-derived tar for hydrogen production with $K_2CO_3/NiO/\gamma$-Al_2O_3 catalyst. Korean J Chem Eng. 2012 Nov;29(11):1525–30.

[5] Samain L, Jaworski A, Edén M, Ladd DM, Seo D-K, Javier Garcia-Garcia F, et al. Structural analysis of highly porous γ-Al_2O_3. J Solid State Chem. 2014 Sep;217:1–8.

[6] Hosseini SA, Niaei A, Salari D. Production of γ-Al_2O_3 from Kaolin. Open J Phys Chem. 2011;01(02):23–7.

[7] Wu X, Shao G, Cui S, Wang L, Shen X. Synthesis of a novel Al_2O_3–SiO_2 composite aerogel with high specific surface area at elevated temperatures using inexpensive inorganic salt of aluminum. Ceram Int. 2016 Jan;42(1):874–82.

[8] Munasir M, Triwikantoro, Zainuri M, Darminto. Synthesis of SiO_2 nanopowders containing quartz and cristobalite phases from silica sands. Mater Sci-Pol. 2015;33(1):47–55.

[9] Morsy FA, El-Sheikh SM, Barhoum A. Nano-silica and $SiO_2/CaCO_3$ nanocomposite prepared from semi-burned rice straw ash as modified papermaking fillers. Arab J Chem. 2014 Nop: http://www.sciencedirect.com/science/article/pii/S1878535214002962.

[10] Luo Z, Xu Y, Ye Q. Effect of nano-SiO_2-LDPE packaging on biochemical, sensory, and microbiological quality of Pacific white shrimp Penaeus vannamei during chilled storage. Fish Sci. 2015 Sep;81(5):983–93.

[11] Liang J, Li X, Zhu Y, Guo C, Qian Y. Hydrothermal synthesis of nano-silicon from a silica sol and its use in lithium ion batteries. Nano Res. 2015 May;8(5):1497–504.

[12] Fu D, Luan B, Argue S, Bureau MN, Davidson IJ. Nano SiO_2 particle formation and deposition on polypropylene separators for lithium-ion batteries. J Power Sources. 2012 May;206:325–33.

[13] Barbé C, Bartlett J, Kong L, Finnie K, Lin HQ, Larkin M, et al. Silica Particles: A Novel Drug-Delivery System. Adv Mater. 2004 Nov 4;16(21):1959–66.

[14] Kiyohara PK, Santos HS, Coelho ACV, Santos PDS. Structure, surface area and morphology of aluminas from thermal decomposition of $Al(OH)(CH_3COO)_2$ crystals. An Acad Bras Ciênc. 2000 Dec;72(4):471–95.

[15] Salahudeen N, Ahmed AS, Al-Muhtaseb AH, Dauda M, Waziri SM, Jibril BY. Synthesis of

gamma alumina from Kankara kaolin using a novel technique. Appl Clay Sci. 2015 Mar;105–106:170–7.
[16] Samadhi TW, Subagyo S, Lismana KR, Fuadi K. Synthesis of gamma-Al_2O_3 Catalyst Support from Kaolin of Indonesian Origin. J Eng Technol Sci. 2011;43(2):113–26.
[17] Temuujin J, Jadambaa T, Mackenzie KJD, Angerer P, Porte F, Riley F. Thermal formation of corundum from aluminium hydroxides prepared from various aluminium salts. Bull Mater Sci. 2000 Aug;23(4):301–4.
[18] Trueba M, Trasatti SP. γ-Alumina as a Support for Catalysts: A Review of Fundamental Aspects. Eur J Inorg Chem. 2005 Sep;2005(17):3393–403.
[19] Abdollahifar M, Zamani R, Beiygie E, Nekouei H. Synthesis of micro-mesopores flowerlike γ-Al_2O_3 nano-architectures. J Serbian Chem Soc. 2014;79(8):1007–17.
[20] Santos PS, Santos HS, Toledo SP. Standard transition aluminas. Electron microscopy studies. Mater Res . Materials Research, 2000, 3(4), 104-114.
[21] Candela L, Perlmutter DD. Pore structure and kinetics of the thermal decomposition of $Al(OH)_3$. AIChE J. 1986 Sep;32(9):1532–45.
[22] Zhou S, Antonietti M, Niederberger M. Low-Temperature Synthesis of γ-Alumina Nanocrystals from Aluminum Acetylacetonate in Nonaqueous Media. Small. 2007 May 4;3(5):763–7.
[23] Mardkhe MK, Huang B, Bartholomew CH, Alam TM, Woodfield BF. Synthesis and characterization of silica doped alumina catalyst support with superior thermal stability and unique pore properties. J Porous Mater. 2016 Apr;23(2):475–87.
[24] Gaweł B, Øye G. Hierarchical γ-alumina monoliths with macro- and meso porosity prepared by using cross-linked dextran gel beads as the template. Mater Lett. 2013 Mar;95:86–8.
[25] Sifontes ÁB, Gutierrez B, Mónaco A, Yanez A, Díaz Y, Méndez FJ, et al. Preparation of functionalized porous nano-γ-Al2O3 powders employing colophony extract. Biotechnol Rep. 2014 Dec;4:21–9.
[26] Violante A. Formation Mechanism of Aluminum Hydroxide Polymorphs. Clays Clay Miner. 1993;41(5):590–7.
[27] Liu X, Xiong X, Liu C, Liu D, Wu A, Hu Q, et al. Preparation of Biodiesel by Transesterification of Rapeseed Oil with Methanol Using Solid Base Catalyst Calcined K_2CO_3/γ-Al_2O_3. J Am Oil Chem Soc. 2010 Jul;87(7):817–23.
[28] Parida KM, Pradhan AC, Das J, Sahu N. Synthesis and characterization of nano-sized porous gamma-alumina by control precipitation method. Mater Chem Phys. 2009 Jan;113(1):244–8.
[29] Nishihara H, Mukai SR, Fujii Y, Tago T, Masuda T, Tamon H. Preparation of monolithic SiO_2-Al_2O_3 cryogels with inter-connected macropores through ice templating. J Mater Chem. 2006;16(31):3231–6.

IOP Conf. Series: Journal of Physics: Conf. Series **1093** (2018) 012016 doi:10.1088/1742-6596/1093/1/012016

Microstructure and Electroconductivity using Silver Nano Particles of *Pterocarpus Indicus W.* Leaf Extracts Films

M Diantoro[1,2,*], S Rakhmah[1], M Khoirudin[1], D W Windasari[1], C I Yogihati[1,2]

[1]Department of Physics, Faculty of Mathematics and Natural Sciences, Universitas Negeri Malang, Jl. Semarang 5 Malang 65145, Indonesia
[2]Central Laboratory of Minerals and Advanced Materials, Universitas Negeri Malang, Jl. Semarang 5 Malang 65145, Indonesia

*Corresponding author's email: markus.diantoro.fmipa@um.ac.id

Abstract: Indonesian is one of the enormous biodiversity countries in the world. Various types of tropical plants exist in Indonesia, one of which is Pterocarpus Indicus W (PIW). PIW is one of the ornamental plants and has a wide range of compound content including flavonoids. Flavonoids are natural polymer compounds with a chemical formula of C_6-C_3-C_6. There are many kinds of flavonoids which could exist in various plants, one of which is quercetin. Quercetin is a unique bioflavonoid that has antioxidant activity. We have performed an extraction of Flavonoid using the maceration method which subsequently doped with several amounts of silver nanoparticles (Ag-NPs). It is suggested that transforming PIW -AgNPS in the form of the film will increase the crystallinity compare to its pristine form. The growth of this crystallinity will enhance the expected properties of the material. The films were prepared based on the content of Ag NPs, i.e., for 0.01, 0.02, 0.03, 0.04, and 0.05 g. It was found that the microstructure of PIW extract showed that flavonoids belonged to flavonoid with quercetin type. The XRD data analyses showed an increase in the crystallinity of the flavonoid PIW-Ag film indicated by the peak at $2\theta = 38.51^{\circ}$ of (111) Bragg's plane influenced by increase of Ag NPs. From the SEM image, we found that AgNPs exist in the sample and also further revealed by TEM analysis. The conductivity of the flavonoid PIW-Ag film shows a linear increase along with the addition of Ag doping.

Keywords: Silver nanoparticles, PIW, flavonoid leaves, microstructure, electrical conductivity.

1. Introduction

Indonesia possesses a significant number of biodiversity. Various kinds of tropical plants grow in Indonesia, one of them is Angsana or *Pterocarpus indicus willd* (PIW) plant. PIW is one of the ornamental plants and has a wide range of compound content including flavonoids. Flavonoids are natural polymer compounds with chemical structure C6-C3-C6. Flavonoids are a class of phenolic compounds with chemical structure C6-C3-C6 [1-4]. Flavonoids have a conjugate bond, there are carbon atoms that are single-bonded (C-C) and double (C = C) alternately [5].

There are more than 100 kinds of flavonoids that exist in various plants, one of which is quercetin. Quercetin is a unique bioflavonoid that has antioxidant activity. Hydrolytic and oxidative enzymes as well as anti-inflammatory measures [6]. Many researchers have studied the content of flavonoids in various plants, such as the flavonoid content of the neriifolia leaf Euphorbia [7-8], from tomatoes [9-12].

IOP Conf. Series: Journal of Physics: Conf. Series **1093** (2018) 012016 doi:10.1088/1742-6596/1093/1/012016

Flavonoid extraction was performed by using the maceration method. In our previous report, we used latex liquid as a source of flavonoid with a simple treatment [3]. We propose to use the leaf of PIW as a raw material using the maceration method. To modify electronics properties of the films we used silver nanoparticles (AgNPs). The inducing AgNPS together with transforming into a film expected to increase the crystallinity as well as their electrical conductivity of flavonoid PIW-AgNPs.

2. Materials and Methods

The initial stages of extracting PIW were started by preparing 800 g of Pterocarpus indicus Willd leaf powder which was mixed with 3 L methanol in a large bottle and shaken and kept for a week. Subsequently, it was filtered using a Buckner funnel under pressure with Whatman's nozzle paper 01. The filtered yield then put in a rotary evaporator to obtain a rough methanol extract. 25 grams of methanol extract was introduced into the separating funnel in which the mixture was manually shaken with 50 mL of n-hexane solvent for 30 minutes. Once separated, the two layers of this process were repeated with 50 mL of n-hexane. The hexane phase in the treatment was separated and the sum of the second procedure is combined and evaporated to obtain a crude hexane extract. The general procedure was applied to obtain extracts of chloroform (100 mL), ethyl acetate (100 mL) and butanol (100 mL), respectively.

We separately also prepared Ag nanoparticles which were synthesized using a chemical reduction method [15]. We used Sodium Borohydride ($NaBH_4$) as a reducing agent according to the following mechanisms.

$$AgNO_3 + NaBH_4 + H_2O \rightarrow Ag + \tfrac{1}{2} H_2 + \tfrac{1}{2} B_2H_6 + NaNO_3 \qquad (1)$$

Mercaptosuccinic acid (MSA) was used as a stabilizer, while aquades and methanol were used as solvents. The first step was mixing 2.34 mL of MSA $C_4H_6O_4S$ with 800 mL of methanol at 5-10 °C with stirring speed of 700 rpm. The next step was to apply 0.68 gram of $AgNO_3$ into 13.548 mL of aquadest to achieve a yellowish solution. Subsequently, we add 1.52 grams of $NaBH_4$ in 200 mL of aquades until the solution is blackish brown with the help of a burette. During this process, the temperature was maintained at 5-10 ° C. Furthermore the solution was precipitated and filtered with 600 mL of methanol three times. The filtrated powder was annealed on a hot plate at 50 °C. To confirm whether the phase of AgNPs crystal is already formed or not, we performed an XRD examination.

The procedure to synthesize flavonoids-AgNPs is briefly described. The initial step was mixing of AgNPs and acetone. This mixture was mixed with flavonoid extract - CSA in beaker glass. The next steps were homogenizing using a magnetic stirrer at 80 °C as shown in Figure 1.

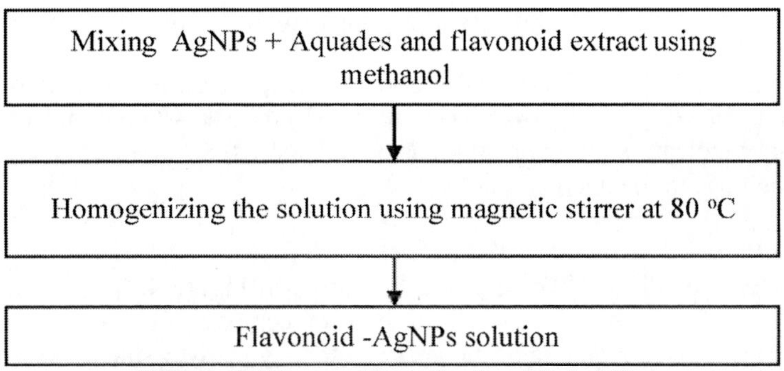

Figure 1. Flavonoid synthesis with AgNPs.

This stage aims to make a film from the synthesis of AgNPs nano flavonoid using a spin coating method. The initial step of sticking the nickel substrate on the spin coater holder. Then drop an AgNPs nano flavonoid solution on the glass substrate and set the spin coater at 1500 rpm for 10 seconds as depicted in Figure 2.

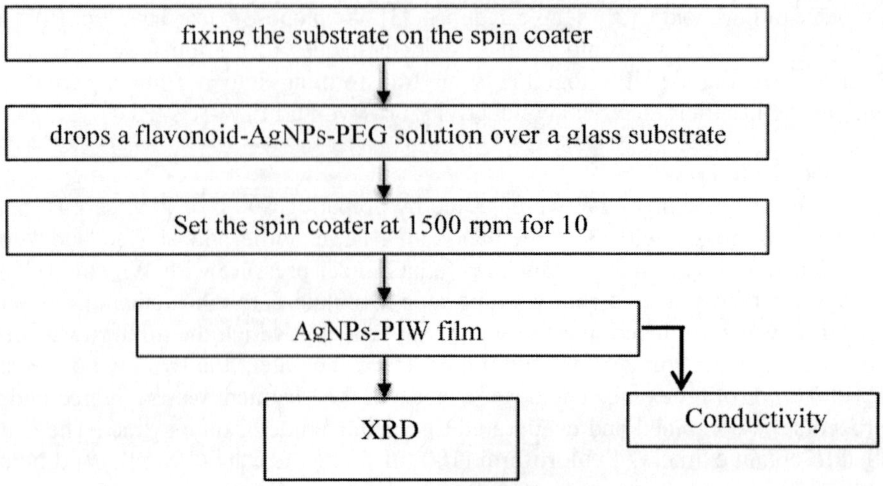

Figure 2. Flavon Ag film process.

3. Results and Discussion

The results of FT-IR spectrophotometric characterization of PIW leaves extract are represented by a wave number and its transmittance shown in Figure 3.

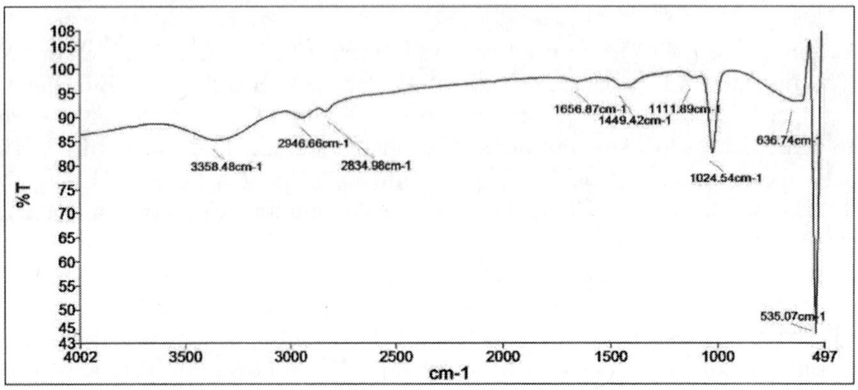

Figure 3. FTIR spectra of PIW leaf extract.

Based on the above data the extract of PIW leaves shows a peak transmittance of 3358.48, 2946.66, 2834.98, 1656.87, 1449.42, 1111.89, 1024.54, 636.74, 535.07, and 474.01 cm^{-1}wave numbers. These peaks are the primary indicators of quercetin kinds of flavonoids. So we inferred that the leaf extract of PIW in this work belongs to the quercetin [13].

3.1. Diffraction Pattern of Flavonoid PIW-Ag Films

The diffraction patterns of Flavon PIW-Ag/glass is shown in Figure 4. It is obtained that the main three prominent peaks indicate the existing of silver nanoparticles on the composite film. Briefly, we observed that the increase of AgNPs gives rise to the rise of the peaks' intensity showing the relative increase of AgNPs phase in the films.

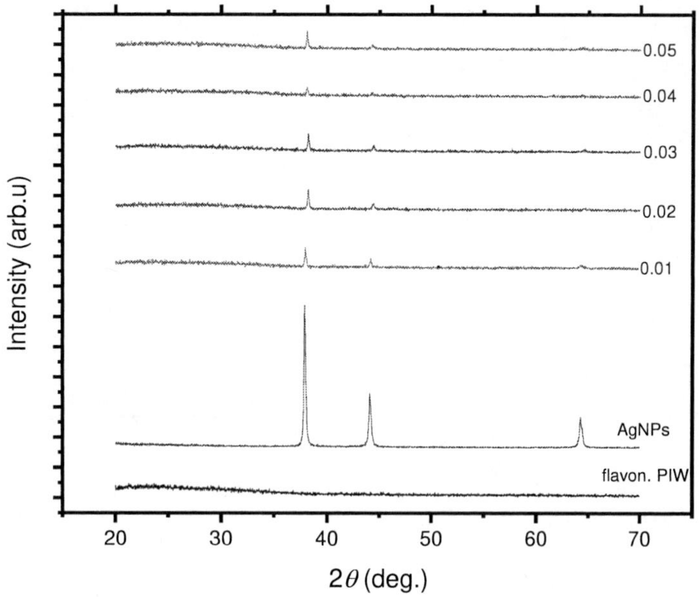

Figure 4. Diffraction pattern of XRD at different AgNPS in the flavon PIW-Ag film

Further detail analysis showed that there is a small shift of main peak of AgNPs. The film with AgNPs concentration of 0.01 g in the PIW-AgNPs flavonoid showed a peak at 2θ = 38.03°. For Flavonoid PIW-Ag 0.02 g produces a peak at 2θ = 38.18°. As for Flavonoid PIW-AgNPs of 0.03, 0.04, and 0.05 g produces a peak at 2θ of 38.12°. We have also found on our previous work that there is a small shift of the XRD patterns due to the introduction of AgNPs on natural flavonoids [3,14].

The X-ray diffraction patterns were analyzed by using the Gaussian fitting model. The silver phase on the whole of the X-ray diffraction patterns was performed under AMCSD 9011608 as a reference. Base on this model the peaks observed at 2θ = 38.51°, 44.51°, 64.69°, 77.78°, and 81.79°. It was found that the crystal volume falls to 66.8 Å^3 with a lattice parameter of $a = b = c = 4.058$ Å under cubic spinel system. The data was used for refinement of Ag NPs for main Bragg's plane of (111), (002), (022), (113), and (222). By using Scherrer's formula, we found that the crystal size is 45.5, 54.5, 52.2, 45.0, and 59.7 nm respectively for 0.01, 0.02, 0.03, 0.04, and 0.05 g of AgNPs content in the films. The average calculated of crystal size falls to 38.45 nm. The crystallinity of AgNPs in the film is depicted in Figure 5.

Although the crystallinity of AgNPs in the films does not linearly increase with the increase of AgNPs, we could see that the trend is increasing. It is also possible that in the higher amount of AgNPs mixed in the films could lower the crystallinity as shown by the 0.05 sample. This trend is generally in line with the Ag-doped PANI.

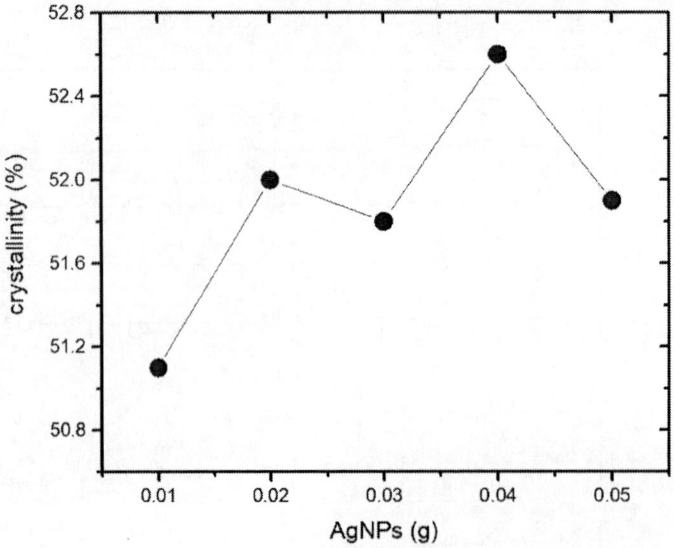

Figure 5. The crystallinity of AgNPs in the flavonoids PIW films.

3.2. Electrical Conductivity of PIW-AgNPs film
The four probe method was implemented to measure the electrical conductivity of flavonoid PIW-AgNPs thin film. The electrical conductivity of AgNPs-Flavonoids film is displayed in Figure 6.

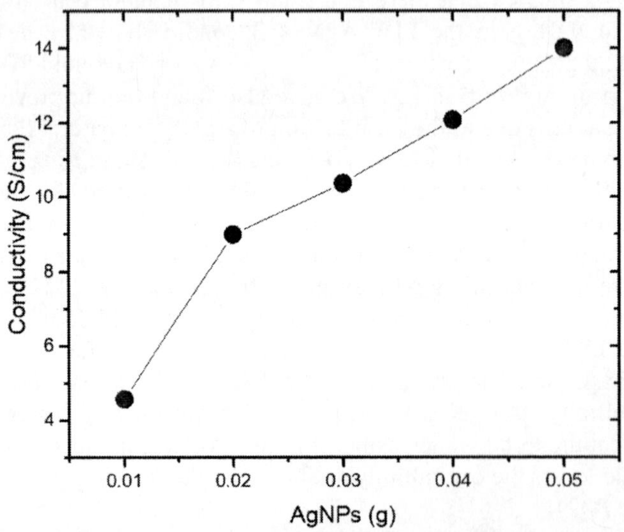

Figure 6. The electrical conductivity of PIW-AgNPs film

AgNPs doped of PIW flavonoid will cause a significant increase in its conductivity. This dependency of electrical conductivity on AgNPs content looks linearly. This trend is also reported previously silver doped natural flavonoids [3], silver doped PANI [16], Fe and Co-doped PANI [18], AgNPs on various materials [20]. We may predict that by introducing a lower conductivity of metal oxide as SnO_2 will reduce the electrical conductivity [19]. We found interesting evidence found that the conductivity of the film depends on the content of AgNPs. Flavonoids are closely a kind of conducting polymer. The typical range of its conductivity lays around 10^{-3} to 10^{-1} S/cm as other conductive polymers. A pure silver shows about 10^7 S/cm. A linear relation of AgNPs on its electrical conductivity could be used to control in what range of films desired. At lower range of AgNPs content, the electrical conductivity dependent of

IOP Conf. Series: Journal of Physics: Conf. Series **1093** (2018) 012016 doi:10.1088/1742-6596/1093/1/012016

AgNPs may also be affected by its crystallinity. The AgNPs content and crystallinity dependent of conductivity could be observed by comparing Figure 5 and Figure 6. This fact gives us a way to control the band gap using AgNPs concentration for solar cell application [21].

3.3. SEM analysis of PIW-AgNPs Film

The results of Flavonoid PIW-Ag film micrograph from SEM characterization and EDX elemental analysis is presented in Figure 7.

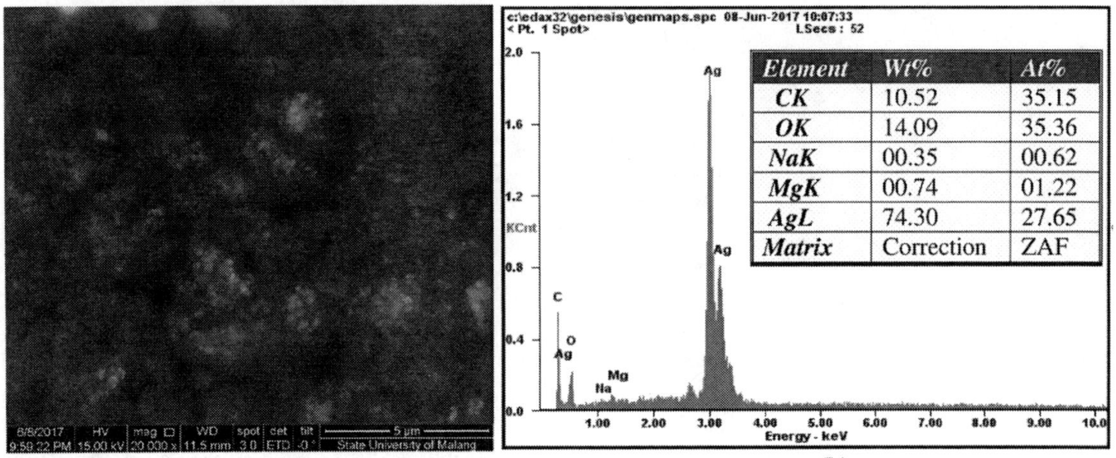

(a) **(b)**

Figure 7. The morphological structure (a), and elemental composition (b) of PIW-AgNPs Film.

A microscopic agglomeration of flavonoids PIW-Ag films is observed. In this image, we may infer that the polymeric flavonoids take a significant role in the agglomeration. From the SEM image, we could not observed where and how the nanosilver lays in the polymeric nets. According to the EDAX analyses, it could be obtained that silver nanoparticles are in the films. Other metallics, as well as organics elements, also observed in the films is come from the substrate and the flavonoids of PIW. Of course that the concentration and the distribution of the elements can be influenced by various factors including the concentration of precursors, reductants, temperatures, and preparation methods [14, 15, 17, 20]. To look for detail nanostructural of Ag-nanoparticles in the flavonoids PIW is better to be observed using TEM.

3.4. TEM Image of PIW-AgNPs

The results of the Flavonoid PIW-Ag image based on the result of TEM characterization is displayed in Figure 8.

Based on the image of TEM image can be presented that the size of Ag nano at TEM Figure 8a is in the range of 30 nm to 50 nm with an average size of 38.45 nm which corresponds to the initial discussion on XRD analysis. Figure 8b shows the result that flavonoids and Ag nanoparticles are inhomogeneously distribute. It is also obtained that the silver nanoparticles are composited rather than in the form of the compound system. This composite system is expected to exist since the procedure which we used is mixing and blending without any further chemical routes of synthesis. It should also depend on several factors including solvents, concentrations, and preparation methods.

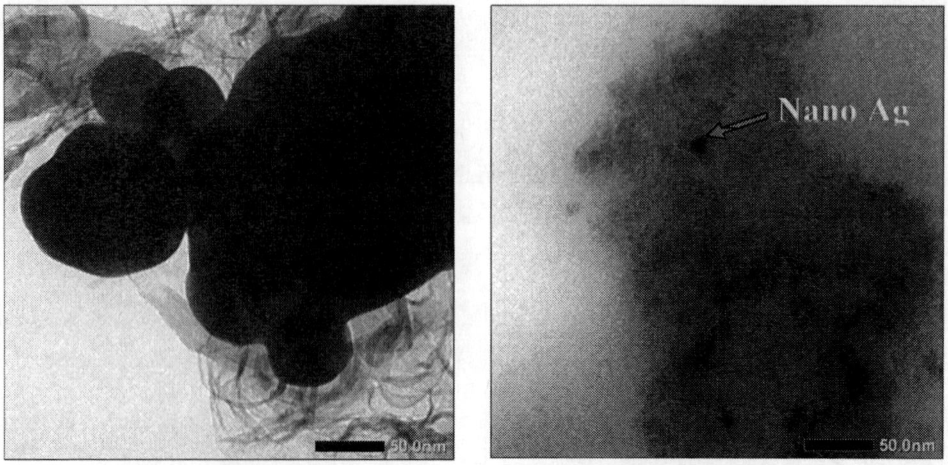

(a) (b)

Figure 8. TEM image of (a) Ag NPs, and b) Flavonoid-Ag NPs

4. Conclusion

We conclude that the flavonoids extract from PIW leaf using maceration method belongs to quercetin type. The grain size of AgNPs obtained in this work is of 38.45 nm. The addition of AgNPs in the PIW extract films increases the crystallinity. The electrical conductivity of PIW-AgNPS is increase with the increase of AgNPs. The SEM results of flavonoid PIW-AgNPs showed that the size of the AgNPs exhibits in the range of 30 nm to 50 nm which is good agreement with XRD data. The TEM analysis of AgNPs showed that the grain size in AgNPS was around 50 nm. We further analyze that AgNPs are not homogeneously distributed in the PIW extract. What the flavonoids extract from PIW leaf is potentially be used as an electronic application by modifying using AgNPs.

Acknowledgments

We thank to the Ministry of Research and Higher Education of Indonesia through the funding grant of this research for MD.

References

[1] Hart, D.J., Scott, K.J., 1995. Development and evaluation of an HPLC method for the analysis of carotenoids in foods, and the measurement of the carotenoid content of vegetables and fruits commonly consumed in the UK. Food Chem. 54, 101–111.

[2] Radke, M., Sittardt, H.G., Welte, D.H., 1978. Removal of soluble organic matter from rock samples with a flow-through extraction cell. Anal. Chem. 50, 663–665.

[3] Markus Diantoro, Nisfi Nahari Sani Hidayati, Rodatul Latifah, Abdulloh Fuad, Nasikhudin, Sujito, and Arif Hidayat. Electrical conductivity modification using silver nanoparticles of Jatropha Multifida L. and Pterocarpus Indicus w. extracts films, AIP Conference Proceedings 1719, 030034 (2016); doi: 10.1063/1.4943729

[4] Sharma, V., Janmeda, P., n.d. Extraction, isolation, and identification of flavonoid from Euphorbia neriifolia leave. Arab. J. Chem. doi:10.1016/j.arabjc.2014.08.01

[5] Merken, H.M., Beecher, G.R., 2000. Measurement of food flavonoids by high-performance liquid chromatography: a review. J. Agric. Food Chem. 48, 577–599.

[6] Mun, J.-H., Park, C.-W., 1995. Flavonoid chemistry ofPolygonum sect. Tovara (Polygonaceae): A systematic survey. Plant Syst. Evol. 196, 153–159.

[7] Bovy, A., Schijlen, E., Hall, R.D., 2007. Metabolic engineering of flavonoids in tomato (Solanum lycopersicum): the potential for metabolomics. Metabolomics 3, 399–412.

[8] Chen, S., Xing, X.-H., Huang, J.-J., Xu, M.-S., 2011. Enzyme-assisted extraction of flavonoids from Ginkgo biloba leaves: Improvement effect of flavonol transglycosylation catalyzed by

Penicillium decumbens cellulase. Enzyme Microb. Technol. 48, 100–105. doi:10.1016/j.enzmictec.2010.09.017

[9] De Sousa Araújo, T.A., Alencar, N.L., de Amorim, E.L.C., de Albuquerque, U.P., 2008. A new approach to study medicinal plants with tannins and flavonoids contents from the local knowledge. J. Ethnopharmacol. 120, 72–80. doi:10.1016/j.jep.2008.07.032.

[10] Gazal, R.M., Blanche, C.A., Carandang, W.M., 2004. Root growth potential and seedling morphological attributes of Narra (Pterocarpus indicus Willd.) transplants. For. Ecol. Manag. 195, 259–266.

[11] Guo, B., Liu, Y.-G., Yan, Q., Liu, C.-Z., 2007. Spectral composition of irradiation regulates the cell growth and flavonoids biosynthesis in callus cultures of Saussurea medusa Maxim. Plant Growth Regul. 52, 259– 263. doi:10.1007/s10725-007-9192-0.

[12] Häkkinen, S.H., Kärenlampi, S.O., Heinonen, I.M., Mykkänen, H.M., Törrönen, A.R., 1999. The content of the flavonols quercetin, myricetin, and kaempferol in 25 edible berries. J. Agric. Food Chem. 47, 2274–2279.

[13] Jeong Mi Kim and Hye Sook Yun-Choi, Anti-platelet Effects of Flavonoids and Flavonoid-glycosides

[14] from Sophora japonica, Arch Pharm Res Vol 31, No 7, 886-890, 2008, DOI 10.1007/s12272-001-1242-1

[15] M. Diantoro, R. Fitrianingsih, N. Mufti, and A. Fuad, "Synthesis of silver nanoparticles by chemical reduction at a various fraction of MSA and their structure characterization," 2014, pp. 257–261.

[16] Ravichandran, V., Vasanthi, S., Shalini, S., Shah, S.A.A., Harish, R., 2016. Green synthesis of silver nanoparticles using Artocarpus altilis leaf extract and the study of their antimicrobial and antioxidant activity. Mater. Lett. 180, 264– 267.

[17] Markus Diantoro, Ika Nur Fitriana, Faradilla Parasmayanti, Nasikhudin, Crystallinity And Electric Conductivity: The Role of Ultrasonic And Silver Doped of Pani/Ni Film, IOP Conf. Series: Materials Science and Engineering 202 (2017) 012005 doi:10.1088/1757-899X/202/1/012005

[18] Diantoro, Markus and Chasanah, Alif Fitria and Mufti, Nandang and Fuad, Abdulloh Fabrication of Silver Nanoparticles and its Films and their Characterization of Structure and Electrical Conductivity}, Advanced Materials Research, Trans Tech Publications,896,341-346, 2014,

[19] Markus Diantoro, Devy Purwaningtyas, Nazilah Muthoharoh, Arif Hidayat, Ahmad Taufiq, Abdulloh Fuad, The Influence of Iron- and Copper- doped of PANi Thin Film on their Structure and Dielectric Properties. American Institute of Physics, 1454. ISSN 0094-243X, p268-271. 2012.

[20] M. Diantoro, Kholid, A. A. Mustikasari, and Yudiyanto, The Influence of SnO2 Nanoparticles on Electrical Conductivity, and Transmittance of PANI-SnO2 Films, IOP Conf. Series: Materials Science and Engineering 367 (2018) 012034 doi:10.1088/1757-899X/367/1/012034

[21] Markus Diantoro, Thathit Suprayogi, Ulwiyatus Sa'adah, Nandang Mufti, Abdulloh Fuad, Arif Hidayat and Hadi Nur (July 18th, 2018). Modification of Electrical Properties of Silver Nanoparticle, Silver Nanoparticles, Khan Maaz, IntechOpen, DOI: 10.5772/intechopen.75682.

[22] Markus Diantoro, Thathit Suprayogi, Arif Hidayat, Ahmad Taufiq, Abdulloh Fuad, and Risa Suryana, Shockley's Equation Fit Analyses for Solar Cell Parameters from I-V Curves, International Journal of Photoenergy, Volume 2018, Article ID 9214820, 7 pages https://doi.org/10.1155/2018/9214820

IOP Conf. Series: Journal of Physics: Conf. Series **1093** (2018) 012017 doi:10.1088/1742-6596/1093/1/012017

Vertical Distribution of Bacteria in Various Lakes of East Java, Indonesia

Sitoresmi Prabaningtyas*, **Agung Witjoro, Endang Suarsini, Dhiyaudin Aridowi, Nafizatuzamrudah, Afifah Nur Aini, Adji Purnomo, Yuniar Indra Permana, Meirinda, Dina Aribah, Yanis**

Department of Biology, Faculty of Mathematics and Natural Science, Universitas Negeri Malang, Jl. Semarang 5 Malang 65145, Indonesia

*Corresponding author's email: sitoresmi.prabaningtyas.fmipa@um.ac.id

Abstract: Ecosystems are composed of a large number of diverse organisms that make up very complex interactions. Several studies have proven the influence of abiotic factors on the distribution of bacteria in the environment. Bacteria play an important role in biogeochemical cycles and support the productivity of ecosystems. The important role of bacteria in the environment causes the need for research on the vertical distribution of bacteria. The purpose of this research was to determine the vertical distribution of bacteria in some lakes in eastern Java. The samples were taken from 3 lakes in East Java namely, Ranu Grati, Ranu Regulo, and Ranu Pane. The bacterial samples were taken at 5 stations in each lake. The samples, vertically taken from 4 points, the upper, middle and lower surface according to the depth of the Secchi disc and the bottom of the lake, were isolated and identified. It was found 14 species of bacteria from Ranu Pane, 24 species of bacteria from Ranu Regulo, and 36 species of bacteria from Ranu Grati. The most dominant and abundant bacteria in Ranu Pane, Ranu Regulo, and Ranu Grati were coded with A, B, G3 respectively. The bacteria in all the lakes were spread over all stations and were vertically scattered at all points. The abiotic factors measured have a positive effect on the vertical distribution of bacteria.

Keywords: Vertical distribution, bacteria, lake, abiotic.

1. Introduction

The presence of bacteria in the environment is greatly influenced by abiotic factors [1]. In a lacustrine (lakes) ecosystem, microorganisms play a pivotal role in driving the biogeochemical processes [2]. First, the photosynthetic and chemolithotrophic microorganisms are the main contributors to the primary productivity. Second, the chemoorganotrophic and chemoheterotrophic microorganisms are actively involved in the decomposition of organic matters. Third, appreciable amounts of microorganisms are the foodstuff for filter-feeding organisms and form the foundation of the aquatic food web [3,4]. Therefore, it is of great interest to study the bacterial communities in Ranu Grati, Ranu Pane, and Ranu Regulo lakes.

Lakes are ecosystems that are composed of a large number of diverse organisms that make up very complex interactions. Interactions can occur in the lake ecosystem between microbes and environment [5]. Bacteria play an important role in biogeochemical cycles and support the productivity of ecosystems [6]. However, the profiles of the distribution bacteria in Ranu Pane, Ranu Regulo, and

Ranu Grati lakes remain unknown. The important role of bacteria in the environment causes the need for research on the relationship of vertical distribution of bacteria with abiotic factors. Thus, it is of great interest to study the microbial communities in Ranu Grati, Ranu Pane, and Ranu Regulo lakes.

In this study, the vertical distribution of bacterial was investigated in 2017. There are about 25 lakes in East Java. The famous lakes in east Java among them are Ranu Pane, and Ranu Regulo located at the border areas of Pasuruan, Lumajang, and Malang at 2100 MASL and Ranu Grati Lake, is located at Pasuruan, with an altitude of about 91m above sea level. The area of Ranu Pane is approximately 1 hectare and the area of Ranu Regulo is about 0.75 hectares. Ranu Grati area is around 198 hectares. The physicochemical parameters of the lake water were examined as well. In addition, the correlation of vertical distribution bacteria with abiotic factors of the lake and the structure of the bacterial communities in the lake were discussed.

2. Experimental Methods

2.1. Study site and sampling stations

Ranu Pane, Ranu Regulo, and Ranu Grati are eutrophic lakes, with the depths about (0.5 - 7 m), (0.5 - 50 m) and (7 - 150 m) respectively. Five sampling stations in the Ranu Pane, Ranu Regulo, and Grati lakes were selected. The five sites were distributed over the lake at the deeps varies between 0.5 m and 91 m.

2.2. Bacterial samples collection

Bacterial samples were collected from water at 5 stations in the three lakes (Figure 1). It was vertically taken from 4 points i.e. the surface, middle, lower surface (according to the depth of Secchi disc) and the bottom of the lake. The bacterial samples were isolated and identified.

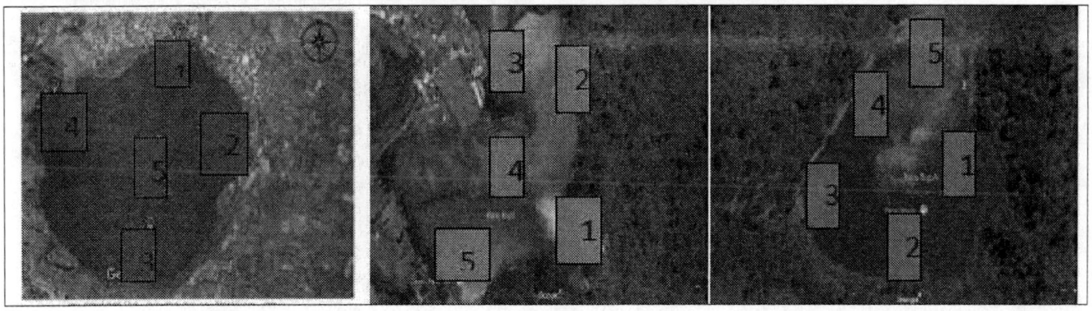

(A) (B) (C)

Figure 1. The conditions of each sampling site in Ranu Grati (A), Ranu Pane (B) and Ranu Regulo (C). The codes of the station taken as samples were (1, 2, 3, 4, 5)

2.3. Bacteria Collection

For collecting the bacteria, 100 ml water from each point of each station was taken using sterile "water sample bottle", and then transferred to a sterile bottle. The water samples from all station were transported in an ice box and stored at about 4°C with ice gel [7].

2.4. Analysis of physicochemical parameters of water

Water transparency was determined in situ using the Disc method. The salinity, concentrations of DO (dissolved oxygen), temperature, and pH were measured using Refractometer ATAGO S-28E, DO meter Lutron 5510, pHmeter Lutron 5509 respectively [8]. The next analysis was conducted in the laboratory. The concentration of total nitrogen was analyzed using the Total Kjeldahl method. The content of dissolved inorganic phosphorus (DIP) was determined using the phosphorus molybdenum blue colorimetric method [9,10].

2.5. Data analysis
Principal coordinate analysis (PCoA) plots were generated based on the weighted Unifrac pairwise distance matrices and on the Bray-Curtis coefficient matrix in order to evaluate vertical distribution in the microbial community over the points of the samplings. Canonical correspondence analysis (CCA) was also carried out to determine the correlation between physicochemical parameters and the relative abundance of bacteria in the microbial community. Multivariate analyses were done using Paleontological Statistics software (PAST, version 2.15). The heat map representing the mean relative abundance of microbial genera was generated by a hierarchical cluster analysis of different genera mean prevalence Paleontological Statistics software (PAST, version 2.15). Moreover, the species richness indices (Chao1 and ACE) and Shannon's diversity index were computed with the software Paleontological Statistics software (PAST, version 2.15).

3. Results and Discussion
The water salinity of the three lakes was around 0.00% – 0.4% and the pH was around 6.83–9.40 at the surface layer. Dissolved Oxygen (DO) of the lake water at the surface layer was between 4.2 and 11.6 mg/L. The transparency of the lake water was between 50 to 130 cm (Table 1, 2 and 3).

Table 1. The Physicochemical Water Parameters of Ranu Regulo Lake

Point	DO(mg/l)	pH	Salinity (‰)	Transparency (cm)
1	8.9	9.38	0.3	75
2	8	8.65	0.2	75
3	8.4	8.65	0.2	80
4	11.3	9.3	0	75
5	11.6	9.4	0.2	75

Table 2. The Physicochemical Water Parameters of Ranu Pane Lake

Point	DO(mg/l)	pH	Salinity	Transparency (cm)
1	4.2	7.24	0.4	80
2	5	6.83	0.3	82
3	6	6.95	0.2	100
4	11.6	7.3	0.4	50
5	6.7	6.88	0.4	50

Table 3. The Physicochemical Water Parameters of Ranu Grati Lake

Point	DO(mg/l)	pH	Salinity	Transparency (cm)
1	7.8	8.82	0	130
2	7.1	7.1	0	100
3	8.4	8.4	0	100
4	7.2	8.8	0	110
5	8.5	8.82	0	110

Among the three lakes observed, the water of Ranu Grati Lake had the salinity of 0‰. Based on the morphology of bacterial colonies, it was found 14 species of bacteria from Ranu Pane, 24 species of bacteria from Ranu Regulo, and 36 species of bacteria from Ranu Grati. The most dominant and abundant bacteria in Ranu Pane, Ranu Regulo, and Ranu Grati were coded with A, B, G3 respectively. Based on the characteristics of bacterial physiology the species G3 is *Enterobacter gergoviae.* The Vertical Distribution of bacteria in Ranu Pani was scattered in all point (Figure 2).

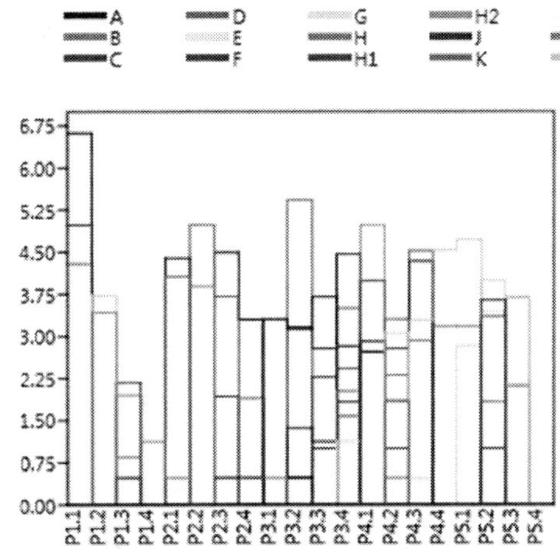

Figure 2. The Vertical Distribution of Bacteria in Ranu Pani. A, B, C and so on are the codes of bacterial species. P1.1, P2.2 and so on are codes of sampling stations.

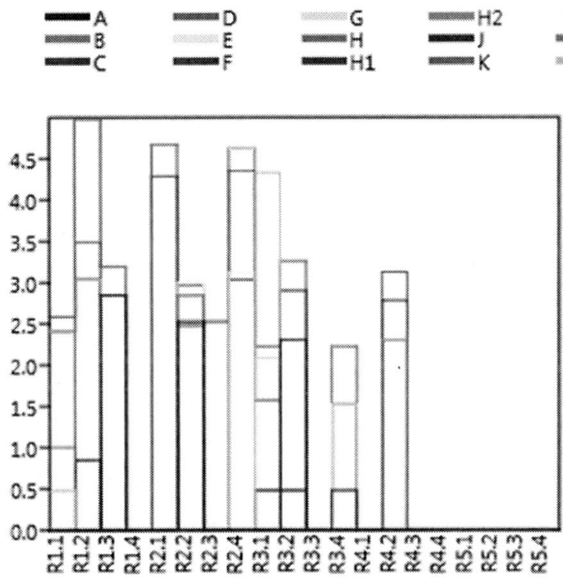

Figure 3. The Vertical Distribution of Bacteria in Ranu Regulo. A, B, C and so on are codes of bacterial species. R1.1, R2.2 and so on are codes of sampling stations.

The diversity profile of bacteria in Ranu Grati**,** Ranu Pane, and Ranu Regulo was computated with the software Paleontological Statistics software (PAST, version 2.15). The results showed that bacteria profile were scattered in another depth of samples (Figure 1, 2, and 3). The bacteria in the lakes were spread over all stations and were vertically scattered at all points. The abiotic factors measured have a positive effect on the vertical distribution of bacteria.

The bacteria were distributed evenly across the sampling point. Abiotic factors measured did not show any significant difference. Certain species exists only at a certain point. Bacteria have a tendency to grow at a particular station and depth. The differences in this tendency are influenced by the type of bacteria, the type of nutrients and the possible range of environments for these types of bacteria. The physicochemical variables measured in this research were transparency, salinity, OD, pH, and turbidity.

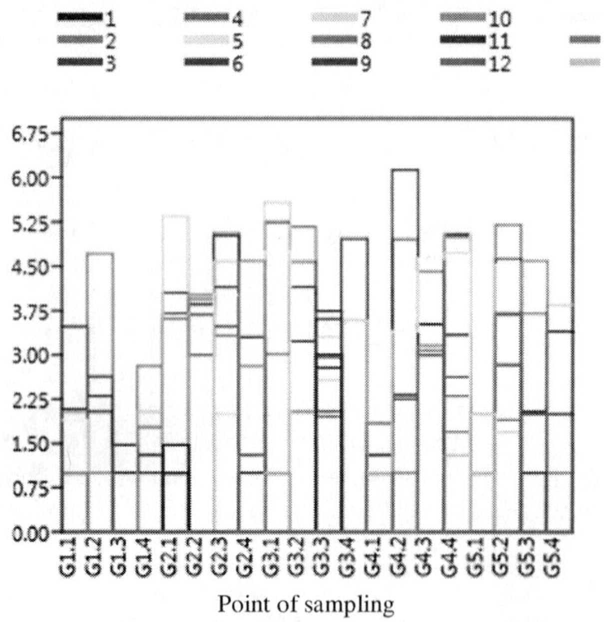

Figure 4. Vertical Distribution of Bacteria in Ranu Grati. . 1, 2, 3 and so on are codes of bacterial species. R1.1, R2.2 and so on are codes of sampling stations

In this research, these physicochemical variables were likely among the dominant factors driving (directly or indirectly) the changes in bacterial community structure. The distribution of bacteria was influenced by its growth which was controlled by various biological and biochemical processes (including the reproduction rate, diffusion of chemical species, microbial species interaction, gene mutation, evolutionary adaptation, and species formation). The rise in the temperature will elevate the rate of biological processes in microorganisms and enhance the growth of bloom-forming Cyanobacteria from the littoral zones [10]. The bacteria in the lake were spread over all stations and vertically scattered at all points.

The composition of bacterial communities appeared to be relatively homogeneous among depth samples, as indicated by UPGMA dendrogram (see Fig. 3). The nutrient availability is considered to be a primary factor that can change the microbial communities in lakes. In our study, the types of lakes seemed to have little effect on the whole bacterial communities in the lakes. Although Ranu Pane, Ranu Regulo, and Ranu Grati lakes was a closed lake, its lake water kept being an active material exchanging and energy transformation with the environment through atmospheric circulation, rainwater precipitation, etc. Consequently, the lake water became an open system and its bacterial communities were determined by the environment. In contrast, the bottom sediments could not freely communicate with the environment and therefore were separated as exclusive niches to preserve the unique properties (including microorganisms) of this lake. The bacteria inhabiting in the sediments of lakes, especially those unidentified, might be a reserve of ancient bacteria which were of geological value.

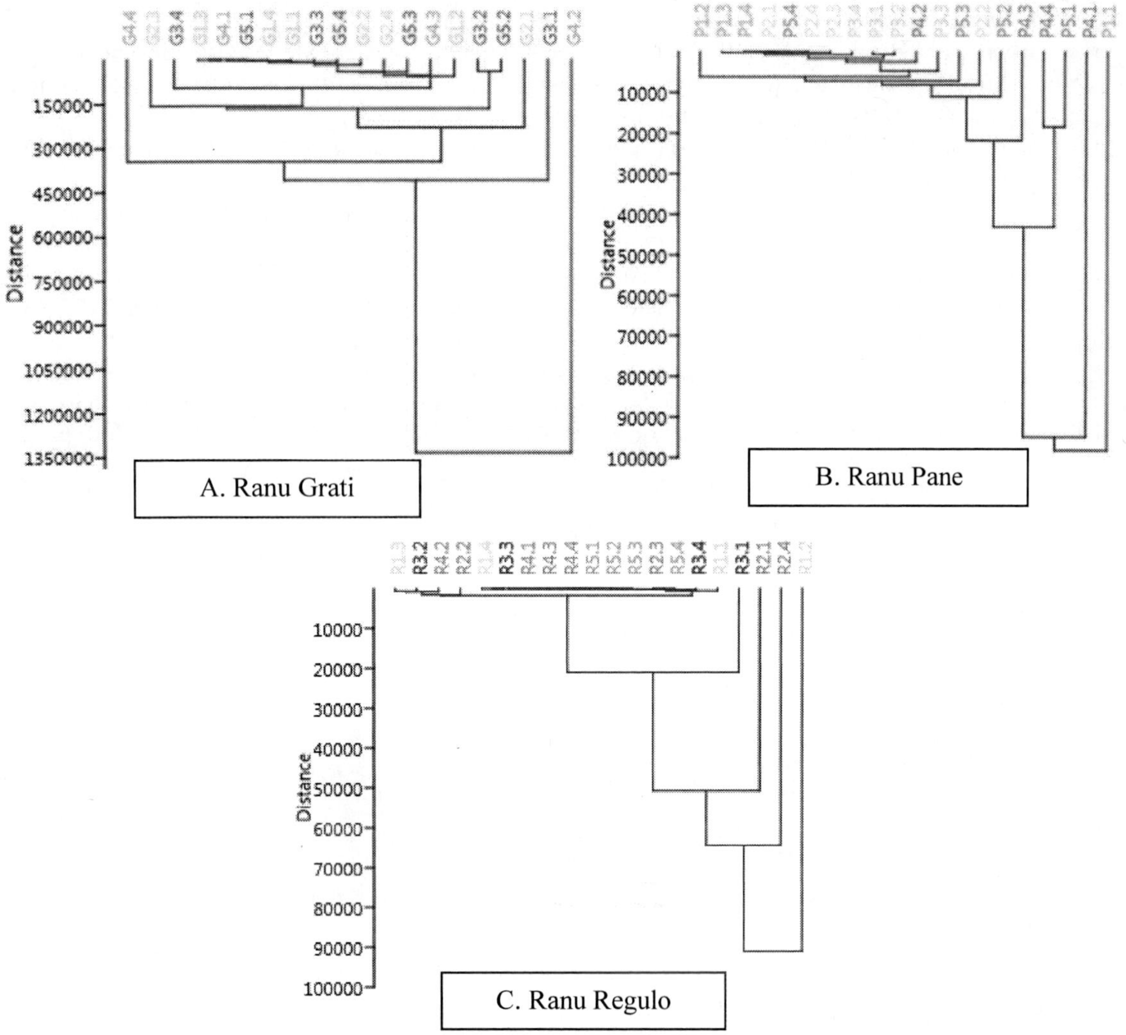

Figure 5. Bacterial spread at the sampling station. The bacteria spread in all stations and there was no difference between the sampling stations.

4. Conclusion

Based on the results and discussion, we conclude that the bacteria in Ranu Pane, Ranu Regulo, and Ranu Grati lakes are spread over all stations and vertically scattered at all points. The physicochemical variables were likely among the dominant factors driving directly or indirectly the changes in bacterial community structure. Furthermore, the abiotic factors measured have a positive effect on the vertical distribution of bacteria.

Acknowledgments

This work was financially supported by DRPM research grant from the Ministry of Research Technology and Higher Education.

References

[1] Madigan M T 2012 *Brock biology of microorganisms* (San Francisco: Benjamin Cummings)
[2] Gualtieri P and Barsanti L 2006 *Algae: anatomy, biochemistry, and biotechnology* (Boca Raton: Taylor & Francis)
[3] Sigee D 2005 *Freshwater microbiology: biodiversity and dynamic interactions of*

microorganisms in the aquatic environment (John Wiley & Sons)

[4] Avery S V, Codd G A and Gadd G M 1998 Microalgal Removal of Organic and Inorganic Metal Species from Aqueous Solution *Wastewater Treatment with Algae* ed Y-S Wong and N F Y Tam (Berlin, Heidelberg: Springer Berlin Heidelberg) pp 55–72

[5] Ramanan R, Kim B-H, Cho D-H, Oh H-M and Kim H-S 2016 Algae–bacteria interactions: Evolution, ecology and emerging applications *Biotechnol. Adv.* **34** 14–29

[6] Cooper M B and Smith A G 2015 Exploring mutualistic interactions between microalgae and bacteria in the omics age *Curr. Opin. Plant Biol.* **26** 147–53

[7] Sugita H, Kuruma A, Hirato C, Ohkoshi T, Okada R and Deguchi Y 1994 The vitamin-B12-producing bacteria in the water and sediment of a carp culture pond *Aquaculture* **119** 425–31

[8] Gao X and Song J 2005 Phytoplankton distributions and their relationship with the environment in the Changjiang Estuary, China *Mar. Pollut. Bull.* **50** 327–35

[9] Paytan A and McLaughlin K 2007 The Oceanic Phosphorus Cycle *Chem. Rev.* **107** 563–76

[10] McKelvie I Phosphate. Handbook of Water Analysis (New York: Marcel Dekker, Inc) pp 273–95

IOP Conf. Series: Journal of Physics: Conf. Series **1093** (2018) 012018 doi:10.1088/1742-6596/1093/1/012018

Effect of Single Garlic Oil for Homeostasis of $CD4^{+}CD25^{+}$ Immuno-regulatory T Cells Controlling Hypercholesterolemia

Sri Rahayu Lestari[1,*] and Muhaimin Rifa'i[2]

[1]Department of Biology, Faculty of Mathematics and Natural Sciences, Universitas Negeri Malang, Jl. Semarang 5 Malang 65145, Indonesia
[2]Department of Biology, Faculty of Sciences, Brawijaya University, Jl. Veteran 42-44 Malang, East Java 65145, Indonesia

*Corresponding author's email: srirahayulestari@um.ac.id

Abstract. This study aims to explore the alternation expression of NFkB, activation degree of naïve T cells, and role of regulatory T cell. Hypercholesterolemia mouse model was created by giving a high-fat diet (HFD) for 45 days, and then they were treated by 12.5 mg/kg BW, 25 mg/kg BW and 50 mg/kg BW SGO for 30 days. The end of the treatment, the mouse was sacrificed, and the spleen was isolated to analyze cell surface molecules and regulatory T cells using flow cytometry. The data were analyzed by Cell Quest software and tested by statistical analysis One Way ANOVA Analysis. The results showed that a high-fat diet decreased the number of regulatory T cell and also decreased the expression of NFkB on $CD4^{+}$ T cell in hypercholesterolemic mice. Moreover, the SGO did not show a significant effect on increasing the number of naïve T cells expressing CD62L molecule. Thus, the administration of SGO might inhibit activation of the NFkB pathway and decrease the stimulation of effector cell development that plays a role in the inflammatory pathway.

Keywords: Hypercholesterolemia, Immunoregulatory, Single garlic oil, Homeostasis, CD4+CD25+

1. Introduction

Atherosclerosis is an arterial chronic inflammatory disease by producing vessel wall plaques is driven by systemic hypercholesterolemia and local accelerated immune responses [1-5]. The atherosclerotic plaque is characterized by an accumulation of vascular smooth muscle cells (VSMCs), foam cell formation, inflammatory cells recruitment, and production of reactive oxygen species (ROS) [6,7,2]. The early pathogenesis of atherosclerosis was indicated by expression of vascular cell-adhesion molecule 1 (VCAM1) on the endothelium as a response to bioactive lipids ox-LDL that have been trapped within the arterial wall [3,7]. Inflammation in atherosclerosis is caused by primarily oxidized low-density lipoprotein (LDL) directly interacting with pro-inflammatory signal pathways in vascular cells and macrophages [8]. Oxidized-LDL (ox-LDL) can be engulfed by aortic macrophages, which become sub-endothelial accumulations of cholesterol-engorged macrophages forming lipid-laden foam cells [9,10,7].

The initiation and progression of atherosclerosis are associated with components of both innate and acquired immune systems [11]. The immune system is activated in the process of atherosclerosis progression, and immune cells, such as T cells, B cells, and dendritic cells (DCs) are present within atherosclerotic lesions and contribute to disease development [10]. Macrophages constitute the dominant cell population on atherosclerotic plaques; they can generate oxygen radicals and proteases and can intake non-specific immune responses mediated by scavenger receptors [10,12]. To initiate specific immune responses, macrophages can present exogenous antigens to T lymphocytes ($CD4^+$ T cells and $CD8^+$ T cells) [12]. Development, activation, and differentiation of T lymphocytes are regulated by the nuclear factor-kappa B (NF-kB) [13]. The activation of the NFkB pathway can mediate the induction of a multitude of responsive genes that contribute toward the inflammatory phenotype role in atherogenesis [1,14].

Garlic has pharmacological effects attributed to the presence of pharmacologically active sulfur compounds including thiosulfinate (allicin), diallyl trisulfide, allyl methyl trisulfide, diallyl disulphide, ajoene, and others, which show anticancer, antioxidation, anti-inflammation, immunomodulatory, antimicrobial, hypoglycemic, cardiovascular protection, and preventing other chronic diseases associated with aging [14-16]. Garlic compound, especially allicin has a protective effect of reducing lipid content in the arterial wall on atherosclerosis by suppressed ox-LDL [17]. Garlic is capable of preventing oxidative modification of DNA, protein, and lipids, such as ox-LDL by scavenging reactive oxygen species (ROS) and increasing the cellular antioxidant enzymes [16,18]. The present study was designed to investigate the effect of garlic compound on immunological function in hypercholesterolemia mice models with the aim to explore the alternation expression of NFkB, activation degree of naïve T cells, and role of regulatory T cell.

2. Experimental Methods

2.1. Animal

Twenty four mice (*Mus musculus*) of male Balb C strain with aged three months old and 38-gram bodyweight were placed in the standard cage and maintained in the free pathogen room. The administration of daily intake was done by *ad libitum* and regular diet treatment.

2.2. Induction of Hypercholesterolemia with High Fat Diet (HFD)

After a one week adaptation period, four mice were given a normal diet, while the other twenty were given a high-fat diet (HFD) for 45 days. HFD feed consists of feed of Hi-Grow Medicated 551 30%+ duck egg yolk 10% + coconut oil 30% + wheat flour 5% + corn 24.9% + cholic acid 0.1%.

2.3. Oral Treatment with Single Garlic Oil

Hypercholesterolemic mouse model was divided into six experimental groups, i.e., standard chow, mice treated by HFD (negative control), positive control with HFD + simvastatin, HFD + SGO dose 12.5 mg/kg BW, HFD + SGO dose 25 mg/kg BW, and HFD + SGO dose 50 mg/kg BW. The administration of this HFD combined with SGO was performed for 30 days. In the next step, the mice were dissected, and the spleen organ was stored in the fridge for further laboratory test. All procedures were approved by the ethics committee of Universitas Brawijaya Indonesia with certificate number 880-KEP-UB.

2.4. Isolation of Lymphoid Cells and Flow Cytometry Analysis

Spleen was washed with sterile Phosphate Buffer Saline (PBS) and put in a petri dish containing sterile PBS. The spleen was pressed by using a syringe holder. Single cell solution was filtered with a sterile wire and put into a 15 mL conical tube. Suspension in the conical tube was added with PBS up to 10 mL and then put in a centrifuge, centrifugation in 2500 rpm, 4^0C, for 5 minutes. The supernatant was removed, and the obtained pellet was resuspended in one mL of sterile PBS. Single cell suspension stained with FITC-conjugated anti-mouse CD4, PE-conjugated anti-mouse CD8, PE-conjugated anti-mouse CD25, and PE-conjugated anti-mouse CD62L. For intracellular staining, the cells were stained with PE/Cy5-conjugated anti-mouse NFk-B, and the staining was performed with a Cytofix/Cytoperm Kit. The cells were ready for analysis according to the parameters that had been set in the flow cytometry FACS Calibur machine.

2.5. Statistical Analysis
The obtained data were analyzed by Cell Quest software and tested by statistical analysis One Way ANOVA Analysis with $P < 0.05\%$. All results were presented as a mean of ± SD values in each group.

3. Results and Discussion
High-fat diet (HFD) can trigger the formation of atheroma plaque in mice, with higher plasma cholesterol. HFD can increase body weight and circulating triglyceride concentrations with the extreme degrees of hypercholesterolemia. Hypercholesterolemia animals are useful models for studies on Atherosclerosis and treatment as well [19-21].

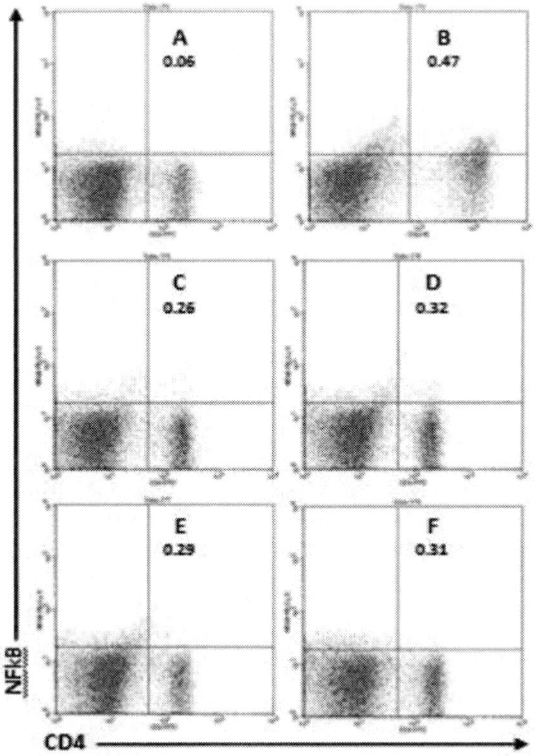

Figure 1. The expression of NFkB in CD4+. A) Normal mice diet. B) Mice with HFD as a negative control. C) Mice with diet HFD treated by simvastatin as a positive control. D) Mice with HFD + 0.5% garlic extract). E) Mice with HFD + 1% garlic extract. F) Mice with HFD + 2% garlic extract.

The administration of HFD showed an increase in NFkB expression at each treatment level compared with normal mice. Figure 1 shows that there was no statistically significant difference in NFkB expression in all of the treatments with HFD. The transcription factor NF-kB is one of the critical regulators of inflammation, immune responses, and cell survival by increasing the expression of specific cellular genes. NFkB can mediate the induction of various inflammatory cytokines, such as IL-1, IL-2, IL-6, IL-8, and TNF-α, many of which have played a role in atherogenesis. Activation of NF-κB can induce activation of iNOS (inducible nitric oxide synthase) catalyzes the formation of a significant amount of NO (nitric oxide), which plays a vital role in the pathogenesis of a variety of inflammatory diseases, such as atherosclerotic [1,14].

The administration of SGO showed the decrease in NFkB expression at all treatment levels compared with negative control. The lowest decrease in NFkB expression occurred in mice with diet HFD treated by 1% garlic extract from 0.47% to 0.29% (Figure 1). The active sulfur compound in garlic has the potential to be anti-inflammatory, which can inhibit the expression of NFkB by downregulation of NF-kB signaling [14, 15].

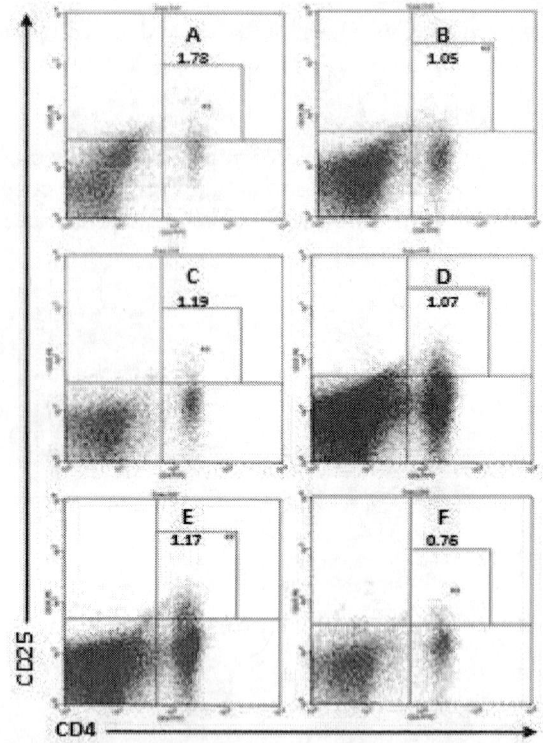

Figure 2. The profile of CD4+/CD25+ T Cells Increase in normal mice and a decrease in the mice with HFD. A) Normal mice diet. B) Mice with HFD as a negative control. C) Mice with HFD + simvastatin as a positive control. D) Mice with HFD + 0.5% garlic extract. E) Mice with HFD + 1% garlic extract. F) Mice with HFD + 2% garlic extract.

$CD4^{+}CD25^{+}$ T regs play crucial roles in the suppression of atherosclerosis by down-regulating inflammatory reactions [22]. In this study, CD4+CD25+ T regs decreased in all groups of mice with HFD administration. Figure 2 shows a decrease in $CD4^{+}CD25^{+}$ T regs in normal mice compared with mice model hypercholesterolemia from 1.78% to 1.05%. In general, natural T regs contains in much lower amounts in atherosclerotic plaques (1%–5% of all T cells) than in normal arterial tissue or inflammatory skin lesions (~25% of all T cells), suggesting that atherosclerotic plaques disturbed the local tolerance protection [8,23].

$CD8^{+}$ T cells are the dominant leukocyte infiltrate in atherosclerotic lesions [4], but the role of $CD8^{+}$ T cells in atherosclerotic is less clear [3,24]. In this study, the administrator of garlic extract in atherosclerotic mice model showed the increase in the activation of $CD8^{+}$ $CD62L^{-}$ T cells, but it was not significant. The percentage of naïve T cells decreased in mice with the HFD diet compared with the normal mice from 4.48% to 4.18%. The amount continued to decline after the treatment with garlic extract at the level of 0.5% and 2%, with the values of 3.57% and 4.16% (Figure 3).

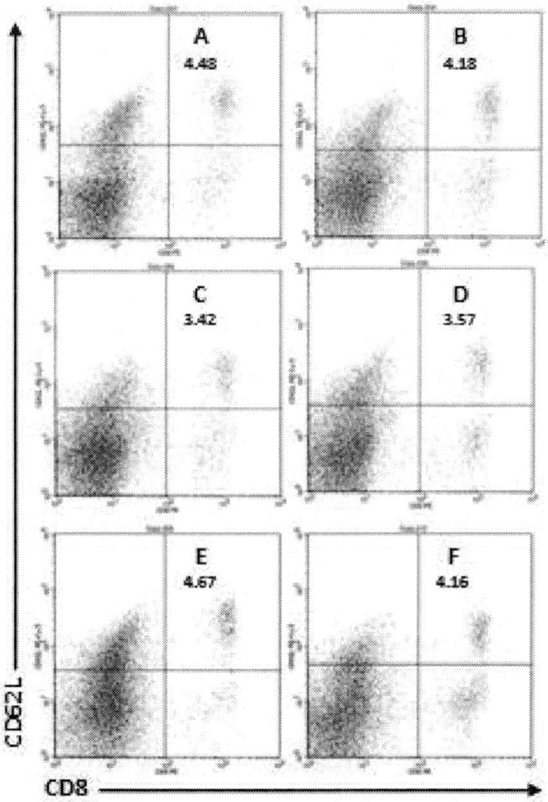

Figure 3. Profile of Naïve T Cells decrease in the mice with diet HFD which was not significant. A) Normal mice diet. B) Mice with HFD as a negative control. C) Mice with HFD + simvastatin as a positive control. D) Mice with HFD + 0.5% garlic extract. E) Mice with HFD + 1% garlic extract. F) Mice with HFD + 2% garlic extract.

4. Conclusion

The oral administration of SGO to hypercholesterolemia mouse model decreased the number of regulatory T cell and might inhibit the activation of the NFkB signaling pathway on CD4+ T cell. Also, SGO treatment decreased the stimulation of effector cell development that plays a pivotal role in the inflammatory pathway related atherogenesis.

Acknowledgments

We would thank Bambang Pristiwanto for assisting flowcytometry analysis. This research is funded by The Ministry of Research, Technology, and Higher Education of Republic Indonesia no. Contract 3.4.8/UN 32.14/LT/2017.

References

[1] Kanters E, Pasparakis M, Gijbels M, Vergouwe M, Partouns-Hendriks I, Fijneman R, et al. Inhibition of NF-κB activation in macrophages increases atherosclerosis in LDL receptor-deficient mice. J Clin Invest [Internet]. 2003;112(8):1176–85. Available from: http://www.jci.org/cgi/content/abstract/112/8/1176?etoc

[2] Karasawa T, Takahashi M. Role of NLRP3 Inflammasomes in Atherosclerosis. 2017;3:443–51.

[3] Andersson J, Libby P, Hansson GK. Adaptive immunity and atherosclerosis. Clin Immunol [Internet]. 2010;134(1):33–46. Available from: http://dx.doi.org/10.1016/j.clim.2009.07.002

[4] Kyaw T, Winship A, Tay C, Kanellakis P, Hosseini H, Cao A, et al. Cytotoxic and proinflammatory CD8+T lymphocytes promote the development of vulnerable atherosclerotic

plaques in ApoE-deficient mice. Circulation. 2013;127(9):1028–39.
[5] Amir S, Binder CJ. Experimental immunotherapeutic approaches for atherosclerosis. Clin Immunol [Internet]. 2010;134(1):66–79. Available from: http://dx.doi.org/10.1016/j.clim.2009.07.009
[6] Wang JC, Bennett M. Aging, and atherosclerosis: Mechanisms, functional consequences, and potential therapeutics for cellular senescence. Circ Res. 2012;111(2):245–59.
[7] Wang L, Huang Z, Huang W, Chen X, Shan P. Inhibition of epidermal growth factor receptor attenuates atherosclerosis via decreasing inflammation and oxidative stress. Nat Publ Gr. :1–14.
[8] Nilsson J, Wigren M, Shah PK. Regulatory T Cells and the Control of Modified Lipoprotein Autoimmunity-Driven Atherosclerosis. Trends Cardiovasc Med [Internet]. 2010;19(8):272–6. Available from: http://dx.doi.org/10.1016/j.tcm.2010.02.010
[9] Manuscript A. NIH Public Access. 2010;407(6801):233–41.
[10] Gil-Pulido J, Cochain C, Lippert MA, Schneider N, Butt E, Zernecke A. Deletion of Batf3-dependent antigen- presenting cells does not affect atherosclerotic lesion formation in mice. 2017;1–14.
[11] Yamashita T, Sasaki N, Kasahara K, Hirata K Ichi. Anti-inflammatory and immune-modulatory therapies for preventing atherosclerotic cardiovascular disease. J Cardiol [Internet]. 2015;66(1):1–8. Available from: http://dx.doi.org/10.1016/j.jjcc.2015.02.002
[12] Huang Y, Wang S, Zhang C, Xu Z, Shen J, Du X, et al. Experimental study of the anti-atherosclerotic effect of demethylzeylasteral. Exp Ther Med. 2017;13(6):2787–92.
[13] Manuscript A. NIH Public Access. 252(1):41–51.
[14] Ban JO, Oh JH, Kim TM, Kim DJ, Jeong H. Research article Anti-inflammatory and arthritic effects of thiacremonone , a novel sulfur compound isolated from garlic via inhibition of NF-κ B. 2009;11(5):1–13.
[15] Zeng Y, Li Y, Yang J, Pu X, Du J, Yang X, et al. Therapeutic Role of Functional Components in Alliums for Preventive Chronic Disease in Human Being. 2017;2017.
[16] The I, Asia M. Characterization of Active Compounds of Different Garlic (. 2010;68(1).
[17] Banerjee SK, Maulik SK. Effect of garlic on cardiovascular disorders : a review. Nutr J. 2002;14(1):1–14.
[18] Sedighi M, Bahmani M, Asgary S, Beyranvand F, Kopaei MR. R eview A rticle A review of plant-based compounds and medicinal plants effective on atherosclerosis. 2004;
[19] Pettersson US, Waldén TB, Carlsson P-O, Jansson L, Phillipson M. Female Mice are Protected against High-Fat Diet-Induced Metabolic Syndrome and Increase the Regulatory T Cell Population in Adipose Tissue. PLoS One [Internet]. 2012;7(9):e46057. Available from: http://dx.plos.org/10.1371/journal.pone.0046057
[20] Hartvigsen K, Binder CJ, Hansen LF, Rafia A, Juliano J, Hörkkö S, et al. A diet-induced hypercholesterolemic murine model to study atherogenesis without obesity and metabolic syndrome. Arterioscler Thromb Vasc Biol. 2007;27(4):878–85.
[21] Otunola GA, Oloyede OB, Oladiji AT, Afolayan AA. Effects of diet-induced hypercholesterolemia on the lipid profile and some enzyme activities in female Wistar rats. African J Biochem Res [Internet]. 2010;4(6):149–54. Available from: http://www.academicjournals.org/journal/AJBR/article-abstract/5D7026911395
[22] Regulatory T Cell Immunity in Atherosclerosis Regulatory T Cell Immunity in Atherosclerosis. 2017;(April).
[23] Tse K, Tse H, Sidney J, Sette A, Ley K. T cells in atherosclerosis. Int Immunol. 2013;25(11):615–22.
[24] Zhou J, Dimayuga PC, Zhao X, Yano J, Lio WM, Trinidad P, et al. Biochemical and Biophysical Research Communications CD8 + CD25 + T cells reduce atherosclerosis in apoE (À / À) mice. Biochem Biophys Res Commun. 2014;443(3):864–70.

The Effect of Sonication Duration on the Characteristics of Nano Hydroxyapatite-Silica (nHAp/SiO_2) Composite and its Mechanical Properties

Hartatiek[1,*], Reri DS[1], Yuanita Amalia Hartanto[1], Nurul Hidayat[1], Yudyanto[1], Joko Utomo[1], Sunaryono[1,2]

[1] Department of Physics, Faculty of Mathematics and Natural Sciences, Universitas Negeri Malang, Jl. Semarang 5 Malang 65145, Indonesia
[2]Research Center of Minerals and Advanced Materials, Faculty of Mathematics and Natural Sciences,Universitas Negeri Malang, Jl. Semarang 5 Malang 65145, Indonesia

*Corresponding author's email: hartatiek.fmipa@um.ac.id

Abstract. In this present study, we report the fabrication of nano-hydroxyapatite/silica (nHAp/SiO_2) composite with the sonochemical method from Bojonegoro natural minerals, i.e. onyx stone and quartz sand. The nano-HAp was synthesized from Ca $(OH)_2$ and H_3PO_4 solutions. The nHAp was successfully composed with silica by using sonication role. The sonication process was done in 1, 2, 3, and 4-hour variations at 20 % wt filler composition of Silica. The crystal structure properties of nHA were evaluated by X-ray diffraction (XRD). The ratio of Ca/P was evaluated by scanning electron microscopy and energy dispersive X-ray spectroscopy (SEM-EDX). The microstructural characterization of the sample was captured by scanning electron microscopy (SEM). Vickers hardness test was also conducted to identify the hardness (mechanical properties) of the sample. On the other hand, the porosity of the sample also was reported. The results of this study show that hydroxyapatite has a size of 30.57 nm and the ratio of Ca/P was 1.57. The variations of sonication affect the mechanical properties of nHA-SiO_2 composite. The increasing sonication duration influences the hardness properties and porosity. The porosity of HAp (31.58 %) is still suitable to be applied for biomedical applications.

Keywords: Nano-hydroxyapatite, silica, hardness, porosity, natural material.

1. Introduction

Hydroxyapatite (HAp) has been extensively investigated and used in the bone clinical application for more than four decades. The increasing interest in HAp is due to its similar chemical composition to that of the inorganic component of natural bone [1] on account of its outstanding properties like biocompatibility, bioactivity, osteocoductivity, non-toxicity and non-inflammatory nature [2].

Synthetic HAp is often stochiometric with a chemical formula of $Ca_{10}(PO_4)_6(OH)_2$, and *a* specific atomic Ca/P molar ratio is 1.67. HAp is generally highly crystalline with the lattice parameter of a= 0.95 nm and c = 0.68 nm and display a hexagonal symmetry typically (S.G. $P6_3/m$) with the preferred orientation along the *c* axis [3]. HAp exists as nanocrystals with dimensions of 4 x 50 x 50 nm whereby the nanocrystals are embedded in organic collagen fiber matrix which comprises 90 % of the protein content [4]. HAp is manufactured in many forms that can be prepared as a dense ceramic [5], powder [6], ceramic coating [7] and porous ceramic [8] as required for particular applications.

The chemical composition, crystallinity, size, and morphology of the HAp crystals and their aggregates play critical roles in determining their properties and potential applications [9-13]. Nano-hydroxyapatite (nHAp) crystals possess excellent sintering ability due to their high surface energy [14]. HAp nano-ceramics with enhanced mechanical properties can be fabricated using HAp nano-powder as raw material [15].

Even though HAp is biocompatible, the mechanical properties of HAp such as brittleness, low tensile strength, and poor impact resistance have restricted its compatibility in load-bearing applications [16]. This study proposes two methods to improve the mechanical properties of HAp. First, HAp was fabricated in nano scale because the nano-sized ceramic is bigger than the micro-sized one [17]. Moreover, Hap nano-bioceramics exhibit better bioactivity than those in micro-scale sizes [15]. Second, nHAp is composed of other materials that have better mechanical properties of silica. Silica (SiO_2) is biocompatible, non-toxic and has good mechanical properties, i.e. strong, hard and tough. In addition, silica also offers the advantages of its ability to increase the density of the structure and to exhibit excellent hardness [18]. Silica results in stronger and harder composites [19] so silica is selected to be a composite material with Hap. In this study, the nHAp-SiO_2 composites were synthesized through the sonochemical method by varying the duration of sonication to examine their effect on mechanical properties.

2. Materials and Methods

2.1. Synthesis of Nano-Hydroxyapatite from Onyx Stone
Onyx stone from Bojonegoro East Java was selected as a source of calcium to produce HAp. Onyx stone was grounded and sieved, and then it was used further in the laboratory. The powder was calcined at 500 °C for 1 h to obtain CaO powder. The CaO was then dissolved in distilled water to obtain $Ca(OH)_2$. The solution was stirred at a rate of 700 rpm and 50 °C for 60 mins to be subsequently incorporated in an ultrasonic bath and disonicated at 50 °C for 60 mins. During the course of the sonication, H_3PO_4 6M was dripped. NH_4OH was dripped to control the pH of the solution (9 ~ 10). The produced HAp was then filtered and dried at 100 °C for 24 h to remove H_2O, and then crushed again to obtain homogeneous powder. The powder was characterized by XRD, SEM, and EDX to identify the phase, surface morphology and Ca/P ratio, respectively. Vickers hardness test and porosity were measured to determine the mechanical properties (hardness) and porosity of the samples.

2.2. Synthesis of nHA/SiO2 Composite with Sonochemical Methods
The powder of nHAp and silica was dissolved in DI water. The solution was stirred at a rate of 1000 rpm at 50 °C for 1 h. Furthermore, the solution was incorporated in an ultrasonic bath and was sonicated for various durations of 1, 2, 3 and 4 h at 50 °C. The solution was filtered and dried in a furnace at a temperature of 100 °C for 24 h. The powder was characterized by XRD and SEM-EDX to study the phase and surface morphology of composite, respectively. Vickers hardness test and porosity test were also measured to determine the mechanical properties (hardness) and porosity of nHA/SiO_2 composite.

3. Results and Discussion
The result of XRF test on onyx stone is shown in Table 1. According to the Table 1, the onyx stone contains a dominant element of calcium, which reached 94.33 % of the compound, and other elements such as Ti, Mn, Fe, Cu, Sr and Ba present in a low percentage within the compound.

Table 1. Result of XRF on onyxstone

Compound	Ca (%)	Ti (%)	Mn (%)	Fe (%)	Ni (%)	Cu (%)	Sr (%)	Ba (%)
Conc Unit	94.33	0.09	1.65	0.40	0.939	0.091	2.3	0.2

The XRD pattern of HAp analyzed by Highscores Plus software shows a good match with the crystallographic open database code of AMCSD 0002299. The presence of $Ca_3(PO_4)_2$ or whitlockite impurities in HAp and whitlockite phase shows a ratio of 94.26 % : 5.74 %. This result indicates that $Ca_3(PO_4)_2$ (tri calcium phosphate) is biocompatible and biodegradable [20]. We obtained the crystal size of HAp by using the Scherrer equation of 30.57 nm. The size of these crystals shows a compatibility of the sample with the size of crystals within the human bones of 20-80 nm [21]. Furthermore, the EDX result shows a Ca/P ratio of about 1.57 that represented the non-stoichiometric sample and this corresponds with the apatite bone characteristic.

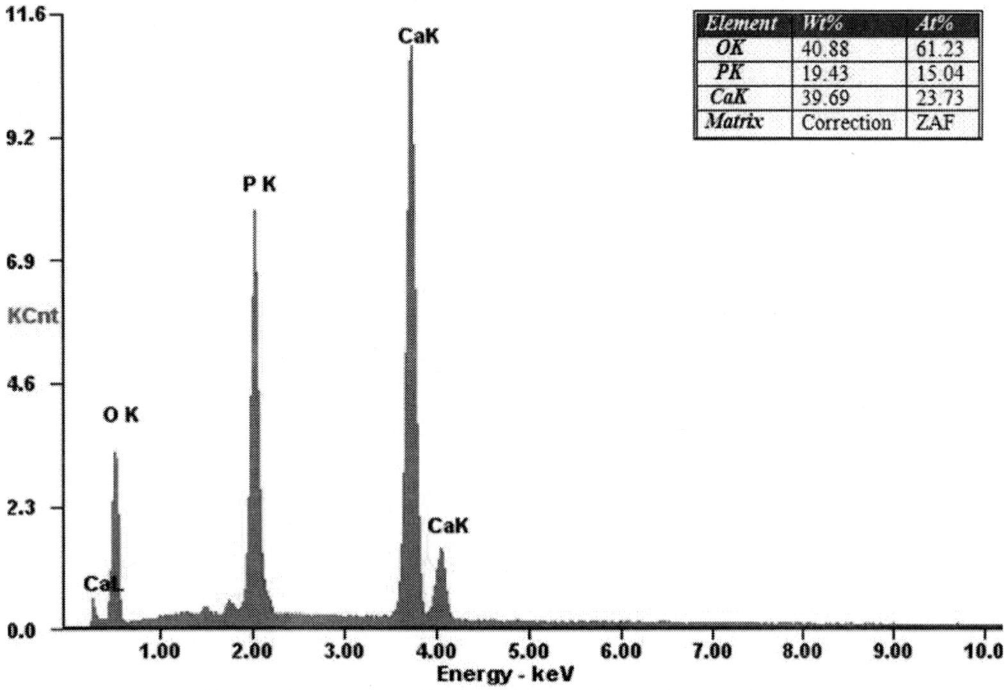

Element	Wt%	At%
OK	40.88	61.23
PK	19.43	15.04
CaK	39.69	23.73
Matrix	Correction	ZAF

Figure 1. SEM-EDX elemental analysis of pure HAp

The XRD pattern for HAp and HAp-SiO_2 composites with various sonication durations can be seen in Figure 2. The PCW analysis for HAp-SiO_2 composite produced the same diffraction pattern as that of the HAp with the 002297 AMCSD code. The peak of SiO_2 is not clearly observed on the XRD pattern of HAp-SiO_2 composite. These results show that silica has amorphous properties. The HAp-SiO_2 composite did not affect the crystal structure of Hap but impacted the intensity of each peak. The high and sharp peak could be attributed to the crystallinity of the samples. The peak intensity and FWHM increased along with the increasing duration of sonication. This implies that the size of the HAp crystals in the composite is decreased while the duration of sonication is prolonged. This result is in line with a previous report [22].

Table 2. The lattice parameters and crystal size of HAp-SiO_2

Duration of sonication	Lattice Parameters (A)			The size of crystal (nm)
	a	*b*	*c*	
1 h	9.4301	9.4301	6.8314	36.35
2 h	9.4352	9.4352	6.8205	30.24
3 h	9.4771	9.4771	6.8513	29.96
4 h	9.4404	9.4404	6.8297	29.72

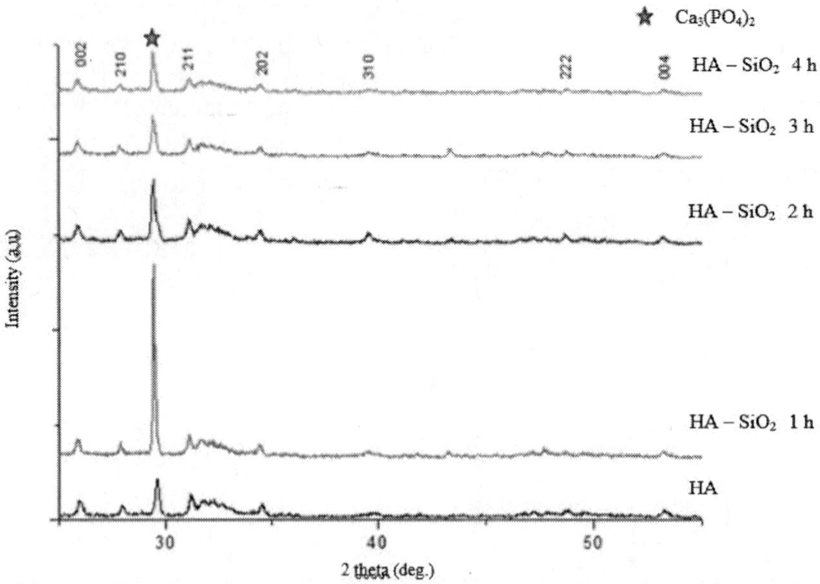

Figure 2. XRD pattern of HAp and HAp-SiO_2 with various sonication durations

Table 2 illustrates the values of lattice parameters and crystal size of nHAp-SiO_2 composite for different sonication durations. According to Table 2, the experimental lattice parameter of nHA-SiO_2 composite is in good agreement with the theoretical lattice parameter, which is $a = 9.432$ A, $c = 6.881$ A [22-24]. On the other hand, Figure 3 shows the hardness of nHAp-SiO_2 composite for various sonication durations. Sonication duration linearly increase along with the increasing hardness of nHAp-SiO_2 composites with a high correlation rate $R^2 = 0.995$.

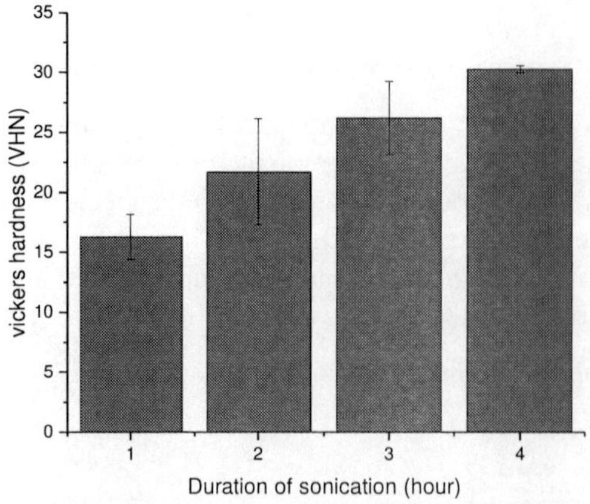

Figure 3. Graph of the hardness of nHA-SiO_2 composite in various sonication durations

IOP Conf. Series: Journal of Physics: Conf. Series **1093** (2018) 012019 doi:10.1088/1742-6596/1093/1/012019

To characterize of hardness of the HA and HA-SiO_2, a material of HA and HA-SiO_2 was synthesized via sonochemical method within ranges of 26-30 HVN or 0.26-0.3 GPa. This value is significantly too high for implant application which requires 3.43 GPa but this is suitable for human tooth dentin application with the hardness standard is 0.25 -0.8 GPa [19]. The different result of hardness of HA and HA-SiO_2 materials might be caused by too low sintering temperature used, i.e. 500 °C for 1-hour detention. Lim et al. [24] reported that the sintering temperature affects the hardness of HAp-Zirconia.

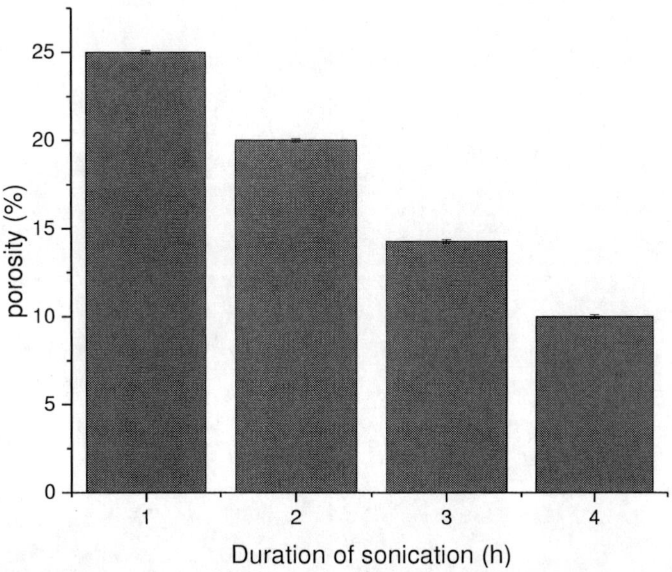

Figure 4. Graph of the porosity of nHA-SiO_2 composite for various sonication durations

Figure 4 shows the effect of sonication duration on the porosity of nHAp-SiO_2 composite. The porosity of nHA-SiO_2 for sonication duration of 1, 2, 3, and 4 hours were 25, 20, 14.29, and 10 %, respectively. The sonication duration linearly decreases the porosity of nHA-SiO_2 composite with the correlation $R^2 = 0.997$. The porosity of nHAp-SiO_2 is lower than the porosity of HAp which is 31.58 %. These results implied that the addition of SiO_2 (20 % of composite weight) as a filler can close the pores of HA. Whereas, the porosity of bioceramic materials for biomedical applications can be tolerated in a range of 10-40 % [25] so the obtained porosity of HAp 31.58 % is still suitable for biomedical applications.

Figure 5 shows the photographs of HA-SiO_2 with the variations of sonication duration, i.e. 1 to 4 hours. Based on Figure 5, all of the samples have high homogeneity. These results confirmed that the porosity of the sample with the 4-hour sonication duration has the highest porosity and is well confirmed by the graphs of the sample porosity illustrated in Fig 4.

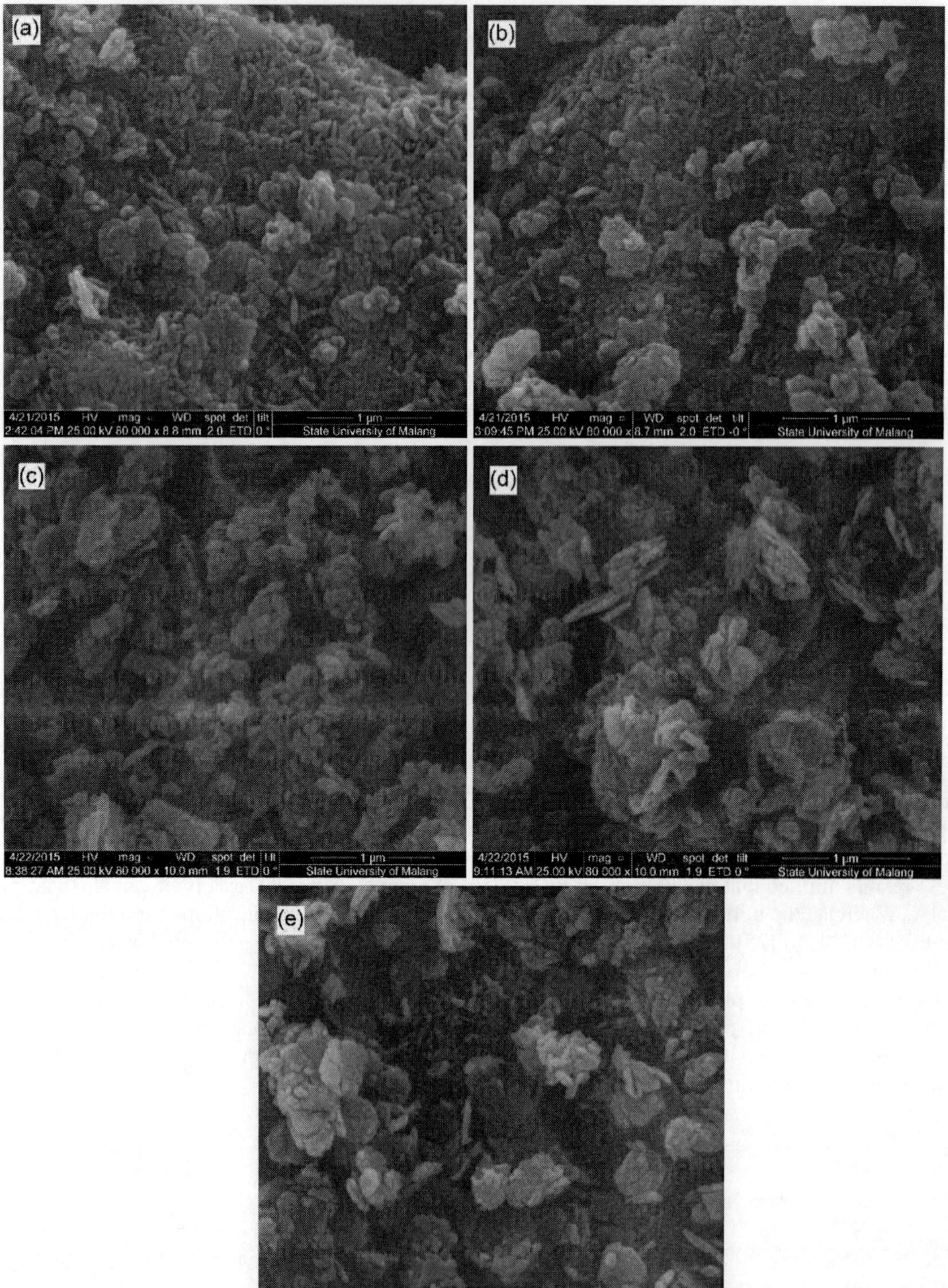

Figure 5. SEM photographs of HA-SiO_2 (sonication duration a=1 h, b=2 h, c=3 h, and d=4 h) and (e) pure HA

4. Conclusion

This work has successfully synthesized nano-hydroxyapatite by using naturally Onyx stone from Bojonegoro East Java as the source of calcium with sonochemistry method. Sonication duration influences the hardness and porosity of nHA-SiO_2. The hardness of nHA-SiO_2 is significantly increased as along with the increasing the sonication duration. On the other hand, the porosity of nHA-SiO_2 is slightly decreased along with the increasing sonication duration. It means that natural materials are potential to synthesize more benefits from biomaterials. Moreover, the porosity of HAp (31.58 %) that shown by this research is suitable biomedical applications.

References

[1] R.A. Ramli , R. Adnan, M.A. Bakar, and S.M. Masudi, Synthesis and characterization of pure nanoporous hydroxyapatite, J Phys Sci 22, 25-37 (2011).
[2] C. Liu ,W. Wang, W. Shen, T. Chen , L. Hu, and Z. Chen, Evaluation of the biocompatibility of a ceramic hydroxyapatite. J Endod 23, 490 (1997).
[3] M. Vallet-Regi, J. Pena, and I. Izquierdo-Barba, Multiclear solid-state NMR studies of ordered mesoporous bioavtive glasses, Solid State Ionics 172, 445-439 (2004).
[4] S. Weiner and H.D. Wagner, Annual Review of Material Science 28, 271-298 (2002)
[5] Yudyanto, Diantoro, M. dan Hartatiek, 2014. Sintesis Nano-Hydroxyapatite Berbasis Batuan Calcite Druju Malang Sebagai Biomaterial Fungsional Pengganti Tulang dengan metode Sonokimia. Proseding Seminar Nasional Fisika di Jurusan Fisika Universitas Negeri Makasar 14 November 2014.
[6] Yudyanto, Diantoro, M. dan Hartatiek, 2014. Pengaruh lama Pengedapan pada Proses Kopresipitasi Sintesis Nano-Hydroxyapatite Berbasis Batuan Calcite Druju terhadap Kristalinitas dan Kekerasan. Proceding Seminar Nasional Fisika (Pertemuan HFI -XXI), 16-17 Oktober 2014, Jurusan Fisika Universitas Udayana Bali.
[7] B. Ben-Nissan, A. Milev, and R. Vago, Morphology of sol-gel derived nano-coated coralline hydroxyapatite. Biomaterials 25, 4971-5(2004).
[8] N.E. Ozgur, and A. Cuneyt Tas. Effort for strength improvement was made on porous HA ceramic in several aspects. J Euro Ceram Soc19, 2569 (1999).
[9] L. Xia, K. Lin, X. Jiang, Y. Xu, and M. Zhang, Enhanced osteogenesis through nano-strutured surface design of macroporous hydroxyapatite bioceramic scaffolds via activation of ERK and p38 MAPK signaling pathways. J.Mater. Chem B1, 5403-5416 (2013).
[10] K. Lin, L. Wei, Z. Zou, W. Zhang, and Y. Qiqn, Biocompatible glasses: from bone regeneration to cancer treatment., Chem. Eng. J. 222, 49-59 (2013)
[11] L. Wu, Y. Dou, K. Lin, W. Zhai, W. Cui, and J. Chang, Hierrachically structured nanocrystalline hydroxyapatite assembled hollow fibers as a promosing protein delivery system. Chem. Commun., 47, 11674-11676 (2011).
[12] N. Zhang, D. Zhai, L. Chen, Z. Zou, K. Lin, and J. Zhang, Phosphate chemical conversi coating on metallic substrates for biomedical application: A review. Mater. Biol. Appl. 37, 281-291 (2014).
[13] Y. Shen, J. Liu, K. Lin, and W. Zhang, Synthesis of strontium substituted hydroxyapatite whiskers used as bioactive and mechanical reinforsement material. Mater. Lett., 70, 76-79 (2012).
[14] K. Lin, L. Chen, and J. Zhang, Hydroxyapatite nanobioceramics wit enhanced mechanical properties via two-step sintering process. Int. J. Appl. Ceram. Technol, 9, 479-485 (2012).
[15] M. Sadat-Shojai, M.T. Khorasani, E. Dinpanah-Khoshdargi, and A. Jamshidi, Sinthesis methods for nanosized hydroxyapatite with diverse structures. Acta Biomater. 9, 7591-7621 (2013)
[16] R.Z. LeGeros, Bioactive bone graft substitute. Clinical orthopaedic and Related Research 395, 81-90 (2002).
[17] J. Wang, and L.L. Shaw, Nanocrystalline hydroxyapatite with simultaneous enhancements in hardness ang toughness. Biomaterials, 30, 6565-6572 (2009).
[18] I.A. Rahman, et al., One-pot synthesis of hydroxyapatite-silica nanopowder composite for

hardness enhancement of glass ionomer cement (GIC). Bull. Mater. Sci, 37, 213–219 (2014).
[19] J. Park, 2008. Bioceramics Properties, Characterizations, and Applications. University of Iowa USA :springer.com.
[20] C. Ruiz-Aguilar, E.A. Aguilar-Reyes, M. Flores-Martinez, C.A. Leon-Patino, and R.E. Nunez-Anita, Advances in Applied Ceramic. Accepted. (2017).
[21] G.M. Cunnife, F.J. O'Brien, S. Partap, T.J. Levingstone, K.T. Stanton, and GR. Dickson,
[22] The synthesis and characterization of nanophase hydroxyapatite using a novel dispersant-aided precipitation method .Journal of Biomedical Materials Researc, 95(4):1142-9 9 (2010).
[23] Hartatiek, Yudyanto, R.Y. Windari, S.D. Ratnasari, and N. Hidayat, " Karakteristik Hidroksiapatit Doping Silicon $Si_{0.5}$HA Hasil Reaksi Keadaan Padat dan potensinya sebahai Material Biomedis" presented at the Seminar Fisika dan Pembelajarannya 2016, Malang, 2016.
[24] E. Kusrini et al. Preparation of Hydroxyapatite from Bovine Bone by Combination Methods of Ultrasonic and Spray Drying. InternationalConference on Chemical, Bio-Chemical and Environmental Sciences.(ICBEE'2012).
[25] K.F. Lim et al. Asian Journal of Scientific Risearch. 7 (4) : 609-615 (2014)
[26] M.W. Lukman, Yudyanto dan Hartatiek. 2013. SintesisBiomaterial Komposit CaO-SiO_2 Berbasis Material Alam(Batu Kapur dan Pasir Kuarsa) dengan Variasi Suhu Pemanasan dan PengaruhnyaTerhadap Porositas, Kekerasan dan Mikrostruktur. Tesis. Malang : Jurusan Fisika FMIPA Universitas Negeri Malang.

Effect of Template on Structural and Band Gap Behaviors of Magnetite Nanoparticles

Muzammil[1,2], Ahmad Taufiq[1,*], Defi Yuliantika[1], Yuanita Amalia Hariyanto[1], Sunaryono[1], Arif Hidayat[1], Syamsul Bahtiar[1], Nandang Mufti[1], Nurul Hidayat[1]

[1]Department of Physics, Faculty of Mathematics and Natural Sciences, Universitas Negeri Malang, Jl. Semarang 5 Malang 65145, Indonesia
[2]Pondok Pesantren Nurul Jadid, Jl. K.H. Zaini Mun'im Karanganyar Paiton, Probolinggo 67293, Indonesia

*Corresponding author's email: ahmad.taufiq.fmipa@um.ac.id

Abstract. Magnetite nanoparticles (MNPs) were prepared via the coprecipitation-sonochemical route. Polyethylene-glycol as a template was employed with volume variations of magnetite/polyethylene-glycol of 1:0; 2:1; and 1:1. The structural and band gap characters of the samples were studied using XRD, SEM, FTIR spectrometer, and UV-Vis spectrometer. The data analysis presented that the magnetite formed spinel structure with particle size ranging from 20 to 23 nm. This result was confirmed by SEM characterization showing a particle size of lower than 25 nm in a spherical form. The FTIR spectra presented that the samples had vibrational peaks originating from the magnetite particles at wavenumbers in the range of 445 to 600 cm^{-1}. Furthermore, the data of the UV-Vis characterization showed band gap values of 2.52- 2.61 eV for respective samples. It can be concluded that the polyethylene-glycol was effective to produce the magnetite nanoparticles via the coprecipitation-sonochemical route.

Keywords: Magnetite, template, nanoparticle, spinel, coprecipitation-sonochemical route.

1. Introduction

Magnetite is a material with a unique magnetic characteristic since its bulk form has a ferrimagnetic property while its reduced size to a nanometric order would have a superparamagnetic property [1]. The superparamagnetic phenomenon of magnetite could be seen when the particle is in a single domain. Hence, the morphology and particle size of magnetite could be controlled by a synthesis method that is appropriate for specific applications.

To date, several synthesis methods for producing magnetite nanoparticles have been developed. For example, ball milling [2], hydrothermal [3], coprecipitation [4], microwave [5], sonochemical [6] method, and so on. Each of the developed methods has its advantages and disadvantages. The previous research showed that the coprecipitation-sonochemical method, which is a combination of coprecipitation and sonochemical methods, is useful for reducing magnetite particle size until it reaches below 10 nm and it has a single domain character [7].

A single-domain particle shows a high magnetic response when induced by an external magnetic field. However, the particle's tendency to agglomerate would be higher due to the van der Walls force among the magnetic particles. Therefore, it is necessary to employ a method to control the particle size such as using a polymer-based template. One of the polymers that could be utilized to form while controlling particle size and distribution is polyethylene-glycol [8]. In this case, polyethylene-glycol functions to prevent the formation of a more extensive particle aggregation to enable the production of uniform magnetite particles. Nonetheless, right molecule length and number are necessary for the polyethylene-glycol to function.

In their research, Radoń *et al.* utilized polyethylene-glycol polymer, PVP, and dextrin to form uniform and water-soluble magnetite particles and citrate as a template to form a soluble surface in an organic acid solution [9]. The polyethylene-glycol polymer is biocompatible, and an active layer that has a hydrophilic property contained in the particle surface enables the particle to disperse in water efficiently [10]. Thus, magnetite particle layering with polyethylene-glycol is excellent for further applications such as water-based ferrofluid fabrication for various biomedical applications. The research trend in the past one year presented several uses of polyethylene-glycol in producing biocompatible particles for biomedical applications such as preclinical imaging [11], MRI and cancer ablation [12], drug delivery [13], MRI contrast agent [14], photocatalyst [15], and so forth.

Based on the explanation above, the magnetite nanoparticles in this research were synthesized via coprecipitation-sonochemical method using polyethylene-glycol as a template. The magnetite nanoparticles with different polyethylene-glycol volumes were characterized by XRD to identify its crystal structure, by FTIR to know its functional group, by SEM to analyze its morphology, size, and particle distribution, and by UV-VIS to identify its structural and energy band gap (*Eg*).

2. Materials and Methods

Iron sands extracted by a permanent magnet were dissolved in HCl through a stirring process using a magnetic stirrer. The solution was then filtered to produce $FeCl_2$ and $FeCl_3$ contents. Such processes were in accordance with our previous studies [16–18]. The polyethylene-glycol solutions in water with 0.75 and 15 ml compositions were added into the $FeCl_2$ and $FeCl_3$ solutions while stirred by using a magnetic stirrer and wisely dropped by NH_4OH. After that, the solution produced by such coprecipitation method was sonicated at 40 kHz frequency for 1 hour to produce black sediment. Next, the sediment was washed with distilled water until it reached a neutral pH and continued by the final process of sediment heating at 100 °C for 1 hour in order to obtain magnetite particles.

3. Results and Discussion

Figure 1 exhibits the XRD patterns of magnetite nanoparticles with polyethylene-glycol composition variations. The data identification shows some diffraction peaks at the angle of 2θ namely 30.2895°; 35.7527°; 43.4787°; 53.7999°; and 57.4081°, respectively in accordance with the *hkl* fields of (220), (311), (422), (511), and (440) based on AMCSD model code 0002402. The *hkl* field indicates that the sample has a cubic-spinel structure as reported by Baltazar *et al.* [19]. The XRD patterns obtained in this research also provide information of the magnetite nanoparticle crystallinity and the absence of a new phase that can cause impurities. It evidenced that magnetite nanoparticles with a high level of purity have been successfully synthesized through coprecipitation-sonochemical method. The highest intensity peaks of all samples were found at (311) *hkl* field. A quantitative analysis of the (311) *hkl* field via Scherer equation produced particle sizes ranging from 21.3 – 25.7 nm. The diffraction peak intensity of the crystal field qualitatively related to the crystallinity level of the particles. An addition in the polyethylene-glycol volume resulted in an increasing number of polyethylene-glycol layers on the magnetite surface. Thus, the peak intensity value of the (311) *hkl* field was decreasing since polyethylene-glycol is a polymer with an amorphous characteristic [20].

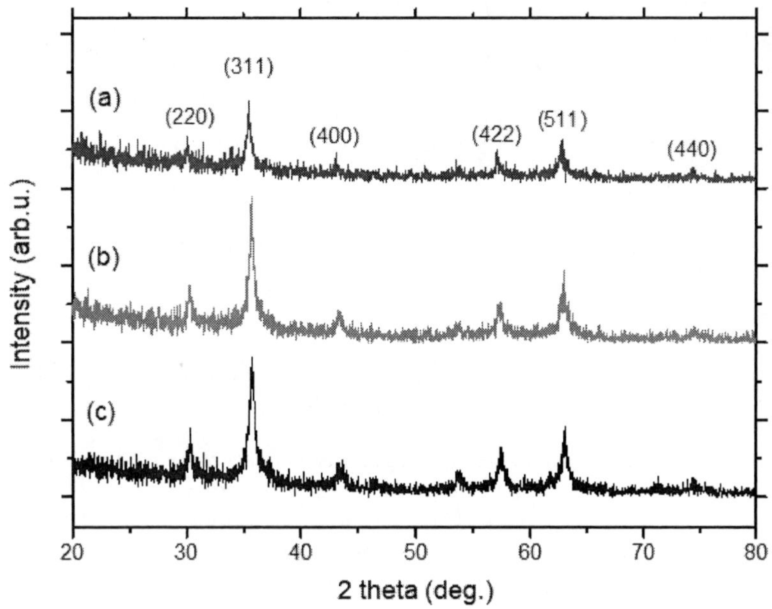

Figure 1. XRD patterns of the magnetite particles with polyethylene-glycol content of (a) 0 ml, (b) 7.5 ml, and (c) 15 ml

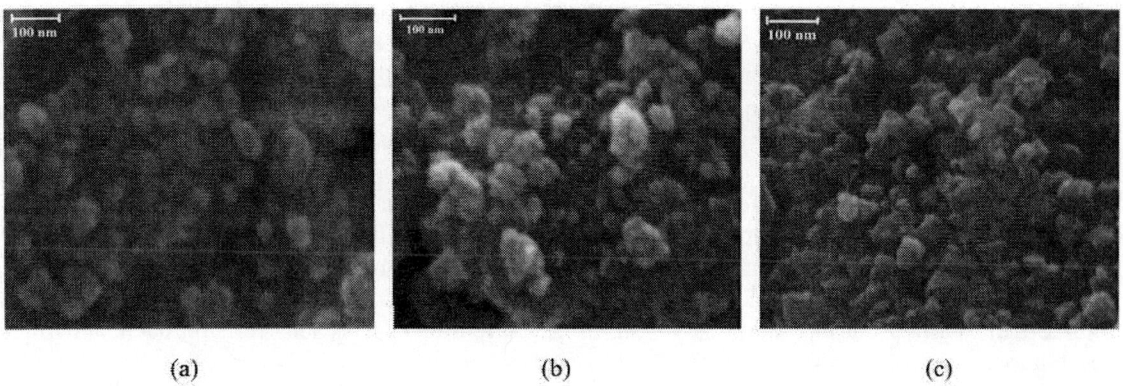

Figure 2. SEM images of the magnetite particles with polyethylene-glycol content of (a) 0 ml, (b) 7.5 ml, and (c) 15 ml

Figure 2 provides information about the particle morphology based on the SEM test. The qualitative analysis via SEM image observation showed that the particles have spherical forms with different shapes and sizes in accordance with the polyethylene-glycol variations. The SEM test results were analyzed using *ImageJ* software resulting in different particle sizes ranging from 22.5–27.3 nm of all polyethylene-glycol variations. The morphology of all of the samples is spherical resulting from the nucleation rate per unit between isotopes on the surface of the magnetite nanoparticles.

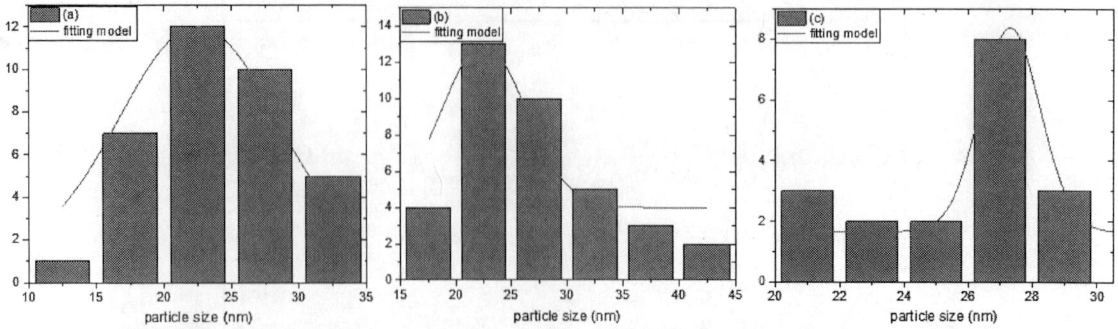

(a) (b) (c)

Figure 3. Particle size distribution of the magnetite particles with polyethylene-glycol content of (a) 0 ml, (b) 7.5 ml, and (c) 15 ml

Complete information about the particle size homogeneity is analyzed based on the histogram fitting of particle size distribution in Figure 3. Such histogram fitting provides information of the magnetite particle size distribution with polyethylene-glycol volume variations shown by the value of the inhomogeneity level obtained from the size deviation towards the average value in percentage unit [21]. It is known that the level of particle size inhomogeneity decreased along with the addition of polyethylene-glycol volume. It was possible due to the polyethylene-glycol percentage that significantly affects the polyethylene-glycol chain's ability to bind and to cover the magnetite nanoparticles [22]. Hence, the chain ability to wrap would be higher, and the homogeneity of the particle size and shape will be uniform. The average size was smaller compared with the previous research where the polyethylene-glycol-200 produced 128 nm particle size [23]. Meanwhile, a single reaction polyol method employed on 80 ml polyethylene-glycol produced magnetite nanoparticles in the size of 32.3 nm [24]. It indicated that the employment of the coprecipitation-sonochemical synthesis method via polyethylene-glycol-1000 as the template has successfully produced smaller-sized particles and an addition of polyethylene-glycol composition resulted in more uniform magnetite particle sizes.

Table 1 presents different particle sizes based on the results of XRD and SEM analysis. The particle size analysis using Scherrer approach resulted in a lower value than SEM. It was due to the inhomogeneity strain and instrumental effect that have an insignificant effect on the value of the diffraction peaks during the calculation using Scherrer's equation [25] and the thin polyethylene-glycol layer on the particle surface resulted in different particle sizes via Scherrer approach and direct observation.

Table 1. Fitting results of magnetite nanoparticle sizes.

Sample	Particle size (nm)		Inhomogeneity level (%)
	Scherrer equation	SEM	
Polyethylene-glycol (0 ml)	21.3	22.5 ± 1.9	8.4
Polyethylene-glycol (7.5 ml)	23.6	22.5 ± 1.8	8.0
Polyethylene-glycol (15 ml)	25.7	27.3 ± 0.8	2.9

The functional group of the magnetite nanoparticles with polyethylene-glycol variations was characterized by using FTIR via electromagnetic wave radiation ranging from 400 – 4000 cm^{-1} as shown in Figure 4. The functional group of the metal oxide at the octahedral position refers to the diffraction peak with a value of ~445 cm^{-1} [26]. Meanwhile, the bonds at the wavenumbers of ~587 cm^{-1} and ~1290 cm^{-1} were metal oxide bonds in a tetrahedral position [27]. There was a large reduction of the peak at the wavenumbers of ~1620 cm^{-1} and ~3460 cm^{-1} that were bending and stretching bonds from the hydroxyl [28]. A very weak spectrum was found at ~ 2350 cm^{-1} was $-CH_2$ stretching vibration group by polyethylene-glycol [10]. The results of this research showed that the polyethylene-glycol wrapped the magnetite particles although it was heated at a temperature of 100 °C. Similar results were shown by the research conducted by Jayanti *et al.* stating that the polyethylene-glycol layer stayed on the surface of the magnetite particles after being heated up to 120 °C [10].

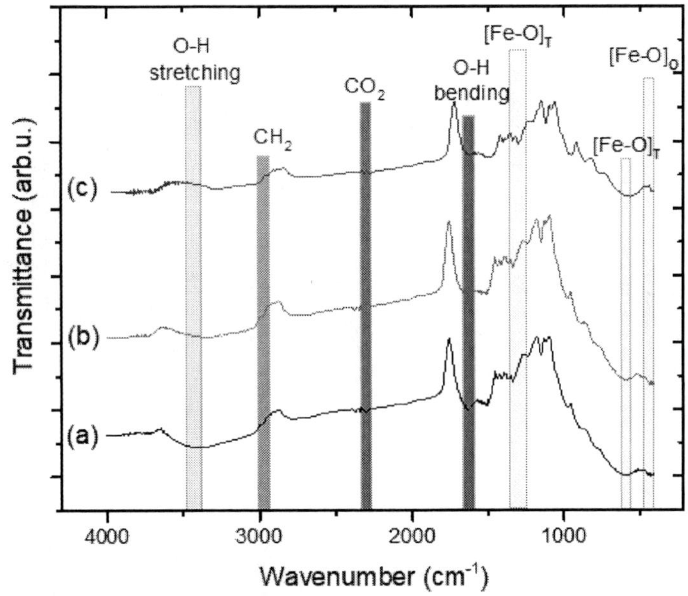

Figure 4. FTIR spectra of the magnetite particles with polyethylene-glycol contents of (a) 0 ml, (b) 7.5 ml, and (c) 15 ml

Figure 5 presents the UV spectrum patterns from the UV-Vis analysis on the magnetite nanoparticles. The absorption spectrum of the magnetite nanoparticles was measured on the particle sizes ranging from 420-800 nm. Hence, the magnetite nanoparticle spectra peak was only found at the wavelength ranging 430-550 nm. The UV-Vis spectroscopy provided information about the band gap, particle size distribution, and absorption spectrum so that the absorption spectrum shows the optical property and band gap of the magnetite nanoparticles. Figure 5 illustrates that the spectra peak moved to a higher wavelength along with the increasing polyethylene-glycol concentration used. Lambert-Beer law was employed to identify the absorbance value. According to the Lambert-Beer law, the absorbance value of a dissolved material is directly proportional to the path length passed by the monochromatic light and concentration of absorbing molecules [30]. Hence, the absorbance value showed the maximum wavelength that could be absorbed by the material. The higher the absorbance value of the magnetite nanoparticles, the higher the monochromatic wavelength that could be absorbed [27].

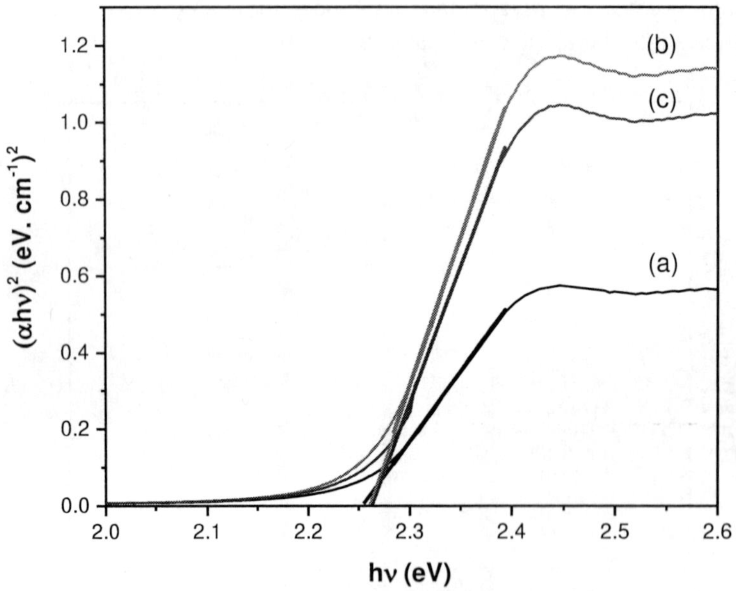

Figure 5. Band gap of the magnetite particles with polyethylene-glycol content of (a) 0 ml, (b) 7.5 ml, and (c) 15 ml

UV-Vis characterization results were analyzed using the Tauc plot model to identify the direct band gap value of the samples. It is possible since the given treatment is not in the forms of doping or annealing process that could change the crystal structure and lattice strain value [31]. The use of polyethylene-glycol as a template on magnetite particles has an insignificant effect on the direct band gap values changes, namely starting from 2.52, 2.57, and 2.61 eV for 0, 7.5, and 15 ml polyethylene-glycol concentrations. In general, such changes in values could relate to the magnetite particle size. A similar tendency was shown in the research performed by Deotale *et al.* stating that both the direct and indirect band gap values increased along with the increase in particle size [32]. Meanwhile, the same energy gap was presented in the results of a study conducted by Gopal *et al.* that employed polyethylene-glycol molarity variations where the optical linear character of the material shown by the energy gap value did not show any significant changes [33].

4. Conclusion

The magnetite nanoparticles synthesis via coprecipitation-sonochemical method assisted by polyethylene-glycol as the template has been successfully conducted. With variations of polyethylene-glycol additions, the magnetite nanoparticles have uniform shapes. The additions of polyethylene-glycol compositions result in a larger magnetite particle size and higher homogeneity level of the particle. The crystallinity of the magnetite particles decreased due to the influence of polyethylene-glycol polymer that has an amorphous characteristic. The functional group of the metal oxide was shown at the absorption peak of ~445 cm^{-1}, and metal oxide bonds in a tetrahedral position were identified at the wavelengths of ~587 cm^{-1} and ~1290 cm^{-1}. Further, the optical characteristic shows that the polyethylene-glycol additions enhanced the absorption value of the magnetite nanoparticles, but it could not significantly change the direct band gap value of approximately 2.52 – 2.61 eV.

Acknowledgments
DRPM KEMENRISTEKDIKTI RI supported this work via "*Penelitian Strategis Nasional*" for AT.

References

[1] I. Karimzadeh, M. Aghazadeh, T. Doroudi, M.R. Ganjali, P.H. Kolivand, Superparamagnetic Iron Oxide (Fe3O4) Nanoparticles Coated with PEG/PEI for Biomedical Applications: A Facile and Scalable Preparation Route Based on the Cathodic Electrochemical Deposition Method, Adv. Phys. Chem. 2017 (2017) 1–7. doi:10.1155/2017/9437487.

[2] D. Chen, S. Ni, Z. Chen, Synthesis of Fe3O4 nanoparticles by wet milling iron powder in a planetary ball mill, China Particuology. 5 (2007) 357–358. doi:10.1016/j.cpart.2007.05.005.

[3] A. Hu, X. Chen, Q. Tang, B. Zeng, Hydrothermal controlled synthesis of Fe3O4 nanorods/graphene nanocomposite for high-performance lithium ion batteries, Ceram. Int. 40 (2014) 14713–14725. doi:10.1016/j.ceramint.2014.06.060.

[4] T. Xia, X. Xu, J. Wang, C. Xu, F. Meng, Z. Shi, J. Lian, J.-M. Bassat, Facile complex-coprecipitation synthesis of mesoporous Fe3O4 nanocages and their high lithium storage capacity as anode material for lithium-ion batteries, Electrochimica Acta. 160 (2015) 114–122. doi:10.1016/j.electacta.2015.02.017.

[5] Y. Wang, Z. Peng, W. Jiang, Controlled synthesis of Fe3O4@SnO2/RGO nanocomposite for microwave absorption enhancement, Ceram. Int. 42 (2016) 10682–10689. doi:10.1016/j.ceramint.2016.03.180.

[6] S. Wu, A. Sun, F. Zhai, J. Wang, W. Xu, Q. Zhang, A.A. Volinsky, Fe3O4 magnetic nanoparticles synthesis from tailings by ultrasonic chemical co-precipitation, Mater. Lett. 65 (2011) 1882–1884. doi:10.1016/j.matlet.2011.03.065.

[7] A. Taufiq, S. Bahtiar, N. Hidayat, A. Hidayat, N. Mufti, M. Diantoro, A. Fuad, R. Rahmawati, W.A. Adi, S. Pratapa, Preparation of Superparamagnetic Zn0.5Mn0.5Fe2O4 Particle by Coprecipitation-Sonochemical Method for Radar Absorbing Material, in: IOP Conf. Ser. Mater. Sci. Eng., IOP Publishing, 2017: p. 012024.

[8] S.-K. Na, S.-A. Jung, H.-W. Yun, J. Ryu, M.-J. Lee, E.M. Song, S.-E. Kim, K.-N. Shim, H.-K. Jung, T.H. Kim, S.Y. Yi, K. Yoo, I.H. Moon, Sa1395 A Randomized Controlled Trial Comparing 2 L-Polyethylene Glycol Solution Containing Ascorbic Acid With 2L-Polyethylene Glycol Plus Bisacodyl Tablets for Colonoscopy Bowel Preparation, DDW 2014ASGE Program Abstr. 2014ASGE Program Abstr. 79 (2014) AB196. doi:10.1016/j.gie.2014.05.014.

[9] A. Radoń, A. Drygała, Ł. Hawełek, D. Łukowiec, Structure and optical properties of Fe3O4 nanoparticles synthesized by co-precipitation method with different organic modifiers, Mater. Charact. 131 (2017) 148–156.

[10] S.A. Jayanthi, D. Sukanya, A.J.A. Pragasam, P. Sagayaraj, The influence of PEG 20,000 concentration on the size control and magnetic properties of functionalized bio-compatible magnetic nanoparticles, Pharma Chem. 5 (2013) 90–102.

[11] A.P. Khandhar, P. Keselman, S.J. Kemp, R.M. Ferguson, P.W. Goodwill, S.M. Conolly, K.M. Krishnan, Evaluation of PEG-coated iron oxide nanoparticles as blood pool tracers for preclinical magnetic particle imaging, Nanoscale. 9 (2017) 1299–1306.

[12] Z. Wang, R. Qiao, N. Tang, Z. Lu, H. Wang, Z. Zhang, X. Xue, Z. Huang, S. Zhang, G. Zhang, Active targeting theranostic iron oxide nanoparticles for MRI and magnetic resonance-guided focused ultrasound ablation of lung cancer, Biomaterials. 127 (2017) 25–35.

[13] S. Kumar, V.K. Meena, P.P. Hazari, R.K. Sharma, PEG coated and doxorubicin loaded multimodal Gadolinium oxide nanoparticles for simultaneous drug delivery and imaging applications, Int. J. Pharm. 527 (2017) 142–150.

[14] F. Xiong, K. Hu, H. Yu, L. Zhou, L. Song, Y. Zhang, X. Shan, J. Liu, N. Gu, A Functional Iron Oxide Nanoparticles Modified with PLA-PEG-DG as Tumor-Targeted MRI Contrast Agent, Pharm. Res. (2017) 1–10.

[15] S. Shamaila, T. Bano, A.K.L. Sajjad, Efficient visible light magnetic modified iron oxide photocatalysts, Ceram. Int. 43 (2017) 14672–14677.

[16] A. Taufiq, Sunaryono, N. Hidayat, A. Hidayat, E.G.R. Putra, A. Okazawa, I. Watanabe, N.

Kojima, S. Pratapa, Darminto, Studies on Nanostructure and Magnetic Behaviors of Mn-Doped Black Iron Oxide Magnetic Fluids Synthesized from Iron Sand, Nano. 12 (2017) 1750110. doi:10.1142/S1793292017501107.

[17] A. Taufiq, Sunaryono, E.G. Rachman Putra, A. Okazawa, I. Watanabe, N. Kojima, S. Pratapa, Darminto, Nanoscale Clustering and Magnetic Properties of MnxFe3-xO4 Particles Prepared from Natural Magnetite, J. Supercond. Nov. Magn. 28 (2015) 2855–2863. doi:10.1007/s10948-015-3111-9.

[18] Sunaryono, A. Taufiq, E.G.R. Putra, A. Okazawa, I. Watanabe, N. Kojima, S. Rugmai, S. Soontaranon, M. Zainuri, Triwikantoro, S. Pratapa, Darminto, Small-Angle X-Ray Scattering Study on PVA/Fe3O4 Magnetic Hydrogels, Nano. 11 (2015) 1650027. doi:10.1142/S1793292016500272.

[19] Á. de Jesús Ruíz-Baltazar, S.Y. Reyes-López, R. Pérez, Magnetic structures synthesized by controlled oxidative etching: Structural characterization and magnetic behavior, Results Phys. 7 (2017) 1828–1832. doi:10.1016/j.rinp.2017.05.001.

[20] M. Khalkhali, K. Rostamizadeh, S. Sadighian, F. Khoeini, M. Naghibi, M. Hamidi, The impact of polymer coatings on magnetite nanoparticles performance as MRI contrast agents: a comparative study, DARU J. Pharm. Sci. 23 (2015) 45.

[21] D.X. Li, W.J. Thomson, Mullite formation kinetics of a single-phase gel, J. Am. Ceram. Soc. 73 (1990) 964–969.

[22] A. Masoudi, H.R.M. Hosseini, M.A. Shokrgozar, R. Ahmadi, M.A. Oghabian, The effect of poly (ethylene glycol) coating on colloidal stability of superparamagnetic iron oxide nanoparticles as potential MRI contrast agent, Int. J. Pharm. 433 (2012) 129–141.

[23] C. Iacovita, R. Stiufiuc, T. Radu, A. Florea, G. Stiufiuc, A. Dutu, S. Mican, R. Tetean, C.M. Lucaciu, Polyethylene glycol-mediated synthesis of cubic iron oxide nanoparticles with high heating power, Nanoscale Res. Lett. 10 (2015) 391.

[24] M. Abbas, B.P. Rao, S.M. Naga, M. Takahashi, C. Kim, Synthesis of high magnetization hydrophilic magnetite (Fe3O4) nanoparticles in single reaction—surfactantless polyol process, Ceram. Int. 39 (2013) 7605–7611.

[25] B.D. Cullity, S.R. Cullity, S.R. Stock, Elements of X-ray Diffraction, 2001.

[26] N. Wiriya, A. Bootchanont, S. Maensiri, E. Swatsitang, Magnetic properties of Zn1−xMnxFe2O4 nanoparticles prepared by hydrothermal method, Microelectron. Eng. 126 (2014) 1–8. doi:10.1016/j.mee.2014.03.044.

[27] J. Zhang, S. Rana, R.S. Srivastava, R.D.K. Misra, On the chemical synthesis and drug delivery response of folate receptor-activated, polyethylene glycol-functionalized magnetite nanoparticles, Acta Biomater. 4 (2008) 40–48.

[28] B. Tang, G. Wang, L. Zhuo, J. Ge, L. Cui, Facile route to α-FeOOH and α-Fe2O3 nanorods and magnetic property of α-Fe2O3 nanorods, Inorg. Chem. 45 (2006) 5196–5200.

[29] A. Gatelytė, D. Jasaitis, A. Beganskienė, A. Kareiva, Sol-Gel Synthesis and Characterization of Selected Transition Metal Nano-Ferrites, 17 (2011).

[30] B. Tanhaei, A. Ayati, M. Lahtinen, M. Sillanpää, Preparation and characterization of a novel chitosan/Al2O3/magnetite nanoparticles composite adsorbent for kinetic, thermodynamic and isotherm studies of Methyl Orange adsorption, Chem. Eng. J. 259 (2015) 1–10.

[31] S.S. Nair, M. Mathews, M.R. Anantharaman, Evidence for blueshift by weak exciton confinement and tuning of bandgap in superparamagnetic nanocomposites, Chem. Phys. Lett. 406 (2005) 398–403.

[32] A.J. Deotale, R.V. Nandedkar, Correlation between Particle Size, Strain and Band Gap of Iron Oxide Nanoparticles, Mater. Today Proc. 3 (2016) 2069–2076.

[33] S.V. Gopal, S. Chitrambalam, I.H. Joe, Coherent source interaction, third-order nonlinear response of synthesized PEG coated magnetite nanoparticles in polyethylene glycol and its application, Opt. Laser Technol. 98 (2018) 84–91.

IOP Conf. Series: Journal of Physics: Conf. Series **1093** (2018) 012021 doi:10.1088/1742-6596/1093/1/012021

Electrolysis Synthesis of MnO_2 in Acidic Environment and Its Electrochemical Performance for Supercapacitor

Mahmudi[1,2], W. Widiyastuti[2], Puspita Nurlilasari[2], Samsudin Affandi[2], Heru Setiawan[2,*]

[1]Department of Chemistry, Faculty of Mathematics and Natural Sciences, Universitas Negeri Malang, Jl. Semarang 5 Malang 65145, Indonesia.
[2]Department of Chemical Engeenering, Faculty of Engeenering, Institut Teknologi Sepuluh Nopember, Jl. Arif Rahman Hakim Surabaya 60111, Indonesia

*Corresponding author's email: sheru@chem-eng.its.ac.id

Abstract. Herein, the simple electrolysis synthesis route to produce α-MnO_2 is reported. The phase and crystal structure were evaluated by means of X-ray diffraction data collection and analysis. The morphological microstructure of as-prepared samples were captured by scanning electron microscope. The diameter distribution was studied by particle size analyzer. The electrochemical performance was systematically investigated by means of cyclic voltametry test. The produced MnO_2 was also prepared to form composite with carbon. The results revealed that the MnO_2/C composites perform outstanding characteristics mainly in terms of their maximum specific capacitance of 922.67 mF/g, with maximum power density of 90.150 W/kg, and maximum energy density of 0.063 Wh/kg. Therefore, the MnO_2/C composite meets the criteria of the prospective applicability for supercapacitor.

Keywords: MnO_2, electrolysis, electrochemistry.

1. Introduction

Among the all diverse crystal morphologies of manganese dioxides (MnO_2), α-MnO_2 and ß-MnO_2 have attracted a great deal of attention due to their (2×2) tunnel structure that provide the incorporation of lithium ions [1]. MnO_2 is considered as vital component for rechargeable lithium batteries and it is regarded as the most appealing catholic material for the near future invention of lithium ion-based batteries. As the implication, various methods have been conducted to produce MnO_2, such as sonochemical [2], electrostatic self-assembly [3], electrodeposition [4,5], chemical bath deposition [6], hydrothermal [7–9], precipitation [10], microwave-assisted reflux [11] , sol-gel, and conventional ceramic or solid state diffusion [12–14]. However, those methods are less effective in terms of cost and time consuming than electrosynthesis method that is introduced in this present study. Previous works that succeeded in producing MnO_2 by electrolysis approach have been reported by [15,16].

The synthesis of MnO_2 nanoparticles can be conducted by electrolyzing $KMnO_4$ solution. In this electrolysis process, the oxidation number of MnO_4^- is +7, which is the highest oxidation number for atom. Hence, the Mn atom in MnO_4^- ion cannot be further oxidized, the only possible change must be reduction. The corresponding chemical reactions for MnO_4^- are

$$MnO_4^- + 3e + 4\,H^+ \rightarrow MnO_2 + 2H_2O \quad E^0 = +1{,}67 \text{ Volt}$$
$$MnO_4^- + 2H_2O + 3e \rightarrow MnO_2 + 4OH^- \quad E^0 = +0{,}588 \text{ Volt}$$

Synthesis of MnO_2 by electrolysis in $KMnO_4$ solution can be explained as follows: when two carbon-based electrodes are immersed in the mixture of $KMnO_4$ and H_2SO_4 solution, and electrical current is given to the electrodes under a DC potential of E_{app} = 2 V, H^+ and K^+ ions will move to cathode as well as MnO_4^- and SO_4^{-2} will move to anode.

The migration of ions occurs during the electrolysis process. An electrolysis will obtain MnO_2, if MnO_4^- ions are controlled on the cathode surface. It is due to the oxidation number of Mn atom (+7) in MnO_4^- ions transforms into Mn (+4) in MnO_2, indicating a reduction process. It requires that the process should take place on the surface of cathode. If MnO_4^- is on the anode surface, MnO_2 will never be produced. Therefore, to synthesize MnO_2 by electrolysis in $KMnO_4$ solution, MnO_4^- ions must be controlled on the cathode surface.

There are three ways that $KMnO_4$ solution in sulfuric acid condition can produce MnO_2. Firstly, the sulfuric acid solution is electrolyzed (E_{app} = 2 V, carbon-based electrodes) for few seconds and $KMnO_4$ solution is dropped wisely on the back part of cathode. If the it is done on the front part of cathode, MnO_2 cannot be formed. Secondly, $KMnO_4$ solution (in a 600 mL glass, completed with the electrodes) is stirred slowly and electrolyzed with E_{app} = 2 V for also few seconds before sulfuric acid is added. Thirdly, electrolysis of $KMnO_4$ was done with nafion membrane between cathode and anode. The membrane is introduced to make MnO_4^- control in cathode. Electrolytes in cathode region are $KMnO_4$ and sulfuric acid solutions, meanwhile electrolyte in anode region is $KMnO_4$ solution only.

From the preliminary study [17], it is revealed that the synthesis of MnO_2 nanoparticles by electrolysizing $KMnO_4$ solution in acidic environment leaded to unpure MnO_2 product. To overcome that problem, it is then introduced a modification process during the electrolysis and the modification will be represented in the next section as well as its discussion.

2. Materials and Methods

The materials that were used in this research were $KMnO_4$ (99.5%, UNI-chem A.R), H_2SO_4 (97%, Merck A.R); aquadest; and electrodes (carbon). 5 gram $KMnO_4$ in 400 mL aquadest was stirred and carbon-based electrodes was immersed into it with DC applied potential of 2 Volt. Electrolysis for 30 mins was done to carbon-based electrodes (with dimension of 5 x 4 x 0,3 cm^3). The immersed electrode was 2 cm and the cathode-anode separation was also 2 cm. Depth of the immersed electrode was changed by introducing 100 mL H_2SO_4 to $KMnO_4$ solution until the final depth was 2.5 cm. The produced precipitate was well washed and dried at 150 °C. X-ray diffraction (Philips: Expert Pro) measurement was conducted to reveal the phase and crystal structure of the sample. The transmission electron microscope (JEOL JEM TEM 1400) was performed to study the crystalline properties; particle size analyzer was used to characterize the particle size.

3. Results and Discussion

Figure 1 represents the XRD characteristics of the obtained MnO_2 with varying electrolysis potential. It is clearly revealed that the all XRD profiles are in a good agreement with α-MnO_2 (JCPDS card no. 44-0140) showing the same diffraction peaks at around 12, 37, and 53. An additional peak is found at diffraction angle of 26° which belongs to carbon from the anode.

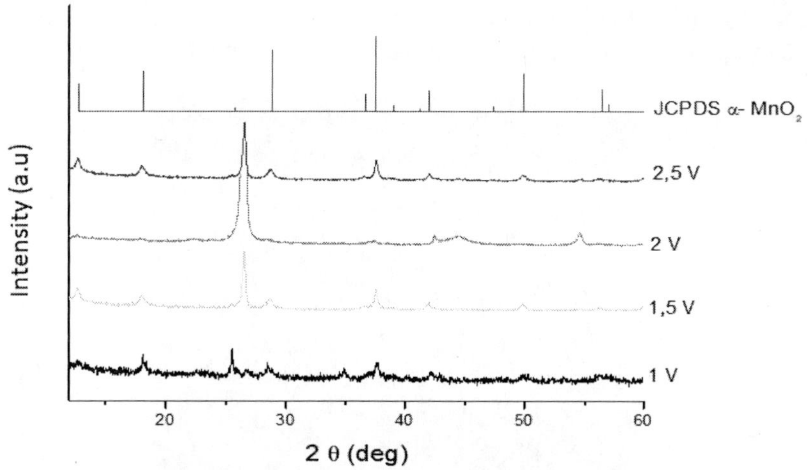

Figure 1. XRD Profiles for MnO_2 Produced by Electrolysis in 0.01 M $KmnO_4$ with H_2SO_4 Addition.

The particle size of MnO_2 after particle size analyzer (PSA) test is given in Figure 2 (a-b). It is found that the average particle diameter of the synthesized MnO_2 is 493.22 nm. The sample with 2.0 V applied electrolysis potential (voltage) shows uniform distribution of particle size being compared with other potentials; 1.0 V, 1.5 V, or 2.5 V. However, there are two peaks of the particle diameter for MnO_2 with 2 V potential (voltage).

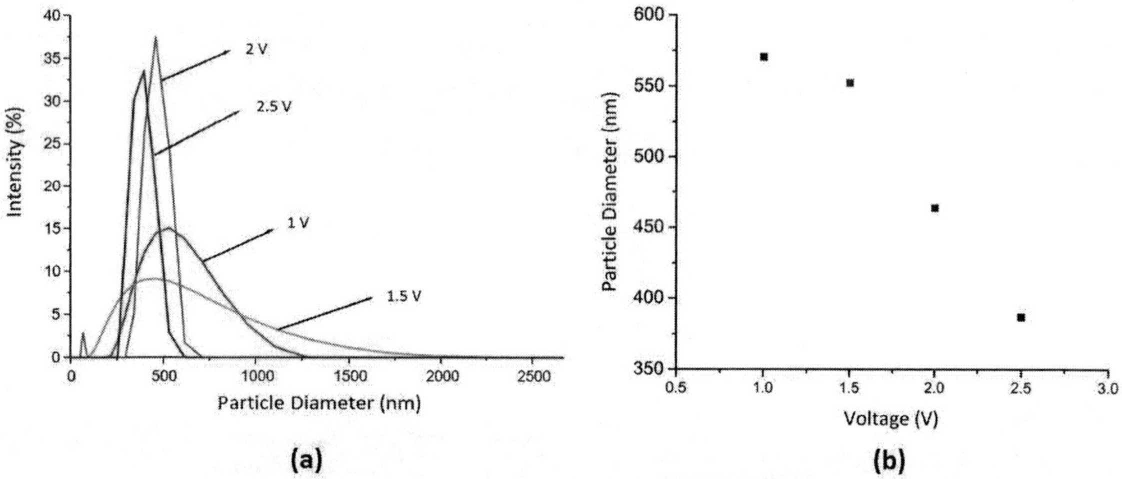

Figure 2. (a) Distribution of MnO_2 Particle Diameter and (b) MnO_2 Particle Diameter with Varying Applied Electrolysis Potential (voltage)

The PSA data is supported by SEM photography, in which the electrolysis-synthesized MnO_2 has rod-like uniform morphology and uniform size as depicted in Figure 3.

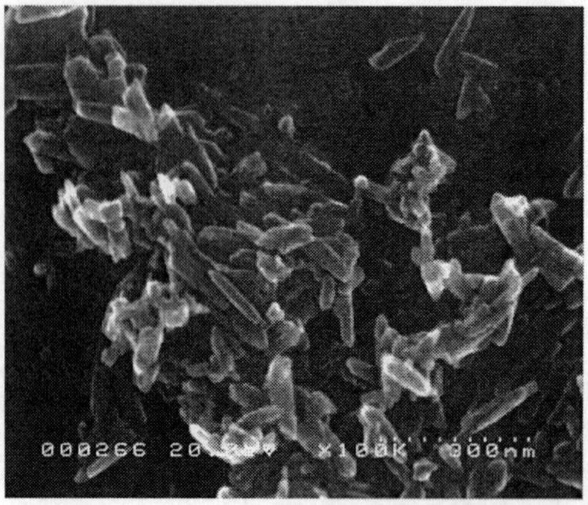

Figure 3. SEM Photography of the MnO_2 Produced by Electrolysis in 0.01 M $KMnO_4$ and Addition of 2 M H_2SO_4 with 2 V Applied Potential.

The PSA data can be compared with the SEM, particularly for sample with 2 V potential. After the comparison, the congruity in terms of particles size characteristic is detected. The rod-like morphology of the MnO_2 imply two kinds of particles size, i.e. (1) the length of the rod and (2) the diameter of the rod. This is what causes the presence of two peaks whe PSA test was conducted.

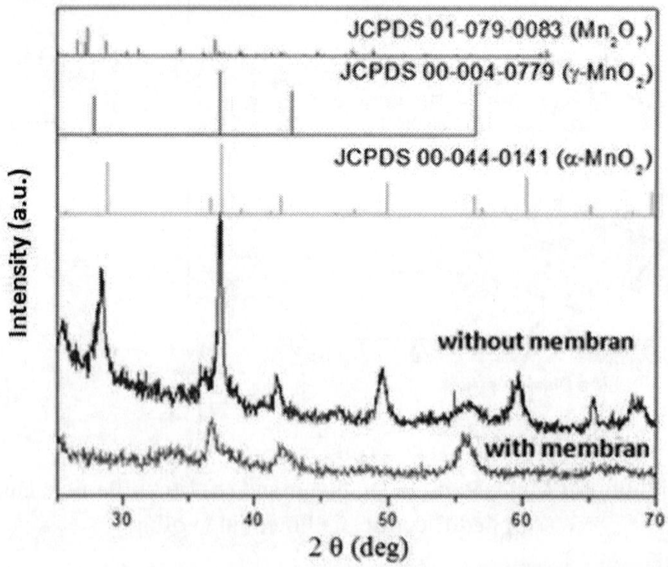

Figure 4. XRD Patterns for MnO_2 Produced by Electrolysis with and without Nafion Membrane.

Synthesis of MnO_2 from electrolysis in $KMnO_4$ solution without stiring the electrolyte was completed by placing nafion membrane between the two electrodes. The effect of the use of the nafion membrane can be observed from the XRD data of the produced samples as given in Figure 4.

The addition of the nafion membrane is to increase the productivity of the MnO2 synthesis. Further analysis by PSA shows the different characteristics MnO_2 particles size due to the nafion membrane. As we can see in Figure 5, the size distribution of the MnO_2 with nafion membrane shows sharper peak than that of without nafion membrane. It indicates the addition of nafion membrane may give rise to particle size uniformity.

The average diameter size of MnO_2 with and without addition of nafion membrane are, respectively, (472.5 ± 130.3) and (362.2 ± 105.1) nm. This result also informs that the MnO_2 with addition of nafion membrane is bigger than that without nafion membrane and it is probably because of the agglomeration is easier to happen by the presence of the nafion membrane.

Furthermore, cyclic voltametry (CV) test was also employed. CV is regarded as an ideal method to investigate the electrochemical performance of material. The CV test was conducted to the MnO_2/C composites as various compositions using scan rate of 10 mV/s and potential range of 0 V to 0.7 V in 0.1 M Na_2SO_3 electrolyte. 1:2 ratio of MnO_2:C is found as the best composition in terms of the electrochemical performance because it has maximum electrical characteristic, see Figure 6.

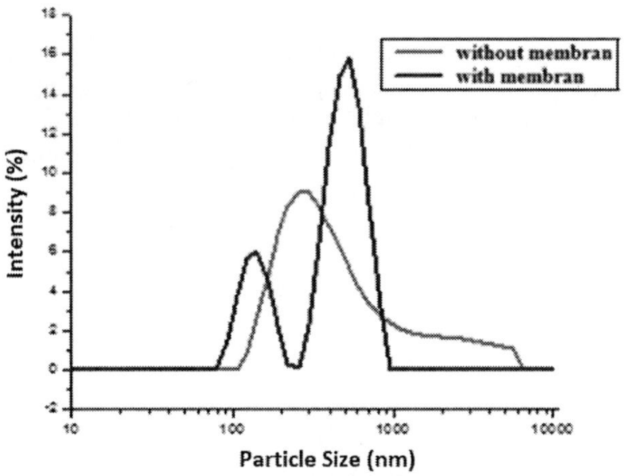

Figure 5. PSA Patterns for MnO_2 Produced by Electrolysis with and without Nafion Membrane.

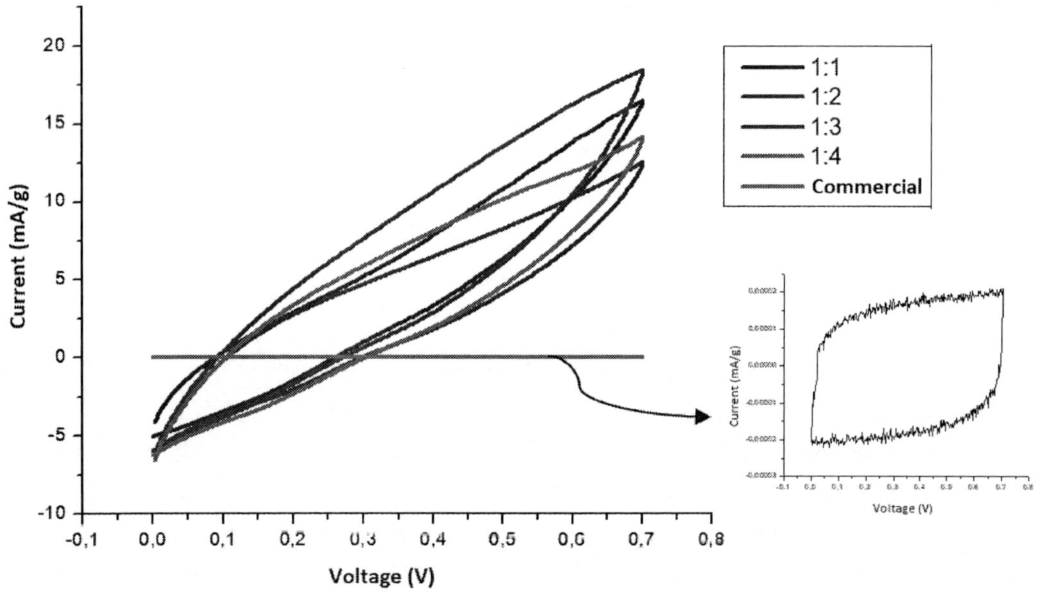

Figure 6. CV Curves for MnO_2:C Composites ratio of 1:1, 1:2, 1:3, and 1:4 at Scan Rate of 10 mV/s in 0.1 M Na_2SO_3 Electrolyte.

IOP Conf. Series: Journal of Physics: Conf. Series **1093** (2018) 012021 doi:10.1088/1742-6596/1093/1/012021

The specific capacitance of the measured sample was calculated by the following mathematical expression:

$$C_{sp} = \frac{i}{sm}$$

Where i is the average cathodic current (mA), s is the scan rate (V/s), and m is the total mass of the electrodes (g). C_{sp} is in unit of mF/g.

The spesific capacitances of the samples with MnO_2:C ratio of 1:1, 1:2, 1:3, and 1:4 are 673.80 mF/g, 922.67 mF/g, 564.78, and 702.11, respectively. Hence, the maximum spesific capacitance is 922.67 mF/g. As a metter of fact, the commercial capacitor has an capacitance of 10 mF and when the same characterzation is applied to that commercial capacitor, we would find value of its spesific capacitance of 0.0017 mF/g. Therefore, the capacitor with MnO_2/C would have specific capacitance 500.000 times larger than that of the commercial capacitor. This provides an evidence that the capacitor with MnO_2/C electrodes are able to store much more electrical charges compared with the commercial capacitor.

Figure 7 shows the impedance plot of MnO_2/C by means of electrochemical impedance spectra (EIS) test. The test was done at AC amplitude of 10 mV and frequency range of 50 kHz – 10 mHz with 0.1 M Na_2SO_3 electrolyte.

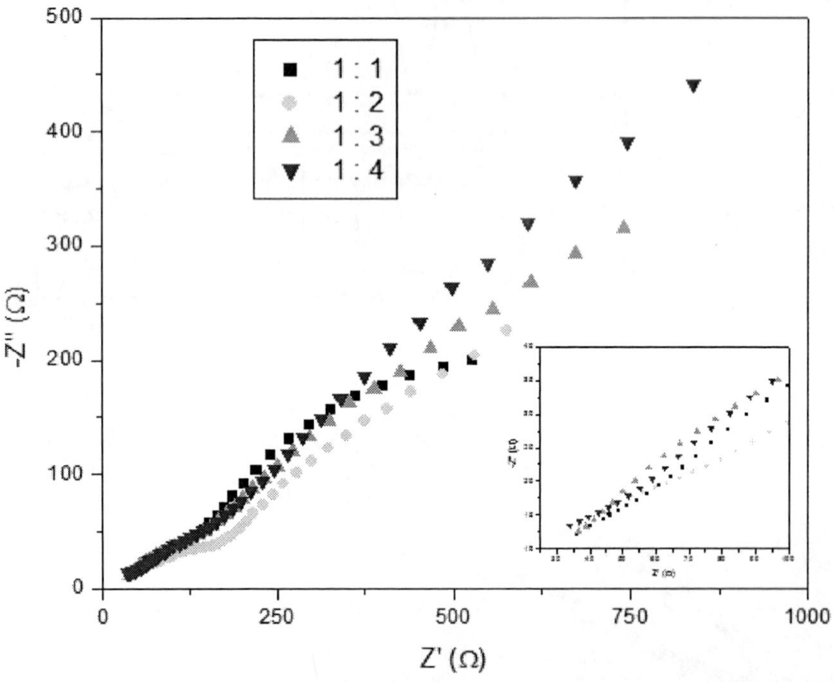

Figure 7. Nyquist Plot for MnO_2/C after EIS Measurement in AC amplitude of 10 mV and in 0,1 M Na_2SO_3 Electrolyte.

From Figure 7, the equivalent series resistance (ESR) for samples with MnO_2:C ratio of 1:1, 1:2, 1:3, and 1:4 are, respectively, 36.17, 45.92, 36.55, and 33.97 Ω. The maximum power density for the samples are 84.678, 65.059, 83.787, and 90.150 W/kg for 1:1, 1:2, 1:3, and 1:4 ratio of MnO_2:C composites respectively. Meanwhile, the maximum energy are 0.046, 0.063, 0.038, and 0.048 Wh/kg for the samples in the same order.

Based on the Ragone plot, it is found that the MnO_2/C composite, in particular, with 1:2 ratio is in the chatagory of supercapacitor. The Ragone plot is a graph that can be used to compare the performance of numerous energy storage materials in terms of energy density (Wh/kg) and power density logaritmically.

To end this discussion, the chemical reactions for the formation of MnO_2 from $KMnO_4$ in acidic environment are given as follows (Equation 1-2):

$$2KMnO_{4(aq)} + H_2SO_{4(aq)} \rightarrow 2MnO_{2(s)} + K_2SO_{4(aq)} + H_2O_{(l)} \quad (1)$$

or

$$MnO_4^- + 4H^+ + 3e \rightarrow MnO_2 + 2H_2O \quad (2)$$

The use of $KMnO_4$ solution with high concentration (0.1 M) to synthesize MnO_2 nanoparticles by electrolysis is discovered to be effective and efficient.

4. Conclussion

Rod-like α-MnO_2 with uniform diameter size was successfully prepared by means of simple electrolysis approach in acidic environment. The α-MnO_2 was then proposed as the electrode material in the form of MnO_2/C for supercapacitor application. Furthermore, the composite shows excellent electrochemical performance such as large maximum specific capacitance of 922.67 mF/g, with maximum power density of 90.150 W/kg, and maximum energy density of 0.063 Wh/kg.

References

[1] Yang Y, Xiao L, Zhao Y and Wang F 2008 Hydrothermal synthesis and electrochemical characterization of a-MnO2 nanorods as cathode material for lithium batteries *Int. J. Electrochem. Sci.* **3** 67–74

[2] Nayak P K and Munichandraiah N 2012 Rapid sonochemical synthesis of mesoporous MnO2 for supercapacitor applications *Mater. Sci. Eng. B* **177** 849–54

[3] Cheng H, Long L, Shu D, Wu J, Gong Y, He C, Kang Z and Zou X 2014 The supercapacitive behavior and excellent cycle stability of graphene/MnO 2 composite prepared by an electrostatic self-assembly process *Int. J. Hydrog. Energy* **39** 16151–61

[4] Jacob G M and Zhitomirsky I 2008 Microstructure and properties of manganese dioxide films prepared by electrodeposition *Appl. Surf. Sci.* **254** 6671–6

[5] Rusi and Majid S R 2014 Controllable synthesis of flowerlike α-MnO2 as electrode for pseudocapacitor application *Solid State Ion.* **262** 220–5

[6] Chodankar N R, Dubal D P, Gund G S and Lokhande C D 2015 Flexible all-solid-state MnO2 thin films based symmetric supercapacitors *Electrochimica Acta* **165** 338–47

[7] Pang S-C, Anderson M A and Chapman T W 2000 Novel Electrode Materials for Thin-Film Ultracapacitors: Comparison of Electrochemical Properties of Sol-Gel-Derived and Electrodeposited Manganese Dioxide *J. Electrochem. Soc.* **147** 444

[8] Chen N, Wang K, Zhang X, Chang X, Kang L and Liu Z-H 2011 Ionic liquid-assisted hydrothermal synthesis of β-MnO2 with hollow polyhedra morphology *Colloids Surf. Physicochem. Eng. Asp.* **387** 10–6

[9] Zhao S, Liu T, Shi D, Zhang Y, Zeng W, Li T and Miao B 2015 Hydrothermal synthesis of urchin-like MnO2 nanostructures and its electrochemical character for supercapacitor *Appl. Surf. Sci.* **351** 862–8

[10] Nam H-S, Kwon J S, Kim K M, Ko J M and Kim J-D 2010 Supercapacitive properties of a nanowire-structured MnO2 electrode in the gel electrolyte containing silica *Electrochimica Acta* **55** 7443–6

[11] Zhang X, Miao W, Li C, Sun X, Wang K and Ma Y 2015 Microwave-assisted rapid synthesis of birnessite-type MnO2 nanoparticles for high performance supercapacitor applications *Mater. Res. Bull.* **71** 111–5

[12] Fritsch S, Post J E, Suib S L and Navrotsky A 1998 Thermochemistry of Framework and Layer Manganese Dioxide Related Phases *Chem. Mater.* **10** 474–9

[13] Nemade K R and Waghuley S A 2014 Preparation of MnO2 immobilized graphene nanocomposite by solid state diffusion route for LPG sensing *J. Lumin.* **153** 194–7

[14] Smart L and Moore E 2005 *Solid state chemistry: an introduction* (Boca Raton: CRC Press)

[15] Jadhav P R, Suryawanshi M P, Dalavi D S, Patil D S, Jo E A, Kolekar S S, Wali A A, Karanjkar

M M, Kim J-H and Patil P S 2015 Design and electro-synthesis of 3-D nanofibers of MnO 2 thin films and their application in high performance supercapacitor *Electrochimica Acta* **176** 523–32

[16] Meng Y, Wang S, Cai R, Jiang B and Zhao W 2015 Discrimination and Content Analysis of Fritillaria Using Near Infrared Spectroscopy *J. Anal. Methods Chem.* **2015** 1–8

[17] Mahmudi M, Setiawan H and Affandi S 2014 Studi Pendahuluan MnO2 dengan Metode Elektrolisis *Pros. Semin. Nas. Kim. Dan Pembelajarannya SNKP*

Synthesis of Fe_3O_4 Nanoparticles using PEG Template by Electrochemical Method

Fauziatul Fajaroh[1,*], Nazriati[1], Yahmin[1], Siti Marfu'ah[1], Sumari[1], Adrian Nur[2]

[1]Department of Chemistry, Faculty of Mathematics and Sciences, Universitas Negeri Malang, Jl. Semarang 5 Malang 65145, Indonesia
[2]Department of Chemical Engineering, Universitas Sebelas Maret, Indonesia

*Corresponding author's email: fauziatul.fajaroh.fmipa@um.ac.id

Abstract. One of the nanoparticles which have been developed is magnetite. Due to its magnetism and reactivity, this particle can be used in various fields including technology, environment, and biomedical. One simple method to synthesize magnetite is electro-oxidation of iron in the water. The particle size produced by this method can be adjusted by controlling the electrochemical cell parameters. Unfortunately, this method usually releases polydispersed particles. One solution to overcome this problem is by using an *in situ* PEG in the synthesis process. In this research, PEG 6000 was chosen as a template, so the synthesis process was performed by electro-oxidation of iron in the dilute solution of PEG. Particles with an average diameter of 62.5 nm were obtained. The monodispersity, surface area, and crystallinity of the particles increased in this way. The specific surface area increased from 55.322 to 391.314 m^2/g. The results of XRD and FTIR analysis showed that PEG acted as a template in the synthesis process. In addition, the yield obtained with PEG template was larger than without the template. This method is quite promising as a way of synthesis of magnetite nanoparticles.

Keywords: Magnetite, PEG, nanoparticle, electrochemical method, template.

1. Introduction

Today, the development of nanoparticles, including magnetite nanoparticles, is a very interesting field of research since nanoscale magnetic materials have superior physical and chemical properties [1]. Magnetite nanoparticles are widely used in many areas of technology including environmental, ferrofluid, energy storage, and medical technology (both as diagnostic and therapeutic agents) [1, 2].

To obtain the desired properties due to their specific use, various methods of synthesizing these nanoparticles have been developed, for example, coprecipitation [3], hydrothermal [4], sol gel [5], and electrochemistry [6-8]. Each method has its own advantages and limitations. The advantages of the electrochemical method, in this case, is the electro-oxidation of iron in water are environmentally friendly and the properties of the synthesized product can be set easily by controlling the parameters of electrolysis cells such as cell dimensions, electrolyte type, current density, and temperature. The disadvantage of this method is the inadequacy of the resulted product, but this weakness will be overcome by using many electrode systems and electrochemical cell parameter setting [9].

Some previous studies have obtained a pure magnetite with a fairly uniform particle size distribution and high saturation magnetization [6]. This study was intended to increase the surface area of the particles using templates in the synthesis process. The template was added to the electrolyte before the electro-oxidation was executed. The template acted only as a mold of the particles formed without affecting the reaction process in electrolysis. After the synthesis process ended, the template could be easily removed, for example, through heating. The release of the particles from the template would leave some pores. Thus, the in situ template-assisted electro-oxidation is expected to produce particles with a larger specific surface area and uniform size. One of the substances that meet the above requirements as a template is PEG. Non-ionic PEG will not interfere with the process of magnetite formation. PEG will decompose on the product drying process. The synthesis of magnetite nanoparticles by the other methods has also been done using PEG as a template [10]. The purpose of this study was to examine the differences in crystallinity, monodispersity, and the specific surface area between magnetite synthesized with and without PEG 6000 as an in situ templates.

2. Materials and Methods

The synthesis of Fe_3O_4 nanoparticles with and without the PEG 6000 template was performed by an electrochemical method and followed by characterization with the following equipment: XRD (Panalytical X'pert Pro), SEM (Type Inspect S50, FEI), FT-IR (Shimadzu Type FT-IR 8400S), and BET (Nova 1200, Quantachrome). The materials which were used included filter paper (Whatman-42), universal indicator (Merck), iron electrode, carbon electrode, technical $FeSO_4$, PEG 6000 (Merck), NaOH (Merck), and demineralized water (Hydrobath).

The synthesis of Fe_3O_4 nanoparticles began with the preparation of anode as a source of iron ion through a process of depositing of pure iron film to a commercial iron plate using a technical iron (II)sulfate solution by electroplating. The synthesis of Fe_3O_4 nanoparticles was carried out in an aqueous solution of PEG 6000 0.004 M (for particles with the PEG template) and in demineralization water (for particles without the PEG template) under DC voltage for 4 hours. Both processes produced black deposits which could be magnetized. The generated particles were then filtered, dried at 60 °C, and characterized.

3. Results and Discussion

3.1. The Mechanism of Magnetite Formation

The processes of the synthesis of magnetite nanoparticles with and without PEG were followed by observing the color change in the solution around the anode ranging from t = 0 to t = 240 minutes. The results of the observation are shown in Figures 1 and 2. Based on these two images, there was no fundamental difference between the two synthesis processes. The solution around the anode was initially colorless, then gradually turned into a light brown turbid solution, orange, dark brown, and black at the end. This phenomenon indicated that the addition of PEG in the electrolyte did not affect the reactions. The color change of solution and sediment around the anodes in both processes can be explained in equation (1) to (4) and proceeded with the following process as described in equation (5), if the PEG solution was used as an electrolyte. The PEG template would decompose on a heating process, so magnetite nanoparticles would be obtained as described by equation (6).

Anode (Fe of electroplating result): $Fe \rightarrow Fe^{2+} + 2e^-$ (1)
Cathode (commercial Fe): $2H_2O + 2e^- \rightarrow 2OH^- + H_2$ (2)

The reactions around the anode:
$Fe^{2+} + 2OH^- \rightarrow Fe(OH)_2$ (3)
$3Fe(OH)_2 + \frac{1}{2}O_2 \rightarrow Fe(OH)_2 + 2FeOOH + H_2O$ (4)
$Fe(OH)_2 + 2FeOOH \rightarrow Fe_3O_4 + 2H_2O$
$Fe_3O_4 + PEG \rightarrow Fe_3O_4\text{-}PEG$ (5)
$Fe_3O_4\text{-}PEG \xrightarrow{\Delta} Fe_3O_4$ (6)

A light brown color indicated the formation of $Fe(OH)_2$ and the beginning of FeOOH formation. Orange was a typical color of FeOOH. The transition from orange to black was shown as dark brown. Black color indicated the existence of magnetite. The above reaction mechanism as written at equation (1) to (4) was corroborated by the results of the study of Lozano et al. who compared various mechanisms of magnetite formation by electrolysis method [9]. Figure 1 and 2 show the electrolyte and sediment color changes around the electrode during the electrolysis process.

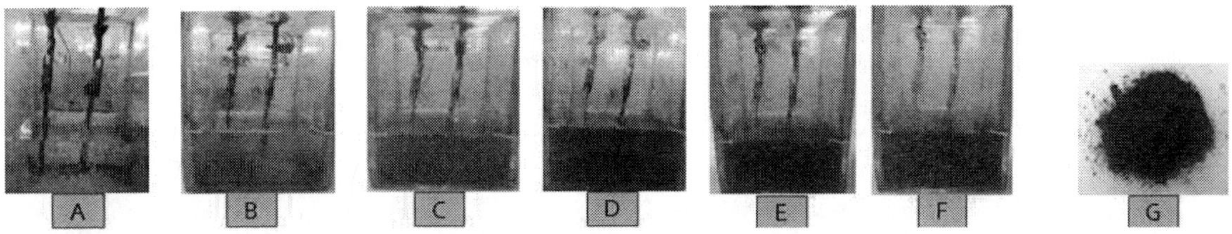

Figure 1. Solution and sediment color changes around the anode in synthesis process of magnetite without template at (A) 0, (B) 30, (C) 60, (D) 120, (E) 180, (F) 240 minutes, and (G) the final product

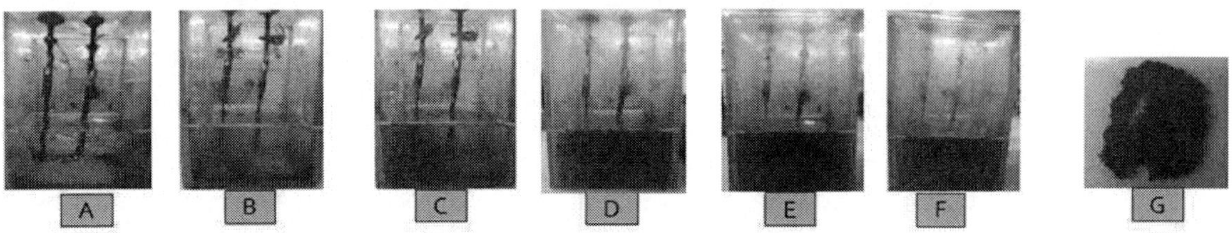

Figure 2. Solution and sediment color changes around the anode in synthesis process of magnetite with the template at (A) 0, (B) 30, (C) 60, (D) 120, (E) 180, (F) 240 minutes, and (G) the final product

In addition, the yield obtained with PEG template was larger than without the template. The presence of PEG around magnetite makes magnetite seem isolated from the reacting system, thus shifting the reaction equilibrium toward the product formation. In other words, the yield will increase.

3.2. The Characteristics of the Particles

Based on the XRD analysis as shown in Figure 3, the XRD spectra of Fe_3O_4 nanoparticles with and without PEG template showed magnetite peaks at $2\theta = 30.3°$, $35.7°$, $43.4°$, $53.8°$, $57.3°$, $62.9°$. These peaks corresponded to a typical magnetite peak in accordance with JCPDS Card No. 19-629 [6]. These spectra did not show typical PEG peaks, i.e. at 2θ: 19.41 and 23.34 according to JCPDS Card No. 49-2097 [11]. This means that the PEG acted as a particle template. The peaks of Fe_3O_4 with PEG template was higher in intensity than Fe_3O_4 without PEG. It also proved that PEG did not coat the magnetite surface, and only acted as a template. Mukhopadhyay, et al. [12] and Mahdavi, et al. [13] found that crystallinity of PEG-coated magnetite was lower than the bare one. In this study, the presence of PEG around magnetite during the synthesis process made the particles longer underwent the maturation process so that the crystallinity was higher. The heating process above PEG decomposition temperature makes the particles free from PEG.

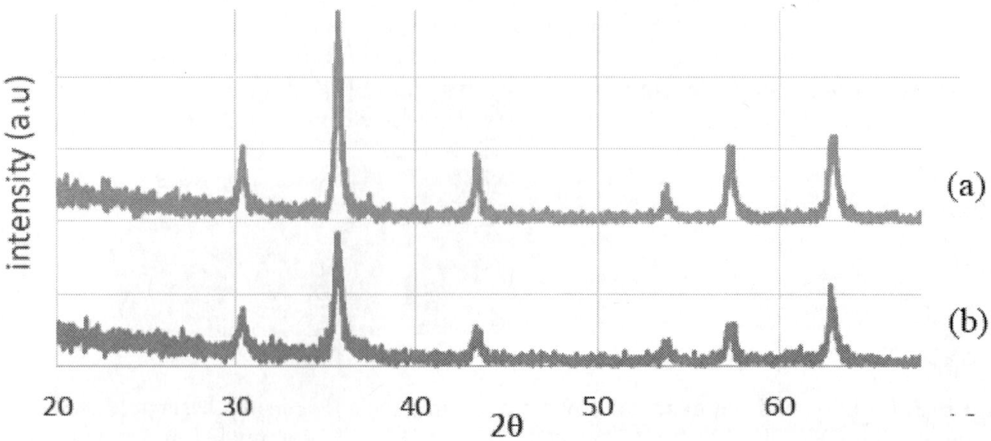

Figure 3. XRD spectrum of Fe_3O_4 (a) with and (b) without template

The FTIR analysis results that are presented in Figure 4 supported that PEG acted as a template. Both synthesized magnetite spectrum showed the presence of a vibrational peak at a region of 560 cm^{-1} indicating that there was a Fe-O bond of Fe_3O_4. However, the absence of the PEG 6000 vibrational peaks [14] on both magnetite spectra, i.e. C-H stretching at 2.890 cm^{-1}, C-O stretching at 1.110 cm^{-1} and O-H stretching at 3.350 cm^{-1} once again proved that PEG had become a mold on this synthesis and had been decomposed.

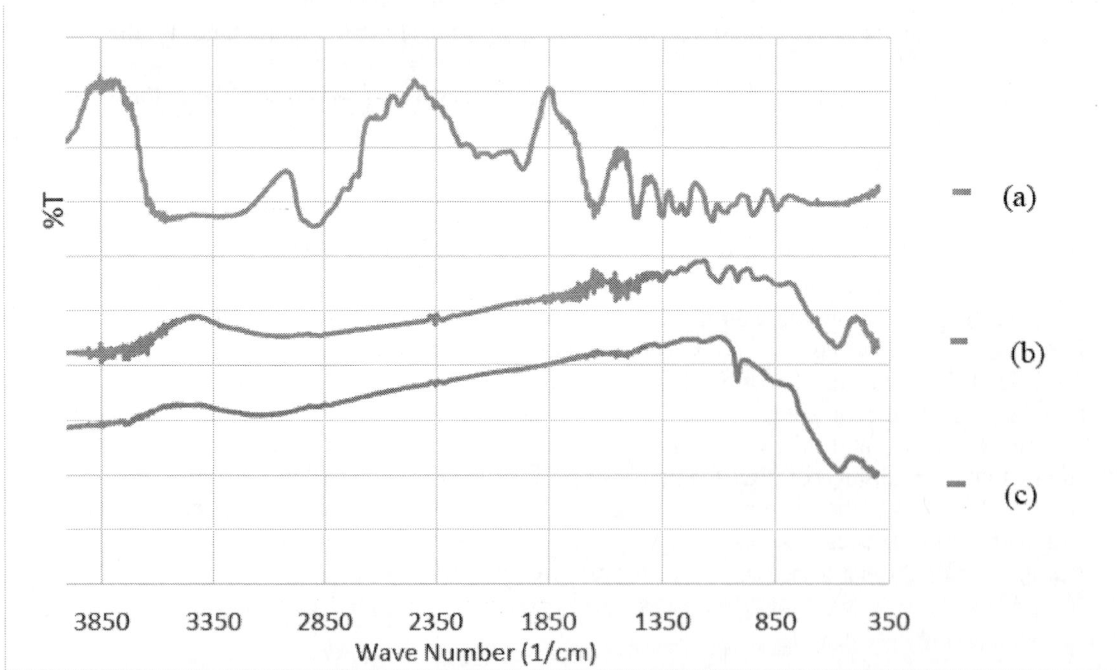

Figure 4. FTIR Spectrum of (a) PEG 6000, Fe_3O_4 (b) with and (c) without template

The specific surface area of Fe_3O_4 with and without the PEG 6000 template obtained from the BET analysis is given in Table 1. The use of PEG 6000 as a template increased the specific surface area.

Particle growth took place in a mold producing smaller particles of the size, causing a larger specific surface area. This increase was also supported by the mold release process of the particles at thermal decomposition. This process left the pores. Similar findings were also obtained by Bukit et al. which synthesized magnetite by coprecipitation methods

Table 1. The specific surface area of the particles

Sample	Specific Surface Area
Fe_3O_4 without PEG template	55.322 m^2/g
Fe_3O_4 with PEG template	391.314 m^2/g

SEM images of magnetite with and without templates are presented in Figure 5. It appeared that the morphology of the particles produced with and without PEG was the same, spherical. The obvious difference of the image was that the magnetite with the template had a more uniform, not agglomerated and smaller size distribution. Furthermore, particles with an average diameter of 62.5 nm were obtained using PEG template process.

(a) (b)

Figure 5. SEM image of Fe_3O_4 (b) with and (c) without template

4. Conclusion

The monodispersed magnetite nanoparticles with high crystallinity and specific surface area of 391.314 m^2/g can be synthesized through an iron electro-oxidation using PEG 6000 as a template. The in situ method was conducted by using a thin solution of PEG as the electrolyte at room temperature. The yield of particles obtained with PEG template was larger than without the template. The particles with PEG showed an average diameter of 62.5 nm which was smaller than the bare particles. This method is quite promising as a way of synthesis of magnetite nanoparticles for many applications.

Acknowledgments
Acknowledgments are conveyed to the Faculty of Mathematics and Natural Sciences, State University of Malang that has funded this research through PNBP funds and to S. Nuraisyah for collecting data.

References

[1] Liu JP, Fullerton E, Gutfleisch O, Sellmyer DJ 2009. Nanoscale Magnetic Material and Application (Springer, New York).
[2] Carlos L, Einschlagm FSG, Gonzales MC, Matire DO 2013. Waste Water Treatment Technologies and Recent Analytical Development (Intech)
[3] Aphesteguy JC, Kurlyandskaya GV, de Cells JP, Safrona AP, Schegoleva NN 2015,

Mat.Chem.Phys.,161, pp 243-319
[4] Daou TJ, Pourroy G, Begin-Colin S, Greneche JM, Ulhaq-Bouilet C, Legare P, Bernhardt P, Leuvry C, Rogez G 2006, Chem.Mat, 18, pp 4399-4404.
[5] Lemine OM, Omri K, Zhang B, El Mir L, Sajieddine M, Alyamani A, Bououdina M 2012, Superlattices and Microstruct, vol. 5, issue 4, pp 793-799
[6] Fajaroh F, Setyawan H, Widiyastuti W, Winardi S 2013, Adv.Powder.Technol., 23, 3.
[7] Setyawan H, Fajaroh F, Pusfitasari MD, Yuwana M, Affandi S 2014, Asia Pac. J. of Chem. Eng. 9 (5), 768-774
[8] Fajaroh F, Sutrisno, Nazriati, Wonorahardjo S 2014, AIP Conference Proceedings 1586 (1), 179-182
[9] Lozano I, Casillas N, Ponce de Leon, Walsh FC, Herrasti P 2017, J. Electrochem. Soc, 164 (4) D184-D191
[10] Cao H, Qi F, Liu R, Wang F, Zhang C, Zhang X, Chaib Y, Zhai Y 2017, RSC Adv., 7, 11244
[11] Biswal S, Sahoo J, Murthy PN, Giradkar RP, Avari JG 2008, AAPS PharmSciTech, Jun, 9(2), pp 563-570
[12] Mukhopadhyay A, Joshi N, Chattopadhyayay K, De Gouten 2012, ACS Appl. Energy Mater 4(1) 142-149
[13] Mahdavi M, Ahmad M, Haron, MJ, Namvor F, Nadi B, Ab Rahman MZ, Amin J 2013, Molecule, 16, 7533-7548
[14] Bukit N, Frida E,, Simamora P, Sinaga T 2015, J. chem. mater. res., Vol.7 No.7.

IOP Conf. Series: Journal of Physics: Conf. Series **1093** (2018) 012023 doi:10.1088/1742-6596/1093/1/012023

Synthesis and Bioactivity of TiO_2/PPy Micelle with *Pachyrhizus erosus* Extract as UV Absorbent Material

Titik Istirohah[1], Dwi Meinita Sari[3], Maqfirotun Ainun Jannah[1], Lisa Amalia Sari[1], Markus Diantoro[1,2,*]

[1]Department of Physics, Faculty of Mathematics and Natural Sciences, Universitas Negeri Malang, Jl. Semarang 5 Malang 65145, Indonesia
[2]Central Laboratory of Minerals and Advanced Materials, Universitas Negeri Malang, Jl. Semarang 5 Malang 65145, Indonesia
[3]Department of Biology, Faculty of Mathematics and Natural Sciences, Universitas Negeri Malang, Jl. Semarang 5 Malang 65145, Indonesia.

*Corresponding author's email: markus.diantoro.fmipa@um.ac.id

Abstract. Indonesia is a tropical country with high intensity and duration of sunlight throughout over year. The condition causes the average UV index in Indonesia is at the level of 9 up to 11+ so that the power of the sun is extreme invade the human body. This condition makes Indonesia one of the countries with the highest UV index in the world. UV rays exposure can be harmful to humans and become one of the most significant factors causing the existing of skin cancer. Researchers have studied many efforts. Unfortunately, the use of oxidic nanoparticles combined with natural products is not yet explored comprehensively. TiO_2 nanoparticles are materials that have excellent absorbability of UV light. In this study the TiO_2 nanoparticles as core and combined with Polypyrrole polymer as shell form a micelle structure. The modification of TiO_2 nanoparticles with Polypyrrole aims to enhance its UV protection, biocompatibility, and nontoxicity on the body. The addition of *Pachyrhizus erosus* extract as an enhancer of the efficacy of micelle as UV absorbent agent because it has the phenolic to skin softening properties. The synthesis of micelle TiO_2/PPy was performed using in situ method, followed by stirring and sonication process to mix the *Pachyrhizus erosus* extract with micelle. The XRD analysis of Micelle synthesis shows that the addition of polypyrrole polymer does not affect the crystallinity of TiO_2 nanoparticles. The results of testing the antibacterial properties of sunblock samples showed a clear zone with glycerine solvent.

Keyword: Micelle TiO_2/PPy, *Pachyrhizus erosus* extract, UV light, sunblock

1. Introduction

Cancer is one type of disease which arises from the abnormal growth of body tissue cells. Cancer has become the number one deadliest disease in the world to date. Skin cancer is one of a kind of cancer, the type of skin cancer itself also several of which one of the most deadly is the kind of melanoma. This type of melanoma skin cancer is the most dangerous, which accounts for about 75 percent of all skin cancer-related deaths. Melanoma cancer influenced by multiple genetic and epigenetic factors, one of the most significant factors causing the growth of skin cancer is UV exposure [1]. Indonesia is a tropical country with sunlight throughout the year. This condition causes the average UV index in Indonesia to

IOP Conf. Series: Journal of Physics: Conf. Series **1093** (2018) 012023 doi:10.1088/1742-6596/1093/1/012023

be at the level of 9 to 11+, or the power of the sun is extreme. Therefore, it makes Indonesia one of the highest UV indexes in the world. To overcome the obstacle, people use natural products as a protective ingredient as well as to beautify the skin. Unfortunately, the natural products may degrade under Sun irradiation.

TiO_2 has been widely developed in both industry and medicine. The TiO_2 nanoparticles have shown an excellent performance in solar cells [2] and cancer therapy [3]. TiO_2 also has a sensitivity to UV light, as evidenced by the study that TiO_2 thin films indicate a sharp transmittance change in the wavelength range of 300 to 400 nm which is the area of ultraviolet wavelength. The change in the value of this transmittance indicates a considerable absorption at that wavelength.

The encapsulating of nano TiO_2 with biocompatible polymer PPy aims to obtain the right choice of sunblock ingredients with an excellent safety for the body. The reason for using PPy is that the PPy polymer itself has proven effective as a safe drug carrier when applied in the human body. So when used for sunblock is not going to irritate, even if absorbed into the body will not be harmful to the body. With the addition of *Pachyrhizus erosus* extract may expect to increase the efficacy of sunblock to maintain the beauty of the skin. TiO_2 nanoparticles wrapped in a PPy polymer with the final step method is by blending *Pachyrhizus erosus* extract on micelle. The resulting sample is a gel that can be identified by the UV absorption, antibacterial, and biocompatibility before applied directly to the body.

2. Materials and Methods

2.1. Synthesis of the TiO_2/Ppy micelle
The fabrication of micelle TiO_2/PPy was performed by using in situ method as previously reported [4,5]. A 1 gram of TiO_2 nanoparticles was suspended in HCl (1.5 M, 50 mL) and followed by ultrasonic radiation for 30 min. The subsequent step was the injection of polypyrrole polymer (PPy) with the ratio of TiO_2: PPy is 100: 1 under constant stirring. We also prepared a solution of $FeCl_3$ (0.1 gram) in HCl (1.5 M, one mL) which was then gently dropped into the previous suspension. The mixed solution was then stirred for 4 hours at 0 °C and followed by centrifugation. The obtained sediment was washed using 1.5 M HCl and DI water until it reached a neutral pH. The washed precipitated materials were then dried at 100 °C to remove the residual solvent.

2.2. Preparation of Pachyrhizus erosus Extract
Preparation of *Pachyrhizus erosus* extract began with the heating of *Pachyrhizus erosus* fruit from 1 kg to reach a dry weight of 1.5 ounces. The dry *Pachyrhizus erosus* was subsequently blended and transformed into flour. A total of 400 grams of *Pachyrhizus erosus* flour was macerated using 800 mL of 96% ethanol. The maceration process took for 1-2 days with interfered by shaking every 6 hours. The macerated materials were then put in a rotary evaporator to separate the extract using the ethanol solvent. The evaporator extract was placed in an open container to remove the remaining solvent.

2.3. Processing of TiO_2/PPy with extract Pachyrhizus erosus sunblock
The micelle TiO_2/PPy powder was dissolved using *Pachyrhizus erosus* extract with a different solvent, PEG-6000, and glycerin. The final product was distilled for 30 min at 50 °C followed by sonication for 30 min.

2.4. Characterization
The samples were then characterized to determine the structure and biophysical properties of the material. The samples' information associated with crystal phase, morphology and grain size were characterized by XRD, FTIR, and SEM. The sample bioactivity included the test of irritation was measured using the mice as the animal test. The anti-bacterial properties tested using the diffusion method on *E.coli*. We also measure the sample absorption using UV-Vis.

3. Results and Discussion

3.1. FTIR Spectra of the TiO_2/Ppy micelle

The results of the characterization of the IR spectra of the micelle TiO_2/PPy are shown in Figure 1. It is obtained that the FTIR peak of TiO_2 nanoparticles exists in the range of 450 to 600 cm^{-1} which indicate the presence of a Ti-O bond vibration in the TiO_2 lattice [6–8]. The peaks between 3100 to 3600 cm^{-1} show vibrations of the hydroxyl (O-H) group [6,7,9] of the remaining solvent water during the synthesis.

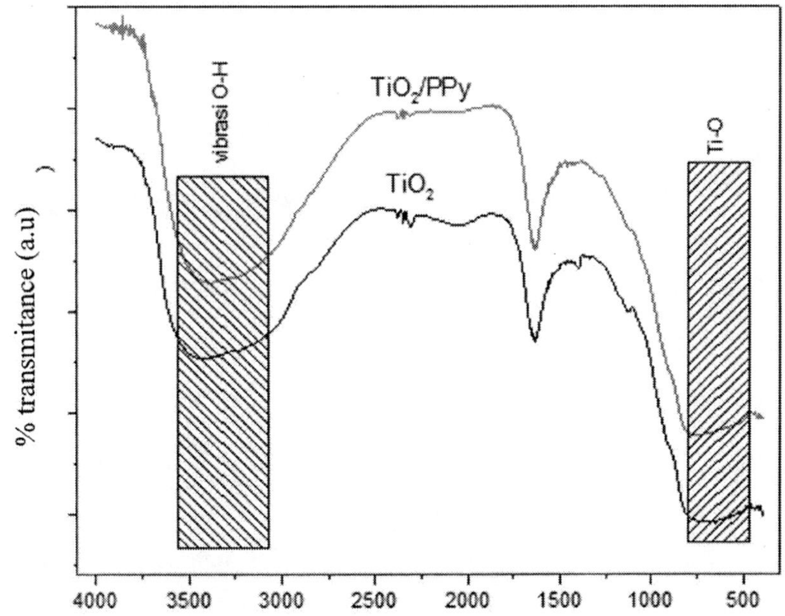

Figure 1. IR spectra of the TiO_2/Ppy *micelle*

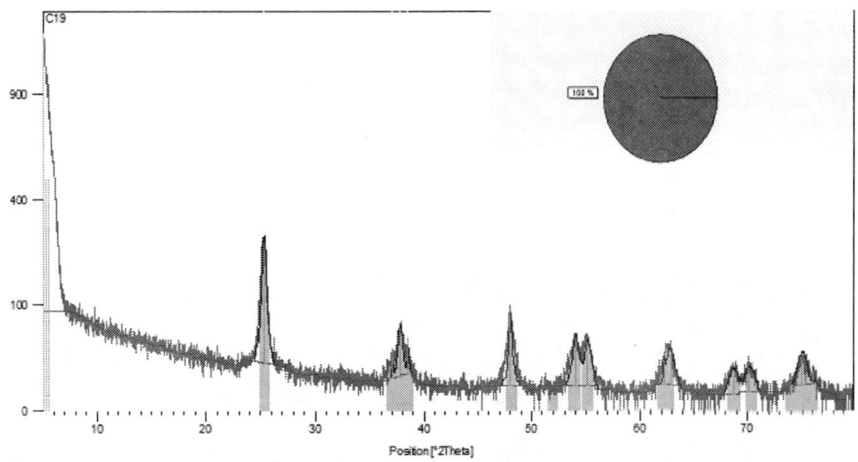

Figure 2. Phase analyses using HighScorePlus of TiO_2 nanoparticles utilizing the model of 01-078-2486

IOP Conf. Series: Journal of Physics: Conf. Series **1093** (2018) 012023 doi:10.1088/1742-6596/1093/1/012023

3.2. Structure of TiO_2/Ppy micelle

The analysis of TiO_2 XRD patterns was performed using Rietica and high score plus analysis (HSP). Figure 2. shows the results of HSP analysis with 100% purity of nanoparticles TiO_2. The matching results of the COD model number 2310710 of the TiO_2 nanoparticles with Rietica software show a matched peak. The good of fitness fitting value (χ^2) reached 1.175 which was obtained from Rietica analysis (Figure 3).

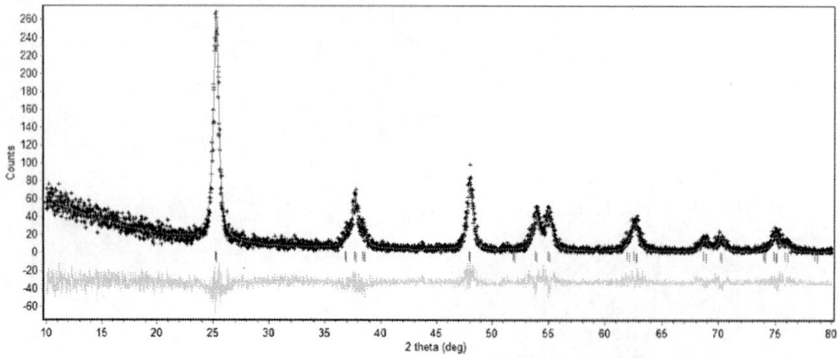

Figure 3. Crystal structure analysis of TiO_2 Nanoparticles.

The X-ray diffraction pattern of the TiO_2/PPy micelle nanoparticles compared to the TiO_2 nanoparticles are shown in Figure 4. From the fitting, we obtain the FWHM values, which were then used to calculate the grain size with the use of the Scherrer equation (Equation 1) [10].

$$D = \frac{k\lambda}{\beta \cos\theta} \qquad (1)$$

where k = *Scherrer constant*, λ = wavelength radiation Cu- Kα (1,5406 Å), β = FWHM (radians), and θ = Bragg's position (radian).

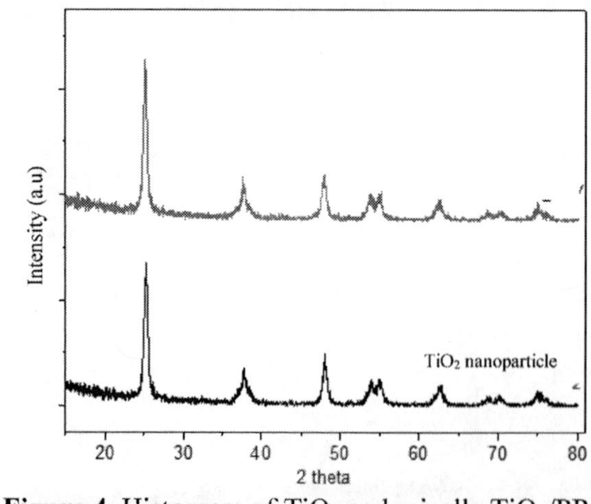

Figure 4. Histogram of TiO_2 and micelle TiO_2/PPy

Figure 4. shows the X-ray diffraction patterns of TiO_2 nanoparticles with TiO_2/PPy micelle. The diffraction peaks have corresponded to the model of the crystal peak of nanoparticles of TiO_2 anatase. The PPy-wrapped nanoparticles exhibit a diffraction pattern similar to those of TiO_2 nanoparticles, indicating that the addition of the PPy polymer as a wrap does not decrease the crystallinity of the TiO_2 nanoparticles. Analysis of the main peak of X-ray diffraction of nanoparticles TiO_2 using Equation 1 yields a grain size of 15.78 nm. This value is associated with the FWHM of 0.57608 under Lorentzian fittings. Using the same way, we obtain that the micelle TiO_2/PPy shows a grain size of 17.38 nm. The micelle TiO_2/PPy grain size shows a greater value than the grain size of TiO_2 nanoparticles. This is because of the effect of coating the particles with polymer PPy.

3.3. Morphology Analysis of micelle TiO_2/PPy using Scanning Electron Microscope (SEM)

The nanomorphology of TiO_2 which was obtained from the SEM analysis is shown in Figure 5. From the morphological results of the SEM of the TiO_2 and TiO_2/PPy nanoparticles, it is seen that after the particles enclosed by the polymer PPy still have a shape like nanoparticles TiO_2. These results support the results of XRD analysis showing that the addition of PPy does not alter crystallinity, which is seen from nearly the same morphology. The elemental composition of the TiO_2 nanoparticles was performed using EDAX is shown in Figure 6.

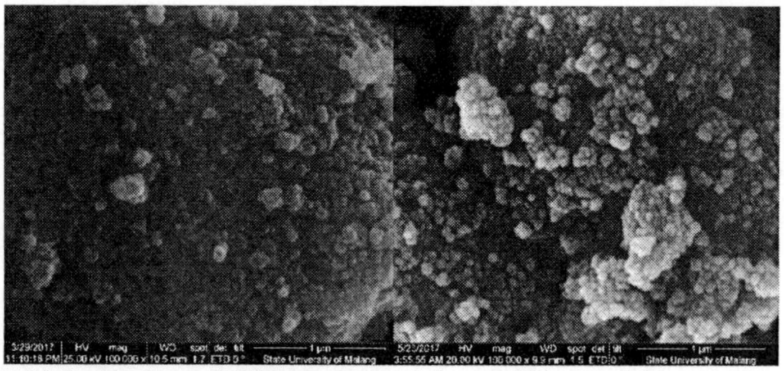

Figure 5. SEM Morphology a) TiO_2 nanoparticles and b) TiO_2/PPy micelle (100.000 x).

Element	*Wt%*	*At%*
OK	37.58	64.32
TiK	62.42	35.68
Matrix	Correction	ZAF

Figure 6. SEM-EDAX result of TiO_2 nanoparticles.

We observed that comparing the ratio of atomic percentage content of oxygen in K-band and titanium in K-band falls as 64.32 to 35.68. This ratio is close to 1 to 2 for respectively for Ti to O. We could conclude that the obtained synthesized material is TiO_2 nanoparticles. The size of nanoparticles has been confirmed from X-RD pattern analysis. Figure 5, as well as Figure 6, show that the TiO_2 shows agglomeration which is also reported for several oxidic nanoparticles [11,12].

3.4. Absorbance performance

The UV-Vis spectra of *Pachyrhizus erosus* extract and various TiO_2 of micelle TiO_2/ Ppy is shown in Figure 6. We found that *Pachyrhizus erosus* extract has a good uptake in the starting range of UV rays of 350-400 nm and have a high absorbance on 400-900 nm. The highest absorbance shows by the lower transmittance [11]. UV-Vis showed the same results from micelle TiO_2/ PPy and sunblock for the sample with 0.05-gram TiO_2/PPy content. Further analysis related to the variation of mass of TiO_2/PPy micelle of 0.05, 0.1, 0.15, 0.20, and 0.25 grams it was revealed that the most appropriate sunblock composition (which has a good uptake) shows by the sample with 0.05 grams micelle.

IOP Conf. Series: Journal of Physics: Conf. Series **1093** (2018) 012023 doi:10.1088/1742-6596/1093/1/012023

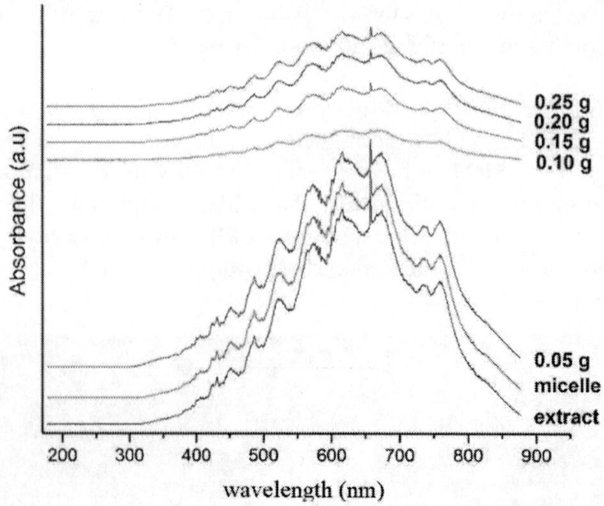

Figure 7. UV-Vis spectra of TiO_2/PPy micelle with *Pachyrhizus erosus* extract.

3.5. Anti-Bacterial Activity of TiO_2/PPy micelle with Pachyrhizus erosus Extract

The result of the antibacterial activity of TiO_2/PPy *micelle* with *Pachyrhizus erosus* extract is shown in Table 1.

Table 1. Antibacterial activity

No	Sample	Active Zone Diameter (cm)
1.	TiO_2-PEG	-
2.	TiO_2-Glycerin	0.9
3.	TiO_2/Ppy-PEG	-
4.	TiO_2/Ppy-Glycerin	2.10
5.	Pachyrhizus extract	-
6.	TiO_2/Ppy-PEG-P.extract	0.225
7.	TiO_2/Ppy- P.extract -Glycerin	0.295
8.	TiO_2/Ppy- P.extract-PEG-Glycerin	-

The observation of *E.Coli* bacterial inhibition zone was performed on the different sample. The width of the inhibitory region indicates the antibacterial power of the material. The wider area exhibits, the more significant of the antibacterial strength. Antibacterial of various samples showed that the inhibitory zone diameter of the TiO_2/Ppy-Glycerin sample shows the highest drag zone. The hygroscopic property of glycerin may provide better absorption of water molecules allow a more effective interaction of the sample. The hygroscopic nature of glycerin causes the sample to diffuse easily into the medium part of Nutrient Agar. The micelle TiO_2/PEG particle diffuses to the maximum in the medium to have more significant contact with the bacterial wall. Glycerin has been widely used in the industry of producing soap, detergent, and cosmetics as a suitable and safe solvent.

3.6. Irritation test of TiO_2/PPy micelle with Pachyrhizus erosus extract

We performed a skin irritation test on mice. We used six young mice which one of them as a control. We found that the skin of mice which were treated of all samples indicate no difference in skin color with the control sample. These results suggest that the sample is not irritating to the skin of test animals. A good performance as antibacterial as well as the nonirritating materials gives the TiO_2/Ppy micelle a great potential to develop and further investigation.

4. Conclusion

We conclude that the TiO_2/PPy micelle samples with *Pachyrhizus erosus* extract have been successfully synthesized. The samples show excellent potential for sunblock prevention of skin cancer that works by optimizing UV light absorption. It is shown that TiO_2/PPy micelle samples with *Pachyrhizus erosus* extract exhibit anti-bacterial properties and nonirritating.

Acknowledgments

One of us (MD) thanks the Ministry of Research and Higher Education of Indonesia through the DRPM decentralized UM of the 2017-2018 grant.

References

[1] Gómez-Moyano E, de Ramón E, Martínez-García S, Gómez R, Crespo-Erchiga V, Crespo A and Ojeda A 2010 Impact of Melanoma Diagnosis on Sun Protection *Actas Dermo-Sifiliográficas Engl. Ed.* **101** 506–16
[2] Egerton T A 2014 UV-absorption—the primary process in photocatalysis and some practical consequences *Molecules* **19** 18192–18214
[3] Ivanković S, Gotić M, Jurin M and Musić S 2003 Photokilling squamous carcinoma cells SCCVII with ultrafine particles of selected metal oxides *J. Sol-Gel Sci. Technol.* **27** 225–233
[4] Luo Q, Li X, Wang D, Wang Y and An J 2011 Photocatalytic activity of polypyrrole/TiO_2 nanocomposites under visible and UV light *J. Mater. Sci.* **46** 1646–1654
[5] Wang D, Wang Y, Li X, Luo Q, An J and Yue J 2008 Sunlight photocatalytic activity of polypyrrole–TiO_2 nanocomposites prepared by 'in situ'method *Catal. Commun.* **9** 1162–1166
[6] Ba-Abbad M M, Kadhum A A H, Mohamad A B, Takriff M S and Sopian K 2012 Synthesis and catalytic activity of TiO_2 nanoparticles for photochemical oxidation of concentrated chlorophenols under direct solar radiation *Int J Electrochem Sci* **7** 4871–4888
[7] Hamadanian M, Reisi-Vanani A and Majedi A 2010 Sol-gel preparation and characterization of Co/TiO_2 nanoparticles: application to the degradation of methyl orange *J. Iran. Chem. Soc.* **7** S52–S58
[8] Shehap A M and Akil D S 2016 Structural and optical properties of TiO_2 nanoparticles/PVA for different composites thin films
[9] Beranek R and Kisch H 2008 Tuning the optical and photoelectrochemical properties of surface-modified TiO_2 *Photochem. Photobiol. Sci.* **7** 40–48
[10] Holzwarth U and Gibson N 2011 The Scherrer equation versus the "Debye-Scherrer equation" *Nat. Nanotechnol.* **6** 534–534
[11] Diantoro M, Kholid, Mustikasari A A and Yudiyanto 2018 The Influence of SnO_2 Nanoparticles on Electrical Conductivity, and Transmittance of PANI-SnO_2 Films *IOP Conf. Ser. Mater. Sci. Eng.* **367**
[12] Diantoro M, Sari L A, Istirohah T, Kusumawati A D, Nasikhudin and Sunaryono 2018 Control of Dielectric Constant and Anti-Bacterial Activity of PVA-PEG/x-SnO_2 Nanofiber *IOP Conf. Ser. Mater. Sci. Eng.* **367** 012012

Fabrication of PAN/ZnO Nanofibers by Electrospinning as Piezoelectric Nanogenerator

Alfan Jamil[1], Nandang Mufti[1,2*], Siti Maryam[1], Arif Hidayat[1], Ahmad Taufiq[1,2], Sunaryono[1,2],

[1]Department of Physics, Faculty of Mathematics and Natural Sciences, Universitas Negeri Malang, Jl. Semarang 5 Malang 65145, Indonesia
[2]Centre of Advanced Materials for Renewable Energy, Universitas Negeri Malang, Jl. Semarang 5 Malang 65145, Indonesia

*Corresponding author's email: nandang.mufti.fmipa@um.ac.id

Abstract. Piezoelectric nanogenerator is a material that is used for converting mechanical energy to electrical energy. This research aimed to study the piezoelectric nanogenerator properties in PAN/ZnO nanofibers layered on the stainless-steel substrate. ZnO nanoparticles that were used in this work were synthesized by coprecipitation method. The ZnO nanoparticles were mixed with PAN dissolved with DMF. Fabrication of PAN-ZnO nanofibers was done using the electrospinning method on the stainless-steel substrate. The formed PAN/ZnO nanofibers were then characterized using XRD, SEM, and FTIR. To test the piezoelectric nanogenerator properties, PAN/ZnO nanofibers were combined to PAN nanofibers and coated on the stainless-steel substrate to form piezoelectric nanogenerator device. This device was then connected to an electrometer and an oscilloscope to measure the current and voltage resulted after bending. The results of XRD of ZnO nanoparticles had the wurtzite crystal structure with the size of about 46 nm. Meanwhile, the PAN/ZnO had an amorphous structure. The test results of piezoelectric nanogenerator properties showed the value of voltage and current of 7.22 V and 47.48 μA, respectively. PAN/ZnO nanofibers on the stainless-steel substrate are potential to be the material of piezoelectric nanogenerators in general.

Keywords: PAN/ZnO nanofiber, electrospinning, piezoelectric nanogenerator, energy harvesting

1. Introduction

Nowadays, the availability of fossil energy in nature decreases due to the continual exploration to fulfill the needs of energy in the world. This is because the use of renewable energy from such sources as mechanical vibration, noise source, thermal energy, wave motion, earth heat, wind energy, and diesel fuel remains to be developed [1]. Mechanical energy becomes one of the renewable energies which can be converted into electrical energy using the piezoelectric material. This case is because the source of mechanical energy is broadly available in nature and now there is no one who utilizes it optimally. In some recent years, the development of piezoelectric-based nanogenerators (PENGs) has caught the researchers' attention since they are potential to be a self-powered device to fulfill the energy needs in the electronic portable that still continually develops [1–3].

PENGs are usually fabricated from piezoelectric materials that have a non-centrosymmetric crystal structure such as ZnO, PZT, and $BaTiO_3$ [4]. Zinc oxide (ZnO) is one of the materials that the researchers are interested in since it has good piezoelectric, pyroelectric, and optical properties [5]. Besides, ZnO also has a relatively wide band gap of 3.37 eV [6], is non-toxic [7], relatively low cost, easily accessible [8], stable, and environmentally friendly [9]. ZnO-based PENGs generally have the output with low power [10,11] and it is less stable [12]. One of the efforts to increase the power and stability of PENGs is by using one-dimensional (1D) piezoelectric nanostructures such as nanowire, nanorods, and nanofibers [9,13,14]. PENGs in the form of nanofiber have various advantages such as having a good piezoelectric characteristic, high flexibility and it can be coated on a flexible substrate [3,7]. Besides, nanofibers can absorb mechanical energy better compared to spherical nanogenerator as nanofibers can continually distribute the mechanical force from one fiber to another fiber continually in a long time [3]. Selecting the appropriate polymer is one of the keys to form nanofibers successfully. The polymer that is commonly used is polyvinyl acetate (PVA) but the output of voltage and power resulted by ZnO/PVA nanofibers is still relatively small [3,15]. This research studied the nanogenerator properties of ZnO nanofibers by using PAN (Polyacrylonitrile) as the composite polymer for PAN molecule has a very polarized functional group [16] so that it is expected to improve the performance of PENGs in PAN/ZnO nanofibers.

2. Materials and Methods

2.1. Synthesis of ZnO nanoparticles

Fabrication of PAN/ZnO nanofibers were conducted in two steps. The first stage was a synthesis of ZnO nanoparticles using the coprecipitation method. 10 gram of zinc acetate dihydrate was dissolved into 50 ml DI water using a magnetic stirrer with the velocity of 600 rpm at the room temperature (27 °C) for 60 minutes. After that, 25 ml of NaOH 3M was dropped to Zinc acetate solution so that milky suspension was formed with pH about 13 at 90 °C, then it was let until precipitating. The deposit was then washed until the pH was neutral. Subsequently, the obtained deposit was heated at the temperature of 100 °C for 60 minutes. The formed ZnO nanoparticles were ready to be utilized in the next process.

Synthesis of PAN/ZnO Nanofibers

The second step was synthesis of PAN/ZnO nanofiber by electrospinning method. PAN/ZnO solution was prepared by weighing 0.28 g of PAN (Polyacrylonitrile) dissolved in 3.72 g of DMF and then stirred at the temperature of 80 °C with the velocity of 1000 rpm for 60 minutes. After that, 0.056 ZnO nanoparticles of the synthesis results were mixed to PAN/DMF solution and stirred for 60 minutes at the temperature of 70 °C and the velocity of 1000 rpm. This solution was then used as a PAN/ZnO nanofiber material. One milliliter PAN/ZnO composite solution was infused to 10-millimeter syringe pump. The 24G-sized syringe needle was set to the syringe and connected to the positive terminal horizontally so that the positive terminal tip (needle) and aluminum collector plate (negative terminal) had the space of 8 cm. The stainless-steel substrate (SS) was set to the collector plate that had been covered with foil aluminum. The applied debit was 10 μL/minute with the voltage of 14 kV. The formed fiber was in the form of PAN/ZnO composite fiber coated on the stainless-steel substrate. Similarly, the synthesis of PAN nanofibers was also carried out on the stainless-steel substrate (SS).

2.2. Characterization

The sample of PAN/ZnO nanofiber composite was characterized using FTIR and XRD by sloughing the composite off from the stainless-steel substrate firstly. FTIR was employed to determine the functional groups of PAN/ZnO composite. The crystal structure of ZnO nanoparticles powder and PAN/ZnO fibers were tested through XRD (X-Ray Diffraction). In the test, the nanogenerator performance was undertaken by combining two stainless steel substrates facing each other so that it formed one device. On the stainless-steel substrate, the wire was set connected to the oscilloscope and electrometer to measure the voltage and current produced from this device.

3. Results and Discussion

The diffraction pattern of ZnO nanoparticles was characterized using XRD and then analyzed using Rietica software shown in Figure 1.a. The XRD pattern ZnO nanoparticles showed the hexagonal crystal structure (wurtzite) with the space group of P63mc which is in accordance with the JCPDS standard (No. 36-1451). The XRD of ZnO nanoparticles was then analyzed by refinement using Rietica software to determine the value of lattice parameter and grain size. The refinement resulted in lattice parameter of ZnO nanoparticles having a,b = 3.2522 Å, c = 5.2095 Å, with the comparison of c/a= 1.6 and the grain size of 46 nm. The lattice parameter which was obtained was in accordance with the lattice parameter of ZnO that previously has been reported [14,17]. In figure 1.b displays the XRD pattern of PAN/ZnO nanofibers. In the diffraction pattern, it can be seen that PAN/ZnO nanofiber had an amorphous structure from PAN polymer. The XRD pattern of ZnO nanoparticles cannot be seen since their concentrations were smaller than that of PAN polymer.

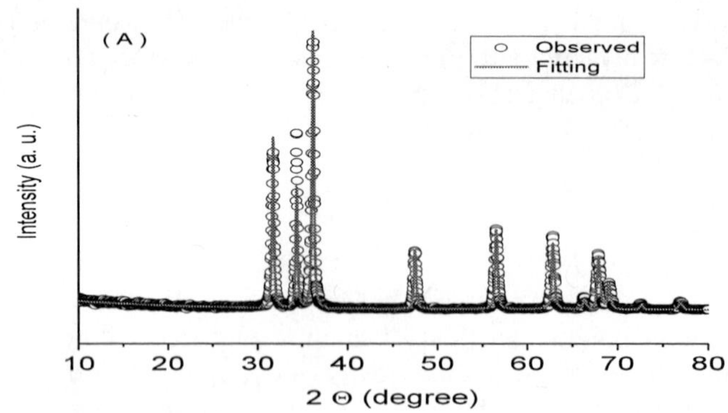

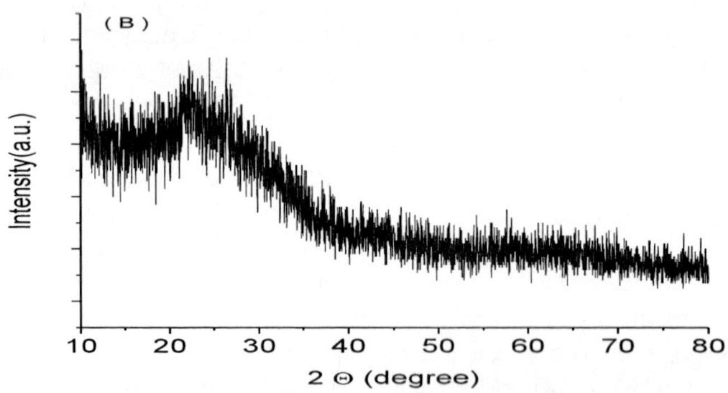

Figure 1. X-Ray Diffraction Pattern of (a) ZnO nanoparticles and (b) PAN/ZnO nanofibers

Figure 2 shows the morphology of PAN/ZnO nanofibers gained from SEM-EDX results with the 20,000 times magnification. In the picture, this SEM shows that PAN/ZnO nanofibers had been formed with the diameter between 0.5 and 1 micrometer. Besides, there was an agglomeration of ZnO nanoparticles in some parts of PAN/ZnO nanofibers. From EDX test results, the synthesized PAN/ZnO nanofibers contained about 3.67% atom of ZnO elements.

Subsequently, the FTIR test was carried out to analyze the resulted PAN/ZnO nanofiber functional groups. The spectrum pattern of PAN/ZnO nanofibers is shown in Figure 3. From the result of the FTIR spectrum, it can be seen that Zn – O stretching band appeared at the wavelength of 451. This case is in line with the research conducted by Zamiri et al. (2014) reporting that the strong absorption band of ZnO was in the range of 400 to 500 cm^{-1} [18]. Besides, there was a vibration pattern at the wavelength of 3643.5, 2939.5, 2243.2, 1666.4, 1452.3, 972.1, and 451. This result was in line with FTIR pattern reported by Yar et al. (2017) for PAN nanofibers, where the vibration appeared at the wavelength of about 2934, 1414, and 1093 cm^{-1} that were the vibration of CH, CH_2, and CH functional groups.

Meanwhile, the vibrations at 2241 cm^{-1} and 1662 cm^{-1} were CN and CO stretching in acrylamide [19]. FTIR pattern showed that PAN/ZnO nanofibers had been formed.

Figure 2. Photo of PAN/ZnO nanofibers SEM with 20,000 times magnification

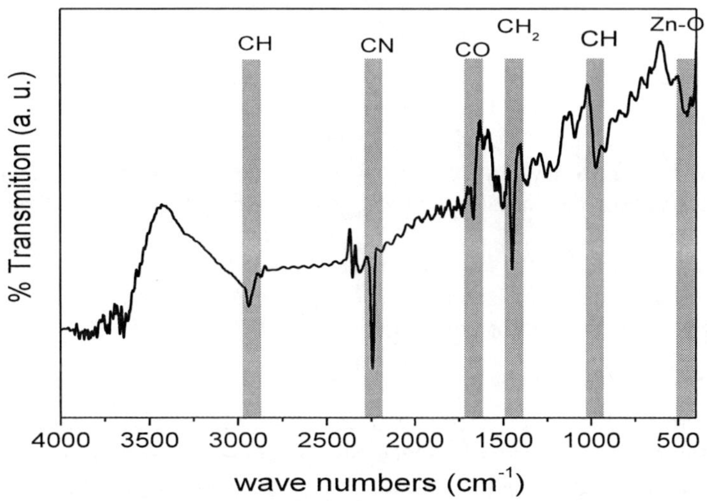

Figure 3. FTIR Spectrum pattern of PAN/ZnO nanofibers

Testing the piezoelectric nanogenerator properties was performed by measuring the current and voltage using electrometer and oscilloscope. This measuring was carried out by bending in the device consisting of two stainless steel substrates coated by composite of PAN/ZnO nanofibers and PAN nanofibers facing each other. Figure 4 shows current and voltage as the time function when the device was bent and released. The average current and voltage resulted from the test were 47.48 μA and 7.22 V, respectively. The resulted voltage was positive and negative indicating the resulted voltage was AC voltage obtained from this research which was bigger than that attained from ZnO/PVA reported in the previous research [3].

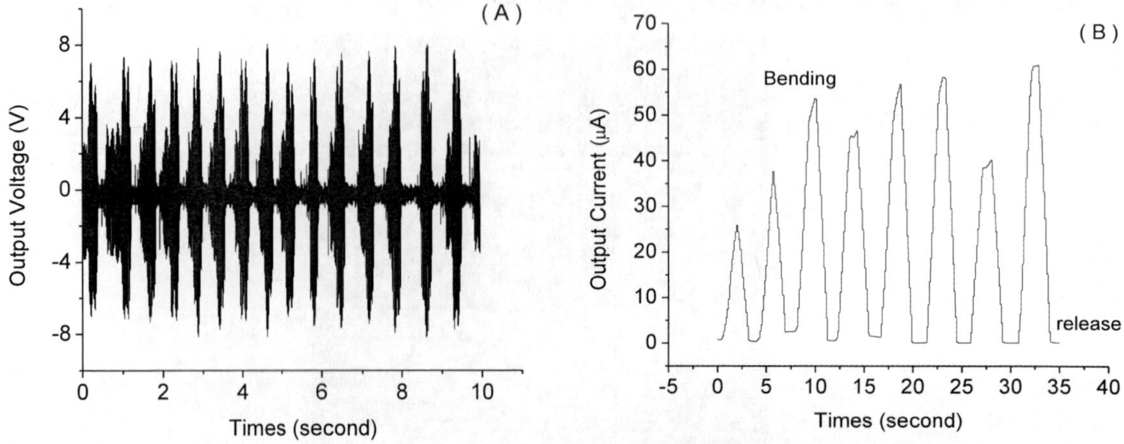

Figure 4. Times dependence of (a) voltage and (b) voltage of PAN/ZnO nanofibers device.

4. Conclusion

ZnO nanoparticles that were used in PAN/ZnO nanofibers had hexagonal crystal structure (wurtzite) with the lattice parameter of a,b = 3.2522 Å, c = 5.2095 Å, with the comparison of c/a = 1.6 and grain size of 46 nm. Meanwhile, PAN/ZnO nanofibers obtained have the diameter size between 0.5 to 1 micrometer. The test results of piezoelectric nanogenerators properties showed the values of voltage and current were 7.22 V and 47.48 μA, respectively. These results indicated that PAN/ZnO nanofibers in the stainless-steel substrate are potential to be piezoelectric nanogenerators material.

Acknowledgments

The authors thank to the Ministry of Research, Technology, and Higher Education of the Republic of Indonesia for the PBK research funding for NM.

References

[1] Saravanakumar B, Mohan R, Thiyagarajan K and Kim S-J 2013 Fabrication of a ZnO nanogenerator for eco-friendly biomechanical energy harvesting *RSC Adv.* **3** 16646–56
[2] Suyitno, Mukhadis A, Suhartadi S, Sutadji E, Saroso and Puspitasari P 2017 Effectiveness of professional practice work with discovery learning methods in engineering program D IV automotive safety *Adv. Sci. Lett.* **23** 1154–7
[3] Suyitno S, Purwanto A, Hidayat R L L G, Sholahudin I, Yusuf M, Huda S and Arifin Z 2014 Fabrication and characterization of zinc oxide-based electrospun nanofibers for mechanical energy harvesting *J. Nanotechnol. Eng. Med.* **5** 011002
[4] Wang X and Shi J 2012 Piezoelectric nanogenerators for self-powered nanodevices *Piezoelectric Nanomaterials for Biomedical Applications* (Springer) pp 135–72
[5] Lee D Y, Cho J-E, Cho N-I, Lee M-H, Lee S-J and Kim B-Y 2008 Characterization of electrospun aluminum-doped zinc oxide nanofibers *Thin Solid Films* **517** 1262–7
[6] Chang Y-T, Chen J-Y, Yang T-P, Huang C-W, Chiu C-H, Yeh P-H and Wu W-W 2014 Excellent piezoelectric and electrical properties of lithium-doped ZnO nanowires for nanogenerator applications *Nano Energy* **8** 291–6
[7] Samanta P, Bagchi S and Mishra S 2015 Synthesis and Sensing characterization of ZnO nanofibers prepared by Electrospinning *Mater. Today Proc.* **2** 4499–502
[8] Nguyen Q, Kim B H and Kwon J W 2015 Paper-Based ZnO Nanogenerator Using Contact Electrification and Piezoelectric Effects *J. Microelectromechanical Syst.* **24** 519–21
[9] Park J-A, Moon J, Lee S-J, Lim S-C and Zyung T 2009 Fabrication and characterization of ZnO

nanofibers by electrospinning *Curr. Appl. Phys.* **9** S210–2

[10] Fang T-H and Kang S-H 2010 Physical properties of ZnO: Al nanorods for piezoelectric nanogenerator application *Curr. Nanosci.* **6** 505–11

[11] Ko Y H, Nagaraju G, Lee S H and Yu J S 2014 PDMS-based Triboelectric and Transparent Nanogenerators with ZnO Nanorod Arrays *ACS Appl. Mater. Interfaces* **6** 6631–7

[12] Cheng Y-Y, Chou S-C and Chang J-A 2011 Development of flexible piezoelectric nanogenerator: Toward all wet chemical method *Microelectron. Eng.* **88** 3015–9

[13] Nechibvute A, Chawanda A and Luhanga P 2012 Piezoelectric Energy Harvesting Devices: An Alternative Energy Source for Wireless Sensors *Smart Mater. Res.* **2012** 1–13

[14] Idiawati R, Mufti N, Taufiq A, Wisodo H, Laila I K R, Fuad A and Sunaryono 2017 Effect of Growth Time on the Characteristics of ZnO Nanorods IOP Conference Series: Materials Science and Engineering vol 202

[15] Subagiyo D and Suyitno 2017 Enhancing Output Voltage of Piezoelectric-Based Nanogenerators Using Zinc Oxide Material Codoped with Aluminum and Cobalt *IOP Conf. Ser. Mater. Sci. Eng.* **267** 012037

[16] Chae D W and Kim B C 2006 Effects of zinc oxide nanoparticles on the physical properties of polyacrylonitrile *J. Appl. Polym. Sci.* **99** 1854–8

[17] Mufti N, Arista D, Diantoro M, Fuad A, Taufiq A and Sunaryono 2017 The Effect of Thickness of ZnO Thin Films on Hydrophobic Self-Cleaning Properties IOP Conference Series: Materials Science and Engineering vol 202

[18] Zamiri R, Rebelo A, Zamiri G, Adnani A, Kuashal A, Belsley M S and Ferreira J M F 2014 Far-infrared optical constants of ZnO and ZnO/Ag nanostructures *RSC Adv.* **4** 20902–8

[19] Yar A, Haspulat B, Üstün T, Eskizeybek V, Avcı A, Kamış H and Achour S 2017 Electrospun TiO 2/ZnO/PAN hybrid nanofiber membranes with efficient photocatalytic activity *RSC Adv.* **7** 29806–14

Phase Transition of SiO_2 Nanoparticles Prepared from Natural Sand: The Calcination Temperature Effect

Munasir[1,2,*], Zainul Arifin Imam Supardi[1,2], Mashadi[1], Zumrotun Nisa[1], Diah Hari Kusumawati[1,2], Nugrahani Primary Putri[1,2], Ahmad Taufiq[3], Sunaryono[3], Nurul Hidayat[3], Darminto[4]

[1]Department of Physics, Faculty of Mathematics and Natural Sciences, Universitas Negeri Surabaya (Unesa), Kampus Ketintang, Jl. Ketintang, Surabaya, 60231, Indonesia
[2]Research Center of Advanced Materials, Faculty of Mathematics and Natural Sciences, Universitas Negeri Surabaya (Unesa), Kampus Ketintang, Jl. Ketintang, Surabaya 60231.
[3]Department of Physics, Faculty of Mathematics and Natural Sciences, Universitas Negeri Malang, Jl. Semarang 5 Malang 64145, Indonesia
[4]Department of Physics, Faculty of Mathematics and Natural Sciences, Institut Teknologi Sepuluh Nopember Surabaya (ITS), Kampus ITS Sukolilo, Surabaya 60111, Indonesia

*Corresponding author's email: munasir_physics@unesa.ac.id

Abstract. In this paper, we systematically report the synthesis of nano silica powder from natural silica sand by a continuous process. Extraction of sodium silicate from silica sand was through a hydrothermal process, then sodium silicate was ready to be used as a precursor in the form of silicate slurry by coprecipitation process. The silicate slurry of $SiO_2.XH_2O$ was then dried in a furnace at 150 °C for 4 hours until obtaining a pure white SiO2 powder. At a calcination temperature of 900 °C, the sample was then tested by XRD to analyze the transformation of the crystal phase of SiO_2 nanoparticles. Identification of functional group absorption was undertaken by FTIR test, and the particle grain microstructure was analyzed by SEM. At calcination temperature of 900 °C, the SiO_2 silica nanoparticles experienced a change of phase from amorphous to crystal or amorphous-cristobalite phase; a shift in stiffness for positions of functional group absorption of Si-O stretching or LO functional groups, Si-O stretching or TO, Si-O-Si bending and Si-O-Si rocking and OH-functional groups occurred; the growth of particle grains happened with a spherical and oval trend, with a larger size. SiO_2 nanoparticles were successfully synthesized by a continuous method and 900 °C calcination temperature had a significant effect on structural phase changes, the formation of siloxane functional groups, silanol, and grain growth on SiO_2 nanoparticles.

Keywords: Silica, nanoparticle, natural sand, calcination, amorphous, cristobalite.

1. Introduction

Quartz sand in nature contains many oxide elements such as SiO_2, Fe_2O_3, Al_2O_3, TiO_2, CaO, MgO and K_2O [1-3]. Silica has crystalline polymorphs namely quartz, cristobalite and tridymite, stishovite, and coesite [4-6] structures. Amorphous silica can be synthesized from natural materials (minerals), such as quartz sand [1], siliceous sands [3], and Douiret sands [7]. The synthesis method which can be used is by a dry method or known as an alkali-fusion process, reacting with an alkaline compound in the furnace; KOH with temperature of 350 °C [8], NaOH with temperature of 500 °C [1], and Na_2CO_3 with temperature of 1100 °C [7]. It can also be done by wet or hydrothermal or chemical processes [9] such as the sol-gel process [10], coprecipitation, etc. [11].

Silica is widely used in various applications such as a mixture of concrete [12,13], tire industry, rubber nanocomposite [14], wall-paint emulsion [15], improved lotions and cosmetics, increased ingredient cosmetic efficacy [16], packaging and anti-bacterial [17,18]. Modification of hydrophobic and hydrophilic properties of nano silica also provides many applications in technology, such as superhydrophobic textiles and ultra-violet yarns [19,20], and super-hydrophobic semi-transparent thin films [21,22]. Nano silica can also be applied as a base material for Ion-Lithium batteries: electrode [23], electrolyte [24,25] and the separator and the like [26].

In nature, the stable silica structure is a quartz phase, so that in macro and micro sizes the silica particles have a crystal structure phase while the nanometer-size silica particles are generally amorphous structured [11]. In various processes of extraction and synthesis of nano silica from the natural (organic or inorganic) material, an amorphous structure is obtained [11,27,28]. Recently, nano-silica synthesized from quartz sand using the coprecipitation method obtained nano silica that encapsulated SiO2-cristobalite and tridymite [11], the amorphous-crystal phase. Nano silica that was processed from organic rice husk through the process of combustion at 900 °C, could also attain tridymite and cristobalite nano silica powder [29].

From the previous research reports, the transformation mechanism of the crystal phase of SiO_2 was influenced by the temperature and pressure. In the crystal phase of SiO_2, the things that possibly happen are α-*quartz* (trigonal), β-*quartz* (hexagonal), α-*tridymite* (orthorhombic), β-*tridymite* (hexagonal), α-*cristobalite* (tetragonal), and β-*cristobalite* (cubic). The phase transformation of α-*quartz* ↔ β-*quartz* occurred at 573 °C; β-*quartz*↔β-*tridymite* occurred at 870 °C; *α-tridymite* ↔*β-tridymite* occurred at 117 °C; *β-tridymite* ↔*β-cristobalite* occurred at 1470 °C, and *β-cristobalit*→*liquid silica* occurred at temperature ≥1710 °C [30,31,32]. The next is how to explain the mechanism of transformation of silica nanoparticles that commonly have an amorphous structure to be crystalline structure. Generally, the raw materials of silica sand have a crystal structure of α-quartz. After the hydrothermal process in the autoclave at the pressure above 1 atmosphere and temperature of ~350-1100 °C, the silicate sodium was produced; generally, amorphous silica or amorphous-crystal was created through coprecipitation process [11]. After that is how the silica nanoparticle properties and the process of phase transformation continue to the crystal.

This research concerned on natural sand purifying process and producing silica powder in nanometer using hydrothermal and coprecipitation methods. The effect of calcination temperature on the silica nanoparticles (amorphous) samples and the phase transformation of amorphous-crystalline were analyzed using X-ray Diffraction (XRD) and the functional groups were analyzed using FTIR spectroscopy wave absorption, and the particle form and the growth of particle grain were analyzed by SEM test.

2. Materials and Methods

The raw materials in this study were as follows: Natural silica sand, HCl 37%, NaOH 99%, and de-mineralized water. The synthesis employed a continuous method (hydrothermal-coprecipitation). The first step was the formation of silicate sodium precursor through hydrothermal. It was then mixed with NaOH 5 M solution and stirred wisely for 2 hours. The second stage was the coprecipitation process; the sodium silicate solution was dried at ambient condition for 24 hours and followed by stirring and adding HCl (2 M) to reach a neutral pH (8 - 7). The solution was looked muddy and formed a white gel. The precipitated gel was then filtrated and cleaned by de-mineralized water for several times. The precipitated $Si(OH)_2$ was dried in the oven with the temperature of 200 °C for 2 hours. The silica gel (SiO_2) was finally characterized by means of X-ray diffraction (XRD) to identify p structure and crystal phase. The analysis of functional groups in SiO_2 (Silanol, siloxane) used Fourier Transform Infra-Red (FTIR) test. The morphology of the particles (the form of particles) and the formation of the particles were observed using Scanning Electron Microscope (SEM).

3. Results and Discussion

3.1. XRD Analysis

The data of diffraction pattern of SiO_2 powder that had been synthesized using continuous method (coprecipitation-hydrothermal) for the hydrothermal time of 10 hours (Figure 1 (a)) showed the single peak at the position of (2θ) 21.6°. The diffraction pattern indicated that this silica powder was in an amorphous condition (Shinohara et al., 2004). If the calcination was done to the samples (silica powder) of the synthesis results at the temperature of 900 °C in the furnace for 10 hours, it would result in the diffraction pattern with the position of angle hump of (2θ) 22.01°. The diffraction pattern of XRD began to show the formation of tridymite crystal phase (Figure 1 (b)). Lastly, the samples calcinated at the temperature of 1000 °C (Figure 12(c)) showed the phase transformation of amorphous SiO_2 to polycrystal phase signed by the formation of some peaks on the crystal structures. The dominant crystal phases were Quartz, Tridymite, and Cristobalite. The XRD pattern showed one characteristic of Quartz phase forming the peak at 26.68° in accordance with Quartz (database 96-900-5018), the tridymite phase at 21.53° and 35.71° (database 96-901-3493), and the cristobalite phase formed the peak at (2θ) 22.05°, 28.49°, 31.43°, and 35.99° (database 96-900-8111) [30] . the analysis results showed multi-spaces with the majority of cristobalite phase (77.4%), a little bit tridymite phase (14.2%), and quartz phase (8.4%). If we see the transition process, the samples transformed from the Quartz phase to cristobalite phase and from cristobalite to tridymite. This case is a little bit different from the phase diagram of silica showing that the phase transformation from tridymite phase to cristobalite phase occurred at 1470 °C with the low pressure (below 20 kbar). The gel silica with the calcination at the temperature of 900 °C will form β-cristobalite form signed at (111), (110) and (120) and it will be more obviously formed at the temperature of 1400 °C [32]. The phase change from α-quartz to β-quartz or the other way around, from β-quartz to β-cristobalite was extremely influenced by the various temperatures [32].

The phase transformation process was different from the previous results for the raw material of SiO_2 crystal quartz (material from nature) with the particle size of micrometer as explained by Marthe Erdal Kjelstadli: quartz → tridymite (870 °C); tridymite → cristobalite (1470 °C) and cristobalite → liquid silica (1710 °C) [33]. This result strengthens the previous finding of Arianit A. Reka, et al. stating that the phase transformations were β-quartz ↔ β-tridymite at 870 °C, β-tridymite ↔ β-cristobalite at 1470 °C, and β-cristobalite → vitreous glass at 1705 °C [34].

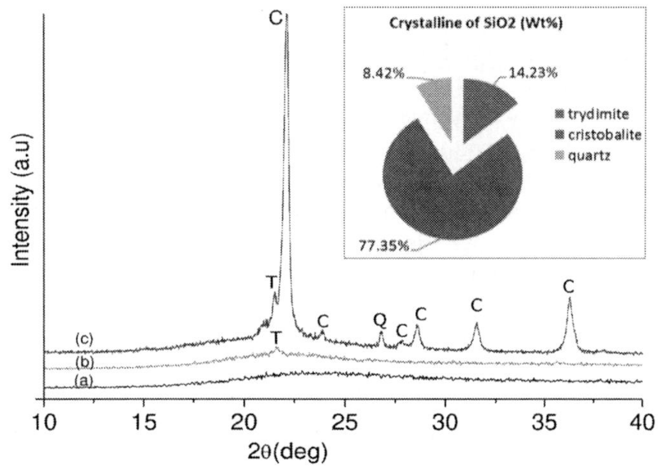

Figure 1. XRD pattern of the samples without (a) and with calcination at (b) 900 °C and (c) 1100 °C

3.2. FTIR Analysis

FTIR characterization aims to know the pattern of wave absorption in the samples. The pattern of infrared wave absorption of SiO sample for the condition before and after the calcination at 900 °C is shown in Figure 3 (a-b). The analysis results and matching of the pattern of infrared absorption in the functional groups of the sample as displayed in Figure 3, at the wavelength of 465-475 cm^{-1}, 800-870 cm^{-1}, and 3000-4000 cm^{-1} showed the absorption pattern of the Si-O functional group, –OH group of SiO, and O-H group. Such absorptions were in accordance with the characteristics of the absorption pattern of silica obtained from reference [34,35]. The absorption pattern of the Si-O-Si group (siloxane) as the synthesis result was located at the wavelength of 1098 cm^{-1}. Such result corresponded to the reference having the range between 1050 and 1115 cm^{-1}. The pattern of O-H functional group (water molecule) had the absorption at the wavelength of 1639 cm^{-1} [11, 30, 36]. The absorption pattern of Si-O-Si (siloxane) group as the synthesis result took place at the wavelength of 1095cm^{-1}. Such result was in line with the reference having the range of 1050-1115 cm^{-1} [29, 35, 37]. The pattern of O-H (water molecule) functional group had the absorption at the wavelength of 1639 cm^{-1} [11, 29, 35, 37].

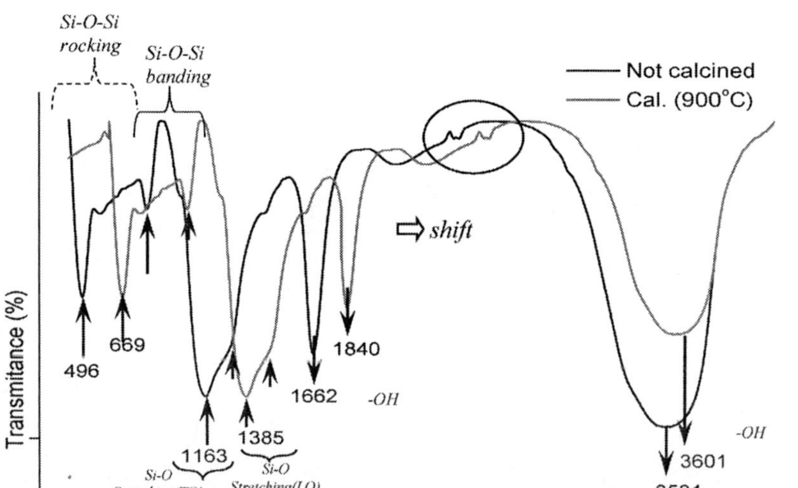

Figure 2. IR Spectroscopy of SiO_2 without (a) and with calcination at 900 °C

The absorption change of O-H functional group was shorter. This indicated that the water (OH) content was more eliminated while the Si-O-Si functional group was still dominant (the vibration was better), and for Si-O group was also getting dominant. This showed that the crystal characteristic of SiO_2 was also better (the crystal phase was formed). The results of FTIR characterization clearly showed the phenomena of the loss of the absorption pattern on O-H (water molecule) group as the calcination temperature used was getting higher [11,37,38]. Figure 2 respectively shows the absorption pattern of silica without calcination, and the absorption pattern (with calcination at 900 °C) indicating the decrease in the absorption pattern of O-H (water molecule) functional group. Besides, the phenomenon of shifting to the right (the wavelength was getting higher). The absorption position of silica functional group: Si-O-Si rocking, Si-O-Si banding, Si-O stretching transversal optic (TO) and Si-O stretching longitudinal optic (LO) [21, 22, 39].

Table 1. Corresponding the functional groups of the results of FTIR with reference

Wavelength (cm^{-1}) experimental data (a)	(b)	Functional Group	Wavelength literature (cm^{-1})	Reference
469	470	Group vibration Si-O	465 - 475	[11, 29, 37]
800	800	OH group vibration of Si-O	800 - 870	[11, 19, 29, 40]
1098	1095	Si-O-Si group asymmetry vibration	1050 - 1115	[11, 19, 37, 40]
1639	1639	O-H (water molecule) group vibration	1639	[11, 37, 38]
3463	3456	O-H group vibration	3000 - 4000	[17, 28, 38, 40, 41]

3.3. SEM Analysis

In this research, the morphology of silica as the synthesis result using hydrothermal-coprecipitation methods without calcination, with calcination at 900 °C (SEM data) is displayed in Figure 3. It is shown that the particle size was smaller and spherical which tended to be oval; the grain border was clear; the estimation showed the particle size of ~ 72.85 nm ($\leq$ 100 nm). The sample with the calcination temperature of 900 °C for 10 hours had shown the growth of its particles so that it formed the bigger size (Figure 3 (b)) so that the particle size was getting bigger with the more dense particle structure, the getting fewer grain borders indicated by the space between particles which disappeared. The particle size resulted was greater; the growth of the particle size caused the grain border between particles disappeared. When the calcination temperature was getting higher (started at 900 °C), the agglomeration process happened to form the bigger particles. The particles looked like a ball and crystal box as shown in Figure 3 (c).

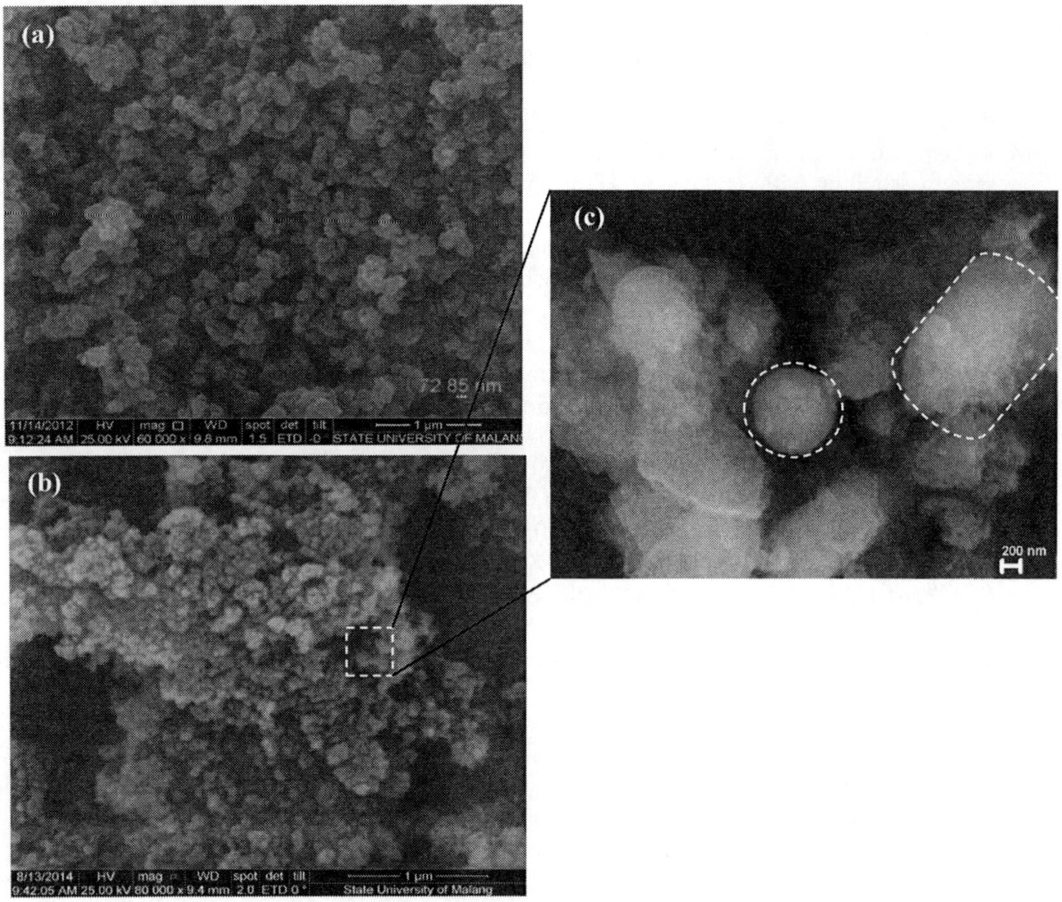

Figure 3. Micrograph SEM of SiO_2 nanoparticles: (a) without calcination, (b) after calcination at 900 °C for 10 hours, and (c) the form of sample particles (b)) in the position of taking the different picture with the magnification: like-spherical and crystal box.

4. Conclusion
Finally, it can be obviously stated that the continuous method has been successfully applied to produce silica (SiO_2) powder nano-crystalline with submicron particle size. The silica from natural sand can be separated from its contaminant and the particle size of the silica can be reduced to the nanoscale. The final product was in the form of SiO_2 nanoparticles powder with the amorphous phase with the size of ~72.85 nm. At 900 °C calcination temperature for 10 hours, the phase transformation happened from the amorphous phase to the polycrystalline phase with the domination of cristobalite crystal structure, tridymite, and quartz. The calcination process definitely influenced the change of the amorphous phase to the crystal and the particle grain size of SiO_2 was bigger and the particle grain border disappeared.

Acknowledgments
This research was supported by the Ministry of Research and Technology and Higher Education of Indonesia.

References

[1] Munasir, A. Sulton, Triwikantoro, M. Zainuri, Darminto, Synthesis of silica nanopowder produced from Indonesian natural sand via alkalifussion route, in: AIP Conf. Proc., AIP Publishing, 2013: pp. 28–31. doi:10.1063/1.4820986.

[2] M. Taxiarchou, D. Panias, I. Douni, I. Paspaliaris, A. Kontopoulos, Removal of iron from silica sand by leaching with oxalic acid, Hydrometallurgy. 46 (1997) 215–227. doi:10.1016/S0304-386X(97)00015-7.

[3] A. Sdiri, T. Higashi, S. Bouaziz, M. Benzina, Synthesis and characterization of silica gel from siliceous sands of southern Tunisia, Arab. J. Chem. 7 (2014) 486–493. doi:10.1016/j.arabjc.2010.11.007.

[4] M. Dapiaggi, L. Pagliari, A. Pavese, L. Sciascia, M. Merli, F. Francescon, The formation of silica high temperature polymorphs from quartz: Influence of grain size and mineralising agents, J. Eur. Ceram. Soc. 35 (2015) 4547–4555. doi:10.1016/j.jeurceramsoc.2015.08.015.

[5] C.S. Marians, L.W. Hobbs, Network properties of crystalline polymorphs of silica, J. Non-Cryst. Solids. 124 (1990) 242–253. doi:10.1016/0022-3093(90)90269-R.

[6] L. Guttman, Ring structure of the crystalline and amorphous forms of silicon dioxide, J. Non-Cryst. Solids. 116 (1990) 145–147. doi:10.1016/0022-3093(90)90686-G.

[7] W. Trabelsi, M. Benzina, S. Bouaziz, Physico-chemical characterisation of the Douiret sand (Southern Tunisia): Valorisation for the production of Silica Gel, Phys. Procedia. 2 (2009) 1461–1467. doi:10.1016/j.phpro.2009.11.117.

[8] H. Mori, Extraction of silicon dioxide from waste colored glasses by alkali fusion using potassium hydroxide, J. Mater. Sci. 38 (2003) 3461–3468. doi:10.1023/A:1025100901693.

[9] U. Zulfiqar, T. Subhani, S.W. Husain, Synthesis and characterization of silica nanoparticles from clay, J. Asian Ceram. Soc. 4 (2016) 91–96. doi:10.1016/j.jascer.2015.12.001.

[10] Q. Guo, D. Huang, X. Kou, W. Cao, L. Li, L. Ge, J. Li, Synthesis of disperse amorphous SiO2 nanoparticles via sol–gel process, Ceram. Int. 43 (2017) 192–196. doi:10.1016/j.ceramint.2016.09.133.

[11] Munasir, Triwikantoro, M. Zainuri, Darminto, Synthesis of SiO2 nanopowders containing quartz and cristobalite phases from silica sands, Mater. Sci.-Pol. 33 (2015) 47–55. doi:10.1515/msp-2015-0008.

[12] A.M. Said, M.S. Zeidan, M.T. Bassuoni, Y. Tian, Properties of concrete incorporating nano-silica, Constr. Build. Mater. 36 (2012) 838–844. doi:10.1016/j.conbuildmat.2012.06.044.

[13] M. Heikal, S. Abd El Aleem, W.M. Morsi, Characteristics of blended cements containing nano-silica, HBRC J. 9 (2013) 243–255. doi:10.1016/j.hbrcj.2013.09.001.

[14] Y. Chen, Z. Peng, L.X. Kong, M.F. Huang, P.W. Li, Natural rubber nanocomposite reinforced with nano silica, Polym. Eng. Sci. 48 (2008) 1674–1677. doi:10.1002/pen.20997.

[15] T. Mizutani, K. Arai, M. Miyamoto, Y. Kimura, Application of silica-containing nano-composite emulsion to wall paint: A new environmentally safe paint of high performance, Prog. Org. Coat. 55 (2006) 276–283. doi:10.1016/j.porgcoat.2005.12.001.

[16] C. Fruijtier-Pölloth, The toxicological mode of action and the safety of synthetic amorphous silica—A nanostructured material, Toxicology. 294 (2012) 61–79. doi:10.1016/j.tox.2012.02.001.

[17] S. Wang, W. Hou, L. Wei, H. Jia, X. Liu, B. Xu, Antibacterial activity of nano-SiO2 antibacterial agent grafted on wool surface, Surf. Coat. Technol. 202 (2007) 460–465. doi:10.1016/j.surfcoat.2007.06.012.

[18] Z. Luo, Y. Xu, Q. Ye, Effect of nano-SiO_2-LDPE packaging on biochemical, sensory, and microbiological quality of Pacific white shrimp Penaeus vannamei during chilled storage, Fish. Sci. 81 (2015) 983–993. doi:10.1007/s12562-015-0914-3.

[19] Q. Zhu, Q. Gao, Y. Guo, C.Q. Yang, L. Shen, Modified Silica Sol Coatings for Highly Hydrophobic Cotton and Polyester Fabrics Using a One-Step Procedure, Ind. Eng. Chem. Res. 50 (2011) 5881–5888. doi:10.1021/ie101825d.

[20] M.D. Teli, B.N. Annaldewar, Superhydrophobic and ultraviolet protective nylon fabrics by modified nano silica coating, J. Text. Inst. 108 (2017) 460–466. doi:10.1080/00405000.2016.1171028.
[21] P.-H. Chen, C.-C. Hsu, P.-S. Lee, C.-S. Lin, Fabrication of semi-transparent super-hydrophobic surface based on silica hierarchical structures, J. Mech. Sci. Technol. 25 (2011) 43–47. doi:10.1007/s12206-010-1013-0.
[22] B. Qiao, Y. Liang, T.-J. Wang, Y. Jiang, Surface modification to produce hydrophobic nano-silica particles using sodium dodecyl sulfate as a modifier, Appl. Surf. Sci. 364 (2016) 103–109. doi:10.1016/j.apsusc.2015.12.116.
[23] Z. Favors, W. Wang, H.H. Bay, Z. Mutlu, K. Ahmed, C. Liu, M. Ozkan, C.S. Ozkan, Scalable Synthesis of Nano-Silicon from Beach Sand for Long Cycle Life Li-ion Batteries, Sci. Rep. 4 (2014). doi:10.1038/srep05623.
[24] X. He, Q. Shi, X. Zhou, C. Wan, C. Jiang, In situ composite of nano SiO_2–P(VDF-HFP) porous polymer electrolytes for Li-ion batteries, Electrochimica Acta. 51 (2005) 1069–1075. doi:10.1016/j.electacta.2005.05.048.
[25] J.T. Park, K.J. Lee, M.-S. Kang, Y.S. Kang, J.H. Kim, Nanocomposite polymer electrolytes containing silica nanoparticles: Comparison between poly(ethylene glycol) and poly(ethylene oxide) dimethyl ether, J. Appl. Polym. Sci. 106 (2007) 4083–4090. doi:10.1002/app.26951.
[26] D. Fu, B. Luan, S. Argue, M.N. Bureau, I.J. Davidson, Nano SiO_2 particle formation and deposition on polypropylene separators for lithium-ion batteries, J. Power Sources. 206 (2012) 325–333. doi:10.1016/j.jpowsour.2011.10.130.
[27] S. Affandi, H. Setyawan, S. Winardi, A. Purwanto, R. Balgis, A facile method for production of high-purity silica xerogels from bagasse ash, Adv. Powder Technol. 20 (2009) 468–472. doi:10.1016/j.apt.2009.03.008.
[28] U. Kalapathy, A. Proctor, J. Shultz, An improved method for production of silica from rice hull ash, Bioresour. Technol. 85 (2002) 285–289. doi:10.1016/S0960-8524(02)00116-5.
[29] I.A. Rahman, V. Padavettan, Synthesis of Silica Nanoparticles by Sol-Gel: Size-Dependent Properties, Surface Modification, and Applications in Silica-Polymer Nanocomposites—A Review, J. Nanomater. 2012 (2012) 1–15. doi:10.1155/2012/132424.
[30] M. A Akl, H. Aly, H. M.A. Sholiman, A.M.. A. Abd-Elrahman, A.I. Abd-ElRahman, Preparation and Characterization of Silica Nanoparticles by Wet Mechanical Attrition of White and Yellow Sand, J. Nanomedicine Nanotechnol. 04 (2013). doi:10.4172/2157-7439.1000183.
[31] F.M. Wahle, R.E. Grim, R.B. Grab, Phase Transformation in Silica as Examined by Continous X-Ray Difraction, The American Mineralogist, 46(1961), 196-208.
[32] M.E. Kjelstadli, Kinetic and Mechanisme of Phase Transformation from Quartz to Crystobalite, NTNU, Norwegian University Science and Technology, Juni 2016.
[33] A. A. Reka, B. Palovski, T.Anovski, Phase Transformations of Amorphous SiO_2 in Diatomite at Temperature age 1000-1200°C, Geologia Macedonica, 29(1) (2015), 87-92.
[34] C. Jiang, Y. Zhang, Q. Wang, T. Wang, Superhydrophobic polyurethane and silica nanoparticles coating with high transparency and fluorescence, J. Appl. Polym. Sci. 129 (2013) 2959–2965. doi:10.1002/app.39024.
[35] R.K. Nariyal, P. Kothari, B. Bisht, FTIR Measurements of SiO_2 Glass Prepared by Sol-Gel Technique, Chem. Sci. Trans. (2014). doi:10.7598/cst2014.816.
[36] N. Dhiman, B.P. Singh, A.K.G. Gathania, Synthesis and characterization of dye-doped TiO2-SiO2 core-shell composite microspheres, J. Nanophotonics. 6 (2012) 063511. doi:10.1117/1.JNP.6.063511.
[37] A.N. Nittaya Thuadaij, Preparation of Nanosilica Powder from Rice Husk Ash by Precipitation Method, Chiang Mai J. Sci. 35 (2008) 206–211.
[38] I.A. Rahman, V. Padavettan, Synthesis of Silica Nanoparticles by Sol-Gel: Size-Dependent Properties, Surface Modification, and Applications in Silica-Polymer Nanocomposites—A Review, J. Nanomater. 2012 (2012) 1–15. doi:10.1155/2012/132424.

[39] P.K. Jal, M. Sudarshan, A. Saha, S. Patel, B.K. Mishra, Synthesis and characterization of nanosilica prepared by precipitation method, Colloids Surf. Physicochem. Eng. Asp. 240 (2004) 173–178. doi:10.1016/j.colsurfa.2004.03.021.
[40] S. Basu, ed., Crystalline Silicon - Properties and Uses, InTech, 2011. doi:10.5772/844.
[41] K.S. Rao, K. El-Hami, T. Kodaki, K. Matsushige, K. Makino, A novel method for synthesis of silica nanoparticles, J. Colloid Interface Sci. 289 (2005) 125–131. doi:10.1016/j.jcis.2005.02.019.

The Attenuation of Physical-Physiological Stresses through Musical-High Intensity Exercise Co-Treatment in Non-Athlete Individual

Sugiharto[1,*], Desiana Merawati[1], Rias Gesang Kinanti[1], Hendra Susanto[2], Ahmad Taufiq[3], Sunaryono[3]

[1]Department of Sports Science, Faculty of Sports Science, Universitas Negeri Malang, Jl. Semarang 5 Malang 65145, Indonesia.
[2]Department of Biology, Faculty of Mathematics and Natural Sciences, Universitas Negeri Malang, Jl. Semarang 5 Malang 65145, Indonesia.
[3]Department of Physics, Faculty of Mathematics and Natural Sciences, Universitas Negeri Malang, Jl. Semarang 5 Malang 65145, Indonesia.

*Corresponding author's email: sugiharto@um.ac.id

Abstract. Exercise is a very effective strategy to improve people's health. However, the unwell managed exercise potentially causes distress and decreases the health. Therefore, this research aimed to know the correlation of exercise combined with music to control stress/tissue injury by using physical-physiological neurohormonal paradigm approach. This study used randomized control group pre-test and post-test design with the sample of 30 non-athlete individuals randomly selected with the age range of 19-23 years old, the body mass index (BMI) of 19-23, VO_2 Max of 46-56 ml/kg/BB, and they do not smoke. The experiment group was divided into high-intensity exercise by listening to hard beat music, high intensity-exercise by listening to slow beat music, high-intensity exercise without listening to the music. The high intensity, in this research, means the intensity with 85% load of maximum working capacity with the multilevel load test using ergometer. The measurement of distress level was performed by taking the sample of 10 ccs of the blood of cube vein and the measurement of heart rate before and after doing an exercise. The distress level was determined using a biomaterial secretion indicator in the form of cortisol hormone, endorphin, cytokine interleukin-6 (IL-6), and the heart work. The method that was used to check the blood sample to know the cortisol, endorphin, and IL-6 content was ELISA (Enzyme-Linked Immunosorbent Assay), while the measurement of heart work was carried out using polar heart rate monitor. The research results showed that there was a significant difference in circulating level of endorphin and IL-6 ($P < 0.05$) after treatment by listening to slow beat music compared to the other experimental groups. Nevertheless, there was no significant difference in the circulating levels of cortisol ($P > 0.05$). In summary, giving co-treatment of high-intensity exercise with slow beat music could decrease the physical-physiological stress of non-athlete individual. It is suggested that the treatment could decrease the inflammatory within the skeletal muscle and improve the decrease in physiological stress by modulating the anti-physiological stress neurohormonal secretion.

Keywords: High-intensity exercise, music, stress, muscle, non-athlete, endocrine.

1. Introduction

Physical exercise is a fundamental effort to improve the health of children, adult, and elderly people. Exercise supports the increase in fat oxidation and the ability of muscle in using the fat as the energy source [1]. Therefore, exercise is a very effective strategy to improve people's health, prevent and treat metabolic syndrome [2], and prevent as well as decrease the stress since it is related to the increase in physical and spiritual fitness [3]. However, the high-intensity exercise is a stressor possibly causing physical, physiological, and psychological stresses as well as oxidative stress [4]. Interestingly, until now, the success of the exercise is in line with the getting higher exercise intensity given so that it is riskier to the increase in muscular stress [5]. Hence, the change of physical-physiological interaction and internal stress with a physio-neurohormonal indicator is vital. However, the information regarding this case is very rarely available so that an in-depth study is necessary through a molecular approach.

The previous research results explained that exercise is an essential thing in enhancing and maintaining the happiness through the healthy lifestyle [6]. Besides, the high-intensity exercise is regarded as beneficial to manage obesity since it can improve the metabolism rate and lipolysis as well as enhance the insulin sensitivity [1]. In contrary, exercise is also a stressor for a body that can impact on physical, physiological, and psychological stresses. This case is indicated by the increase in cortisol secretion and the decrease in testosterone hormone, endorphin hormone, and the other hormones [7]. In the high-intensity exercise, the muscle will be more active, need much more energy, improve the metabolism, and the inflammatory cytokine secretion results in the inflammatory of muscle due to the increase in IL-6 [8,9]. Importantly, exercise can influence the neurohormonal function related to the effort to face the exercise stressor [10]. The exercise with 60% intensity of VO_2 will improve the cortisol secretion and the higher the exercise intensity; the cortisol secretion will be higher as well. It is proven by the anaerobic sport which can enhance the cortisol secretion [11].

The research results in the last decades showed that listening to the music could prevent the cortisol hormone secretion during the exercise and avert the stress [12]. The exercise while listening to the music could enhance the mood, motivation, reduce the anxiety, stress response, blood pressure, heart rate, and stress hormone [13]. The rhythm, beat, and color of music physiologically influenced the convenience during exercise and could distract the attention of tiredness [14]. As a consequence, the physical performance increased, the tiredness sensation decreased, and people could enjoy and avoid distress [15]. The effect of music on the body was influenced by not only the exercise intensity but also the suitability of exercise intensity for the music beat, habitual factor, culture, and the music rhythm [16]. Classical music that was listened to during exercise had been proven to be able to provide relaxation compared to rock music [12]. Interestingly, the exercise while listening to the rock music for 30 minutes could increase the heart rate, blood pressure, and cortisol compared to listening to classical music [17]. Even though several previous studies have been done with the preliminary investigation related to musical-exercise combination model, the basic molecular profile linked cytokine, and hormonal alteration in high-intensity exercise combined with serial beat musical treatment are still not fully elucidated.

2. Experimental Methods

This study was an experimental research aiming at uncovering the physical-physiological response and muscle inflammatory to the musical beat combined with the high-intensity exercise. This research used a randomized control group pre-test and post-test design, with three experimental groups consisting of high-intensity exercise with hard beat music, high-intensity exercise with slow beat music, and high-intensity exercise without music. Slow beat music was decided about 130-140 beats/minute while hard beat music was selected on 150-170 beats/minute.

The human samples were selected from 30 healthy subjects with several criteria namely male, 19-23 years old, normal body mass index or it is about 19-23, heart rate, normal rest, VO_2 Max 46-56 ml/kg/BB, not smoking, non-athlete, and do not have the habit of drinking alcohol beverage, normal blood pressure, and willing to be the sample (informed content). The ethical clearance has been approved by ethic committee of Faculty of Medicine, Universitas Brawijaya, Indonesia No.

100/EC/KEPK/04/2018. The data collection was conducted by asking people to do warming up for five minutes by pedaling an ergometer bicycle with low intensity continued by pedaling the ergometer bicycle for 15 minutes with 85 % intensity of maximum working capacity and ended by the cooling for five minutes with low intensity. The measurement of stress level was by using marker of neuroendocrine and cytokine included cortisol, endorphin, and IL-6. The heart rate performance was measured pre and post treatment. Taking 10 ccs blood sample was to measure the cortisol hormone, endorphin, and IL-6 of the cube vein, while the measurement of heart performance used polar heart rate monitor set in the chest and the monitor set in the ergometer bicycle. Checking the blood sample was to measure the cortisol hormone, endorphin, and IL-6 using the ELISA method; the data analysis was by one way ANOVA with a 5 % significance level.

3. Results and Discussion

The research results of the physical condition of the samples related to anthropometric and baseline characteristics are presented in Table 1.

Table 1: Baseline characteristics of population

Groups	Age (years)	Resting Heart Beat (beats/minute)	BMI (kg/m^2)	Sistolic Blood Pressure (mmHg)	Diastolic Blood Pressure (mmHg)	VO_2 max (L)
HIE + HBM (n = 10)	20	74	21	110	90	51
HIE + SBM (n = 10)	20	69	22	120	90	52
HIE (n = 10)	19	68	21	120	90	49

HIE + HBM (High-Intensity Exercise + Hard Beat Music), HIE + SBM (High-Intensity Exercise + Slow Beat Music), HIE (High-Intensity Exercise); BMI (Body Mass Index)

Based on Table 1, the physical condition from the anthropometric aspect and physiological condition of each group were good, and the physical and physiological conditions of each group were not significantly different ($P < 0.05$). The research subjects were then given high-intensity exercise using music, and their stress level was measured using heart performance indicator, and the results are presented in Table 2. The high-intensity exercise without listening to the music caused the enhancement of heart performance than the high-intensity exercise while listening to the music. To know the stress level was by using a marker of cortisol hormone, endorphin, and IL-6 presented in Table 3. Based on the ANOVA (Table 3), there was a significant difference in endorphin content and interleukin-6 (IL-6) of the group given high-intensity exercise with slow beat music compared to control group ($p < 0.05$). However, there was no significant difference in circulating cortisol content in all experimental groups ($p > 0.05$).

Table 2. The cardiac performance of population

Groups	Heart Beat (beats/minute)	
	Resting	Post Exercise
HIE + HBM (n = 10)	73.	189
HIE + SBM (n = 10)	69	185
HIE (n = 10)	67	194

HIE + HBM (High-Intensity Exercise + Hard Beat Music), HIE + SBM (High-Intensity Exercise + Slow Beat Music), HIE (High-Intensity Exercise)

Physio-neurohormonal, music can increase the enjoyable feeling, mood, pleasure, decrease the emotion, as well as physical, physiological, and psychological stresses [18]. Besides, music can also reduce the perception of tiredness, mood, and pain suppression [19]. Exercise is a stressor that potentially disturbs body homeostatic and causes distress [11]. Hence, the well and right, as well as regulated and measured sports management, were by shifting the stress perception by using music is very important to improve the coping mechanism in the body in building the body adaptation. Music is a non-pharmacological approach to decrease the stress related to central nerve regulation in the hypothalamus-pituitary-adrenal axis (HPA axis) and sympathetic nervous system (SNS). Treatment using music could enhance the enjoyable feeling for the condition related to physical stress, emotional/mental stress, distress change to eustress through control of psychological and physiological aspects [20]. The stress level related to physical and physiological changes in this research used the marker of cortisol hormone, endorphin, cytokine IL-6, and heart performance. The heart performance used heart rate parameter as the distress indicator [5]. The research results showed that the high-intensity exercise without listening to the music caused the heart performance higher than the high-intensity exercise while listening to the music. Based on the variant test, there was a significant difference ($P < 0.05$). The improvement of higher heart performance in high-intensity exercise without listening to the music was the effect of sympathetic nerve stimulation that was more active than parasympathetic nerve. The research of [21] concluded that listening to the music during exercise could increase the parasympathetic nerve activity stimulation giving effect on the stress decrease in heart muscle. Moreover, the increase in heart rate was found in the exercise without music compared to the exercise with music [18]. This case was because music contributes to control and reduce the sympathetic nerve stimulation and increase the parasympathetic nerve stimulation. The parasympathetic nerve stimulation could decrease the stress of heart so that the heart performance decreased. Besides, music could also control the emotion level and decrease the anxiety level decreasing the heart performance [21]. The decrease in heart performance in the exercise with music was also influenced by the stressor level. The exercise with music caused the stress hormone activity lower so that that the anxiety level reduced, the relaxation and enjoyment while doing exercise increased [13]. The decrease in heart performance in the exercise while listening to music was caused by the music listened to is a voice wave of which stimulation arrives at the auditorious cortex in celebri cortex 41, stimulates to limbic system through cerebral cortex ring (limbic cortex), to the hypo campus, then Hypothalamus does the secretion of some hormones related to the emotional behavior setting [22]. Besides, the voice wave of music continues to formati retikularis giving the impulse to the autonomy nerve namely sympathetic and parasympathetic nerves. Both nerves are responsible for the heart performance setting [23].

Table 3. The circulating levels of physiological stress marker in subjects after high exercise treatment

Groups	Endorphin (ng/mL)		Cortisol (ng/mL)		IL-6 (ng/mL)	
	Pre HIE	Post HIE	Pre HIE	Post HIE	Pre HIE	Post HIE
HIE + HBM (n = 10)	301.0384	244.538	908.700	764.4000	359.7500	327.5000
HIE + SBM (n = 10)	192.4615	233.9230*	819.3000	1113.200	337.00	298.000*
HIE (n = 10)	212.3077	251.385	790.600	878.800	287.7500	294.2500

HIE + HBM (High-Intensity Exercise + Hard Beat Music), HIE + SBM (High-Intensity Exercise + Slow Beat Music), HIE (High-Intensity Exercise). *Significant difference with $p < 0.05$

The increase in heart rate in the high-intensity exercise with hard beat music was higher than the high-intensity exercise with slow beat music. The difference in heart performance of both groups was possibly caused by the difference in music beat, sympathetic and parasympathetic nerves. The hard beat music caused the sympathetic nerve oscillator stimulation in the brain more active. The stimulation level of sympathetic and parasympathetic nerves system to the respiration system as the consequence of hard beat music could increase the respiration frequency of 1:8, while the respiration frequency of the slow beat music was only 1:4 [24]. This case was assumed that listening to the hard beat music enhance the sympathetic nerve response and stimulate the high contraction of heart muscle so that the heartbeat content increased. Such enhancement was also caused by the active sympathetic nerve in doing secretion of catecholamine hormone of medulla adrenal, or the increase in venous return through frank starling mechanism. Besides, doing exercise while listening to the hard beat music also could improve the skeletal muscle performance influencing the increase in energy need. To fulfill the energy need, the heart performance was enhanced in order there is no homeostatic disturbance impacting on the distress increase. On the contrary, the slow music decreases the physiological and psychological responses, increase the resilience, motivation, decrease the tiredness, become more relax, reduce the stress of muscle, and prevent the working perception [25].

Based on the results of this research, the secretion of cortisol hormone in the group of high-intensity exercise with slow beat music and the high-intensity exercise without music tended to increase but the increase was responded by the increase in endorphin secretion. On the contrary, the high-intensity exercise with the hard beat music, the cortisol decrease occurred that was also responded by the decrease in endorphin secretion. Endorphin is an opied hormone secreted from adrenal pituitary gland to face distress, while if the body undergoes distress, it will stimulate corticotrophin releasing hormone (CRH) from hypothalamus and also stimulate the endorphin secretion and Adrenocorticotrophic Hormone (ACTH) as well as α-Melanocyte Stimulating Hormone (α-MSH) from pituitary anterior gland as the stimulator of the increase in cortisol hormone secretion [26]. Interestingly, the enhancement of cortisol secretion was responded by the endorphin increase. This case was caused by the exercise intensity done, the training level, and the type of music used. Exercise gives an opportunity for body to overcome stress that can cause the pleasure, happiness, but to adapt needs time to face the stress [6]. Therefore, the trained people could be able to do adaptation to the stress caused by exercise so that the exercise can improve the pleasure feeling supported by music that will influence neurohormonal system [27]. Listening to the music during exercise could change the depression level function, cognitive system, and endocrine glands [28]. The previous research proved that the rhythm and music melody influence human's endocrine nerve system that the impulse is started from limbic system, thalamus, and reticular activity system (RAS). However, the hard beat music influences the neuroendocrine, improves the heart performance, noradrenalin hormone secretion, cortisol, and adrenotrophic hormone causing the stress increase, arrhythmia, the high heart rate and blood pressure [27].

The increase in body response to do endorphin secretion when cortisol increases or decreases in all groups was possibly caused by the training level of the research subjects used so that it could prevent the stress of muscle and physiology. The endorphin secretion is an indicator of enjoyable feeling, pleasure, and relaxation in the body system, also as the anti-pain. Exercise can increase the brain-derived neurophic factor (BDNF) [29], while the response of cortisol secretion is related to cortisol hormone function in maintaining the glucose equilibrium in fulfilling the energy needed by muscle to do high-intensity exercise [5]. Cortisol is human's glucocorticoid. It is a catabolic hormone secreted from adrenal cortex because it was responded from exercise and psychological stresses, exercise with 60% intensity of maximum oxygen is one of the stressors that can cause the cortisol discretion. The cortisol secretion influences the metabolism to help the maintenance of blood glucose during exercise. This case is the part of activities in the skeletal muscle of adipose tissue to increase the amino acid and fat metabolism [30].

The high-intensity exercise without listening to the music caused the increase in IL-6 secretion higher than the other two groups. This case showed that exercise without listening to the music, the stress level was higher directing to distress. One of the differences was caused by the effect of music as the anti-inflammatory, anti-pain, as well as more relaxes muscle, and avoided the damage caused by the endorphin effect. Since the music causes the endorphin secretion, while the endorphin is hormone that causes not only enjoyment and relaxation, but also anti-pain hormone or opied system [26]. Endorphin secretion is very dominant with *agonist* μ and δ as the *opied* receptor. Such stimulation impacts on the decrease in Ca^{2+} ion concentration, while the decrease in Ca^{2+} ion causes the decrease in contraction and skeletal muscle relaxation that can prevent the inflammatory and stress of muscle [26]. The increase in inflammatory level indicated the distress of muscle signed by IL-6 secretion in the high-intensity exercise without music was higher than the exercise with listening to the music. This enhancement is related to the muscle performance that also caused the increase in heart performance by the epinephrine hormone and glucocorticoid hormones stimulation secreted during exercise (Pedersen, 2009). This case proved that music decreases the physiological and psychological stresses and increases the resilience ability, motivates to improve the performance, prevents the tiredness, and causes the relaxation so that it can prevent pain, adds the attention, and reduces the stress of muscle [25].

4. Conclusion

Based on the research results, it can be concluded that the high-intensity exercise while listening to the music could prevent the physical-physiology stress confirmed by neurohormonal and marker inflammatory. Giving the co-treatment of high-intensity exercise with music could suppress the muscle inflammatory and physiology stress. Hence, giving regular exercise combined with music could possibly reduce the metabolic disease progression. However, a further research is needed related to the suitability between the type of music, the number of beat, the type/color of music for the exercise in sport and recreation model.

Acknowledgments

We thank KEMENRISTEKDIKTI and Universitas Negeri Malang for Supporting research funding for this work.

References

[1] Jabbour G and Iancu H-D 2017 High-intensity exercise training does not influence body weight but improves lipid oxidation in obese adults: a 6-week RCT *BMJ Open Sport Exerc. Med.* **3** e000283

[2] Blizzard LeBlanc D R, Rioux B V, Pelech C, Moffatt T L, Kimber D E, Duhamel T A, Dolinsky V W, McGavock J M and Sénéchal M 2017 Exercise-induced irisin release as a determinant of the metabolic response to exercise training in obese youth: the EXIT trial *Physiol. Rep.* **5**

[3] Mastorakos G, Pavlatou M, Diamanti-Kandarakis E and Chrousos G P 2005 Exercise and the stress system *Horm. Athens Greece* **4** 73–89

[4] Brondani L de A, Assmann T S, Duarte G C K, Gross J L, Canani L H and Crispim D 2012 The role of the uncoupling protein 1 (UCP1) on the development of obesity and type 2 diabetes mellitus *Arq. Bras. Endocrinol. Metabol.* **56** 215–25

[5] Aizawa K, Nakahori C, Akimoto T, Kimura F, Hayashi K, Kono I and Mesaki N 2006 Changes of pituitary, adrenal and gonadal hormones during competition among female soccer players *J. Sports Med. Phys. Fitness* **46** 322–7

[6] Bhavsar D S D, S. Abhange D R and Afroz D S 2014 Effect of Different Musical Tempo on Post-Exercise Recovery in young adults *IOSR J. Dent. Med. Sci.* **13** 60–4

[7] Jarraya M, Chtourou H, Aloui A, Hammouda O, Chamari K, Chaouachi A and Souissi N 2012 The Effects of Music on High-intensity Short-term Exercise in Well Trained Athletes *Asian J. Sports Med.* **3** 233–8

[8] Gleeson M 2004 Immune function and exercise *Eur. J. Sport Sci.* **4** 52–66

[9] Elkington L J, Gleeson M, Pyne D B, Callister R and Wood L G 2015 Inflammation and Immune Function: Can Antioxidants Help the Endurance Athlete? *Antioxidants in Sport Nutrition* ed M Lamprecht (Boca Raton (FL): CRC Press/Taylor & Francis)

[10] Hackney A C 2006 Exercise as a stressor to the human neuroendocrine system *Med. Kaunas Lith.* **42** 788–97

[11] Minetto M A, Paccotti P, Borrione P, Massazza G, Ventura M, Termine A, Di Luigi L, Pigozzi F and Angeli A 2006 Effects of the training status on the hormonal response and recovery from high-intensity isokinetic exercise: comparisons between endurance-trained athletes and sedentary subjects *J. Sports Med. Phys. Fitness* **46** 494–500

[12] Hebert S, Béland R, Dionne-Fournelle O, Crête M and Lupien S J 2005 Physiological stress response to video-game playing: the contribution of built-in music *Life Sci.* **76** 2371–80

[13] Clark M, Isaacks-Downton G, Wells N, Redlin-Frazier S, Eck C, Hepworth J T and Chakravarthy B 2006 Use of preferred music to reduce emotional distress and symptom

activity during radiation therapy *J. Music Ther.* **43** 247–65

[14] Savitha D, Sejil T V, Rao S, Roshan C J and Roshan C J 2013 The effect of vocal and instrumental music on cardio respiratory variables, energy expenditure and exertion levels during sub maximal treadmill exercise *Indian J. Physiol. Pharmacol.* **57** 159–68

[15] Szabo A, Small A and Leigh M 1999 The effects of slow- and fast-rhythm classical music on progressive cycling to voluntary physical exhaustion *J. Sports Med. Phys. Fitness* **39** 220–5

[16] Bernardi L 2005 Cardiovascular, cerebrovascular, and respiratory changes induced by different types of music in musicians and non-musicians: the importance of silence *Heart* **92** 445–52

[17] Priest D L, Karageorghis C I and Sharp N C C 2004 The characteristics and effects of motivational music in exercise settings: the possible influence of gender, age, frequency of attendance, and time of attendance *J. Sports Med. Phys. Fitness* **44** 77–86

[18] Yamashita S, Iwai K, Akimoto T, Sugawara J and Kono I 2006 Effects of music during exercise on RPE, heart rate and the autonomic nervous system *J. Sports Med. Phys. Fitness* **46** 425–30

[19] Fukui H and Yamashita M 2003 The effects of music and visual stress on testosterone and cortisol in men and women *Neuro Endocrinol. Lett.* **24** 173–80

[20] Jurcău R and Jurcău I 2012 Influenţa terapiei prin muzică asupra stării de anxietate şi a cortizolului salivar, în stresul indus de efortul fizic intens şi de scurtă durată 5

[21] Jia T, Ogawa Y, Miura M, Ito O and Kohzuki M 2016 Music Attenuated a Decrease in Parasympathetic Nervous System Activity after Exercise *PloS One* **11** e0148648

[22] Hall J E 2016 *Guyton and Hall textbook of medical physiology* (Philadelphia, PA: Elsevier)

[23] Stefano G B, Zhu W, Cadet P, Salamon E and Mantione K J 2004 Music alters constitutively expressed opiate and cytokine processes in listeners *Med. Sci. Monit. Int. Med. J. Exp. Clin. Res.* **10** MS18-27

[24] Larsen P D 2005 The sound of silence is music to the heart *Heart* **92** 433–4

[25] Ghavam-Bakhtiar R, Nikbakht H, Ziaee N and Mohammadi M 2012 The effect of relaxing music on changes in blood lactate level during recovery following a maximal exercise session in young female athletes 4

[26] Nandhra T S and Carson R J 2000 beta-endorphin inhibits the production of interleukin-8 by human chorio-decidual cells in culture *Mol. Hum. Reprod.* **6** 555–60

[27] Amaral J A T, Nogueira M L, Roque A L, Guida H L, De Abreu L C, Raimundo R D, Vanderlei L C M, Ribeiro V L, Ferreira C and Valenti V E 2014 Cardiac autonomic regulation during exposure to auditory stimulation with classical baroque or heavy metal music of different intensities *Turk Kardiyol. Dernegi Arsivi Turk Kardiyol. Derneginin Yayin Organidir* **42** 139–46

[28] Satoh M, Ogawa J, Tokita T, Nakaguchi N, Nakao K, Kida H and Tomimoto H 2014 The Effects of Physical Exercise with Music on Cognitive Function of Elderly People: Mihama-Kiho Project ed C J Stam *PLoS ONE* **9** e95230

[29] Yeh S-H, Lin L-W, Chuang Y K, Liu C-L, Tsai L-J, Tsuei F-S, Lee M-T, Hsiao C-Y and Yang K D 2015 Effects of Music Aerobic Exercise on Depression and Brain-Derived Neurotrophic Factor Levels in Community Dwelling Women *BioMed Res. Int.* **2015** 1–10

[30] Brownlee K K, Moore A W and Hackney A C 2005 Relationship between circulating cortisol and testosterone: influence of physical exercise *J. Sports Sci. Med.* **4** 76–83

The Influence of Fire on Biomass Weight of Herbs in Lowland and Highland

Suhadi[1,*], Sueb[1], Sonny Wedhanto[2]

[1]Department of Biology, Faculty of Mathematics and Natural Sciences, Universitas Negeri Malang, Jl. Semarang 5 Malang 65145, Indonesia
[2]Department of Civil Engineering, Faculty of Engineering, Universitas Negeri Malang, Jl. Semarang 5 Malang 65145, Indonesia.

*Corresponding author's email: suhadi@um.ac.id

Abstract. The objectives of this study were to determine the differences of biomass in burning and non-burning sites in upland and lowland areas, to know the types of non-burning plants in burning locations, and to classify fast burning plants in burning sites. The study was done in Baluran National Park representing lowland and Ijen Crater Natural Tourism Park representing highland. The size of the plot is 1×1 m^2, and the repeated measures are four replications. The conclusions are that fire in lowland savanna decreased herb biomass, *Tectona grandis* Linn.f forest fire slightly decreased herb biomass, and fire in highland savanna increased herb soil biomass.

Keywords: Fire, herb biomass, lowland, highland.

1. Introduction

Forest fire frequently sweep over Indonesia. Reportedly, fire hit more than100.000 million hectares of forests and savanna from June until October 2015. It is forecasted to cause severe economic loss of more than US $ 15 billion that is equivalent to IDR 196 trillion. The fire took place in situ areas of Sumatera, Kalimantan, and Java. Another unfavorable impact of fire is the distribution of carbon dioxide emissions. Forest fire in 1997 contributed 2.6 billion tonnes of carbon dioxide emissions to the six billion tonnes total carbon emissions all over the world in the same year [1].

On the other hand, fire can be a tool in the management of savanna ecosystems. When used, the fire has more prolonged effects on vegetation. Fire prevents forest regeneration, and control the growof shrubs and worms that feed on growing grasses. Fire is the fastest way to control overgrown shrubs. Savana fire set by people in the early rainy season is one of the ways to decrease bigger fire risk in the future[2]. In Brazil, fire control on weeds increases pH, Ca, Mg, and Na exchange, decreases Al but the P decreases in the first up to the fourth year. Oxisol with a combination of fire and P fertilizer can improve the soil quality [3].

Forest fire smoke on moss land contains harmful organic carbons. There is 80% N composition in the air, and the rests are CO_2, N_2O, NOx, and CO gases. There are other elements which are more dangerous, i.e., numerous heavy metals presenting a very smooth shape. The particle sizes from 2.5 up to 0.1 microns like Cr, Cd, and Ni [4].

Ijen Crater National Park is located in Banyuwangi and Bondowoso. It covers an area divided into a 92-hectare National Park and 2,468-hectare nature conservation. This area is a transition of Mount Rain Forest at an altitude of 1,000 - 2,500 from the sea level toSub-Alpine Rain Forest altitude of 2,500 - 4,000 m above sea level. The dominant species are Compositae and Ericaceae Families. In the Sub-Alpine rainforest were found montane dry forest and shrubs. This area often burns in dry season since it is dominated by pines (*Casuarina junghuhniana* Miq) and shrubs.

Baluran National Park covers an area of 25,000 hectares with a 40-km coastline. In the middle of Baluran National Park, there is an inactive Baluran mountain with a peak of 900-1247 m above sea level. The topography varies from flat to a 60° slope. Flat until slope areas are found along the beach to the foot of Mount Baluran. Specific forest ecosystems on the island of Java dominate the savanna ecosystem. The forest is made by savanna that occupies 40%, 20% teak forests, and the remaining are mangrove, coastal, swamp, and evergreen forest. There are 444 species of flora and 26 species of mammals and 147 species of birds. It burns every year with the fire concentration mainly in savanna and the *Tectona grandis* Linn.f forest [5]. The Baluran National Park houses 15 dominant species of Polytrias praemosa and *Axonopus compressus* (Sw.) P.Beauv., while the savanna forest consists of 58.46% Gramineae; 24.21% Fabaceae; 2.56% Euphorbiaceae; 8.86% Malvaceae and 5.91% Cyperaceae [6]. *Acacia nilotica* (L.) Willd. ex Delile at Baluran National Park is used for very high firebreaks due to its rapid growth inhibiting the growth of herbs in the savanna [7].

Forest fires decreases soil nutrients in soil surface in number and quality so the growth and development of plants will be hampered. Similarly, heavy metals especially Hg will accumulate in soil and may accumulate in animal tissues, plants and contaminate water areas around the fire. There is a fire in the areas of Baluran National Park and Ijen Crater Park in almost every 3-5 years yet the medium of fire and its cause are indeed unidentified. The objectives of this study were to determine the differences of biomass in burning and non-burning sites in upland and lowland areas, to identify the types of plants in burning and non-burning locations, and to know fast growing plants in burning locations.

2. Experimental Methods

The lowland sampling locations are Briu Savannah and teak forests of Baluran National Park of between 5-20 meter altitudes above sea level, while the highland sampling places are Ijen Crater Natural Tourism Park, which is between 2500-3000 meters above sea level. The samples of lowland are taken at Baluran National Park, and the samples of highland are taken at Ijen Crater Natural Tourism Park. The method used is the quadratic method. The points of the sample point lay on the transect path following the contour line. The distance between plots are 10 meters and 50 meters between transects. The size of the plot is 1×1 m^2, and the measures are repeated in ten replications.

3. Results and Discussion

At Briu Savanna, the biomass of *Aeschynomene elegans* Schlecht&Cham in the fire area reached 15.96 gram.m^{-2} and 13.25 gram.m^{-2}in non-burning area.Besides that were found Genus Aeschynomene, herbaceous perennial with a strong taproot, pubescent stems (sometimes hispidulous) to 1 m long and with 1 - 3 mm in diameter, leaves 5-7 (-9) foliolate, leaflets obovate to elliptical about 6 - 12 mm long, and 2.5 – 4 mm wide, pubescent on both surfaces. Also, there was inflorescence comprising 1 or 2 yellow flowers with a standard 7 - 9 mm long. Next also found a pod curved, 15 - 30 mm long comprising 4 - 8 articles, dehiscing along the scalloped suture at maturity. Finally, there were seeds of 2 mm long, and 1.5 mm wide, commonly dark brown, but varying from light yellowish-brown to almost black; c. 370,000-450,000 seeds/kg. Plants are morphologically similar to A. elegans and A. brevifolia [8,9]. At burned and unburned areas, the growth of *Aeschynomene elegans* Schlecht & Cham is so rapid that its biomass in one season occurs in competition over other plants, especially in the lowlands, whereas a slow growth was found in the non-burning areas at the highlands. In their research [3], Serrao et. al. expressed that fire would increase the pH, Ca, Mg, Na exchange.

Table 1. Lowlands (*Tectona grandis* Linn.f Forest at Baluran National Park)

	No Fire	**Transition**	**Fire**
Mean of biomass weight	49.09 gram.m^{-2}	12.61 gram.m^{-2}	44.57 gram.m^{-2}
The dominant plant	*Mimosa invisa* Mart. ex Colla	*Themeda arguens* (L.) Hack	*Themeda arguens* (L.) Hack

Table 2. Burning and non-burning plants biomass in the lowlands

No	**Species**	**Burning (gram.m^{2})**	**Non Burning (gram.m^{-2})**
1.	*Aeschynomene elegans* Schlecht & Cham	15.96	13.25
2.	*Apluda mutica* L.	0.79	0.00
3.	*Blumea balsamifera* (L) DC	1.68	3.03
4.	*Calopogonium mucunoides* Desv	0.54	6.84
5.	*Commelina diffusa* Burm.f	0.54	0.00
6.	*Desmodium gangeticum* (L.) DC	0.65	0.00
7.	*Elephantopus scaber* L	0.56	0.00
8.	*Eranthemum capense* L	0.65	0.00
9.	*Eudalia amaura* (Buese) Ohwi	10.08	12.23
10.	*Euphorbia hirta* L	0.52	0.00
11.	*Imperata cylindrica* (L.) P.Beauv.	0.00	0.93
12.	*Mimosa invisa* Mart. Ex Colla	3.18	12.84
13.	*Moghania strobilifera* ((L.)J.St.-Hil.	2.40	0.00
14.	*Oplismenus burmani* (Retz.) Beauv	1.84	0.00
15.	*Passiflora foetida* L	0.00	0.79
16.	*Pennisetum purpureum* Schumach	0.23	1.54
17.	*Cyperus rotundus* L	0.00	0.93
18.	*Schlerachne punctata* R. Br	2.84	5.39
19.	*Sida rhombifolia* L	0.81	0.00
20.	*Sorghum nitidum* (Vahl) Pers	0.70	9.87
21.	*Synedrela nodiflora* (L.) Gaertn	0.52	0.00
22.	*Themeda arguens* (L.) Hook	8.81	13.69
23.	*Uraria lagopodoides* (L.) Desv. Ex DC	0.79	3.17
	Total	54.12	84.51

In research[10] and [11], the biological characteristics that make *Imperata cylindrica* (L.) P.Beauv. very successful were reviewed. The weed shows wide genetic variability that allows it to adapt to a wide range of ecological and management conditions. It possesses five taxonomic varieties with var. major in Asia and var. Africana in West African being the most serious. *Imperata cylindrica* (L.) P.Beauv. breed through sexual reproduction of seeds and vegetative by rhizomes. It flowers rapidly after experiencing stress from fire, absorbing soil nutrients high, and drought resistant. It produced 3000 seeds, without dormancy period, and with more than 1 year life [12]. In aggressive and invasive natures, rhizomes that usually concentrate above 15-20 cm above ground can survive for a long period [13]. Rhizome has a high regenerative ability because of the numerous buds that readily sprout into new shoots after fragmentation by tillage or any other forms of disturbances. Rhizome is resistant to fire because of its deep soil burial. Deep burial also makes *Imperata cylindrica* (L.) P.Beauv. very resistant to most control strategies [13]. The ability of rhizome fragments to regenerate decreases with a reduction in length of rhizome segment. Longer rhizomes have better chances of sprouting because they have more carbohydrate reserves than short fragmented rhizomes [14].

Table 3. The biomass of burned and unburned plants in the highlands

No	Species	Burning (gram.m^2)	Non burning (gram.m^{-2})
1.	*Aeschynomene elegans* Schlecht& Cham	14.09	0.50
2.	*Ageratum conyzoides* L	0.00	0.34
3.	*Blume lacera* (Burm.f.) DC	16.08	0.50
4.	*Carex bohemica* L	5.38	56.08
5.	*Centella asiatica* (L) Urban	0.00	29.68
6.	*Cyperus* sp	0.00	2.80
7.	*Desmodium hirtum* Guill. &Perr	17.93	25.59
8.	*Desmodium paniculatum* (L.) DC.	23.89	54.23
9.	*Dryopteris* sp	65.20	0.00
10.	*Equisetum ramosissimum* Desf	25.30	0.00
11.	*Eragrostis brownii* (Kunth) Nees	16.63	4.96
12.	*Eragrostis cilianensis* (All.) Vign ex Janchen	6.96	7.15
13.	*Eupatorium inulifolium* Kunth	25.92	0.03
14.	*Hedychium roxburghii* (Blume) Kuntze	2.57	0.05
15.	*Imperata cylindrica* (L.) P.Beauv.	512.97	52.36
16.	*Ipomoea* sp.	2.57	0.00
17.	*Ipomoea triloba* L.	2.57	0.00
18.	*Oplismenus compositus* (L.) P. Beauv	0.00	38.30
19.	*Oxalis corniculata* L.	0.00	5.25
20.	*Phyllanthus niruri* L.	2.64	0.00
21.	*Plantathera susannae* (L.) Lindl	2.57	0.05
22.	*Polygonum* sp.	4.01	0.00
23.	*Pteris* sp.	0.00	7.89
24.	*Rubus niveus*Thunb.	2.57	0.00
25.	*Sorgum nitidum* (Vahl) Pers	222.99	239.01
26.	*Spilanthes* sp.	5.96	0.11
27.	*Vernonia cinerea* (L.) Less	0.00	1.16
28.	*Debregeasia longifolia* (Burm.f) Wedd	0.00	3.85
	Total	978.80	529.88

Imperata cylindrica (L.) P.Beauv. is a plant that has a rapid growth of biomass at fire area reaching 512.97 grams.m^{-2} on the plateau. Savanna which is dominated by this plant will accelerate the rate of natural succession to the secondary forest that will ultimately shade and suppress the weeds. In order to prevent fire, the growth of *Imperata cylindrica* (L.) P.Beauv. must be suppressed as it causes the risk of fire in the next year [15]. Fire separation by removing leaves, soil processing, and combustion in a controlled manner is necessary to prevent land fires[16].

Biomass in the location of Savana Briu National Park of Baluran is concentrated in lowland areas that have not experienced heavy fire reaching 84.51 gram.m^{-2} while it reached 54.12 gram.m^{-2}at the location of the heavy fire and a biomass of 40.58 grams.m^{-2}in weight transition area. The dominating plants are *Aeschynomene elegans* Schlecht&Cham and *Themeda arguens* (L.) Hack. These types of plants are commonly found in the area of ample water. In the state of lack of water, they were quickly dried up. Therefore, they are easy to burn and grow very fast in the open state. Flammable herbs have a biomass weight of 0.4 - 1.3 gram.m^{-2} [17].

The *Tectona grandis* Linn.f forest in flammable areas has been dominated by the type *Themeda arguens* (L.) Hack.The area that has not burned is dominated by *Mimosa invisa* Mart. ex Colla. This species is easily depressed by water and quickly dries in the dry season. The *Tectona grandis* Linn.f forest area includes deciduous forest type. Deciduous teak leaves add soil surface biomass, so the area is flammable. Fire will increase the pH, Ca, Mg, Na exchange[3] so it will accelerate the growth of weeds in the rainy season. Furthermore, Ijen Tourism Area has an altitude of 2500-4000 m above sea level. Both burned and unburned areas are dominated by *Imperata cylindrica* (L.) P.Beauv. dan *Sorghum nitidum* (Vahl) Press. In severe burning areas, the biomass reached 978.80 grams.m^{-2}. High herbaceous biomass is prone to fire. *Imperata cylindrica* (L.) P.Beauv. and *Sorghum nitidum* (Vahl) Pres species are easily depressed by water.

4. Conclusion

Based on the results and discussion, it can be concluded that fires in lowland savanna decrease herb biomass. *Tectona grandis* Linn.f forest fire decreases little herb biomass, and fires in highland savanna increase herb soil biomass. *Imperata cylindrica* (L.) P.Beauv. in the areas of highland fires have rapid growth.

Acknowledgments

Thanks to the Directorate General of Higher Education of the Ministry of Research of Technology and Higher Education and LP2M Universitas Negeri Malang who financed this research.

References

[1] Badan Pengendalian Dampak Lingkungan. dan Canadian International Development Agency (CIDA) - Collaborative Environmental Project in Indonesia (CEPI). Data 1997/1998 dari Asian Development Bank (ADB) . Departemen Kehutanan (2016). 1997.

[2] MacKinnon JR, IUCN Commission on National Parks and Protected Areas., International Union for Conservation of Nature and Natural Resources., United Nations Environment Programme., Global Environmental Monitoring System., World Congress on National Parks (1982 : Bali I. Managing protected areas in the tropics. International Union for Conservation of Nature and Natural Resources; 1986.

[3] Serrao, E.A.S., I.C. Falesi JBDV &J. FTN. Productividad de praderas cultivadas en suelos de baja fertilidad de la Amazonia del Brasil. Prod Pastos En Suelos Acidos Los Trop 1978;CIAT, Cali:211–43.

[4] Tempo. Ini Komposisi Berbahaya Kabut Asap Kebakaran Hutan - Nasional Tempo.co 2015. https://nasional.tempo.co/read/714629/ini-komposisi-berbahaya-kabut-asap-kebakaran-hutan (accessed August 3, 2018).

[5] Baluran BTN. Review Rencana Pengelolaan Taman Nasional Baluran. Proyek Pengembangan Taman Nasional Baluran Dengan Dana Reboisasi Tahun Anggaran 1994/1995. p. 20. 1995.

[6] Raharjo Y. Komposisi, Kelimpahan dan Diversitas Komunitas Tumbuhan Hutan Savana di Taman Nasional Baluran Banyuwangi. Universitas Brawijaya Malang; 1995.

[7] Alikodra HS. Tanaman Eksotik Akasia (Acacia Nilotica) dan Masalahnya bagi Ekosistem Savanna di Taman Nasional Baluran. Duta Rimba 1987;13:30–4.

[8] Cook B. Aeschynomene falcata (Poiret) DC. In : 't Mannetje, L. and Jones, (R.M. (eds) Plant Resources of South-East Asia No.4 Forages. Pudoc Scientific Publishers, Wageningen The Netherlands; 1992.

[9] CLEMENTS R, Grasslands GB-T, 2000 undefined. Sown pastures in subcoastal south-eastern Queensland: pasture composition, legume persistence and cattle liveweight gain over 10 years. TropicalgrasslandsInfo n.d.

[10] Townson J. Imperata cylindrica and its control. Weed Abstr 1991;40: 457-46.

[11] Terry PJ, Adjers G, Akobundu IO, Anoka AU, Drilling ME, Tjitrosemito S, et al. Herbicides and mechanical control of Imperata cylindrica as a first step in grassland rehabilitation. Agrofor Syst 1996;36:151–79. doi:10.1007/BF00142872.

[12] Santiago A. Studies on the autecology of I. cylindrical (L.) Beauv. Proc 9th Int Grasslands Congr Sao Paulo, Brazil 1965;pp. 499-50.

[13] Ivens GW. Imperata cylindrica (L.) Beauv. in West African agriculture. Proc BIOTROP Work Alang-Alang, Bogor, 27-29 July 1976 1980:149–56.

[14] Ivens GW. Studies on Imperata cylindrica (L.) Beauv. and Eupatorium odoratum L. Weed Research Project R 2552, 1971-1973. Stud Imp Cylind Beauv Eupatorium Odoratum L Weed Res Proj R 2552, 1971-1973 1975.

[15] Anonim. Imperata management for smallholders: an extensionist's guide to rational Imperata management for smallholders 1996.

[16] Wibowo A, Suharti M, Sagala APS, Hibani H, Van Noordwijk M. Fire management on Imperata grasslands as part of agroforestry development in Indonesia. Agrofor Syst 1996;36:203–17. doi:10.1007/BF00142874.

[17] Reis SM, de Oliveira EA, Elias F, Gomes L, Morandi PS, Marimon BS, et al. Resistance to fire and the resilience of the woody vegetation of the "Cerradão" in the "Cerrado"–Amazon transition zone. Brazilian J Bot 2017;40:193–201. doi:10.1007/s40415-016-0336-1.

IOP Conf. Series: Journal of Physics: Conf. Series **1093** (2018) 012028 doi:10.1088/1742-6596/1093/1/012028

The Effect of ZnO Nanorods Morphology on Electrical Properties of Perovskite Solar Cells

Nandang Mufti[1,2,*], Riris Idiawati[1], Hari Wisodo[1], Yoyok Adisetio Laksono[1], Abdulloh Fuad[1], Markus Diantoro[1,2]

[1]Department of Physics, Faculty of Mathematics and Natural Sciences, Universitas Negeri Malang, Jl. Semarang 5 Malang 65145, Indonesia
[2]Centre of Advanced Materials for Renewable Energy, Universitas Negeri Malang, Jl. Semarang 5 Malang 65145, Indonesia

*Corresponding author's email: nandang.mufti.fmipa@um.ac.id

Abstract. Perovskite solar cells have attracted the researchers' attention because they are one of the most promising candidates that are cheap, easy to fabricate and high-efficiency solar cell. ZnO nanorods (ZnO NRs) can be used as an electron and hole recombination barrier to improve the perovskite solar cell performance. This research aimed to investigate the effect of growing time on the characteristics of ZnO NRs and its application for perovskite solar cell device. The ZnO NRs was synthesized on ITO substrate by hydrothermal method at the different growing times of 4, 6, and 8 hours. The $CH_3NH_3PbI_3$ perovskite solar cells were deposited on the top of ZnO NRs/ITO thin films by a two-stage spin coating method. The solar cell devices were characterized by using XRD, SEM-EDX, FTIR, UV-vis, and LCR meter. The results showed that ZnO NRs/ITO thin films were formed homogeneously and vertically oriented on all ITO substrates with hexagonal structures of zincite phases. When the growing time of ZnO NRs increased, the degree of homogeneity, density, and length of the rods enhanced as well. All the samples have the bandgap around 3.05 eV. The FTIR results showed the absorbent functional forces at 870 cm^{-1} and 1250 cm^{-1} related with H-C-N wavelengths, and ZnO wavelength ranges from 400 to 515 cm^{-1}. The $CH_3NH_3PbI_3$ perovskite solar cell device still contained PbI_2 as impurities phase. The longer the growing time, the rods resulted by ZnO NRs had the wider diameter and the longer length and they were more homogenous. Meanwhile, the effect of the number of PbI_2 impurities resulted in the higher dielectric and smaller current.

Keywords: ZnO nanorods, perovskite solar cell, $CH_3NH_3PbI_3$ and hydrothermal method.

1. Introduction

In some recent decades, many researchers are interested in conducting the research on perovskite solar cells (PSCs) [1–3]. This is because PSCs have the potential to be produced with low cost, relatively easy fabrication and result in high efficiency [3]. The organic-inorganic hybrid compounds ($CH_3NH_3PbX_3$, X=I, Cl, Br) with the perovskite crystal structure is a material that can absorb the light well [4]. Especially the $CH_3NH_3PbI_3$ (methyl-ammonium lead halide) perovskite material has resulted in 19% efficiency [5]. The $CH_3NH_3PbI_3$ has a direct bandgap of 1.55 eV [6]. The organometal perovskite material of halide group has the high absorption coefficient about 10^5 cm^{-1} with the absorption wavelength above 800 nm and it also has unique electrical characteristic [7].

One of the factors determining the efficiency of perovskite solar cells is the active layer morphology namely $CH_3NH_3PbI_3$ film. The $CH_3NH_3PbI_3$ film must have the big crystal size in the range of 500 - 1000 nm and high crystallinity with the cubic and homogeny form to be able to result in perovskite solar cells with the high efficiency [1,2,4]. One of the weaknesses of PCSs is its low stability where the PCs stability decreases to the duration of UV ray radiation. Solving such thing needs certain material that functions as perovskite electron transport with semiconductor materials such as TiO_2 [8] or ZnO [9]. ZnO has some benefits since it is cheaper, its band gap is wide so it is easily grown up on the substrate, it is non-toxic and environmentally friendly [9,10].

ZnO synthesis with any kinds of morphology has been conducted by many researchers such as nanorod [11], nanowire [12] nanoflow [13], nano-comblike [14], nanosheet [15], and nanoforest [16]. ZnO with nanorod morphology has the benefit since it is vertically oriented to the substrate so that it can well improve the effectiveness of the electron transport on PSCs [9,17]. ZnO nanorods (NRs) can enhance the electron diffusion on the photo-electrode layer by providing the conduction path directly (direct pathway) so that it reduces the charge recombination level usually occurring in conventional polycrystalline nanoparticles randomly oriented [18].

Synthesis of ZnO NRs that have been carried out were such as chemical bath deposition (CBD) method at the low temperature [19], and the hydrothermal method [20]. Meanwhile, $CH_3NH_3PbI_3$ perovskite coating is commonly undertaken by one stage and two stages-spin coating method [21]. The perovskite coating with two stages could result in good morphology so that it results in the higher photovoltaic performance [22]. PSCs used ZnO NRs with HTM Spiro-OMeTAD resulted in 8.90 % efficiency on FTO substrate and 2.62% on PET/ITO substrate [19]. PSCs with 11% efficiency based on ZnO NRs were successfully done by Son *et al.* (2014) using $CH_3NH_3PbI_3$ perovskite [9]. PSCs efficiency based on ZnO NRs was extremely influenced by the rods morphology besides its perovskite structure and the used HTM. Therefore, this research examined the influence of ZnO NRs morphology resulted from the various growing times to the electrical properties of PSCs.

2. Materials and Methods

This research used Indium Tin Oxide (ITO) substrate coated in the glass. For the synthesis of ZnO seed layer, the main materials of Zinc Acetate Dihydrate, Ethanol, and Monoethanolamine were utilized. Meanwhile, to grow up ZnO NRs, the main materials used were Zinc NitrateTetrahydrate, Hexamethylenetetramine (HMT), and Deionized Water.

Making the ZnO seed layer solution was conducted by dissolving Zinc Acetate Dihydrate with ethanol, then stirred using a magnetic stirrer on the hot plate at the temperature of 70 °C for 45 minutes. Subsequently, Monoethanolamine (MEA) was added wisely with the molar ratio between Zinc Acetate Dihydrate and MEA about 1:1. The solution was untreated for 24 hours at the room temperature then sparingly dropped on the ITO substrate, and then the spin coating was done with the velocity of 2500 rpm for 25 seconds. After that, the preheating was done at the temperature of 150 °C for 10 minutes and the annealing was performed at the temperature of 550 °C for two hours so that the ZnO seed layer film was formed.

Growing up ZnO NRs was carried out by the hydrothermal method. The solution for ZnO NRs growing was done by mixing $Zn(NO_3)_2.4H_2O$ 75Mm with HMT with the molar ratio of 1:1. The solution was then stirred using a magnetic stirrer for 45 minutes at the room temperature. Subsequently, ZnO seed layer film was dipped into the prepared solution. The growing process was undertaken at the temperature of 90 °C with the various growing times of 4, 6, and 8 hours. The obtained sample was rinsed with DI-water, dried, and then annealed at the temperature of 550 °C for two hours in the furnace. The sample was then characterized using XRD, SEM, FTIR, and UV-Vis.

The fabrication of PSCs was performed by two-stage spin coating on ZnO NRs substrate. The first step of PbI_2 coating was from the mixture of 1 mol/L PbI_2 solution and dimethylformamide (DMF) with the rotational speed of 4500 rpm for 15 minutes. The sample was then dried at the temperature of 100 °C for 30 minutes. The next step was the spin coating of 10 mg/mL CH_3NH_3I solution dissolved in IPA with the rotational speed of 2500 rpm for 10 seconds. The sample was then dried in the oven at the temperature of 100 °C for 30 minutes. The sample was characterized by XRD, LCR meter, and solar cell efficiency test.

3. Results and Discussion

The X-ray diffraction pattern of ZnO NRs/ITO thin film with the precursor concentration of 75 mM in the various growing times of 4, 6, and 8 hours using hydrothermal method is shown in Figure 1. The analysis of ZnO NRs phase informed that the formed diffraction pattern was in accordance with the data of XRD JCPDS standard No.36-1451with the hexagonal crystal (zincite) system having the space group of P63mc, with the crystal lattice of *a, b* =3. 2533 Å and *c* =5.2073 Å. The formed phase was suitable for the report of Rai, *et al.* (2014), that ZnO NRs were formed in zincite phase [23]. Figure 1 shows the diffraction peak located at (100), (002), and (101). The XRD pattern with the annealing temperature of 550 °C for two hours at the NRs growing temperature of 90 °C had the crystal orientation at the highest intensity peak on (002), while on (100) and (101) the pattern weakened since the ZnO NRs growing was oriented to c direction. This case is in line with the report of Selvan, *et al.* (2014) [24].

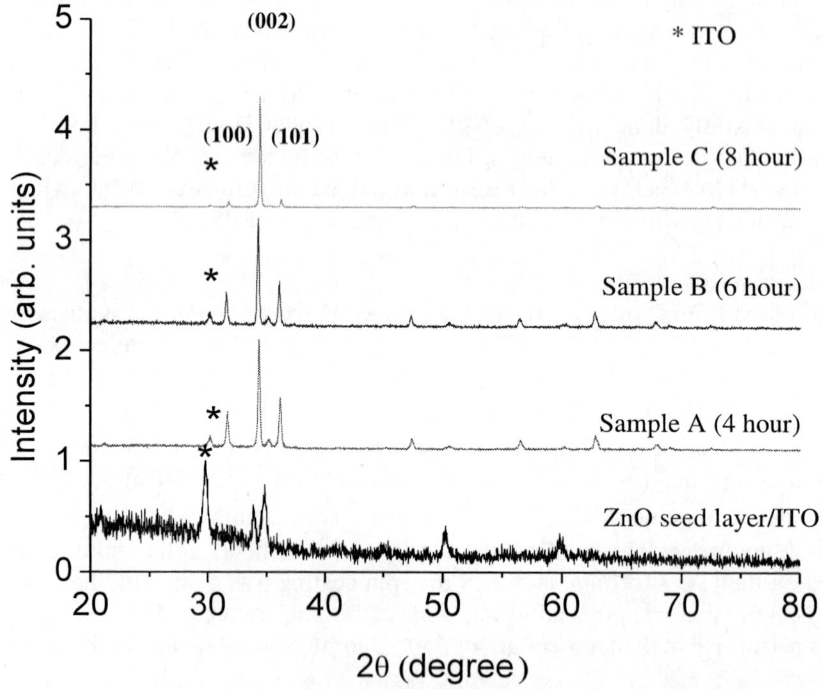

Figure 1. The diffraction pattern of ZnO NRs/ITO as the function of temperature

Figure 2 shows the cross-section and top-section morphology of ZnO nanorods film with the various growing times of 4, 6, and 8 hours. The results of SEM characterization showed the forming of the structure having rod form with a hexagonal structure on all substrate surfaces.

Table 1. The dimension of ZnO nanorods.

Samples	Length (nm)	Diameter (nm)
ZnO NRs 4 hours	835	138
ZnO NRs 6 hours	934	230
ZnO NRs 8 hours	1399	236

The results of the SEM characterization were analyzed and tabulated in Table 1. They showed that the longer the growing time of ZnO NRs, the resulted rods were higher and the diameter was bigger. This case showed that the growing time could increase the diameter and length of ZnO as reported by the previous researchers [25,26]. NRs morphology with the higher density can be seen in ZnO NRs synthesized with the growing time of hydrothermal for 8 hours. The longer the growing time with the hydrothermal method, it influenced the increase in the formed NRs density. This case is in line with the statement of Baruah and Dutta (2009) [27]. The more dense structure of ZnO NRs could enhance the orientation degree to a vertical direction. This case was because each rod supported each other so that they minimized the possibility of the rods to be inclined or collapsed.

Figure 2. SEM image of the ZnO NRs in various growing times (a) 4 hours, (b) 6 hours, and (c) 8 hours

The characterization of UV-Vis spectrophotometer in the wavelength range of 350-800 nm has been carried out to identify the transmittance spectrum and optical properties of ZnO NRs. The transmittance spectrum of ZnO NRs is shown in Figure 3. ZnO NRs in the precursor concentration of 75 mM with the various growing times of 4, 6, and 8 hours showed the strongest excitonic absorption at the wavelength of ~376 nm. This result is in line with the research results of Foo *et al.* (2014) with the strongest excitonic absorption at the wavelength of 378 nm [28]. This case indicated that ZnO NRs had the good optical quality and strong excitonic bonding energy. The strongest peak at the wavelength of ~360 nm was the emission peak of UV from free exciton. The absorption coefficient (α) for the transition direction of the ZnO NRs/ZnO seed layer of ITO is formulated through the Equation 1.

$$\alpha = \ln \frac{\left(\frac{1}{T}\right)}{d} \tag{1}$$

where T is the transmittance of ZnO film and d is the film thickness [28].

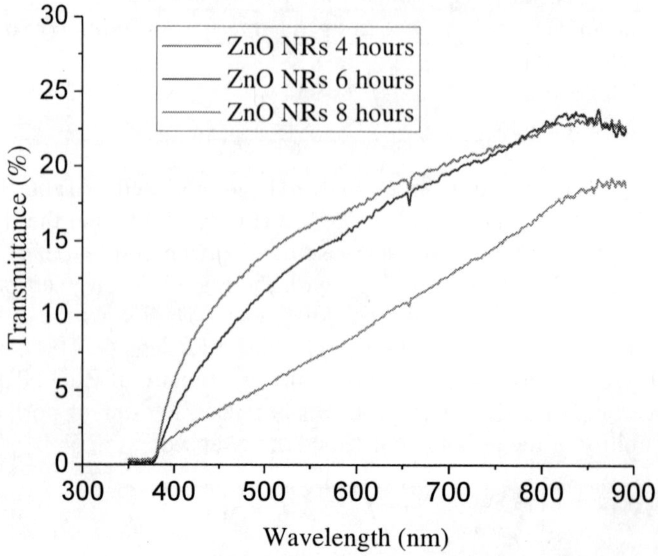

Figure 3. UV-Vis spectra of ZnO NRs as the function of wavelength at the different growing times

Determining the energy of bandgap was based on the Kubelka Munk equation [29] for the semiconductor for the direct type of bandgap described through plot Tauc between $(\alpha h\nu)^2$ versus $h\nu$, as shown in Figure 4. This result is the data interpretation of UV-Vis Spectra in Figure 3. The value of the ZnO NRs bandgap was obtained from the intersection between extrapolation line with x-axis summarized in Table 2.

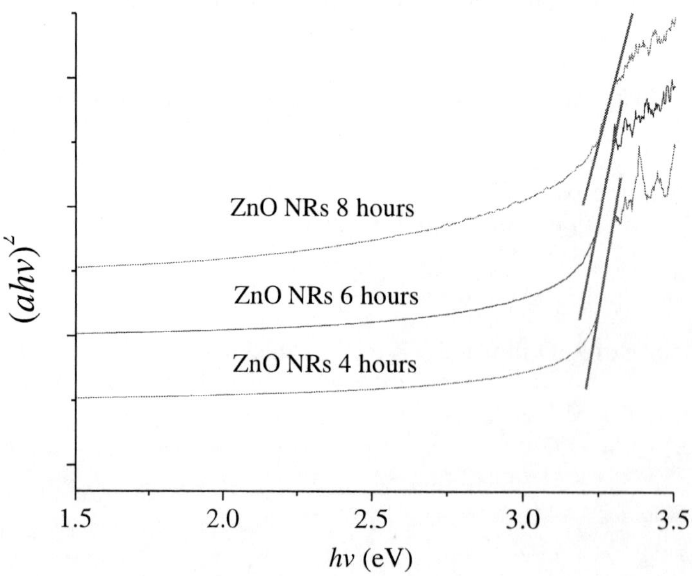

.**Figure 4**. Touch graph of ZnO NRs with various growing times

IOP Conf. Series: Journal of Physics: Conf. Series **1093** (2018) 012028 doi:10.1088/1742-6596/1093/1/012028

Table 2 presents that the value of the ZnO seed layer bandgap was 3.2 e. This is in accordance with the range of the bandgap value of ZnO nanoparticles film namely 3.2-3.5 eV[10]. The bandgap values of ZnO NRs at the precursor concentration of 75 mM for the various growing times of 4, 6, and 8 hours were 3.145 eV, 3.120 eV, and 3.049 eV, respectively. By the increase in the growing time, the bandgap decreased. The ZnO NRs bandgap was smaller than the ZnO seed layer. This was because ZnO NRs film has a wider area compared to ZnO seed layer film so that it could increase the high absorption of light and cause the bandgap smaller. This case corresponds to the research conducted by Selvan, *et al.* (2014) that the longer the growing time, the bandgap value would decrease [24].

Tabel 2. The bandgap of ZnO nanorods with various growing times

Type of Samples	Bandgap (eV)
ZnO seed layer	3.216
ZnO NRs 4 hours	3.145
ZnO NRs 6 hours	3.120
ZnO NRs 8 hours	3.049

Figure 5 shows the results of FTIR of ZnO Seed layer and ZnO NRs with the precursor concentration of 75 mM in the various hydrothermal growing times of 4 hours, 6 hours, and 8 hours in the wavelength range of 4000-500 cm^{-1} at the room temperature. The FTIR results showed three main peaks in the wavelength range of 400-515 cm^{-1}, 870 cm^{-1}, and 1250 cm^{-1}. Generally, ZnO has the absorption at the wavelength range of 400-515 cm^{-1} in accordance with the report of Gayen,*et al.* (2010) [30], while in the wavelength of 870 cm^{-1} and 1250 cm^{-1} there was H-C-N functional group formed from Hexamethylenetetramine (HMT) material with the chemical formula of $C_6H_{12}N_4$, in line with the report of Cottin, *et al.* (2002) [31] and Garcia, *et al.* (2010) [32].

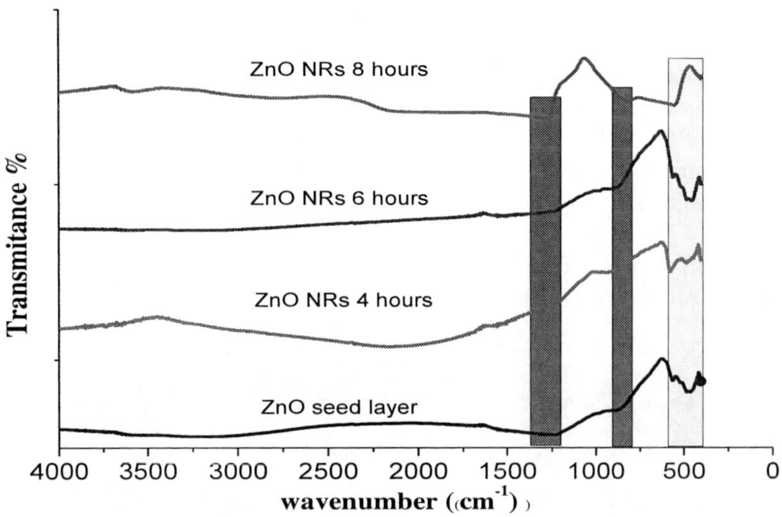

Figure 5: FTIR Spectrum of ZnO NRs with various growing times

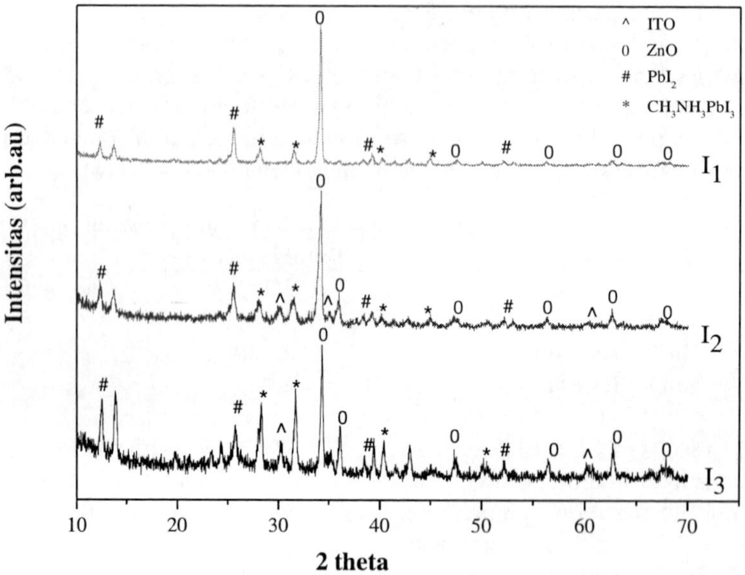

Figure 6. XRD Pattern of $CH_3NH_3PbI_3$/ZnO NRs sample with various growing times

Figure 6 shows XRD pattern of PbI_2 film, CH_3NH_3I powder, and $CH_3NH_3PbI_3$ deposition in ZnO NRs/ITO film with the concentration of 75 mM in 4 hours (sample A), 6 hours (sample B), and 8 hours (sample C) growing times. The XRD pattern of PbI_2 film on the ZnO NRs/ITO substrate showed the crystal growth oriented to c-axis indicated by four main peaks namely in the angle of 12.4°, 25.5°, 38.7°, and 52.5° that were respectively the crystal structure of (001), (002), (003), and (004). This case is in accordance with the research conducted by Burschka *et al.* (2013) showing that the PbI_2 peak was located at the crystal structure of (001), (002), (003), and (004) [33]. The diffraction peaks of perovskite appeared at the angle of 14.0°, 28.2°, 31.6°, 40.1°, and 44.9° that were respectively related to the crystal structure of perovskite tetragonal phase of (110), (220),(310), (224), and (134). The similar research was also undertaken by Zheng *et al.* (2015) showing the perovskite peaks were on (110), (220), (310), (224), and (134) with 2θ at 14.2°, 28.5°, 31.8°, 40.5°, and 44.0° [34]. The percentage of the crystal phase of CH_3NH_3I and PbI_2 in all samples are sorted from the low to the high percentage shown in Table 3 where sample C (sample I_1), sample A(sample I_2), sample B (sample I_3)

Table 3. The percentage of the crystal phase of CH3NH3PbI3/ZnO NRs

Type of Sample	Crystal Phase (%)	
	CH_3NH_3I	PbI_2
I_1	43.01	56.99
I_2	74.67	25.33
I_3	82.32	17.68

The XRD pattern on sample I_1 showed the percentage of the CH_3NH_3I phase of 43.01% and PbI_2 phase of 56.99%. In I_2 sample, the percentage of CH_3NH_3I phase was 74.67% and PbI_2 phase was 25.33%, while in I_3 sample, the percentage of CH_3NH_3I phase was 82.32% and PbI_2 phase was 17.68%. The percentage of CH_3NH_3I phase increased as the decrease in PbI_2 phase. The percentage of the forming of $CH_3NH_3PbI_3$ perovskite device was low. This case was because the percentage of PbI_2 was still high meaning that the PbI_2 phase had not been converted totally became $CH_3NH_3PbI_3$ perovskite. The conversion of PbI_2 to $CH_3NH_3PbI_3$ occurred quickly when the solution was dropped on the film since the heavy metal halide layered structure tended to interact with a small molecule. Thereby, the dense perovskite layer was usually formed on the PbI_2 surface and blocked the $CH_3NH_3PbI_3$ diffusion to the basic layer so that it caused imperfect perovskite conversion. Besides, the influence of the oxygen humidity definitely influenced the reaction of perovskite forming directly. $CH_3NH_3PbI_3$ perovskite has the high sensitivity to the water and will tend to be hydrolyzed due to the humidity. Hence, the fabrication process should be done in the glove box passed by an inert gas.

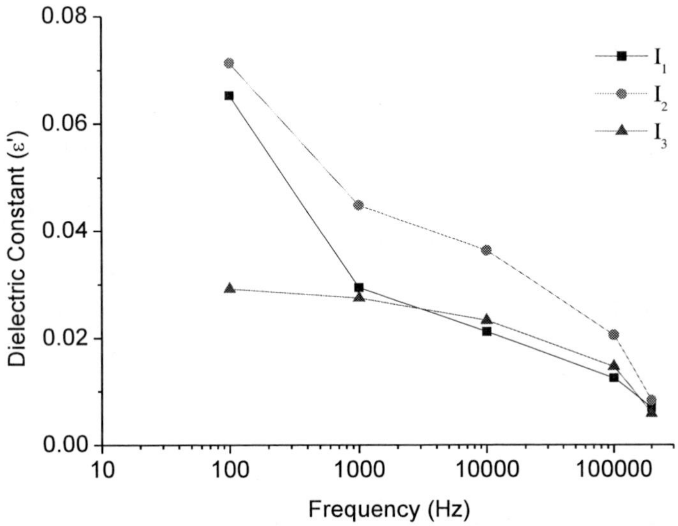

Figure 7. The influence of frequency on the dielectric constant of $CH_3NH_3PbI_3$/ZnO NRs with the variation of PbI_2 impurities

Figure 7 shows the results of the dielectric constant that tended to decrease as the increase in the given frequency. This case is in line with the research of Hoque *et al.* (2016) that studied the effect of polarization and perovskite dielectric that the higher the given frequency, the value of dielectric constant tended to decrease [35]. The lowest dielectric constant was resulted by I_3 sample with the frequency of 100 Hz. Meanwhile, in the high frequency, the value of dielectric was almost similar to all samples. This case showed that the polarization between the grain groups was more dominant in influencing the dielectric characteristic of the material.

The current resulted by $CH_3NH_3PbI_3$/ZnO NRs with the variation of I_1, I_2, and I_3 measured by using multimeter used the light through LED 13 Watt lamp with the various spaces between the test sample and the lamp as presented in Table 4.

Table 4. The results of the measurement of 2 probes of $CH_3NH_3PbI_3$/ZnO NRs

Type of Sample	Current (μA)			
	Intensity (400000 lux)	Intensity (173000 lux)	Intensity (120000 lux)	Intensity (8300 lux)
I_1	0.013(7)	0.001(1)	0.001(1)	0.001(1)
I_2	0.077(6)	0.001(1)	0.001(4)	0.001(1)
I_3	0.096(8)	0.074(3)	0.001(1)	0.001(1)

Table 4 shows that in the $CH_3NH_3PbI_3$/ZnO NRs film, the resulted current from any light intensities was very small. This case might be because the perovskite as the synthesis result was unwell formed due to the humidity factor where the light absorption by perovskite was not optimal. The highest current was resulted by I_3 sample. In general, the smaller the PbI_2 impurity of the sample, the resulted current was bigger. This result was consistent enough in any radiation intensity.

4. Conclusion

ZnO NRs synthesis has been successfully conducted using a hydrothermal method. The longer the growing time, the resulted ZnO NRs had the bigger diameter, and their rod length was longer. The created ZnO NRs had the diameter between 138 and 236 nm and the rod length between 835 nm and 1399 nm. Besides, the growing time increased the homogeneity and density of the produced ZnO NRs, while the bandgap was getting smaller. The electrical properties of $CH_3NH_3PbI_3$/ZnO NRs PSCs were related to the number of PbI_2 impurities as well. The smaller the PbI_2 impurity, the dielectric value becomes smaller at the low frequency, while it was relatively similar at the high frequency. Moreover, the smaller the PbI_2 impurity, the current value was higher in the light source with the high intensity.

Acknowledgments
The authors gratefully express appreciations for the financial support from PNBP Universitas Negeri Malang.

References

[1] Nie W, Tsai H, Asadpour R, Blancon J-C, Neukirch A J, Gupta G, Crochet J J, Chhowalla M, Tretiak S, Alam M A, Wang H-L and Mohite A D 2015 High-efficiency solution-processed perovskite solar cells with millimeter-scale grains *Science* **347** 522–5
[2] Boix P P, Nonomura K, Mathews N and Mhaisalkar S G 2014 Current progress and future perspectives for organic/inorganic perovskite solar cells *Mater. Today* **17** 16–23
[3] Saliba M, Correa-Baena J-P, Grätzel M, Hagfeldt A and Abate A 2018 Perovskite Solar Cells: From the Atomic Level to Film Quality and Device Performance *Angew. Chem. Int. Ed.* **57** 2554–69
[4] Niu G, Guo X and Wang L 2015 Review of recent progress in chemical stability of perovskite solar cells *J Mater Chem A* **3** 8970–80
[5] Zhou H, Chen Q, Li G, Luo S, Song T -b., Duan H-S, Hong Z, You J, Liu Y and Yang Y 2014 Interface engineering of highly efficient perovskite solar cells *Science* **345** 542–6
[6] Nagarjuna P, Narayanaswamy K, Swetha T, Rao G H, Singh S P and Sharma G D 2015 CH3NH3PbI3 Perovskite Sensitized Solar Cells Using a D-A Copolymer as Hole Transport Material *Electrochimica Acta* **151** 21–6
[7] Han G S, Chung H S, Kim B J, Kim D H, Lee J W, Swain B S, Mahmood K, Yoo J S, Park N-G, Lee J H and Jung H S 2015 Retarding charge recombination in perovskite solar cells using ultrathin MgO-coated TiO_2 nanoparticulate films *J Mater Chem A* **3** 9160–4
[8] Li X, Dai S-M, Zhu P, Deng L-L, Xie S-Y, Cui Q, Chen H, Wang N and Lin H 2016 Efficient Perovskite Solar Cells Depending on TiO_2 Nanorod Arrays *ACS Appl. Mater. Interfaces* **8** 21358–65
[9] Son D-Y, Im J-H, Kim H-S and Park N-G 2014 11% Efficient Perovskite Solar Cell Based on ZnO Nanorods: An Effective Charge Collection System *J. Phys. Chem. C* **118** 16567–73
[10] Mufti N, Arista D, Diantoro M, Fuad A, Taufiq A and Sunaryono 2017 The Effect of Thickness of ZnO Thin Films on Hydrophobic Self-Cleaning Properties IOP Conference Series: Materials Science and Engineering vol 202
[11] Li H, Liu H, Li Y and Liu Q 2017 Controllable Growth of the ZnO Nanorod Arrays on the Al Substrate and Their Reversible Wettability Transition *J. Nanotechnol.* **2017** 1–4
[12] Kjeldstad T, Thøgersen A, Nilsen O, Monakhov E and Galeckas A 2017 Controllable template approach for ZnO nanowire growth: Controllable template approach for ZnO nanowire growth *Phys. Status Solidi A* **214** 1600480
[13] Zhang S, Chen H-S, Matras-Postolek K and Yang P 2015 ZnO nanoflowers with single crystal structure towards enhanced gas sensing and photocatalysis *Phys Chem Chem Phys* **17** 30300–6
[14] Li W, Gao S, Li L, Jiao S, Li H, Wang J, Yu Q, Zhang Y, Wang D and Zhao L 2016 Hydrothermal synthesis of a 3D double-sided comb-like ZnO nanostructure and its growth mechanism

analysis *Chem Commun* **52** 8231–4

[15] Demel J, Pleštil J, Bezdička P, Janda P, Klementová M and Lang K 2011 Few-Layer ZnO Nanosheets: Preparation, Properties, and Films with Exposed {001} Facets *J. Phys. Chem. C* **115** 24702–6

[16] Ko S H, Lee D, Kang H W, Nam K H, Yeo J Y, Hong S J, Grigoropoulos C P and Sung H J 2011 Nanoforest of Hydrothermally Grown Hierarchical ZnO Nanowires for a High Efficiency Dye-Sensitized Solar Cell *Nano Lett.* **11** 666–71

[17] Fuad A, Fibriyanti A A, Mufti N, Taufiq A, Maryam S and Hidayat N 2018 Growth of CH3NH3PbI3 Perovskite on Stainless Steel Substrate Layered by ZnO Nanoparticles Using One-Step Spin Coating Route *J. Phys. Conf. Ser.* **1011** 012011

[18] Wanit M, Yeo J, Hong S J, Suh Y D, Ko S H, Lee D and Grigoropoulos C P 2012 ZnO nano-tree growth study for high efficiency solar cell *Energy Procedia* **14** 1093–8

[19] Kumar M H, Yantara N, Dharani S, Graetzel M, Mhaisalkar S, Boix P P and Mathews N 2013 Flexible, low-temperature, solution processed ZnO-based perovskite solid state solar cells *Chem. Commun.* **49** 11089–91

[20] Idiawati R, Mufti N, Taufiq A, Wisodo H, Laila I K R, Fuad A and Sunaryono 2017 Effect of Growth Time on the Characteristics of ZnO Nanorods IOP Conference Series: Materials Science and Engineering vol 202

[21] Ahmed M I, Habib A and Javaid S S 2015 Perovskite solar cells: potentials, challenges, and opportunities *Int. J. Photoenergy* **2015**

[22] Im J-H, Kim H-S and Park N-G 2014 Morphology-photovoltaic property correlation in perovskite solar cells: One-step versus two-step deposition of CH3NH3PbI3 *Apl Mater.* **2** 081510

[23] Rai P, Jeon S-H, Lee C-H, Lee J-H and Yu Y-T 2014 Functionalization of ZnO nanorods by CuO nanospikes for gas sensor applications *RSC Adv.* **4** 23604

[24] selvan P T, Venkatachalam, Saroja M, Gowthaman P, Ravikumar,S and Shankar. S 2014 Influence of Growth Time on Zinc Oxide Nano Rods Prepared By Dip Coating Method **3**

[25] Babu E S, Hong S-K, Vo T S, Jeong J-R and Cho H K 2015 Photoelectrochemical water splitting properties of hydrothermally-grown ZnO nanorods with controlled diameters *Electron. Mater. Lett.* **11** 65–72

[26] Shabannia R 2015 Vertically aligned ZnO nanorods on porous silicon substrates: Effect of growth time *Prog. Nat. Sci. Mater. Int.* **25** 95–100

[27] Baruah S and Dutta J 2009 pH-dependent growth of zinc oxide nanorods *J. Cryst. Growth* **311** 2549–54

[28] Foo K L, Hashim U, Muhammad K and Voon C H 2014 Sol–gel synthesized zinc oxide nanorods and their structural and optical investigation for optoelectronic application *Nanoscale Res. Lett.* **9** 429

[29] Yang L and Kruse B 2004 Revised Kubelka–Munk theory I Theory and application *J. Opt. Soc. Am. A* **21** 1933

[30] Khun K, Ibupoto Z, AlSalhi M, Atif M, Ansari A and Willander M 2013 Fabrication of Well-Aligned ZnO Nanorods Using a Composite Seed Layer of ZnO Nanoparticles and Chitosan Polymer *Materials* **6** 4361–74

[31] Cottin H, Bachir S, Raulin F and Gazeau M-C 2002 Photodegradation of hexamethylenetetramine by VUV and its relevance for CN and HCN extended sources in comets *Adv. Space Res.* **30** 1481–1488

[32] García B B, Liu D, Sepehri S, Candelaria S, Beckham D M, Savage L W and Cao G 2010 Hexamethylenetetramine multiple catalysis as a porosity and pore size modifier in carbon cryogels *J. Non-Cryst. Solids* **356** 1620–1625

[33] Burschka J, Pellet N, Moon S-J, Humphry-Baker R, Gao P, Nazeeruddin M K and Grätzel M 2013 Sequential deposition as a route to high-performance perovskite-sensitized solar cells *Nature* **499** 316–9

[34] Zheng E, Wang X-F, Song J, Yan L, Tian W and Miyasaka T 2015 PbI2-based dipping-controlled material conversion for compact layer free perovskite solar cells *ACS Appl. Mater. Interfaces* **7** 18156–18162

[35] Hoque M N F, Yang M, Li Z, Islam N, Pan X, Zhu K and Fan Z 2016 Polarization and dielectric study of methylammonium lead iodide thin film to reveal its nonferroelectric nature under solar cell operating conditions *ACS Energy Lett.* **1** 142–149

Magnetic Domain Distribution Analysis of Volcanic Material from the 2017 Eruptions of Mount Agung, Indonesia

Siti Zulaikah[1,2,*], **Ika Putri Nurlaily**[1]

[1]Department of Physics, Faculty of Mathematics and Natural Science, Universitas Negeri Malang, Jl. Semarang 5 Malang 65115, Indonesia
[2]Advance Material and Mineral Laboratory, Faculty of Mathematics and Science, Universitas Negeri Malang, Jl. Semarang 5 Malang 65115, Indonesia

*Corresponding author's email: siti.zulaikah.fmipa@um.ac.id

Abstract. Three types of volcanic materials of Mount Agung Bali Indonesia, i.e., volcanic ash, sand, and gravel have been tested magnetic susceptibility using a Bartington MS2B susceptibility meter. Of the three types of materials, volcanic ash has one order lower of the magnetic susceptibility of about 7.7×10^{-6} m^3kg^{-1} than that of sand and gravel which have magnetic susceptibility ranging from 14.1 to 26.8×10^{-6} m^3kg^{-1}. Based on the χ_{lf} Vs. χ_{fd} analysis, it can be seen that magnetic minerals contained in the volcanic ash have dominated the stable single domains (SSD), while magnetic minerals in sand and gravel have dominated by multidomain (MD). The domain distribution of the magnetic mineral in various sizes of volcanic materials from Mount Agung of naturally approximates the exponential form.

Keywords: Magnetic susceptibility, volcanic material, mount agung

1. Introduction

In the past few decades, the Rock Magnetic method has been widely applied and is developing rapidly. Besides the applications for the study of paleomagnetism, environmental magnetism, biomagnetism, magnetoclimatology, in the last decade, the application of rock magnetism methods penetrated in agriculture known as agromagnetism and in the volcanic material to understand more deeply the volcano mechanism, known as vulcanomagnetism. The measurement of magnetic properties, especially magnetic susceptibility of volcanic materials such as basalt, diorite, although for different purposes such as magnetic anisotropy susceptibility (AMS) test for the benefit of paleomagnetism studies has been done, magnetic susceptibility test can also be used to predict the origin of magnetic minerals and processes that affect the time of deposition [1-2]. In an effort to deduce various geophysical and environmental phenomena, geophysical methods are often juxtaposed with a geochemical analysis [3-7]. The application of rock magnetism method is also applied to materials in geothermal environments such as hot springs [8-9].

Vulcanomagnetism is specifically studied with long-term goals, understanding the characteristics of the volcanic material that may tell the characteristics of volcanoes more closely related to temperature and energy of the eruption. Several studies concerning volcanic processes properties have long been initiated by researchers as discussed in the book on the use of paleomagnetic and magnetic magnetism to understand volcanic processes [10]. In this study, the method of rock magnetism and chemical element analysis was used in the volcanic material to provide a detailed description of the magnetic properties of the volcanic material and their usefulness in understanding the characteristics and mechanisms of volcanoes.

2. Experimental Methods

Three types of samples namely volcanic ash, sand and gravel of Mount Agung on November 2019 eruption were taken around the location of the mountain peak on March 4, 2018, precisely in the village of Sebudi, Selat District, Karangasem Regency, Bali. There are three sampling points, the first point at the top, near the parking area, volcanic ash (P1) and gravel (P4) samples were taken. The second sampling point is 500 meters from the top parking area, with sand sample types and coded P2. The third point is 1 km from the top parking are taking two types of samples, namely sand and gravel which are coded P3 and P5 respectively. The samples taken in the field are then taken to the laboratory to prepare measurements of magnetic susceptibility by taking a small portion of sample samples and put them into a standard plastic holder for measuring magnetic susceptibility (plastic cylinders with a diameter of 2 cm and height of 2.2 cm). Each type of sample is prepared 10 holders, for example, volcanic ash samples are coded P1.1 to P1.10. Then the sample was measured by Bartington, MS2B with low and high frequencies and calculated mass-based susceptibility from low and high-frequency measurements and dependent frequency susceptibility. The calculation results are then mapped to see the domain of magnetic mineral distribution [11]. The method is also used in detail can be used to map magnetic top soil susceptibility and soil stability [12-13]. Some of the samples selected in this study were also tested for the content of elements with Panalytical X-Ray fluorescence (XRF). Furthermore, the correlation of element content with magnetic susceptibility values as described in the discussion.

3. Results and Discussion

The results of the magnetic susceptibility test using Magnetic Susceptibility Meter Bartington MS2B generated the volume of susceptibility value (κ), then analyzed and produced a mass-specific susceptibility (χ). The measurement was done with two frequencies namely low frequency and high frequency. From this measurement, we obtained the value of low-frequency magnetic susceptibility (χ_{lf}) and high-frequency magnetic susceptibility value (χ_{hf}). Based on the calculation the value of low-frequency magnetic susceptibility (χ_{lf}) and high-frequency magnetic susceptibility values (χ_{hf}, we obtain magnetic susceptibility frequency dependent (χ_{fd}).

The overall measurement results of the magnetic susceptibility of volcanic material from Mount Agung eruption have a range of low-frequency magnetic susceptibility values (χ_{lf}) with a range of (7.66–26.80) $\times$ $10^{-6}m^3kg^{-1}$ with a dependent magnetic susceptibility value(χ_{fd}) between 0.8%–2.7%. The correlation of the low-frequency magnetic susceptibility value (χ_{lf}) with high-frequency magnetic susceptibility value (χ_{hf}) is shown in Figure 1. Based on the R-value, the measurement of low-frequency susceptibility with high frequency shows high consistency and stability.

At each sampling point also showed a high consistency in the measurement which can prove the distribution of magnetic susceptibility values in Figure 2 looks to collect for specific sampling points. The lowest susceptibility of volcanic ash compared to sand and gravel samples. However, the gravel sample has the widest susceptibility range, namely in the sand area, between 15–20 $\times$ $10^{-6}m^3kg^{-1}$ and 25–30 $\times$ $10^{-6}m^3kg^{-1}$. Based on the χ_{fd} value in each group susceptibility value, it can be seen that for sand and gravel, the magnetic minerals are dominated by multidomain (MD) grains, while volcanic ash is dominated by single domain (SD) to superparamagnetic (SP) grains [11].

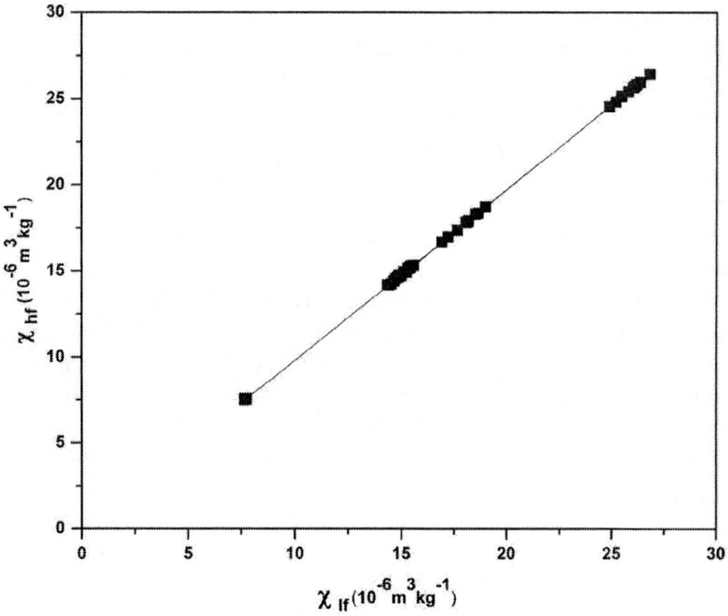

Figure 1. The Correlation of low-frequency χ_{lf} magnetic susceptibility and high-frequency χ_{hf} magnetic susceptibility of samples from Sikka Nusa Tenggara Timur

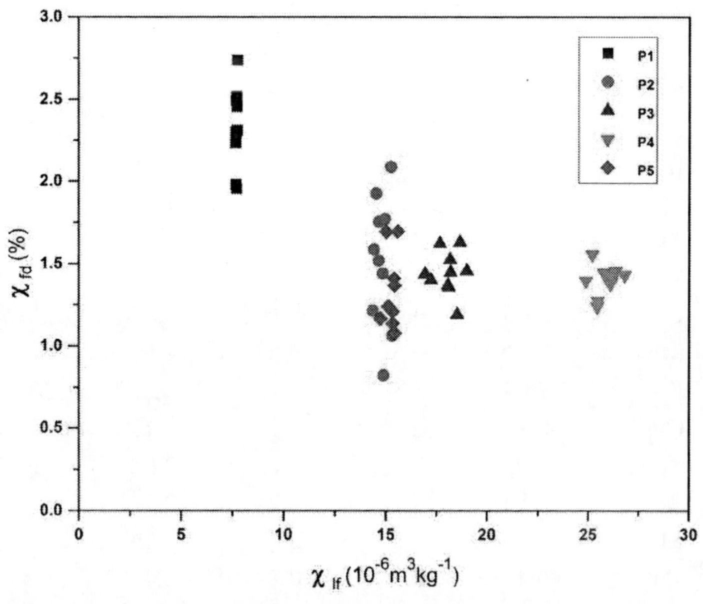

Figure 2. Distribution of χ_{lf} and χ_{fd} of vulcanic material from Agung Mount, Bali P1 is volcanic ash, P2 and P3 are sand, while P4 and P5 are gravel.

The XRF measurement results from each selected representative sample show the contents of each element as shown in Table 1. The dominant elements measured are Fe, Si, Ca and Al. The sample content of the volcanic ash is the biggest among the others, but the Fe content is the lowest compared to the other two types of samples. Sand has a relatively similar content of Si, Al, Ca and Fe in two different picking places. The magnetic susceptibility, in this case, has a good correlation with Fe% content. The higher Fe, the higher the value of magnetic susceptibility. From these data, it can be concluded that magnetic minerals from the three types of samples are sourced from the same place [3]. This characteristic illustrates mount Agung which can be compared with other mountain materials in Indonesia and other countries. At one time, it might be possible to find a relationship between the characteristics of magnetic minerals and the characteristics of volcanoes such as shields or cones or the type of mountain volcano.

In Table 2, the susceptibility value and the content of Wt (%) elements from XRF data are shown from the four dominant elements and their correlation of five representative samples. Based on the data in Table 2, it can be seen that there is a positive correlation between an iron element content of Fe with a magnetic susceptibility, namely R = 0.98. Besides having a positive correlation with Fe, a magnetic susceptibility also has a positive relationship with Ca. Al and Si, on the other hand, show that the relationship is inversely proportional to Fe. This discovery can support previous findings on volcanic material in other places such as alkaline volcanic material such as in Isparta area SW Turkey [4]. This condition does not apply to other natural samples which in general susceptibility is not positively correlated with Fe because natural elements of Fe are not always associated with other elements forming ferromagnetic minerals, but can also form paramagnetic. Therefore, the presence of enough Fe even in natural samples can form paramagnetic with susceptibility that tends to be small [9], [11].

Table 1. XRF test results of Mount Agung Volcanic eruption material

Unsur	**P1.1 (%)**	**P2.9 (%)**	**P3.3 (%)**	**P4.9 (%)**	**P5.7 (%)**
Al	10	9.1	9.3	7.7	10
Si	34.7	30	30.4	24.4	32
S	3.2	1.2	0.78	-	-
P	-	-	-	0,43	-
K	2.43	2.21	2.2	1.88	2.7
Ca	13.8	16	15.9	13.9	17.7
Ti	1.89	2.32	2.26	2.8	2.02
V	0.13	0.16	0.15	0.23	0.12
Cr	0.064	0.073	0.065	0.074	0.074
Mn	0.31	0.51	0.53	0.61	0.58
Fe	28.4	34	34.3	41.1	33
Cu	0.15	0.16	0.15	0.13	0.2
Zn	0.02	0.02	0.03	0.02	0.01
Sr	0,54	0.72	0.74	0.7	0.69
Mo	3.5	2.8	2.6	5.46	-
Eu	0.4	0.3	0.3	0.4	0.4
Re	0.2	0.32	0.2	0.3	0.34
Na	2.1	3.1	3.1	2.9	2.9
Mg	0.3	0.48	0.4	0.56	0.66

Table 2. Magnetic susceptibility and element content dominance of 5 samples representative.

Sample ID	Al	Si	Ca	Fe	χ_{lf} (10^{-6} m^3/kg)
P1.1	10	34.7	13.8	28.4	7.75
P2.9	9.1	30	16	34	15.15
P3.3	9.3	30.4	15.9	34.3	18.72
P4.9	7.7	24.4	13.9	41.1	26.41
P5.7	10	32	17.7	33	15.31

4. Conclusion

From the results of the study, it can be concluded that the volcanic ash has the lowest susceptibility and the size of the magnetic mineral grain tends to be smaller than that of the two other measured materials, sand and gravel. Based on the χ_{lf} Vs. χ_{fd} plot, the domain of magnetic minerals from the mount Agung volcanic ash, varies from MD to SD, while sand and gravel range in MD size. The content of the Fe element is positively correlated with magnetic susceptibility, while Al and Si are inversely proportional to magnetic susceptibility. Further studies are needed regarding the magnetic mineral morphology of the three materials to understand more about the mechanism of the mount Agung.

References

[1] Hatfield, H.G., and Maher, B.A., 2009, Fingerprinting Upland Sediment Sources: particle size-specific magnetic linkages between soils, lake sediments and suspended sediments, Earth Surface Process and Lanforms.

[2] Zong, Y., Xiao, Q. & Lu, S. J Soils Sediments, 2017**,** Magnetic signature and source identification of heavy metal contamination in urban soils of steel industrial city, Northeast China 17: 190. https://doi.org/10.1007/s11368-016-1522-2

[3] Yunginger, R., Bijaksana, S., Dahrin, D., Zulaikah, S., Hafidz, Abd., Kirana, K.H., Sudarningsih, Maryanto, Fajar, S.J., 2018, Lithogenic and Anthropogenic Components in Surface Sediments from Lake Limboto as Shown by Magnetic Mineral Characteristics, Trace Metals, and REE Geochemistry, Geosciences , 8, 116.

[4] Elitok, O., Kamaci, Z., C\Dolmaz, M.N., Yilmaz. K., and Sener, M, 2010, Relationship between Chemical Composition and Magnetic Susceptibility in the Alkaline Volcanic from the Ispanta Area Sw Turkey, J.Earth Syst. Sci 119 no 6 pp 853 – 860.

[5] Murdock, K.J, Wilkie, K., and Brown, L.L., 2013, Rock magnetic properties, magnetic susceptibility, and organic geochemistry comparison in core LZ1029-7 Lake El'gygytgyn, Russia Far East, Climate of The Past 9 467-479.

[6] Azzahro, R., Zulaikah, S., Diantoro, M., and Budi, P.S., and 2017, Magnetic Susceptibility and Elements composition of Mangrove Sediment in Malang, East Java Indonesia, have been presented at GEO-EM2017, Bandung, February 21-24th and accepted in conference proceedings of American Institute of Physics (AIP).

[7] Budi, P.S., Zulaikah, S., Hidayat, A., and Azzahro, R., 2017, Magnetic Susceptibility and dielectric properties of peatland in Central Kalimantan, Indonesia, have been presented at GEO-EM2017, Bandung, February 21-24th and accepted in conference proceedings of American Institute of Physics (AIP).

[8] Pandarinath, K., Shankar, R., Alvarado, I.S.T., Warrier, A.K., 2014, Magnetic susceptibility of volcanic rocks in geothermal areas: application potential in geothermal exploration studies for identification of rocks and zones of hydrothermal alteration, Arabian Journal of Gesciences, V. 7, Issue 7, pp. 2851-2860

[9] Zulaikah, S., Azzahro., R., Mu'alimah, E.S., Munfarikha, N., Dewiningsih, Fitria, W.L., and

Niarta., H.A., 2017, Magnetic Susceptibility and Morphology of Natural Magnetic Mineral Deposit in Vicinity of Human's Living, IOP Conf. series, Materials science and Engineering, 202.

[10] Ort, M.H., Porreca, M., and Geissman, J.W., 2015, The Use Paleomagnetism and Rock Magnetism to understand Volcanic Processes.

[11] Dearing, J.1999. Environmental Magnetic Susceptibility Using the Bartington MS2System. British Library Cataloguing in Publication Data.

[12] Kanu M.O., Meludu O. C. and Oniku S. A. 2014. Comparative study of top soil magnetic susceptibility variation based on some human activities. Geofisica International. 53-4: 411-423

[13] Cerato, A. B. and Miller, G. A. 2013. Determination of Soil Stabilizer Content Using X-Ray Fluorescence. Geotechnical Testing Journal, Vol. 36, No. 5, pp. 781–78.

Quantum Otto Engine based on Multiple-State Single Fermion in 1D Box System

Deny Pra Setyo, Eny Latifah*

Departement of Physics, Faculty of Mathematics and Natural Sciences, Universitas Negeri Malang, Jl Semarang 5 Malang 65145, Indonesia

*Corresponding author's email: eny.latifah.fmipa@um.ac.id

Abstract. In this work, a quantum Otto engine (QOE) based on a multiple-state single fermion particle trapped in an one-dimensional potential well has been studied. In this study, we used an analogical model completed with the first law of thermodynamics for the quantum system. The system satisfied the relativistic Hamiltonian, especially Dirac Hamiltonian. The engine consisted of the quantum isochoric and isentropic processes to form an Otto Cycle. The heat input and output were transferred under isochoric processes. As a result, the higher the number of energies was arranged, the lower the efficiency the system has. However, we could increase the efficiency of QOE by controlling the compression ratio. In addition, we also obtained the correspondence of the efficiency ultra-relativistic limits QOE and non-relativistic limits one.

Keywords: Quantum Otto engine, Dirac Hamiltonian, relativistic limits.

1. Introduction

A heat engine is a device for converting the heat energy to work, such as mechanical work, electrical work, chemical work, etc. The heat energy is absorbed from the heat reservoir and converted into work and partly transferred into a cold reservoir. Classically, the heat engine consists of an ideal gas as a working substance that expands and pushes a piston in a cylinder. The main difference between classical and quantum systems is the working substance and it is concerned with the discrete energy level [1–5]. It causes the quantum heat engine to have unique properties.

In recent years, the study of a quantum heat engine increases significantly. Some studies have been focusing on a working substance [4–10], and others on the dynamic of the source of a reservoir energy [11,12] with some models such as a potential well [4–8,13–15]], oscillator harmonic [11,12,16] etc. It has been also inspecting on the kind of heat engines like Carnot Engine [4,8,17,18], Diesel Engine [19–21], Otto engine [14,19,22]. The study has been done by [11] that used a harmonic oscillator model for irreversible cyclic. They have obtained that the efficiency can reach maximum power at a high temperature and quasi-static operating condition. Another study for a reversible cycle, the efficiency of quantum Otto engine has been similar with the classical one. Moreover, the efficiency has relied on the compression adiabatic ratio. Another study investigated the use of a single Dirac particle system for quantum Carnot engine and they obtained the efficiency of quantum Carnot engine which has the similarly with the classical cycle [7]. The optimization of relativistic Carnot Engine [15] and the internal effect process [8] have also been investigated.

For the generalization of Otto engine, however, it is necessary to study about a relativistic Otto Engine. The relativistic Otto engine which is explored uses a single fermion that fulfills the Dirac Hamiltonian. The study focused on evaluating the efficiency of quantum Otto engine (QOE) using a single fermion system as a working substance, trapped in a potential well. The number of energy arranged will also be investigated in order to find out the effect of level state on the performance of QOE in relativistic approach. The performance of relativistic QOE must be valid in a non-relativistic limit. Thus, we can clarify this result with the previous study.

2. Experimental Methods

In this work, we used an analogical model first law of thermodynamics implementation. The physical system that we chose was a single fermion which was trapped in 1 Dimensional system. The implementation of the first law of thermodynamics was used to redefine the external force act to the 1D box system.

Since we used the fermion system, the motion equation is represented by Dirac equation that satisfied relativistic Hamiltonian. After we obtained the solution of Dirac equation, explicitly expectation value of energy would be found. The expectation value of energy was used to describe the quantum thermodynamic processes.

2.1. A Fermion Trapped in the 1D Box System

The fermion which is trapped in the 1D box system can be described by Dirac Hamiltonian operator

$$\hat{H} = -i\hbar c\hat{\alpha}\cdot\nabla + m_0c^2\hat{\beta} + V(x) \quad (1)$$

Here $\hat{\alpha}$ and $\hat{\beta}$ are Dirac matrices which are expressed by

$$\hat{\alpha}_i = \begin{pmatrix} 0 & \sigma_i \\ \sigma_i & 0 \end{pmatrix},\ \hat{\beta} = \begin{pmatrix} I & 0 \\ 0 & -I \end{pmatrix}.$$

For the case, 1D box system the potential is bound and given by

$$V(x) = \begin{cases} 0, & 0 \le x \le L \\ +\infty, & x < L \text{ and } x > L \end{cases} \quad (2)$$

From that equation, in region $0 \le x \le L$ particle is under the free-states because the particle motion is not influenced by the potential. Thus Dirac Hamiltonian operator is expressed by

$$\hat{H}_0 = -i\hbar c\hat{\alpha}\cdot\nabla + m_oc^2\hat{\beta} \quad (3)$$

In 1D box system, there is no particle can go through and out. It is guaranteed by the current density in boundary ($x = 0$ and $x = L$) is zero. Summarizing the Eigen equation of Dirac Hamiltonian for particle trapped in 1D box system is given by [7]

$$\hat{H}\psi = E\psi, \quad (4)$$

with boundary condition $j(x = 0 \text{ and } x = L) = 0$.
According to equation (3), equation (4) will be able to be expressed by

$$\left(-i\hbar c\left(\hat{\alpha}_i\cdot\nabla\right) + \hat{\beta}m_0c^2\right)\psi(x) = E\psi(x) \quad (5)$$

Since $\hat{\alpha}$ and $\hat{\beta}$ are matrices, so the solution of Dirac equation has the column vector is consist of 4-component spinors [9]. Therefore, the solution is expressed by $\psi(x) = (\varphi_1, \varphi_2, \chi_1, \chi_2)$ [4]. In this notation, $\varphi(x)$ is denoted as the "large" component and $\chi(x)$ is denoted as the "small" component.

The solution of equation (5) is

$$\psi = A\begin{pmatrix} \sin\left(kx-\frac{kL}{2}\right) \\ 0 \\ 0 \\ -\frac{i\hbar ck}{E+m_0c^2}\cos\left(kx-\frac{kL}{2}\right) \end{pmatrix}, \tag{6}$$

with A is $2iA_1e^{i\frac{kL}{2}}$. That equation associated with discrete energy eigenvalue

$$E_n = \pm m_0c^2\left(\sqrt{1+\left(\frac{n\lambda}{2L}\right)^2}-1\right), \tag{7}$$

with $\lambda = 2\pi\hbar/(m_0c)$ is wavelength. The positive and negative sign corresponds with particle and anti-particle solution respectively, however, in this paper, we used a positive solution because by assuming that the particle is trapped in the potential well and hence we would keep the positive eigenvalue [7].

2.2. *Quantum Thermodynamic Processes*

In the previous study, the definition of force is not well defined to construct thermodynamic processes for quantum system especially for isochoric and isobar process [1,4]. We redefined the force of the system based on the phenomenological representation of infinitesimally changing of the expectation value of Hamiltonian [5]. The infinitesimally changing of the expectation value of Hamiltonian is

$$dE = \sum_n E_n dp_n + \sum_n p_n dE_n \tag{8}$$

In the right side of equation (8), the first term indicates the heat absorbed by the system or released to the surroundings and the second term indicates work produced by the system. The second term of equation (8) is the work of the quantum system, $\sum_n p_n dE_n = -dW = -FdL$. Thus, we have the ensemble average force as,

$$F = -\sum_n p_n \frac{dE_n}{dL} = -\sum_n p_n F_n \tag{9}$$

The ensemble average of force F is the average of F_n over all the states represented in the ensemble. According to equation (9) the force act on the wall is

$$F(L) = \left(\frac{m_0c^2\lambda^2}{4L^3}\right)\sum_n p_n \frac{n^2}{\sqrt{1+(n\lambda/2L)^2}} \tag{10}$$

2.2.1 *Adiabatic Process.*

Classically, there is no exchanged heat in the system for the adiabatic process. Therefore, the changing of the internal energy produces work to push or pull the system

$$dU = -PdV \tag{11}$$

In quantum adiabatic, since there is no exchanged of heat, the occupation probability remained constant. Thus, according to the first law of thermodynamics, the internal energy changed is

IOP Conf. Series: Journal of Physics: Conf. Series **1093** (2018) 012030 doi:10.1088/1742-6596/1093/1/012030

$$dE = \sum_n p_n dE_n \tag{12}$$

2.2.2 Isochoric Process.

Classically, there is no changing of the volume of the piston under isochoric process. Consequently, there is no work produced during this process. According to the first law of thermodynamics, all of the heat transferred to the system is used to change the internal energy of the system. In a quantum system, all of the heat transferred to the system is used to change occupation probability. Thus, the particle is exited to the higher level state *n*. The heat that transferred to the system is

$$\delta Q = E(L_f) - E(L_i) \tag{13}$$

3. Results and Discussion

Otto engine consists of adiabatic compression, isochoric, adiabatic expansion, and back to the initial condition by isochoric. Figure 1 describes the quantum Otto cycle for a single fermion system. The process is started by compressing adiabatic. The occupation probability remains constant under this process. After that, the system is thermally contacted with the hot reservoir by an isochoric process. During this process, the particle is excited to the other *n* state. Next, the system is on expansion adiabatic process and the eigenvalue of the system energy decrease along of increasing the width of well. Finally, the cycle of QOE is back to the initial condition by an isochoric process by transferring the residual heat energy to the cold reservoir.

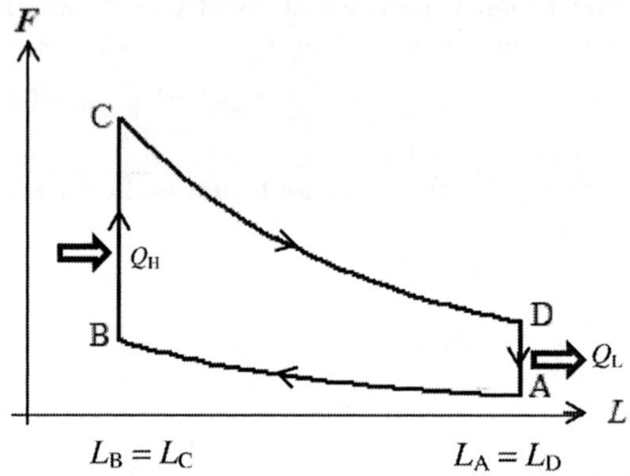

Figure 1. Diagram *F-L* for quantum otto cycle with single fermion system

The heat is transferred from hot reservoir Q_H into the system by an isochoric process. According to the first law of thermodynamics, the heat input to a system is

$$Q_H = \frac{mc^2}{2L_B}\left[\sqrt{4L_B^2+\lambda^2} - \sum_n p_n\sqrt{4L_B^2+n^2\lambda^2} + 2L_B\left(\sum_n p_n - 1\right)\right] \tag{14}$$

Meanwhile, the heat output to the cold reservoir by an isochoric process is

$$Q_L = -\frac{mc^2}{2L_A}\left[\sqrt{4L_A^2+\lambda^2} - \sum_n p_n\sqrt{4L_A^2+n^2\lambda^2} + 2L_A\left(\sum_n p_n - 1\right)\right] \tag{15}$$

According to the conservative law of energy, absorbed heat is converted to work and residual heat is transferred to the cold reservoir. The total work for this cycle is expressed by

$$W = Q_H - Q_L \tag{16}$$

IOP Conf. Series: Journal of Physics: Conf. Series **1093** (2018) 012030 doi:10.1088/1742-6596/1093/1/012030

The efficiency of quantum Otto engine (QOE) is the ratio between work that is produced by system and heat absorbed. According to equation (14) (15) and (16), we obtained the efficiency of QOE with single fermion system as follow

$$\eta_{\mathrm{O}}^{\mathrm{QR}}=1-\left(\frac{L_{\mathrm{B}}}{L_{\mathrm{A}}}\right)\left(\frac{\sqrt{4L_{\mathrm{A}}^{2}+\lambda^{2}}-\sum_{n}p_{n}\sqrt{4L_{\mathrm{A}}^{2}+n^{2}\lambda^{2}}+2L_{\mathrm{A}}\left(\sum_{n}p_{n}-1\right)}{\sqrt{4L_{\mathrm{B}}^{2}+\lambda^{2}}-\sum_{n}p_{n}\sqrt{4L_{\mathrm{B}}^{2}+n^{2}\lambda^{2}}+2L_{\mathrm{B}}\left(\sum_{n}p_{n}-1\right)}\right), \tag{17}$$

with $L_{\mathrm{A}}/L_{\mathrm{B}}$ is ratio of compression and n which is integer number is the number of energy arranged.

From equation (17), the efficiency of QOE based on single fermion system depends on compression ratio and the number of energy arranged. Considering the number of energy arranged are n = 2, 3, 4 (by assuming Fermion excited to the maximum state), we have obtained the efficiency of QOE based on single fermion system which has less number of energy arranged that is the highest efficiency for a certain condition as a Figure 2. However, we can increase the efficiency by controlling the compression ratio. If the system has less number of energy arranged, the compression ratio should be bigger. This result has corresponded with a non-relativistic case [2,4].

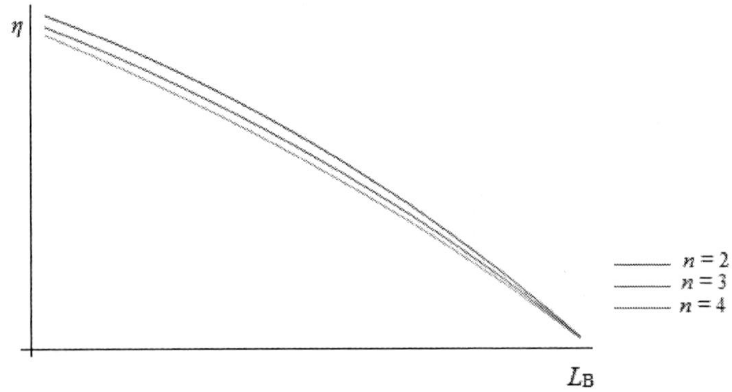

Figure 2. The efficiency of QOE depends on the number of level-state for arbitrary L_{B}

Consider for the ultra-relativistic limit ($\lambda\to\infty$), the efficiency of QOE is

$$\eta_{\mathrm{O}}^{\mathrm{UR}}=1-\frac{L_{\mathrm{B}}}{L_{\mathrm{A}}} \tag{18}$$

That efficiency does not depend on the number of the level state, but it only depends on ratio compression. If we take for a non-relativistic limit ($\lambda\to 0$) the efficiency of QOE is

$$\eta_{\mathrm{O}}^{\mathrm{NR}}=1-\left(\frac{L_{\mathrm{B}}}{L_{\mathrm{A}}}\right)^{2} \tag{19}$$

Equation (19) has also confirmed the previous study [5,20,23]. Therefore, by exploring a relativistic Otto Engine we can reformulate the efficiency of non-relativistic Otto engine.

If we compare this result with the classical one,

$$\eta=1-\left(\frac{V_{\mathrm{B}}}{V_{\mathrm{A}}}\right)^{\gamma-1}, \tag{20}$$

with γ is the heat capacity ratio then we also obtained the heat capacity ratio for a fermion system that was trapped in a 1D box. The value of the heat capacity ratio for ultra-relativistic is 2. These results also correlated with Diesel fermion engine [21]. Meanwhile, for a non-relativistic limit, the ratio of the heat capacity is 3 [5,7,22,23]. This difference of the heat capacity ratio for classical, non-relativistic and ultra-relativistic is corresponding with the properties of the system.

4. Conclusion

In this paper, we have explored quantum Otto engine (QOE) based on a single fermion in a 1D box system. The fermion satisfies Dirac Hamiltonian which is a relativistic equation and we have obtained the eigenvalue of relativistic energy for this system. The system occurred by some processes such as isochoric and adiabatic processes. By using the first law of thermodynamics, we have obtained the efficiency of QOE for single fermion system. For certain condition, the higher the level state of the system, the lower the efficiency of QOE is. However, by controlling of ratio compression we can keep the performance of QOE high. By setting the limit $\lambda \to 0$ and $\lambda \to \infty$, we also obtained the efficiency of QOE in non-relativistic limit and ultra-relativistic limit.

References

[1] Bender C M, Brody D C and Meister B K 2000 Quantum Mechanical Carnot Engine *J. Phys. A* **33** 4427–36
[2] Quan H T, Zhang P and Sun C P 2005 Quantum heat engine with multilevel quantum systems *Phys Rev E* **72** 56110
[3] Quan H T, Liu Y, Sun C P and Nori F 2007 Quantum thermodynamic cycles and quantum heat engines *Phys Rev E* **76** 31105
[4] Latifah E and Purwanto A 2011 Multiple-State Quantum Carnot Engine *J. Mod. Phys.* **2** 1366–72
[5] Latifah E and Purwanto A 2013 Quantum Heat Engines; Multiple-State 1D Box System *J. Mod. Phys.* **04** 1108–15
[6] Bender C M, Brody D C and Meister B K 2000 Quantum mechanical Carnot engine *J. Phys. Math. Gen.* **33** 4427–36
[7] Muñoz E and Peña F J 2012 Quantum heat engine in the relativistic limit: The case of a Dirac particle *Phys. Rev. E - Stat. Nonlinear Soft Matter Phys.* **86** 1–11
[8] Purwanto A, Sukamto H, Subagyo B A and Taufiqi M 2016 Two Scenarios on the Relativistic Quantum Heat Engine *J. Appl. Math. Phys.* **4** 1344–53
[9] Ivanchenko E A and Science N 2015 Quantum Otto cycle efficiency on coupled qudits 1–19
[10] Akbar M S, Latifah E, Wisodo H, Hidayat A, Prasetyo D and Qomariyah S N 2018 Proses Adiabatis dan Isovolume Kuantum Sistem Dua Partikel Simetri *JPSE J. Phys. Sci. Eng.* **2** 55–65
[11] Rezek Y and Kosloff R 2006 Irreversible performance of a quantum harmonic heat engine *New J. Phys.* **8**
[12] Insinga A, Andresen B and Salamon P 2016 Thermodynamical analysis of a quantum heat engine based on harmonic oscillators *Phys Rev E* **94** 012119
[13] Quan H T, Zhang P and Sun C P 2005 Quantum Heat Engine with Multilevel Quatum Systems *Phys Rev E* **72** 056110
[14] Wu F, Yang Z, Yang L, Liu X and Wu S 2010 Work Output and Efficiency of a Reversible Quantum Otto Cycle *Therm. Sci.* **14** 879–86
[15] Pena F J, Ferre M, Orellana P A, Rojas R G and Vargas P 2016 Optimization of a relativistic quantum mechanical engine *Phys. Rev. E - Stat. Nonlinear Soft Matter Phys.* **94** 1–12
[16] Quan H T 2009 Quantum thermodynamic cycles and quantum heat engines. II. 1–10
[17] Quan H T, Liu Y, Sun C P and Nori F 2007 Quantum Thermodynamic Cycles and Quantum Heat Engines *Phys. Rev. E* **76** 031105
[18] Munoz E and Pena F 2012 A single Dirac-fermion Quantum Heat Engine *ArXiv Prepr. ArXiv12076149* **2** 1–8
[19] Latifah E and Purwanto A 2013 Quantum Heat Engines ; Multiple-State 1D Box System **2013** 1091–8
[20] Wang J, Ma Y and He J 2015 Quantum-mechanical engines working with an ideal gas with a finite number of particles confined in a power-law trap *EPL Europhys. Lett.* **111** 20006
[21] Setyo D P, Latifah E, Wisodo H and Hidayat A 2018 Quantum Relativistic Diesel Engine with

Single Massless Fermion in 1 Dimensional Box System *J. Penelit. Fis. Dan Apl. JPFA* **8** 25–32

[22] Wang J, Ma Y and He J 2015 Quantum-Mechanical Engines Working with an Ideal Gas with a Finite Number of Particle Confined in a Power-Law Trap *EPL* **111** 2

[23] Wu F, Yang Z, Chen L, Liu X and Wu S 2010 Work output and efficiency of a reversible quantum otto cycle *Therm. Sci.* **14** 879–86

IOP Conf. Series: Journal of Physics: Conf. Series **1093** (2018) 012031 doi:10.1088/1742-6596/1093/1/012031

Limit of Relativistic Quantum Brayton Engine of Massless Boson Trapped 1 Dimensional Potential Well

Muhammad Syawaluddin Akbar, Eny Latifah*, Hari Wisodo

Department of Physics, Faculty of Mathematics and Natural Sciences, Universitas Negeri Malang, Jl. Semarang 5 Malang 65145, Indonesia

*Corresponding author's email: eny.latifah.fmipa@um.ac.id

Abstract. A quantum Brayton engine using a massless Boson trapped in a one-dimensional potential well as a working substance has been constructed to be studied. We used a modified analogical model by the first law of thermodynamics for a quantum system implementation. The system was built with Klein Gordon equation, relativistic quantum mechanics Hamiltonian, without the mass term of the Hamiltonian and it was trapped in a one-dimensional potential well. Then, it underwent the quantum ideal Brayton cycle process. The quantum Brayton process consisted of adiabatic compression, isobaric expansion, adiabatic expansion, and isobaric compression processes. This relativistic quantum mechanics Brayton engine has found that the efficiency formulation was similar to the classical one. This study obtained the conformity results between the quantum relativistic and classical Brayton engines. By this cycle process, we also invented that the ratio of heat capacity under constant pressure and volume of the system was 2. It could indicate that the efficiency of quantum relativistic Brayton engine is higher than the classical one.

Keywords: Brayton, Klein-Gordon equation, relativistic, quantum.

1. Introduction

A heat engine is a device to convert a heat energy from a high-temperature reservoir to other work [1–3], but the heat energy is not all converted into a mechanical work because some energy is rejected into a low-temperature reservoir, so the efficiency of device is always lower than 100%, same as the thermodynamic second law [4]. Generally, the efficiency of a heat engine has more than 50% [5]. The presence of quantum theory is expected to increase of heat engine efficiency. The performance of the heat engine can be reviewed by the principle of quantum mechanics, which is known as the quantum heat engine[2,3,6–10].

In this cases, a quantum heat engine system is the quantum particle (single mass-less Boson) trapped in a one-dimensional potential well, and one of the dividing wall on the potential-well can be friction-less move, so this system is similar to the piston in the cylinder. The expectation value of the Hamiltonian system is the internal energy of the system. The quantity volume of the cylinder in the classical thermodynamics is replaced by the length of the distance between the dividing walls on the potential well in quantum mechanics system. The Pressure of the piston in the classical thermodynamic system is replaced by the expectancy value of force in dividing the wall potential well [1,6,11]. The quantum heat engine is different from the classical heat engine, the working substance of quantum heat engine is quantum particles so it implies that the energy state of the system is discrete[11].

In previous studies [1,2] it has a disadvantage that does not apply of first law thermodynamics for the quantum system, therefore the expectation value of force and energy are not well defined. Carnot quantum engine for two states of the system has an efficiency value larger than 2 more states of the system, this shows that the more the number of states involved, the lower the efficiency [2]. The efficiency value of a quantum heat engine depends on the ratio of adiabatic expansion process [1,2,6,12] and the other studies for other engines such as Otto, Brayton, and Diesel [3,8,13] has a similar efficiency as the classical system. The study of relativistic Dirac particles as a working substance of quantum Carnot engine has found that the efficiency value depends on the root square of Compton wavelength, the ratio of adiabatic expansion and the number of state system [9,12].

The study of massless boson as a working substance in a quantum heat engine is rarely conducted, so further research is needed. Therefore, a research using relativistic particle such as massless boson as working substance of quantum heat engine is required, especially using Brayton cycle. Thus, we studied about the limit of relativistic quantum Brayton engine of massless boson trapped 1 dimensional well. This study is expected to develop results from previous theories and can be a consideration for a further research development.

2. Experimental Methods

2.1. Single mass-less boson in a one-dimensional potential well

This study considered the case of a classical thermodynamic system which has a proximity analogy with a quantum system. The analogy model implemented the first law of thermodynamics modified for such quantum system, is intended to describe a change of a physical quantity throughout in the process [3]. The potential well or one-dimensional box system in a quantum mechanics is the closest analogy with a piston-cylinder in a classical thermodynamic system. The potential well of the system, in this case, is a particle trapped in a 1-dimensional box system. The case taken was the potential system which was infinite ($V = \infty$) at $x \leq 0$ dan $x \geq 0$, and the zero ($V = 0$) at $0 < x < L$ in 1-dimensional potential well (1-Dimensional box system). The illustration of an analogical mechanical quantum system as a piston-cylinder system is Figure 1 showed.

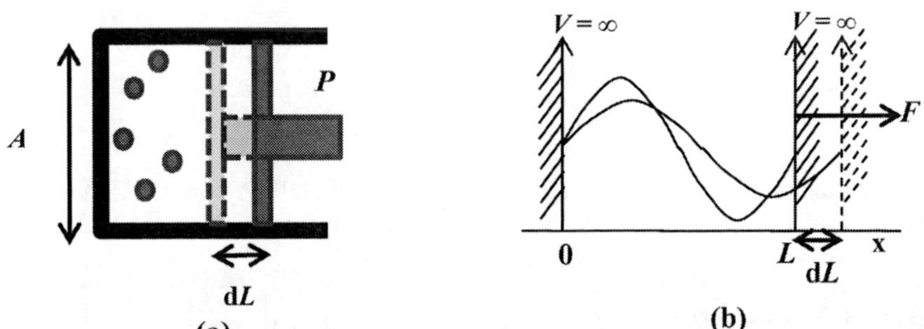

Figure 1. The piston system between the classical thermodynamics and mechanical quantum system; (a) piston inside the cylinder in a classical system (b) the analogy model of a piston in a mechanical quantum.

The generality of a free-state motion equation for a mass-less Boson at Minkowsky space (space-time dimension) is described as Klein- Gordon equation [14]. The Klein-Gordon equation is

$$\hat{p}^{\mu}\hat{p}_{\mu}\Psi = \left(\frac{E^2}{c^2} - \hat{p} \bullet \hat{p}\right)\Psi = m_0^2 c^2 \Psi. \tag{1}$$

In this equation (1), m_0 is a motionless mass of the particle, each momentum operator in Minkowski space is

$$\hat{p}^{\mu} = i\hbar\frac{\partial}{\partial x_{\mu}} = i\hbar\left(\frac{\partial}{\partial(ct)}, -\frac{\partial}{\partial x}, -\frac{\partial}{\partial y}, -\frac{\partial}{\partial z}\right) = i\hbar\left(\frac{\partial}{\partial(ct)}, -\vec{\nabla}\right) = i\hbar\left(p_0, \vec{p}\right) \tag{2}$$

$$\begin{aligned}\hat{p}^{\mu}\hat{p}_{\mu} &= -\hbar^2\frac{\partial}{\partial x_{\mu}}\frac{\partial}{\partial x^{\mu}} = -\hbar^2\left(\frac{1}{c^2}\frac{\partial^2}{\partial t^2} - \left(\frac{\partial^2}{\partial x^2} + \frac{\partial^2}{\partial y^2} + \frac{\partial^2}{\partial z^2}\right)\right) \\ &\cong -\hbar^2\vec{\Delta} = -\hbar^2\left(\frac{1}{c^2}\frac{\partial^2}{\partial t^2} - \vec{\nabla}^2\right)\end{aligned} \tag{3}$$

So with considering the equation (3) and (2), we can rewrite the Klein-Gordon equation as follow

$$\left(\vec{\Delta} + \frac{m_0^2c^2}{\hbar^2}\right)\Psi = \left(\frac{\partial^2}{c^2\partial t^2} - \frac{\partial^2}{\partial x^2} - \frac{\partial^2}{\partial y^2} - \frac{\partial^2}{\partial z^2} + \frac{m_0^2c^2}{\hbar^2}\right)\Psi = 0. \tag{4}$$

Consider the case of massless boson trapped 1-dimensional potential well, we obtain the equation motion is

$$\hat{H}^2\psi(x) = -\hbar^2c^2\frac{\partial^2\psi(x)}{\partial x^2}. \tag{5}$$

Equation (5) is a second order linear differential equation and the general solution of the equation is

$$\psi(x) = A\sin kx + B\cos kx,\ k = \frac{E}{\hbar c}. \tag{6}$$

Consider of applying boundary and normalized condition that probability of a state system is zero at $x = 0$ and $x = L$, and 1 for $0 < x < L$, we obtained the solution which is a wave-function as Eigen-vector is

$$\psi_n(x) = \sqrt{\frac{2}{L}}\sin\frac{n\pi x}{L}. \tag{7}$$

eigen-vector (Eigen-Function) functions as representations of each Eigen-state level. Eigen-function at equation (7) associate with Eigen-energy system, is

$$E_n(L) = \frac{n\pi\hbar c}{L} = nE_1(L). \tag{8}$$

If the system has the state which is unsure or probability of state is not 100% at one of the state systems, then the state of the system is superposition by each other of Eigen-state. The mathematical form of superposition by Eigen-function at stationary condition is

$$\Psi(x,t) = \left(\sum_{n=1}^{\infty} a_n\psi_n\right)\theta(t) = \left(a_1\psi_1(x) + a_2\psi_2(x) + \ldots\right)\exp\left(-i\frac{Et}{\hbar}\right). \tag{9}$$

The expectation value of energy possessed by the particle massless Boson trapped in a one-dimensional potential well is

$$\langle H\rangle = \int_{-\infty}^{\infty}\Psi^*(x,t)\hat{H}\Psi(x,t)\mathrm{d}x = \sum_{m=1}^{\infty}\sum_{n=1}^{\infty}a_m^*a_nE_n\delta_m^n = \left\{\sum_{n=1}^{\infty}\left|a_n(L)\right|^2 n\right\}\frac{\pi\hbar c}{L} \tag{10}$$

where δ_{mn} describes the orthogonal state property of the system.

2.2. *The first law thermodynamics in quantum Brayton engine*

The description of the physical quantity between the classical thermodynamics and quantum mechanics system is different. The classical thermodynamics considers macroscopic quantities (such as pressure, volume, and temperature), while the quantum mechanics and mechanical statistic consider the microscopic quantities (such as internal energy, distribution function, and the probability of state system). However, from the difference between the classical thermodynamics and quantum mechanics system, there is a relation that can attribute to each other, such as the quantity of temperature. The quantity of temperature is associated with the internal energy system, with considering the microscopic for the quantum system shown the energy internal system is manifested of the expectation value of the Hamiltonian for particle trapped in a one-dimensional potential well system. The quantity of Pressure in the classical thermodynamics system is replaced by expectation-value of Force in dividing wall potential well. The relation of the physical quantity of classical thermodynamics and quantum mechanics is shown by table 1. The general form for the First law of thermodynamisc is

$$\mathrm{d}U = \delta Q - \delta W. \tag{11}$$

Previously research has not defined the expectation value of force well since it has not developed and implemented the modified first law of thermodynamics for the quantum mechanics system [1,2,6]. The modified first law of thermodynamics for a quantum mechanics system is

$$\mathrm{d}\langle H\rangle = \sum_{n=1}^{N} E_n(L)\mathrm{d}P_n(L) + \sum_{n=1}^{N} P_n(L)\mathrm{d}E_n(L). \tag{12}$$

Table 1. The association of quantities for classical thermodynamics and quantum mechanics system

No	Quantities of classical thermodynamics	Quantities of quantum mechanics
1	Pressure (P)	Force (F)
2	Volume (V)	Length of the system (L)
3	Internal Energy (U)	Expectation value of Hamiltonian
4	Heat energy (Q)	The change of probability in the expectation value of Hamiltonian
5	Work (W)	The change of Eigen-energy system

The heat energy absorbed to the system or released to the surrounding of the system will imply the change of probability in the expectation value of Hamiltonian (internal energy system), therefore, the quantities of heat energy for quantum mechanics system is

$$\delta Q = \sum_{n=1}^{N} E_n(L)\mathrm{d}P_n(L). \tag{13}$$

The Quantities work of system expanding the dividing wall of the potential well or imposing on the system can be writen as follow

$$\delta W(L) = -\sum_{n=1}^{N} P_n(L)\mathrm{d}E_n(L) \tag{14}$$

2.2.1. *Adiabatic Process*

An adiabatic process for the classical thermodynamic system is characterized by the absence flow of the heat energy between system and environment (reservoir), so the work of system to expand comes from the change of internal energy system. In addition, the adiabatic compression will increase the internal energy system. The first law of thermodynamics for the adiabatic process is

$$\mathrm{d}U = -\delta W = -\vec{F}(L) \bullet \mathrm{d}\vec{L}. \tag{15}$$

The quantum mechanics system for the adiabatic process characterized by the probability of a level state is constant [3]. Thus, the modified first law of thermodynamics for quantum mechanics system is

$$d\langle H\rangle = \sum_{n=1}^{N} P_n(L)dE_n(L). \tag{16}$$

The expectation value of force to moving the dividing wall of potential well for massless Boson as working substance in a quantum heat engine is

$$F(L) = \left\{\sum_{n=1}^{N} P_n(L)n\right\}\frac{\pi\hbar c}{L^2}. \tag{17}$$

The Quantities of work throughout the adiabatic process is obtained by integrating equation (17) to infinitesimal partition of length (d L) from the initial length (L_i) to the final length (L_f), i.e.

$$W_{i\to f} = \int_{Li}^{Lf} F(L)dL = \left\{\sum_{n=1}^{N} P_n(L_i)n\right\}\frac{\pi\hbar c}{L_i}\left(1-\frac{L_i}{L_f}\right). \tag{18}$$

2.2.2. *Isobaric Process*

The isobaric process for the classical thermodynamic system is characterized by absence the change of pressure (the pressure throughout is constant), the system is expanded or compressed in conjunction by the flow of the heat energy between the system and the environment (reservoir). The first law of thermodynamics for the Isobaric process is

$$dU = \delta Q - \delta W. \tag{19}$$

The quantum mechanics system for the isobaric process is characterized by absence the change of expectation-value of force (expectation-value of force throughout is constant). The modified First law of thermodynamics for the isobaric process is

$$d\langle H\rangle = \delta Q - \int_{Li}^{Lf} FdL. \tag{20}$$

The quantities of work throughout the Isobaric process is characterized by expectation-value of force for initial-state and final-state is constant. Thus, for massless Boson as a working substance of a quantum heat engine in the Isobaric process occurs as follow:

$$\begin{aligned} F(L_i) &= F(L) \\ \left\{\sum_{n1}^{N} |a_n(L_i)|^2 n\right\}\frac{\pi\hbar c}{L_i^2} &= \left\{\sum_{n1}^{N} |a_n(L)|^2 n\right\}\frac{\pi\hbar c}{L^2} \\ \left(\frac{L}{L_i}\right)^2 \sum_{n=1}^{N} P_n(L_i)n &= \sum_{n=1}^{N} P_n(L)n. \end{aligned} \tag{21}$$

To find the Quantities of work throughout the Isobaric process is obtained by integrating the expectation-value of force to the infinitesimal partition of length (dL) from the initial to final length. Consider the equation (21) to integrating operation obtain the Quantities of work, i.e.

$$\begin{aligned} W_{i\to f} &= \int_{Li}^{Lf} F(L)dL = \int_{Li}^{Lf}\left\{\sum_{n=1}^{N} P_n(L)n\right\}\frac{\pi\hbar c}{L^2}dL \\ &= \int_{Li}^{Lf}\left\{\sum_{n=1}^{N} P_n(L_i)n\right\}\left(\frac{L}{L_i}\right)^2\frac{\pi\hbar c}{L^2}dL \\ &= \left\{\sum_{n=1}^{N} P_n(L_i)n\right\}\frac{\pi\hbar c}{L_i}\left(\frac{L_f}{L_i}-1\right). \end{aligned} \tag{22}$$

The amount of quantities of heat energy flowing between the system and environment (reservoir) along the isobaric process from initial length to final length condition is obtained by implying the modified of the first law thermodynamics, i.e.

$$Q = \langle H \rangle_f - \langle H \rangle_i + W_{i\to f}$$
$$= \left\{ \sum_{n=1}^{N} |P_n(L_f)|^2 n \right\} \frac{\pi \hbar c}{L_f} - \left\{ \sum_{n=1}^{N} |P_n(L_i)|^2 n \right\} \frac{\pi \hbar c}{L_A} + W_{i\to f}. \quad (23)$$

3. Results and Discussion

3.1. The Cycle of Quantum Brayton Engine

The Brayton engine is an idealization of a turbine engine. However, this cycle is complicated to realize because the turbine machine has not closed process in that cycle, so it is possible for the change of working substance. George Brayton (1870) is a first-engineer to present of the Brayton cycle. The Brayton cycle compared to the Otto and Diesel cycle, the Brayton cycle acts on a larger volume condition but a smaller pressure[15].

The Brayton cycle process consisted of adiabatic compression, isobaric Expansion, adiabatic expansion and isobaric compression. The Diagram F-L and scenario of quantum Brayton engine has shown by Figure 2.

The Initial condition of the Brayton cycle starting by adiabatic compression process from D→A condition, and then the heat energy from surrounding (high-temperature reservoir) will be absorbed by the system through the process of isobaric expansion process (at A→B condition). Afterward, the system undergoes an adiabatic expansion process (at B→C condition). The last cycle process is an isobaric compression where the system will return to the initial condition (at C→D condition).

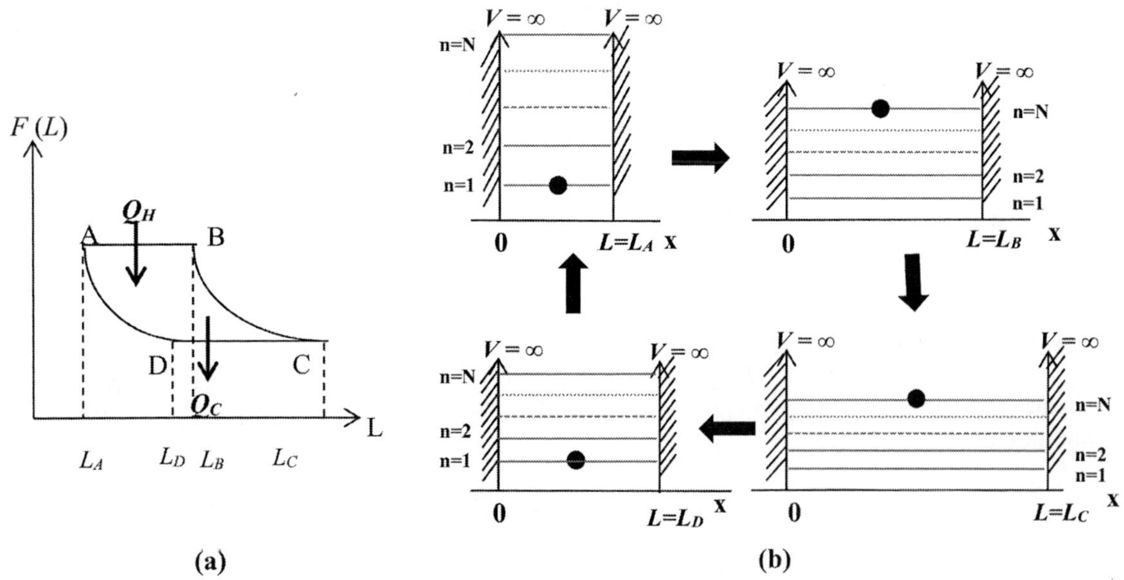

Figure 2. The cycle of quantum brayton engine: (a) diagram F-L of the brayton cycle (b) scenario state of the brayton cycle

3.2. Initial State of System

The initial state for this Brayton cycle is started by D-condition, where the state-probability of the system is 100% at the ground-state (n=1). The wave function to describe the system for the initial condition is

$$\psi_D = \sqrt{\frac{2}{L_D}} \sin\left(\frac{\pi x}{L_D}\right). \tag{24}$$

The expectation value of Hamiltonian in this system for initial condition ($L=L_D$) is

$$\langle H \rangle_D = \left\{ \sum_{n=1}^{N} |a_n(L_D)|^2 n \right\} \frac{\pi \hbar c}{L_D} = \frac{\pi \hbar c}{L_D}. \tag{25}$$

3.3. Adiabatic Compression D→A

The D→A process is adiabatic compression. In this process is characterized by the state-probability of the system is constant, so the state-probability of the system is fixed (100% at the ground state). Nevertheless, the internal energy of the system increasde because the length of the system decreased (compressed). The quantities of work throughout the adiabatic compression process (from D→A condition) is

$$W_{D\to A} = \int_{L_D}^{L_A} \left\{ \sum_{n=1}^{N} |a_n(L_D)|^2 n \right\} \frac{\pi \hbar c}{L^2} \mathrm{d}L = \frac{\pi \hbar c}{L_A}\left(\frac{L_A}{L_D} - 1\right) = \frac{\pi \hbar c}{L_A}\left(\frac{1}{\beta_1} - 1\right). \tag{26}$$

The β_1 coefficient is the ratio of L_D and L_A. The value of work throughout adiabatic compression is minus (-) that shown by $L_D > L_A$ or $\beta_1 > 1$. This proves that the system is subject of work throughout the adiabatic compression process.

3.4. Isobaric expansion A→B

The A→B process is an isobaric process. In this process is characterized by the expectation values constant throughout A→B isobaric expansion. The heat energy from the boundary (high-temperature reservoir) will be absorbed by the system in such a way the state of the system increases, this study assumes that the state probability of the highest state (n=N) system increased until 100%. The initial process of isobaric expansion started at A-condition, the state of state-probability of the system is still 100% at the ground-state. The wave function to describe the system at A condition is

$$\psi_A(x) = \sqrt{\frac{2}{L_A}} \sin\frac{\pi x}{L_A}. \tag{27}$$

The expectation value of the Hamiltonian system at A-condition is

$$\langle H \rangle_A = \left\{ \sum_{n=1}^{N} |a_n(L_A)|^2 n \right\} \frac{\pi \hbar c}{L_A} = \frac{\pi \hbar c}{L_A}. \tag{28}$$

Along the isobaric expansion process, state-probability at the ground state will decrease, however, the state-probability of a higher state increases. An Increase of state-probability in the highest state ($n = N$) reaches 100% when the system at B-condition. The wave function to describe the system at B condition is

$$\psi_B(x) = \sqrt{\frac{2}{L_B}} \sin\frac{N\pi x}{L_B}. \tag{29}$$

The expectation value of the Hamiltonian system at B-condition is

$$\langle H \rangle_B = \left\{ \sum_{n=1}^{N} |a_n(L_B)|^2 n \right\} \frac{\pi \hbar c}{L_B} = N\frac{\pi \hbar c}{L_B}. \tag{30}$$

The quantity of work and heat energy throughout the isobaric expansion process from A→B condition is

$$W_{A\to B} = \int_{L_A}^{L_B} \left\{ \sum_{n=1}^{N} P_n(L) n \right\} \frac{\pi \hbar c}{L^2} \mathrm{d}L = \frac{\pi \hbar c}{L_A}\left(\frac{L_B}{L_A} - 1\right) = \frac{\pi \hbar c}{L_A}\left(\sqrt{N} - 1\right). \tag{31}$$

$$Q_H = \langle H \rangle_B - \langle H \rangle_A + W_{A\to B} = 2\frac{\pi\hbar c}{L_A}\left(\sqrt{N}-1\right). \tag{32}$$

The value of work throughout A→B isobaric expansion process is positive (+) showed by $L_B > L_A$, this proves that the system does the work to expansion. The value of heat energy throughout A→B process is positive (+) shown by $\sqrt{N} > 1$, this proves that the heat energy from surrounding (high-temperature reservoir) is absorbed by the system to increase the state-system. $\sqrt{N}$ is the ratio of L_B and L_A, this relation has obtained by considering the equation (21) that expectation-value of force is constant.

3.5. Adiabatic Expansion B→C

The B→C process is an adiabatic expansion that absent of heat energy flow between the system and environment, so the state-probability of the system is fixed (100% at the N-state). The quantity of work throughout the adiabatic expansion (B→C) is

$$W_{B\to C} = N\frac{\pi\hbar c}{L_B}\left(1-\frac{L_B}{L_C}\right) = \frac{\pi\hbar c}{L_A}\left(\sqrt{N}-\frac{\sqrt{N}}{\beta_2}\right). \tag{33}$$

The β_2 coefficient is the ratio of L_C and L_B. the value of work is positive (+) showed by $L_C > L_B$ or $\beta_2 > 1$, this proves that the system does the work for expansion.

3.6. Isobaric Compression C→D

The C→D process is an isobaric compression. In this process is characterized by the expectation value of force is constant throughout C→D isobaric compression. Heat energy from the system is released to the environment/boundary (low-temperature reservoir), so the state of the system will decrease until the state probability of the system is 100% at the ground-state when the system reaches to D-condition. The state system at C-condition is still in the highest-state (state-probability of the system is 100% at n=N-state). The wave-function to describe the system at C-condition is

$$\psi_C(x) = \sqrt{\frac{2}{L_C}}\sin\frac{N\pi x}{L_C}. \tag{34}$$

The expectation value of the Hamiltonian system at C-condition is

$$\langle H \rangle_C = \left\{\sum_{n=1}^{N}\left|a_n(L_C)\right|^2 n\right\}\frac{\pi\hbar c}{L_C} = N\frac{\pi\hbar c}{L_C}. \tag{35}$$

Along the isobaric compression process, state-probability at N-state will decrease, however, the state-probability at the ground state (n=1) increases up to 100% when the cycle back to D-condition. The wave function to describe the system at D-condition shown by equation (24), and the expectation value of the Hamiltonian system showed by equation (25). The quantity of work and the flow of heat energy between the system and environment throughout the isobaric compression process i.e.

$$W_{C\to D} = N\frac{\pi\hbar c}{L_C}\left(\frac{L_D}{L_C}-1\right) = \frac{\pi\hbar c}{L_A}\left(\frac{1}{\beta_2}-\frac{\sqrt{N}}{\beta_2}\right). \tag{36}$$

$$Q_C = \langle H \rangle_D - \langle H \rangle_C + W_{C\to D} = 2\frac{\pi\hbar c}{L_A\beta_2}\left(1-\sqrt{N}\right). \tag{37}$$

The value of work throughout C→D isobaric compression process is negative (-) showed by $L_C > L_D$, this proves that a system is a subject of work. The value of heat energy throughout C→D isobaric compression process is negative (-) showed $\sqrt{N} > 1$, this proves that the heat energy system is released to environment/surrounding (low-temperature reservoir) and then the internal energy system will decrease until the ground-state.

3.7. The Efficiency of Quantum Brayton Engine

The efficiency is ratio the quantity total value of the working system and absorbed heat energy system by the environment (high-temperature reservoir). The expansion ratio β_1 and β_2 is same-value because $L_B/L_A = L_C/L_D = \sqrt{N}$, so the total value of work and efficiency i.e.

$$W_{total} = \oint F(L)\mathrm{d}L = \frac{\pi\hbar c}{L_A}\left[2\sqrt{N} - 2 + \left(\frac{1}{\beta_1} + \frac{1}{\beta_2}\right) - 2\frac{\sqrt{N}}{\beta_2}\right] = 2\frac{\pi\hbar c}{L_A}\left(\sqrt{N}-1\right)\left(1-\frac{1}{\beta}\right) \tag{38}$$

$$\eta_{Brayton} = \frac{W_{total}}{Q_H} = \left(1-\frac{1}{\beta}\right) = 1 - \frac{L_B}{L_C} \vee 1 - \frac{L_A}{L_D}. \tag{39}$$

3.8. Heat Capacity of Quantum Brayton Engine

The heat capacity is the ratio of heat energy to increase the temperature system. A heat capacity depends on the amount and the characteristic of the working substance system. For a specific property of working substance heat capacity is known as specific heat (C). Specific heat property based on the process in the thermodynamic system divided by two properties, that specific heat at constant volume condition (C_V) and specific heat at constant pressure condition (C_P). Specific heat at constant volume is obtained from the partial differential of internal energy by temperature. Specific heat at constant pressure is obtained from the partial differential of Helmholtz free-energy by temperature [16], the mathematics form of specific heat is

$$C_V = \frac{\partial U}{\partial T} \tag{40}$$

$$C_P = \frac{\partial H}{\partial T} = \frac{\partial(U+PV)}{\partial T}. \tag{41}$$

Specific heat is related to a temperature, therefore it needs to introduce the temperature function approach to statistical mechanics. The working substance particle in this study is only a single mass-less Boson, so the properties of the particle following the Bose-Einstein distribution. Bose-Einstein distribution can be approached by Maxwell-Boltzmann distribution because the working substance is a single particle. The mathematics form of distribution-function is

$$f(E,T) = \left(\exp\left(\frac{E_n(L)}{k_B T}\right) - 1\right)^{-1} \approx \exp\left(-\frac{E_n(L)}{k_B T}\right). \tag{42}$$

Statistical mechanics system present of partition-function. The partition function is the amount of state-partition in the system, the mathematical form of a partition-function is

$$Z(\beta,L) = \sum_{n=1}^{\infty}\exp\left(-\beta E_n(L)\right), \beta = \frac{1}{k_B T} \tag{43}$$

The internal energy of the system (U) is the expectation value of the Hamiltonian system. The internal energy of the system is obtained by considering the Maxwell-Boltzmann distribution and partition-function, (Z(L,β)), i.e.

$$U(L,T) = \langle H\rangle = \frac{\sum_{n=1}^{\infty} E_n(L)\exp\left(-\beta E_n(L)\right)}{\sum_{n=1}^{\infty}\exp(-\beta E_n(L))} = -\left(\frac{\partial \ln Z(\beta,L)}{\partial\beta}\right). \tag{44}$$

The determination of a specific heat at constant volume condition is complicated if we following the equation (44) because the partition-function have sigma notation form. The partition-function is following the amount of geometric series, so mathematic form of partition-function can be re-write as

$$Z(\beta,L)=e^{-\beta E_1}+e^{-2\beta E_1}+e^{-3\beta E_1}+\ldots=\frac{\exp\left(-\beta E_1(L)\right)}{1-\exp\left(-\beta E_1(L)\right)}. \tag{45}$$

The internal Energy of the system is

$$U\left(\beta,L\right)=-\frac{\partial\ln Z\left(\beta,L\right)}{\partial\beta}=\frac{E_1(L)}{1-e^{-\beta E_1(L)}}. \tag{46}$$

Specific heat at constant volume condition in Quantum Brayton Engine can be analogized by specific heat at constant length condition because the quantity of volume in the classical system is replaced by the length of the dividing wall potential-well. Specific heat at constant length condition is

$$C_L=\frac{\partial}{\partial T}\frac{E_1(L)}{\left(1-\exp\left(-\beta(T)E_1(L)\right)\right)}=\frac{E_1^{\,2}(L)e^{-\beta(T)E_1(L)}}{k_BT^2\left(1-e^{-\beta(T)E_1(L)}\right)^2} \tag{47}$$

Specific heat at constant pressure condition in Quantum Brayton Engine can be analogized by a specific heat at the expectation value of force is a constant condition because the quantity of pressure in the classical system is replaced by expectation value of force. Expectation value of force is obtained by considering partition-function, i.e.

$$F\left(\beta,L\right)=\frac{\sum_{n=1}^{\infty}\frac{1}{\beta}\frac{\partial}{\partial L}\exp\left(-\beta E_n\left(L\right)\right)}{\sum_{n=1}^{\infty}\exp\left(-\beta E_n\left(L\right)\right)}=\frac{1}{\beta}\frac{\partial\ln Z(\beta,L)}{\partial L} \tag{48}$$

Substitution of equation (45) to equation (48), obtained the expectation-value of force

$$F(\beta,L)=\frac{1}{\beta}\frac{\partial}{\partial L}\ln\left(\frac{e^{-\beta E_1(L)}}{1-e^{-\beta E_1(L)}}\right)=-\left(\frac{1}{1-e^{-\beta E_1}}\right)\frac{\partial E_1(L)}{\partial L}. \tag{49}$$

Helmholtz free energy is related by specific heat at constant force condition. Helmholtz free energy is an addition of internal energy (expectation value of energy) and the multiplication expectation value of force by the length of the system. The Helmholtz free energy is

$$H=U\left(\beta,L\right)+F\left(\beta,L\right)L=\left(\frac{1}{1-e^{-\beta E_1(L)}}\right)\left(E_1(L)-L\frac{\partial E_1(L)}{\partial L}\right) \tag{50}$$

Specific heat at the expectation value of force is constant condition is

$$C_F=\frac{\partial H}{\partial T}=\frac{E_1^{\,2}(L)}{k_BT^2}\left(1-\frac{L}{E_1(L)}\frac{\partial E_1(L)}{\partial L}\right)\left(\frac{e^{-\beta(T)E_1(L)}}{\left(1-e^{-\beta(T)E_1(L)}\right)^2}\right). \tag{51}$$

The ratio of specific heat at expectation value of force is constant and specific heat at the length of the system is at a constant condition as follow

$$\frac{C_F}{C_L}=\left(1-\frac{L}{E_1(L)}\frac{\partial E_1(L)}{\partial L}\right), \tag{52}$$

substitution of equation (8) to equation (52) is

$$\frac{C_F}{C_L} = \left(1 - \frac{L^2}{\pi\hbar c}\frac{\partial \pi\hbar c L^{-1}}{\partial L}\right) = 2 \tag{53}$$

The general form of efficiency for Brayton engine is

$$\eta_{Brayton} = 1 - \left(\frac{V_A}{V_D}\right)^{\gamma-1}, \gamma = \frac{C_P}{C_V} \tag{54}$$

The general form efficiency of classical thermodynamics [5,15], non-relativistic quantum mechanics [1–3,6,8] and ultra-relativistic quantum mechanics [9,12][13] showed Table 2.

Table 2. The Comparison efficiency of heat engine among classical thermodynamics, non-relativistic quantum mechanics and ultra-relativistic quantum

No	Based on the system	Efficiency of the Brayton Engine
1	Classical Thermodynamic	$1-\left(\frac{V_A}{V_D}\right)^{\gamma-1}$
2	Non-Relativistic quantum heat engine	$1-\left(\frac{L_A}{L_D}\right)^{2}$
3	Ultra-relativistic quantum heat engine	$1-\left(\frac{L_A}{L_D}\right)$

The general form of the Brayton efficiency at the classical thermodynamics have a relation with the efficiency of Ultra-relativistic Quantum Brayton Engine in equation (39), it can be taken correlation that the existence of correspondence between the quantum system and the classical system. Based on Tabel 2, shown that the existence of correspondence between the quantum system and the classical system, the difference is that there is a system which is Laplace constant (γ) or the ratio of specific heat at constant volume and specific heat at constant. The Laplace constant for classical Brayton engine depended by working substance (diatomic or monoatomic). The Laplace constant of non-relativistic quantum Brayton engine from the results of the previous study [3] show the Laplace constant (γ) or the ratio of specific heat at constant pressure and specific heat at constant volume is 3, it showed that the powered of L_A/L_D ratio is 2. The Laplace constant of Ultra-relativistic quantum Brayton engine forms a mass-less Boson as working substance is 2.

4. Conclusion

In this paper, we have explored the limit of relativistic quantum Brayton engine based on the massless boson trapped in a one-dimensional box system. We used the Klein Gordon equation to describe the boson motion. The boundary condition is applied for the system to obtain the Eigenvalue and associated by Eigenstate of the system. The motion walls of a one-dimensional box satisfy the Bryton cycle. The Brayton cycle occurred some of the adiabatic and isobaric processes. The results of the study showed the efficiency of quantum Brayton engine based on massless boson as working substance is larger than the classical Brayton engine but lower than non-relativistic quantum Brayton engine. The efficiency of ultra-relativistic Brayton engine is larger than the classical Brayton engine but lower than non-relativistic Brayton engine can be shown by a ratio of the specific heat at a constant pressure by a specific heat at a constant volume.

References

[1] Bender C M, Brody D C and Meister B K 2000 Quantum Mechanical Carnot Engine *J. Phys. A* **33** 4427–36
[2] Latifah E and Purwanto A 2011 Multiple-state quantum carnot engine *J. Mod. Phys.* **2** 1366
[3] Latifah E and Purwanto A 2013 Quantum Heat Engines; Multiple-State 1D Box System *J. Mod. Phys.* **04** 1108–15
[4] Cápek V and Sheehan D P 2005 *Challenges to the second law of thermodynamics* (Springer)
[5] Zemansky M W and Dittman R 1997 Heat and Thermodynamics: An Intermediate Textbook, International Series in Pure and Applied Physics
[6] Bender C M, Brody D C, Meister B K and A P R S L 2002 Entropy and temperature of a quantum Carnot engine *Proc. R. Soc. A* 1519–26
[7] Bender C M, Brody D C and Meister B K 2000 Quantum mechanical Carnot engine *J. Phys. Math. Gen.* **33** 4427–36
[8] Wu F, Yang Z, Yang L, Liu X and Wu S 2010 Work Output and Efficiency of a Reversible Quantum Otto Cycle *Therm. Sci.* **14** 879–86
[9] Munoz E and Peña F J 2012 Quantum Heat Engine in the Relativistic Limit: The Case of a Dirac Particle *Phys Rev E* **86** 061108
[10] Kosloff R and Levy A 2014 Quantum heat engines and refrigerators: Continuous devices *Annu. Rev. Phys. Chem.* **65** 365–93
[11] Akbar M S, Latifah E, Wisodo H, Hidayat A, Prasetyo D and Qomariyah S N 2018 Proses Adiabatis dan Isovolume Kuantum Sistem Dua Partikel Simetri *JPSE J. Phys. Sci. Eng.* **2** 18–28
[12] Purwanto A, Sukamto H, Subagyo B A and Taufiqi M 2016 Two Scenarios on the Relativistic Quantum Heat Engine *J. Appl. Math. Phys.* **4** 1344–53
[13] Setyo D P, Latifah E, Hidayat A and Wisodo H 2018 Quantum Relativistic Diesel Engine with Single Massless Fermion in 1 Dimensional Box System *J. Penelit. Fis. Dan Apl. JPFA* **8** 25–32
[14] Greiner W 1990 *Relativistic quantum mechanics* vol 3 (Springer)
[15] Ginting R U 1989 Dasar-Dasar Termodinamika Teknik *Jkt. Dep. Pendidik. Dan Kebud. Dir. Jendral Pendidik. Tinggi Proy. Pengemb. Lemb. Pendidik. Tenaga Kependidikan*
[16] Pointon A 1980 introduction to statistical Physics

Band Gap Shift and Electrical Conductivity of (Ag-xSnO_2)NPs-β-Carotene Thin Film

Ulwiyatus Sa'adah[1], Arif Hidayat[1,2], Solehudin[1], Sujito[1], Nandang Mufti[1,2], Markus Diantoro[1,2,*]

[1]Department of Physics, Faculty of Mathematics and Natural Sciences, Universitas Negeri Malang, Jl. Semarang 5 Malang 65145, Indonesia
[2]Center of Advanced Materials for Renewable Energy, Universitas Negeri Malang, Jl. Semarang 5 Malang 65145, Indonesia

*Corresponding author's email: markus.diantoro.fmipa@um.ac.id

Abstract. This research proposed to use β-carotene as a conducting polymer to encapsulate Ag-SnO_2 to enhance the electrical conductivity, so it becomes more new green technology. Initially, the synthesis of Ag and SnO_2 nanoparticles was performed by chemical reduction and a sol-gel method respectively. The (Ag-xSnO_2) nanoparticles with β-carotene composite were synthesized by the blending method and followed by deposition using a spin coating method. Samples of nanoparticles and films were characterized using X-ray Diffraction (XRD) and Scanning Electron Microscopy (SEM) to obtain their structure, morphology, and size. Based on the analysis of XRD pattern, it was shown that the grain size of Ag and SnO_2 are 25.50 nm and 37.79 nm respectively. Optical properties of samples which were investigated by using UV-vis spectroscopy to find out the energy gap based on the Tauc plot method. It was found that that the energy gap of (Ag-xSnO_2)NPs-β-carotene thin film reduce which the average of ~3.7 eV. It was also shown that inducing of SnO_2 nanoparticles could enhance the electrical conductivity of films with the increase of SnO_2. This combination of reducing the band gap and enhancing its electrical conductivity opens this composite for solar cell application.

Keywords: Bandgap shift, electrical conductivity, (Ag-SnO_2)NPs-β-carotene composite, thin film

1. Introduction

Metal-oxide semiconductors such as SnO_2, TiO_2, and ZnO are intensively used in photoelectrochemical applications. Among the semiconductor materials, SnO_2 shows relatively high electron mobility (~250 $cm^2\ V^{-1}\ s^{-1}$), high electron-hole separation ability, and high transport properties [1]. Moreover, SnO_2 as a stable n-type semiconductor with a wide band gap (E_g = 3.6 eV) has been used for various applications, including a solid-state gas sensor, photovoltaic devices, transparent conductive films, and dye-based solar cells [2]. In the case of dye-sensitization, the conduction band edge has to match the lowest excited state of the day molecule to enable effective electron transfer from the molecules excited state to the semiconductor's conduction band [3].

Organic dyes can easily found in nature, such as chlorophyll, anthocyanins, and carotenoid which are contained in any plants and fruits. Carotenoids are lipid-soluble C40 tetraterpenoids. Carotenoid is one of the conducting polymers that can conduct electricity in the range of semiconductors [4]. The hydrocarbon carotenoids which are known as β-carotene is an asymmetrical molecule with a chemical structure of $C_{40}H_{56}$. β-carotene has conjugated bonds; there are single and double bonds as 11 [12]. In conducting polymers, the charge carriers can move easily on the conjugated polymer chain by hopping mechanism. Thus β-carotene is expected to be a good candidate as organic electrical applications. These doped conducting polymers play a key role in the optoelectronic and electrochemical devices. To modify the electrical properties of this material, we can use metallic based materials. Some elements in the transition group such as Mn, Fe, Cu, Zn, Pt, Au, Ag has been used to modifying the material properties [4]. Among metallic nanoparticles, the silver nanoparticles have a typical attractive characteristic, they are high electrical conductivity, stable under the chemical environment, and can be used as a catalyst [5]. So far there is no intensive study related to synthesis (Ag-SnO_2) NPs-β-carotene thin film. In this work, we report the band gap shift and electrical conductivity of the films with the various mass of SnO_2.

2. Materials and Methods

2.1. Synthesis of Ag and SnO_2 Nanoparticles

Silver nanoparticles (AgNPs) were prepared according to the chemical reduction method [5]. The materials used are silver nitrate ($AgNO_3$) as a precursor, sodium borohydride ($NaBH_4$) as a reductor, and mercaptosuccinic acid (MSA) as a stabilizer to prevent the nanoparticle from agglomerating. Tin oxide nanoparticles (SnO_2NPs) have been obtained by the sol-gel method [6] from stannic chloride dihydrate ($SnCl_2.2H_2O$). In a typical synthesis procedure, 8 g $SnCl_2.2H_2O$ was dissolved in 25 ml methanol under continuous stirring at 400 rpm. After 30 minutes, 0.1 M polyethylene glycol (PEG, M_w = 6000) added into the solution. The next step was refluxed at 200 rpm for 5 hours till and the temperature maintained at 120 °C. After the solution transformed into amber one, kept it to obtain the precipitate. Further, the yellowish precipitate was filtered and annealed at 100 °C for 3 hours. Finally, the precipitate calcined for 1 hour at 350 °C resulting the SnO_2 nanoparticles.

2.2. Preparation of (Ag-SnO_2) NPs-β -Carotene Thin Film

Basic of an experimental method for synthesis (Ag-SnO_2)NPs-β-Carotene composite was blending method. Firstly, 0.04 g AgNPs dissolved in 1 mL of phosphoric acid (H_3PO_4) under for 1 hour. SnO_2NPs, as prepared, was added to the solution with the various mass; they were 0.05 g, 0.010 g, 0.015 g, 0.020 g, and 0.025 g. After 2 hours, 0.1 M β-carotene was added into the solution. The important one of this process was sonication to obtain a homogeneous solution with small particles; it ran for 2 hours. For comparison, the control AgNPs without adding of SnO_2NPs was named as 0 g sample. Deposition method used in this current work was a spin coating on the glass substrate. The composite was deposited for 10 s with the rate of 1500 rotation per minute (rpm). A thin film of (Ag-SnO_2)NPs-β-Carotene obtained after annealing at 100 °C for about 4 hours.

2.3. Characterization of (Ag-SnO_2) NPs-β-Carotene Thin Film

The phase identification of the resulting samples investigated by X-ray diffraction (X'Pert Pro) measured with Cu-Kα radiation (λ = 1.540598 Å) in the range of 2θ 10-90° and a step size as 0.02°. The surface morphology and elemental compositions were characterized using SEM-EDAX (FEI INSPECT-S50). Fourier Transform Infrared spectroscopy measurement was recorded by Shimadzu FTIR (IR-Prestige 21). UV-Vis absorption was obtained from (UV-1700 Pharmaspec Shimadzu) UV-Vis spectrophotometer. Electrical properties of the samples were measured by the four-point probe method.

3. Results and Discussion

3.1. XRD Analysis

The phase purity and crystallinity of synthesized products were analyzed by x-ray diffraction (XRD). The diffractogram of silver nanoparticles displayed in Figure 1. Several identified peaks of intensity at 2θ are 38.19°, 44.38°, 64.53°, 77.53° and 81.69°, according to Bragg planes of (111), (002), (022), (113), and (222) respectively. Analyzing was done using Celref with *Fm3m* space group of AMCSD 0011135. The lattice parameter of a = 4.0862 Å as input model. After the refinement process, we obtained that the lattice parameter decreases to a = 4.0834 Å. The crystallite size was calculated using the Scherer equation, where in this case we use (111) plane as the highest plane. The crystallite size of AgNPs falls at 25.20 nm.

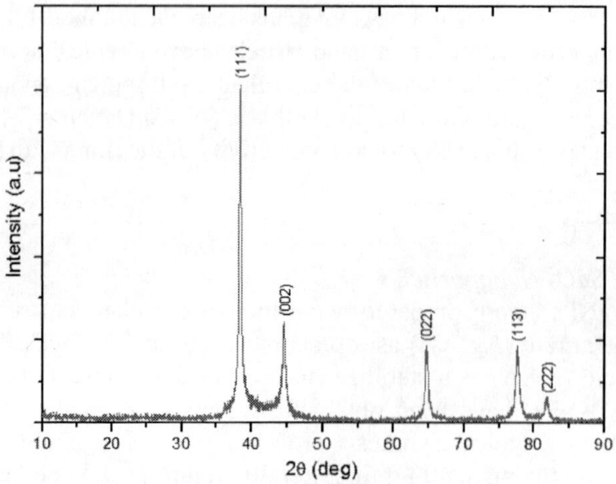

Figure 1. X-ray diffraction pattern of AgNPs

The x-ray diffraction pattern of SnO_2NPs prepared by the sol-gel method shown in Figure 2. There were several main identified peaks indexed using the Celref software at 2θ = 26.60°, 33.90°, 37.94°, 51.75°, 54.77°, dan 64.77° according to Bragg planes of (110), (101), (020), (121), (220), and (112) respectively.

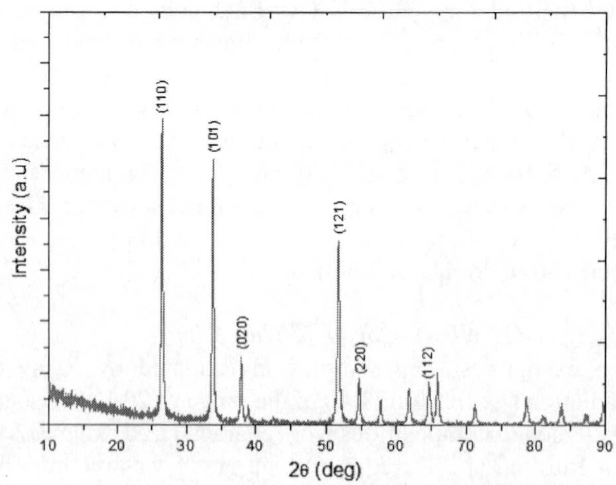

Figure 2. X-ray diffraction pattern of SnO_2NPs

The peaks observed in the XRD patterns are in good agreement with the standard of COD 9007433. It corresponded to the rutile type of SnO_2NPs which has a tetragonal crystal structure. Lattice parameters of refinement give a = 4.7419 Å and c = 3.1881 Å, it was rather difference with input model as a = 4.7370 Å and c = 3.1850 Å. The crystallite size of the prepared samples is 37.79 nm as calculated by Debye Scherrer equation.

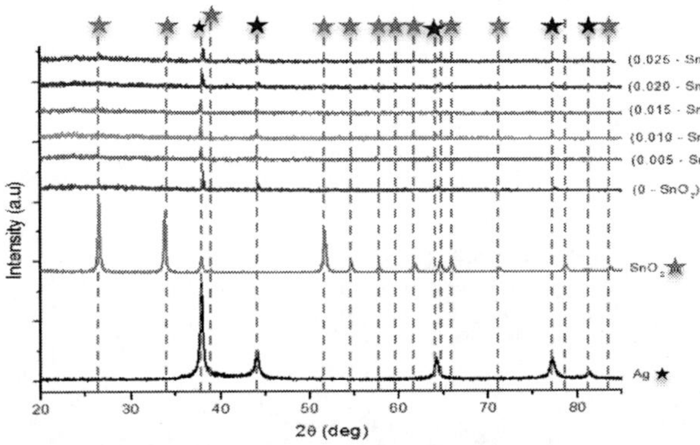

Figure 3. Typical x-ray diffraction patterns of (Ag-SnO_2)NPs-β-Carotene thin films with a various mass of SnO_2NPs.

Figure 3 represents the x-ray diffraction patterns of the (Ag-SnO_2)NPs-β-Carotene thin films. All histogram were rearranged start from 2θ = 10° up to 85°. Generally, it could be observed that two peaks appear at 37.80° and 44.87° belong to silver with low intensity. There are no clear peaks of SnO_2NPs detected in the 0 g, 0.005 g. 0.010 g, and 0.015 g of SnO_2NPs mass since the film is extremely thin. Specifically, several peaks of SnO_2NPs seen in the mass variations of 0.020 g and 0.025 g, two of them are identified at 26.57° and 33.85°, represent two strong peaks of tin oxide. It is revealed that there are small peak shifts in the film. The crystallinity of thin films is obtained by comparing the area of the peaks to the total area under histogram [4]. The crystallinity of the samples relatively low, since the materials contain a polymer which naturally has an amorphous structure.

3.2. FTIR Studies

FTIR measurements were carried out to find the functional groups of films. Some peaks according to the functional groups of chemical bonds are drawn in Figure 4.

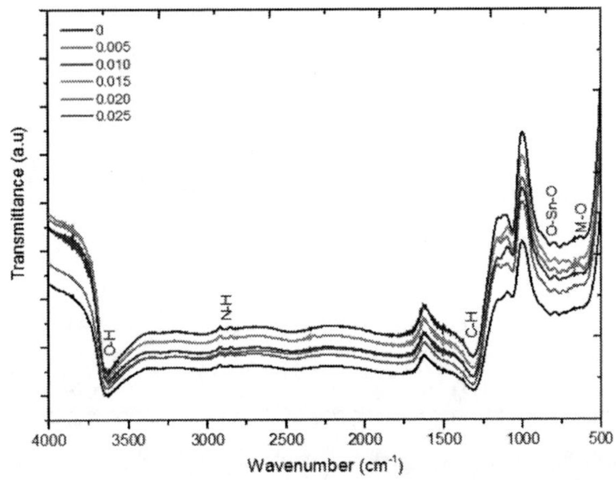

Figure 4. FTIR spectra of (Ag-SnO_2)NPs-β-Carotene thin films with a different mass of SnO_2NPs

The results of the analysis show that some peaks identified as C-H, O-H, N-H, O-Sn-O, and M-O. The peaks at a wave number of 592.15 cm^{-1} belong to the type of (M-O) metallic oxygen bond; it assigns to silver which is bond the oxygen [7]. Stretching modes of the O-Sn-O functional group shown by existing at wave number 827.46 cm^{-1} [8]. The wavenumber about 3496.94 cm^{-1}, which is in the range of 3349-3409 cm^{-1} related to stretching vibrations of the O-H bond [8]. Beside, O-H stretching vibration also identified at 359 cm^{-1} indicating the presence of hydroxyl group. The peak at 1319.31 cm^{-1} suggests the appearance of a C-H bond corresponds to the β-carotene functional group [7]. At wavenumber

2833.43 cm^{-1}, investigated as an N-H functional group [9]. There is no remarkable discrepancy since SnO_2NPs added in a very small portion.

3.3. Morphological Analysis

The morphology and microstructure of film were characterized by SEM as shown in Figure 5.

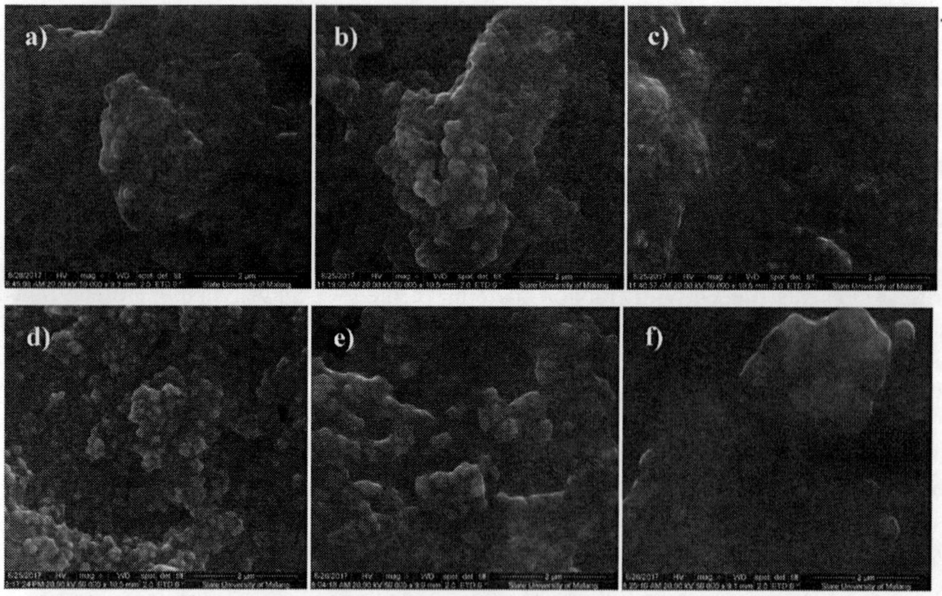

Figure 5. SEM images of (Ag-xSnO_2)NPs-β-Carotene thin films with various mass of SnO_2NPs a) 0, b) 0.005, c) 0.010, d) 0.015, e) 0.020, dan f) 0.025 g

From the SEM images depicted in Figure 5, it seems that silver and tin oxide nanoparticles are not fully dispersed as expected. The materials agglomerated to form clusters and porous. Further, the agglomeration increases at a higher fraction of SnO_2NPs. Surface porosity of (Ag-xSnO_2)NPs-β-Carotene thin films are shown in Table 1.

Table 1. Surface porosity of (Ag-xSnO_2)NPs-β-Carotene thin films at the various mass of SnO_2NPs

SnO_2NPs (g)	**Surface Porosity (%)**
0	5.15
0.005	5.94
0.010	4.80
0.015	4.00
0.020	9.68
0.025	21.60

The results of SEM analysis using ImageJ software show that porosity surface of films in the range of 4 to 6% for 0 to 0.015 g. Percentage of porosity surface increase significantly at 0.020 g, and 0.025 g as 9.68, and 21.60% respectively. Higher porosity of films due to higher SnO_2NPs added on the films which tend to agglomerate with AgNPs and β-Carotene.

3.4. Optical Absorption Properties

The optical and energy band of the film confirmed using UV-vis spectroscopy. Absorption of the light at certain wavelength related to electron energy needed for excitation from valence band to conduction band [10], thus it is necessary to use Tauc plot method to find out the energy gap of the films. Figure 6 shows the curve of $(\alpha h\nu)^2$ as a function of $h\nu$.

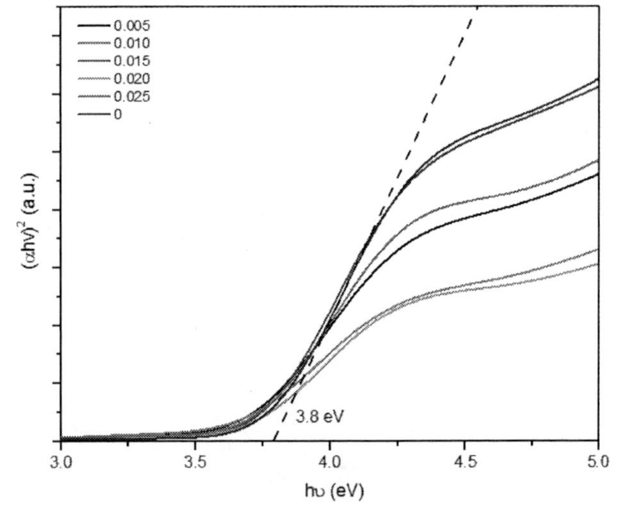

Figure 6. Tauc plot curve of (Ag-SnO_2)NPs-β-Carotene thin films.

From the specific Tauc plot, we further analyzed by using a linear fit of each curve of Figure 6. One example is shown in Figure 6 for the undoped film. The band gap could be obtained from the linear fit cut to *hv* axis. The film of undoped SnO_2NPs exhibit a higher energy gap of 3.8 eV. The results of the bandgap analysis of all samples are shown in Figure 7.

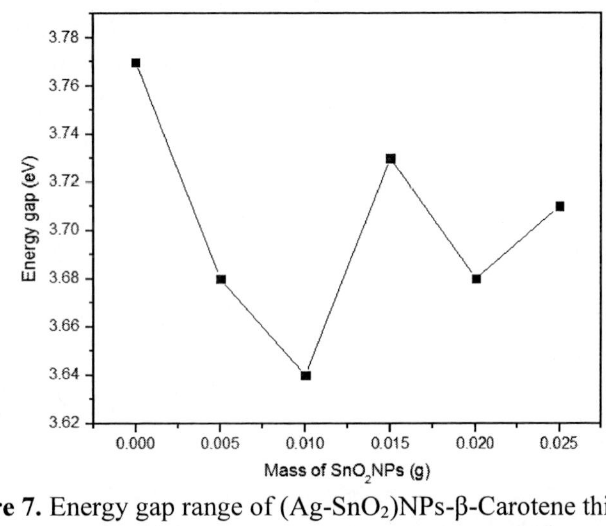

Figure 7. Energy gap range of (Ag-SnO_2)NPs-β-Carotene thin films

Introducing SnO_2NPs on the films provide a shift the bandgap shift. The (Ag-xSnO2)NPs-β-carotene films show an average band gap around ~3.7 eV as depicted in Figure 7. These values approach the energy gap of the tin oxide nanoparticles as 3.68 eV [7]. The energy gap between samples decreases with the various increasing mass of SnO_2NPs, due to several factors as like grain size, morphology, or crystal structure [11]. The wide band gap of this samples corresponds to wavelength absorption in the range of ultraviolet at ~380 nm.

3.5. Electrical Conductivity

Measurement of electrical conductivity has been investigated using four points probe method; it will be provided current and voltage (*I-V*) curve relationship. Figure 8 shows the influence of SnO_2NPs doping on the electrical conductivity of (Ag-SnO_2)NPs- β-Carotene thin films.

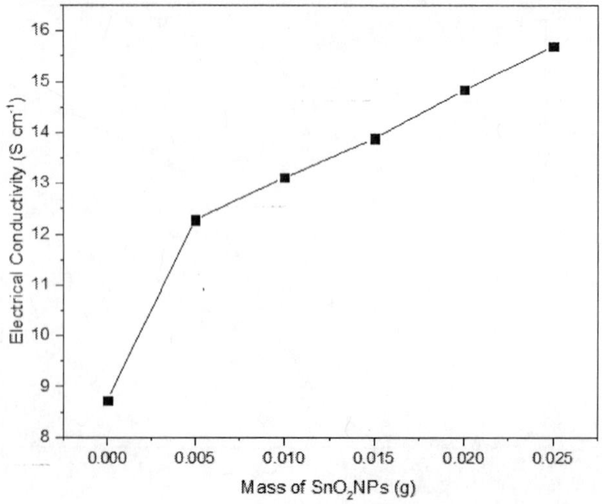

Figure 8. The effect of SnO_2NPs on the electrical conductivity of (Ag-xSnO_2)NPs- β-Carotene thin films.

The addition of tin oxide nanoparticles gives higher electrical conductivity, which is due to the presence of SnO_2NPs which provide higher electron mobility. The undoped sample shows a relatively conductive which provided by Ag NPs [4, 15] and other metallic [13], or metal oxides [14] in conducting polymers. This also results Electrical conductivity properties of the films significantly increases from 12.28 S/cm to 15.70 S/cm with the various mass of SnO_2NPs as 0.005 g to 0.025 g, while the AgNPs-β-carotene film reaches up to 8.72 S/cm. Although the electrical conductivity of the films is relative low, thus it opens a wide range of research to enhance the electrical properties and modify the films for solar cell applications [16].

4. Conclusion

(Ag-SnO_2)NPs-β-carotene thin films have been synthesized using the spin coating method. It is observed that the composite films were clarified according to XRD patterns analysis. Observation using SEM indicates that materials agglomerate and more evident along with the increasing of mass tin oxide nanoparticles. It is found that the film provides the wide energy gap ~3.7 *eV*, it has been decreased from 3.8 *eV* without any tin oxide nanoparticles. Moreover, the electrical properties were investigated using four points probe, and it shows that SnO_2NPs doping gives rise to the electrical conductivity of the films. The decrease of the band gap and high electrical conductivity make this film possess excellent potential for solar cell applications, mainly for photoanode of DSSC device.

Acknowledgments

This work was partly supported by Ministry of Research and Higher Education through PNBP UM.

References

[1] W. M. N. M. B. Wanninayake, K. Premaratne, and R. M. G. Rajapakse, "High Efficient Dye-Sensitized Solar Cells Based on Synthesized SnO 2 Nanoparticles," *J. Nanomater.*, vol. 2016, pp. 1–8, 2016.

[2] M. M. Rashad, I. A. Ibrahim, I. Osama, and A. E. Shalan, "Distinction between SnO2 nanoparticles synthesized using co-precipitation and solvothermal methods for the photovoltaic efficiency of dye-sensitized solar cells," *Bull. Mater. Sci.*, vol. 37, no. 4, pp. 903–909, 2014.

[3] M. Dürr, S. Rosselli, A. Yasuda, and G. Nelles, "Band-Gap Engineering of Metal Oxides for Dye-Sensitized Solar Cells," *J. Phys. Chem. B*, vol. 110, no. 43, pp. 21899–21902, Nov. 2006.

[4] M. Diantoro *et al.*, "Electrical conductivity modification using silver nanoparticles of Jatropha

Multifida L. and Pterocarpus Indicus w. extracts films," 2016, p. 030034.

[5] M. Diantoro, R. Fitrianingsih, N. Mufti, and A. Fuad, "Synthesis of silver nanoparticles by chemical reduction at a various fraction of MSA and their structure characterization," 2014, pp. 257–261.

[6] P. Tripathi, A. Ahmed, T. Ali, and M. Obaidurrahman, "Polyethylene glycol (PEG) assisted size-controlled SnO2 nanoparticles by a sol-gel process," 2016, p. 050067.

[7] K. Saravanakumar and V. Muthuraj, "Fabrication of sphere-like plasmonic Ag/SnO2 photocatalyst for the degradation of phenol," *Opt. - Int. J. Light Electron Opt.*, vol. 131, pp. 754– 763, Feb. 2017.

[8] S. Majumder, "Synthesis and characterization of SnO 2 films obtained by a wet chemical process.," *Mater. Sci. 0137-1339*, vol. 27, no. 1, 2009.

[9] D. Geethalakshmi, N. Muthukumarasamy, and R. Balasundaraprabhu, "Structural and optical study of spin-coated camphorsulfonic acid-doped polyaniline/titanium-di-oxide nanoparticles hybrid thin films," *Superlattices Microstruct.*, vol. 82, pp. 447–460, Jun. 2015.

[10] M. Caglar, S. Ilican, and Y. Caglar, "Influence of dopant concentration on the optical properties of ZnO: In films by a sol-gel method," *Thin Solid Films*, vol. 517, no. 17, pp. 5023–5028, Jul. 2009.

[11] A. H. Kuroda, Y. M. Hassan, and N. M. Ahmed, "Controlling Diameter, Length and Characterization of ZnO Nanorods by Simple Hydrothermal Method for Solar Cells," *World J. Nano Sci. Eng.*, vol. 05, no. 01, pp. 34–40, 2015.

[12] Wang, X.-F., Matsuda, A., Koyama, Y., Nagae, H., Sasaki, S., Tamiaki, H., Wada, Y., 2006. Effects of plant carotenoid spacers on the performance of a dye-sensitized solar cell using a chlorophyll derivative: Enhancement of photocurrent determined by one electron-oxidation potential of each carotenoid. Chemical physics letters 423, 470–475.

[13] Markus Diantoro, Devy Purwaningtyas, Nazilah Muthoharoh, Arif Hidayat, Ahmad Taufiq, Abdulloh Fuad, The Influence of Iron- and Copper- doped of PANi Thin Film on their Structure and Dielectric Properties. American Institute of Physics, 1454. ISSN 0094-243X, p268-271. 2012.

[14] M. Diantoro, Kholid, A. A. Mustikasari, and Yudiyanto, The Influence of SnO2 Nanoparticles on Electrical Conductivity, and Transmittance of PANI-SnO2 Films IOP Conf. Series: Materials Science and Engineering 367 (2018) 012034 doi:10.1088/1757-899X/367/1/012034

[15] Markus Diantoro, Thathit Suprayogi, Ulwiyatus Sa'adah, Nandang Mufti, Abdulloh Fuad, Arif Hidayat and Hadi Nur (July 18th, 2018). Modification of Electrical Properties of Silver Nanoparticle, Silver Nanoparticles Khan Maaz, IntechOpen, DOI: 10.5772/intechopen.75682.

[16] Markus Diantoro, Thathit Suprayogi, Arif Hidayat, Ahmad Taufiq, Abdulloh Fuad, and Risa Suryana, Shockley's Equation Fit Analyses for Solar Cell Parameters from I-V Curves, International Journal of Photoenergy, Volume 2018, Article ID 9214820, 7 pages https://doi.org/10.1155/2018/9214820

Synthesis and Structural Analysis of Silicon Carbide from Silica Rice Husk and Activated Carbon Using Solid-State Reaction

Kusuma Wardhani Mas'udah[1,*], Markus Diantoro[2], Abdulloh Fuad[2]

[1]Faculty of Mathematics and Natural Science, University of Pesantren Tinggi Darul Ulum (Unipdu), PP. Darul'Ulum Tromol Pos 10 Peterongan, Jombang 61481, Indonesia.

[2]Department of Physics, Faculty of Mathematics and Natural Sciences, Universitas Negeri Malang, Jl. Semarang 5 Malang 65145, Indonesia.

*Corresponding author's email: kusumawardhani.m@mipa.unipdu.ac.id

Abstract. Synthesis of Silicon Carbide (SiC) has been performed using a solid-state reaction method. We used silica and activated carbon as raw materials. The silica was synthesized from silica rice husk using an alkali extraction and a sol-gel method. The purified silica was then mixed with the activated carbon at the same ratio, homogenized, and then cold pressed into pellets by adding polyvinyl alcohol to glue them perfectly. The pellets were then sintered in a vacuum of a high-temperature furnace in an inert arc-furnace at 1200, 1300, and 1400 °C for 6 hours. The samples were characterized for their particle size, surface area, phase composition, microstructure, and resistivity. The XRD data analysis showed that the samples are dominated by the SiC phase in the form of 3C-SiC and 6H-SiC, CO (Carbon (II) oxide), and SiO_2 phases. The weight fractions of SiC samples were respectively fallen to 68, 98, and 69% for 1200, 1300 and 1400 °C sintering temperatures.

Keywords: Silicon carbide, silica rice husk, activated carbon, solid-state reaction

1. Introduction

Silicon carbide (SiC) is one of the most widely used non-oxide ceramics for many industrial applications because of its attractive high-temperature properties, such as high strength, high hardness, and high wear and thermal shock resistance. It also possesses an excellent chemical oxidation resistance. The silicon carbide's performance under such extreme conditions is expected to enable significant improvements to a variety of applications. Nanostructured SiC has shown to exhibit superior properties (greater elasticity and strength) compared to the bulk SiC. It also shows the potential applications in light emitting diodes and UV Photodetectors due to a higher light-emission efficiency [1]. There is a significant interest in the synthesis of nanostructured SiC including nanospheres, nanowires, nanorods and so on as novel functional materials for a nanoscale engineering [2]. In normal conditions, the synthesis of SiC is difficult due to oxidation of carbon and silica at a high temperature. In contrast, the synthesis of phase- pure of sub-micron-sized of βSiC powder has been achieved from its constituent elements namely carbon and silicon to form SiC can occur at a temperature as low as 900 °C. The phase-pure SiC can be formed only at the temperatures of 1200 °C and above [3].

Recently, Zhang et al. [4] reported the direct synthesis of nanostructured silicon carbide (SiC) materials using a pulsed laser deposition technique. Kawamura et al. [5] showed that β-type cubic SiC could be prepared from a Si and fullerene powder mixture or a-Si and amorphous carbon (carbon black) powder mixture by the Na flux method at 900-1000 K. Kriener et al. [6] presented a study of mainly hexagonal boron-doped 6H-SiC , without any indication of a cubic 3C phase fraction, by means of X-ray diffraction, and resistivity measurements. Walter et al. [7] used solid-state reactions in the Al-Si-C at a sintering temperature of 1450 °C for 8 hours under a vacuum condition in a graphite furnace. *Andrievski* [8] reported that amorphous SiC nanoparticles have been synthesized by using a low-pressure microwave plasma, inductively coupled plasma to synthesis β-SiC nanopowders, carbothermic reduction to synthesis α-SiC nanocrystals. The amorphous SiC particles were synthesized from tetramethylsilane precursor at room temperature and a low precursor partial pressure using argon as carrier gas. However, all these methods are expensive and require special equipment.

In this paper, we reported a synthesis of SiC using polyvinyl alcohol to prevent sticking the precursors. Due to its scarce availability in nature, SiC ceramics must be synthesized. The industrially high demand of SiC raw material takes intensive attention to produce a renewable material at a low cost. The rich silica content in the rice husk ash is one of prospective low cost and renewable starting materials in synthesizing SiC due to its desired silica and carbon-activated contents. A simple approach to synthesized SiC nanoparticles by using a solid-state reaction at the temperatures of 1200, 1300, and 1400 °C and characterized their mechanical properties. A novel synthesis route resulting high purity SiC but at a low cost is addressed in this study.

2. Materials and Methods

The synthesis of Silicon Carbide was prepared by solid-state reaction methods using the raw material of SiO_2, activated carbon, and PVA. The SiO_2 was synthesized from silica rice husk using an alkali extraction and sol-gel method with a purity of up to 99.45 % [9] as the main material. In this method, solid powders of starting materials with particular molar mass ratios were mixed, homogenized, and then cold pressed into pellets with the addition of polyvinyl alcohol to stick perfectly. The pellets were then sintered in a vacuum of the high-temperature furnace in an inert gas at the temperatures of 1200 °C, 1300 °C, and 1400 °C for 6 hours. The samples were then characterized by using powder X-ray diffractometer (XRD) with Cu $K_{\alpha 1}$ with 2θ between 20° to 80° (wavelength 1.5406 Å), scanning electron microscopy (SEM) with energy dispersive X-ray (EDX), and resistivity by using a four-point probe method (I-V meter).

3. Results and Discussion

The phase identification of as-prepared samples was carried out based on XRD patterns. The most common polytypes of SiC presently being developed for electronics are 3C-SiC, 2H-SiC, 4H-SiC, and 6H-SiC that shown in the schematic cross-section (Figure 1) [10].

Figure 2 shows the typical XRD pattern of samples prepared at 1200, 1300, and 1400 °C for 6 hours. The sintering temperature and duration affect the intensity of the material and tend to reduce the intensity when surpassing its melting point [11]. Therefore, we noted that the crystalline structures and diffraction peaks intensity were mainly depended upon the increasing temperature.

The XRD data analysis showed that the samples have a dominant phase of SiC in the form of 3C-SiC and 6H-SiC structure. We also found a weak intensity of CO (Carbon (II) oxide), and SiO_2 phases. This composition is a good agreement with the EDX data. In the sample sintered at 1200 °C, four phases were observed, i.e., 6H-SiC, 3C-SiC, CO, and SiO_2 with the PDF number of 01-073-6302, 01-073-1708, 01-076-2378, and 01-070-2540 respectively. It is found the appearance of several peaks identified as a 6H-SiC phase. The 6H-SIC peaks appear at diffraction angle 2θ of 25.92°, 27.25°, 35.35°, 45.56°, 49.36°, 56.48°, 59.72°, 62.60°, and 71.56°. We also observed three peaks belong to 3C-

SiC. The 3C-SiC observed at 2θ of 35.35°, 59.72°, and 71.56 °. Minor peaks found showing CO phase at 2θ of 29.62° and 31.53°. We found SiO_2 phase exists at 2θ of 21.57°, 51.92°, and 53.52°.

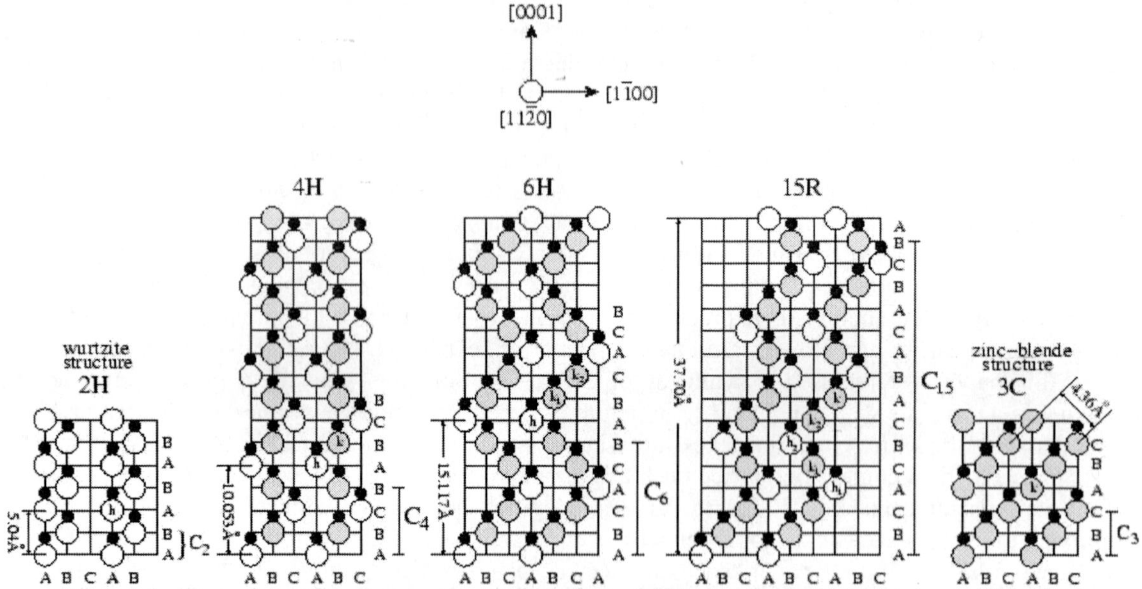

Figure 1. The schematic cross-sectional depiction of SiC atomic crystal structure. After [10].

Meanwhile, the sample sintered at 1300 °C shows three phases exist, i.e., 6H-SiC, 3C-SiC, and CO. It was also observed that two weak peaks represented the CO phase, while reducing SiO_2 intensity. The shoulder is associated with stacking faults, which serves as a common feature of SiC material, demonstrating the crystalline nature of the sample prepared at the temperature of 1200 °C [4]. Increasing temperature up to 1400 °C provides to increase the 6H-SiC and reduce 3C-SiC. On the other hand, a SiO_2 increase significantly. The refinements gave the cell constant falls to a = 4.3480 Å, which was consistent with the reported value in the literature (a = 4.3480 Å, AMCSD, No. 001791). The weight fraction of SiC which were sintered at 1200, 1300, and 1400 °C reached 68, 98, and 69% respectively. All samples have also been analyzed for the particle size, particle morphology, and distribution using a scanning electron microscope (SEM) which is depicted in Figure 3.

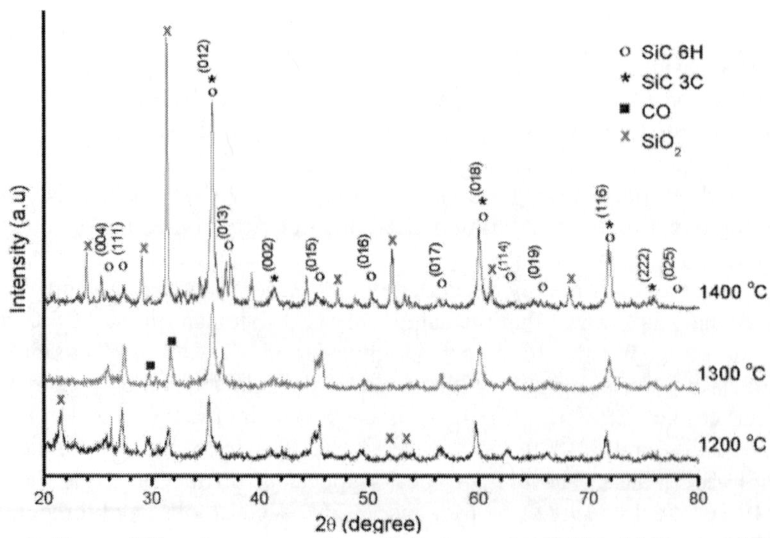

Figure 2. X-ray diffraction pattern from SiC sintered at 1200, 1300, and 1400 °C.

It is seen that activated carbon of charcoal yielded as platelets with the size of ~50 μm (Figure 3a), while purified silica tends to agglomerate as shown in Figure 3b. The size of the agglomerate is ~60μm. A typical SiC plate-likes grain is shown in Figure 3c. On the selected area indicated by red lines, showing that the plate like grains comprises by a stacked vertically thin plates. The particle morphology of SiC were examined using three different pellets. The morphology of the particles change following the increaase of in the sintering temperature. The size of individual thin plates in each case show a relatively constant of ~0.5 μm (Figure 3c).

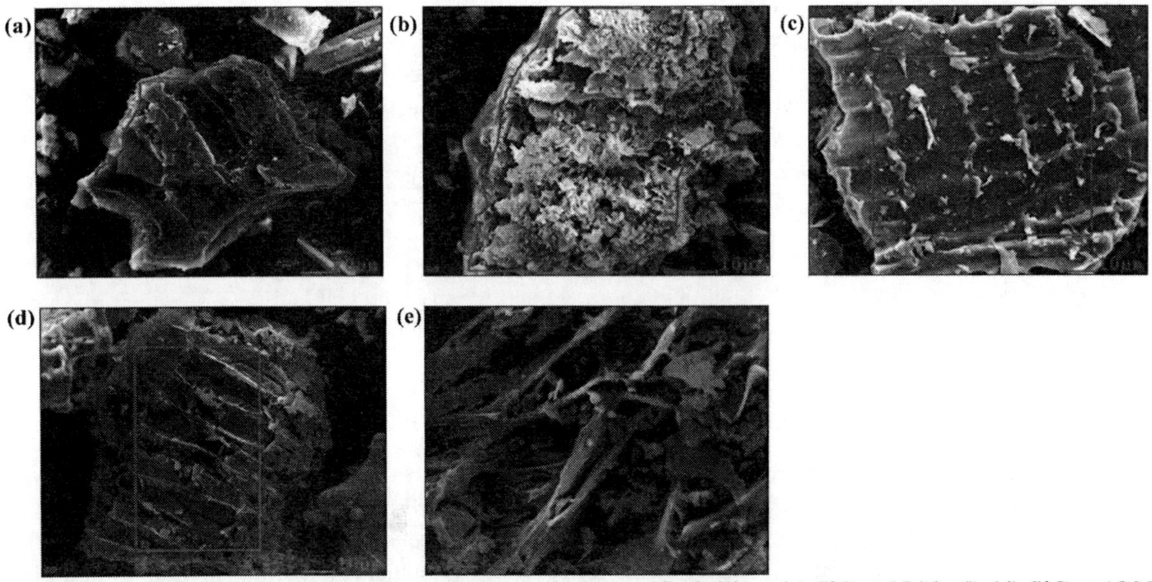

Figure 3. SEM images of (a) activated charcoal, (b) purified silica, (c) SiC at 1200 °C, (d) SiC at 1300 °C, and (e) SiC at 1400 °C

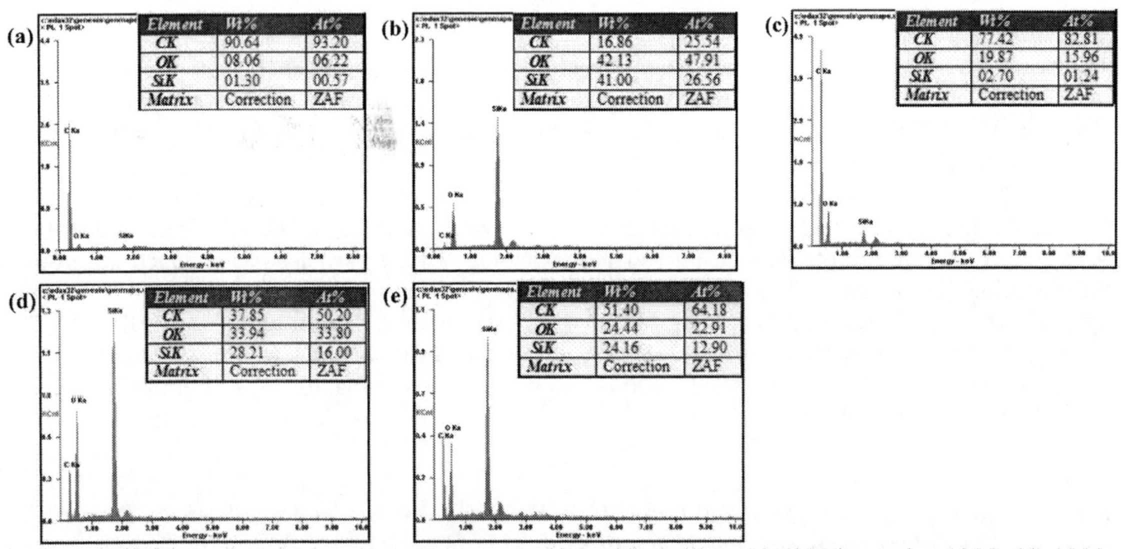

Figure 4. EDX results of (a) activated charcoal, (b) purified silica, (c) SiC sintered at 1200, (d) 1300 °C, and (e) at 1400 °C.

Like those above, the agglomerates also contained whiskers in almost all cases. However, the morphology of the whiskers changed from free, sharp-edged whiskers at temperature 1300 °C and blunt-type whiskers at 1400 °C. The processing temperatures of 1300 and 1400 °C is close to partial sintering. Figure 3e indicates the fusion of many whiskers or individual thin sheets to form a large grain of ~5 μm. The continuous formation of such grains increases the overall grain size. A similar observation was reported for a SiC synthesis by Johnson et al. [12]. These authors reported that as soon as SiC forms on the surface of the carbon primary particles, the primary particles begin to grow by sintering. The SiC particles resemble to the carbon precursor which have a tendency to grow significantly at higher temperatures, because of increased sintering. The similar grains were not observed for other synthesis conditions, confirming that the fusion of particles is uniquely a high-temperature process and not a process-dependent factor. These fused particles are assumed to decrease the surface area of the particles, which eventually results in the reduction of the reactivity.

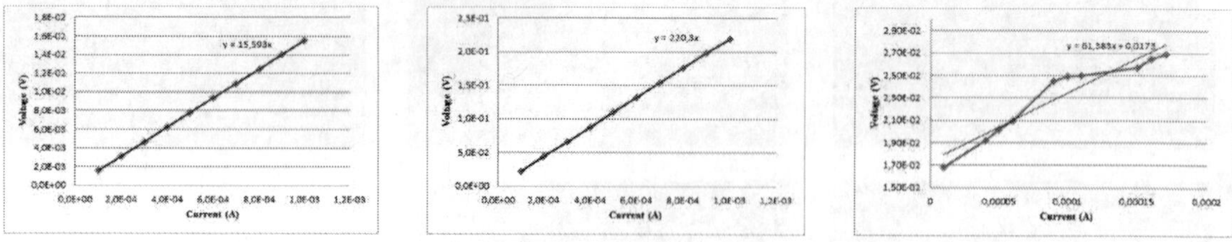

(a) (b) (c)

Figure 5. I-V meter results of SiC at (a)1200, (b)1300, and (c)1400 °C.

The resistivity of SiC which were sintered at 1200 °C (Figure 5a) and 1300 °C (Figure 5b) shows an ohmic relation [13]. The calculated resistivity falls to 9.8 $\Omega.cm$ and 138.3 $\Omega.cm$ respectively. On the other hand, the SiC which sintered at 1400 °C (Figure 5c) exhibit a non-ohmic behavior with average resistivity falls to 38.55 $\Omega.cm$. The resistivity of semiconductor materials such as Si and Ge were 10^{-2}-10^{2} $\Omega.cm$ [13]. The three samples of SiC being studied show a resistivity in the range of a semiconductor. By comparing the purity of the samples obtained from the XRD analysis and resistivity, we may conclude that resistivity is strongly affected by SiC fraction of the samples.

4. Conclusion

The synthesis of Silicon Carbide has been done by using solid state reaction methods using the raw material of SiO_2, activated carbon, and PVA under vacuum condition. The SiO_2 was synthesized from silica rice husk using alkali extraction and sol-gel method with a purity of up to 99.45 %. It was shown that the samples are dominated by SiC in the form of 3C-SiC and 6H-SiC structure. We also found a minor impurity of CO (Carbon (II) oxide), and SiO_2 phases within a good agreement with the EDX characterization. The weight fraction of SiC which were sintered at 1200, 1300, and 1400 °C respectively achieved of 68, 98, and 69%. The morphology of the particles has changed according to the sintering temperature. The EDX data show that the carbon atom concentration (at %) is approximately 82.81 %, 50.20%, and 64.18% and the silicon atom concentration varies from 01.24%, 16.00 and 12.90%. According to the resistivity, the obtained SiC belongs to the semiconductor. The resistivity of SiC at the 1200 °C and 1300 °C show an ohmic behavior ranged from 9.8 $\Omega.cm$ to 138.3 $\Omega.cm$. We noted that the resistivity is strongly affected by its purity of the samples.

References

[1] W. Wong E, E. Sheehan P and Lieber C 1997 *Nanobeam Mechanics: Elasticity, Strength, and Toughness of Nanorods and Nanotubes* vol 277
[2] Biernacki J J and Wotzak G P 1989 Stoichiometry of the C + SiO2 Reaction *Journal of the American Ceramic Society* **72** 122–9
[3] Satapathy L N, Ramesh P D, Agrawal D and Roy R 2005 Microwave synthesis of phase-pure,

fine silicon carbide powder *Materials Research Bulletin* **40** 1871–82

[4] Zhang H X, Feng P X, Makarov V, Weiner B R and Morell G 2009 Synthesis of nanostructured SiC using the pulsed laser deposition technique *Materials Research Bulletin* **44** 184–8

[5] Kawamura F, Yamane H, Yamada T, Yin S and Sato T 2008 Low-Temperature Fabrication of Porous β-SiC Ceramics in Sodium Vapor *Journal of the American Ceramic Society* **91** 51–5

[6] Kriener M, Muranaka T, Kato J, Ren Z-A, Akimitsu J and Maeno Y 2009 Superconductivity in heavily boron-doped silicon carbide *Science and technology of advanced materials* **9** 044205

[7] Walter D and Karyasa I W 2005 Solid State Reactions in the Al-Si-C System *Journal of the Chinese Chemical Society* **52** 873–876

[8] Andrievski R A 2009 Synthesis, structure and properties of nanosized silicon carbide *Rev. Adv. Mater. Sci* **22** 1–20

[9] Fuad A, Mufti N, Diantoro M, Subakti and S. S K 2016 Synthesis and characterization of highly purified nanosilica from pyrophyllite ores p 030020

[10] Zetterling C-M 2000 Process Technology for Silicon Carbide Devices *Docent Seminar, March 21st, 2000*

[11] Kim J and Creasy T S 2004 Selective laser sintering characteristics of nylon 6/clay-reinforced nanocomposite *Polymer Testing* **23** 629–36

[12] Johnson J A, Hrenya C M and Weimer A W 2002 Intrinsic Reaction and Self-Diffusion Kinetics for Silicon Carbide Synthesis by Rapid Carbothermal Reduction *Journal of the American Ceramic Society* **85** 2273–80

[13] Newnham R E 2005 *Properties of materials: anisotropy, symmetry, structure* (Oxford ; New York: Oxford University Press)

Small Scale DSSC Panels Design and Its Performance

Sahrul Saehana[*], Darsikin, Muslimin, and Nasar

Departement of Physics Education, Tadulako University, Jl. Soekarno-Hatta km. 09 Kel.Tondo Palu, Indonesia, 94118

*Corresponding author's email: sahrul_saehana@untad.ac.id

Abstract. This study reports an effort to design dye-sensitized solar cell (DSSC) panels. The panel was proposed using a TiO_2 layer, dye, electrolyte and counter electrode. TiO_2 layer was deposited on FTO glass by using spray methods then it was heated on the furnace for 45 minutes. We applied ruthenium dye (N719), iodolyte and platinum as a counter electrode. Performance of solar cells was characterized by solar illumination using a multimeter. The solar's intensity was measured by using a solar power meter. It was found the efficiency and fill factor of DSSC about 2.08% and 0.80, respectively. This study can be scaled up then applying in daily life.

Keywords: DSSC, panel, efficiency, fill factor and scale up

1. Introduction

The development of solar cell research, which can convert photon energy from the sun into electrical energy, as very rapid alternative energy is done [1-11]. In the first generation, solar cells are made from semiconductor crystals, while the next generation uses thin layer semiconductors (amorphous silicon). In the last decade also developed more economical solar cells, namely organic solar cells.

Dye-sensitized solar cell (DSSC), first discovered by Gratzel et al., is one of the organic solar cells with oxide semiconductor materials, dye, electrodes, electrolyte solutions that attract the attention of many people. The organic material commonly used in DSSC solar cells is titanium dioxide (TiO_2). The TiO_2 material is an n-type semiconductor that has a wide bandgap (about 3.2 eV). Many researchers report good solar cell performance by optimizing the TiO_2 layer, dye, electrolyte and counter electrode. In our previous research [12-15], we found that the efficiency of solar cells is still low at about 1%.

Many obstacles are encountered in the development of DSSC solar cells. One of them is the price of expensive components, not environmentally friendly and complicated fabrication process. However, to date, most of the developed DSSC solar cells are still in a laboratory scale so they can not be applied widely.

In this study, we report on our initial research to create DSSC panels. In this study, we use some simple methods, such as spray so that the resulting panel can be of low cost. We hope that the results of this research can be the basis for the development and application of DSSC solar cells in the community.

2. Materials and Methods

TiO_2 films were deposited on ITO substrates (10 Ω/cm^2) using a spray method. A suspension was prepared by dispersing 5 g of TiO_2 into 10 mL of distilled water, followed by stirring for 45 min. The homogeneous suspension was coated on the ITO surface, which was maintained at a temperature of approximately 150 °C to evaporate the water [10]. The film was then heated at 450 °C in a furnace for

IOP Conf. Series: Journal of Physics: Conf. Series **1093** (2018) 012034 doi:10.1088/1742-6596/1093/1/012034

45 minutes to strengthen the contact between the TiO_2 particles and the contact between the TiO_2 particles and the substrate. Synthetic (N719) dye was used in this research. The heat-treated TiO_2 films were immersed in this dye solution for 24 hours and then cleaned with distilled water before finally being heated at 40 °C for 10 minutes. Prototype solar cells were made by sandwiching HI-30 iodolyte solution (Solaronix, Switzerland) containing I^-/I^{3-} ion between a dye-impregnated TiO_2 film and a platinum-coated ITO electrode.

3. Results and Discussion

Structure of DSSC can be seen in Fig. 1(a) [13]. The mechanism that occurs in DSSC solar cells starts when the photons come on the dye so that the electrons are injected into TiO_2. The electrons then flow in the circuit and move to the counter electrode. The iodine-containing electrolytes then receive the electrons and pass them back to the degenerated dye [14-15].

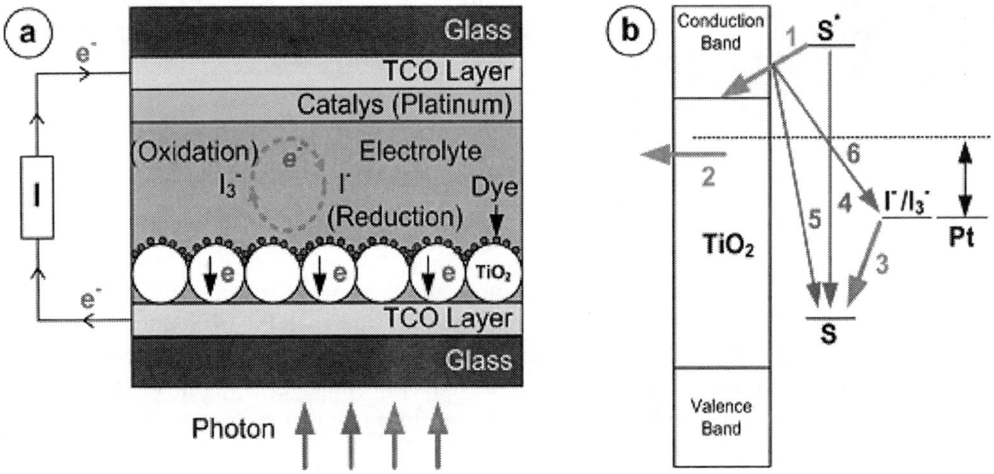

Figure 1. (a) Mechanism and (b) band diagram of DSSC solar cells.

Figure 1b shows the band diagram of DSSC solar cells, where the desired process is marked by the green line (1) electron injection, (2) the collection of electrons, and (3) dye regeneration. On the contrary, it also shows the unexpected process, which is signed by red lines, namely: (4) decay, (5) recombination and (6) recaption reaction [12-15].

In this paper, we proposed a DSSC panel by employing ruthenium dye (N719), iodine electrolyte and platinum. The absorbance spectrum of this synthetic dye can be observed in Figure 2. The maximum peak of N719 was 313 nm, 393 nm, and 533 nm. Thus, the spectrum of sunlight can be well absorbed so that DSSC solar cell performance becomes better.

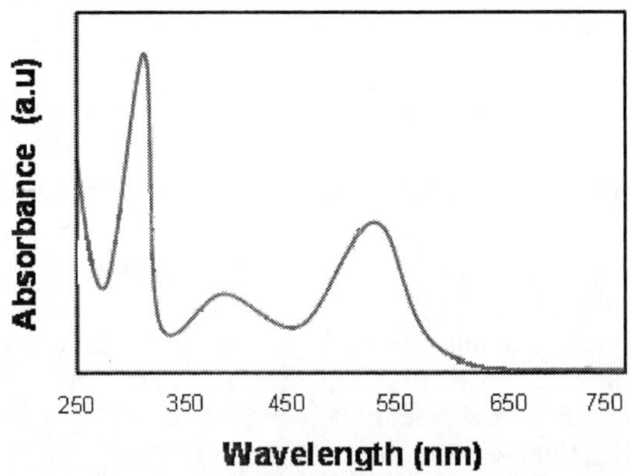

Figure 2. The absorbance spectrum of N719 dye.

The dimension of the DSSC panel was 10 x 10 cm as shown in Figure 3(a). I-V characteristic of the DSSC panel was measured under sunlight illumination. Efficiency and fill factor of the DSSC panel was 2.01% and 0.80, respectively. We suspect that the high internal resistance of solar cells is caused by conductive glass, ionic conductivity, and counter electrodes. However, the results obtained are better when compared with the studies conducted by our previous team [12-15].

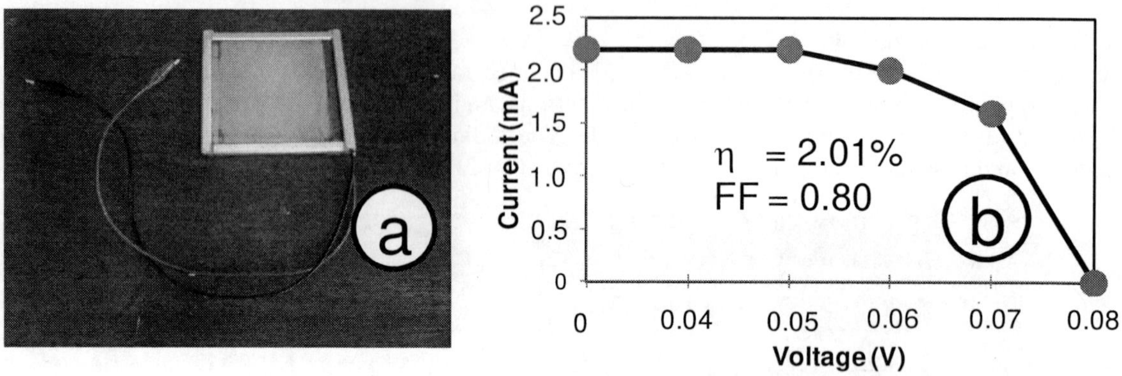

Figure 3. (a) Design of DSSC panel resulted, and (b) I-V characteristic of DSSC panel under sunlight illumination.

In this work, undesirable processes such as recombination still occur so that the performance of solar cells is not optimum. In a subsequent study, we attempted to increase the efficiency of solar cells by optimizing some parameters such as (1) TiO_2 particle size, (2) electrolyte conductivity and (3) better dye absorbance. One of the things that solar cells need to have is long-term stability. Thus, in our future study, we attempted to use an electrolyte polymer that can improve its performance.

4. Conclusion

We have successfully proposed DSSCs panel by using TiO_2 nanoparticle, ruthenium dye, iodolyte, and platinum. It was found the efficiency and fill factor of DSSC about 2.08% and 0.80, respectively. The small DSSC panels (its dimension is about 10 cm x 10 cm) with sandwiched structured are expected to develop in large scale and applied in real life.

Acknowledgments

This work was supported by Riset Terapan Unggulan Perguruan Tinggi Decentralization Research Grant from the Research, Technology and Higher Education of the Ministry of the Republic of Indonesia in the fiscal years of 2017. The researcher also gives appreciation to Dr. Eng. Agus Purwanto, M.Si for supplying material in this research.

References

[1] N.T.R.N. Kumara, A. Lim, C.M. Lima, M.I. Petra, and P. Ekanayake, Renew. Sustain. Energy Rev. **78**, 301–317 (2017).

[2] H. Setyawati, H. Darmokoesoemo, F. Rochman, and A.J. Permana, Mater Renew Sustain Energy, **6** 17 (2017).

[3] P. Guarracinoa, T. Gattia, N. Canevera, M. Abdu-Aguyeb, M. A. Loib, E. Mennaa, and L. Francoa, Phys. Chem. Chem. Phys, 00, 1-3 (2017).

[4] K. Galappaththi, A. Lim, P. Ekanayake, and M.I. Petra, Int. J. Photoen. Vol. 2017.

[5] S. Mayumi, Y. Ikeguchi, D. Nakane, Y. Ishikawa, Y. Uraoka, and M. Ikeguchi, Nanoscale Research Letters **12**, 513 (2017).

[6] J. Anuntahirunrat, Y-M. Sung, and P. Pooyodying, IOP Conf. Ser.: Mater. Sci. Eng. **229**, 012019 (2017).

[7] F. Mendizabal, R. Mera-Adasme, W-H. Xu, and D. Sundholm, RSC Adv., **7**, 42677 (2017).
[8] I.C. Maurya, S. Singh, Neetu, A.K. Gupta, P. Srivastava, and L. Bahadur, J. Electronic Materials (2017).
[9] K.K. Manar, Neetu, Anamika, P. Srivastava, Michael G. B. Drew, and N. Singh, Chemistry Select, **2**, 8301–8311 (2017).
[10] B. Tang, G. Ji, Z. Wang, H. Chen, X. Li, H. Yu, S. Li, and H. Liu, RSC Adv., **7**, 45280 (2017).
[11] M.A.K.L. Dissanayake, H.N.M. Sarangika, G.K.R. Senadeera, H.K.D.W.M.N.R. Divarathna, and E.M.P.C. Ekanayake, J Appl Electrochem, (2017).
[12] S. Saehana, P. Arifin, Khairurrijal, and M. Abdullah, J. Appl. Phys., **111**, 123109 (2012).
[13] S. Saehana, E. Yuliza, Khairurrijal, and M. Abdullah, Mat. Sci Forum, **737**, 43-53 (2013).
[14] S. Saehana, Darsikin, E. Yuliza, P. Arifin, Khairurrijal, and M. Abdulah, J. Sol. Energy Eng., **136,** 044504-1 - 044504-5 (2014).
[15] S. Saehana, Muslimin, M. Abdulah, J. Renew. Sustain. Energy, **6,** 023109-1 - 023109-6 (2014).

IOP Conf. Series: Journal of Physics: Conf. Series **1093** (2018) 012035 doi:10.1088/1742-6596/1093/1/012035

Synthesis of Methyl Ester from Chicken Oil and Methanol Using Heterogeneous Catalyst of CaO-MgO as well as Characterization Its Potential as a Biodiesel Fuel

Aman Santoso*, Sumari, Agus Salim, Siti Marfu'ah

Department of Chemistry, Faculty of Mathematics and Natural Sciences, Universitas Negeri Malang, Jl. Semarang 5 Malang 65145, Indonesia

*Corresponding author's email: aman.santoso.fmipa@um.ac.id

Abstract. The aim of this study was to synthesize methyl ester from chicken oil and methanol using catalyst of CaO-MgO as well as to characterize its potential as a biodiesel fuel. Synthesis methyl ester from chicken oil and methanol using CaO-MgO catalyst conducted through a trans-esterification reaction. Identification the methyl ester generated as a result of synthesis was used GC-MS. Characterization its potential as a biodiesel consisted of density, viscosity, refractive index, and acid number test. The results showed that methyl ester can be synthesized from chicken oil through trans-esterification with methanol using CaO-MgO catalyst with the yield of 81.32 wt%. The types of methyl esters resulted in this reaction were methyl palmitoleic (3.64 wt%), methyl palmitate (37.05 wt%), methyl linoleate (4.75 wt%) methyl oleate (40.90 wt%), and methyl stearate (2.28 wt%). The synthesized methyl ester has a density of 0.88 g/mL, a viscosity of 5.84 cSt, a refractive index of 1.45018, and the acid number of 0.69 mg KOH/g of methyl ester. Therefore, the methyl esters produced using chicken oil as a raw material meets the standard requirements for use as biodiesel fuel and has good prospects for use as a renewable energy source.

Keywords: Methyl ester, chicken oil, trans-esterification, CaO-MgO catalyst.

1. Introduction

The development of the economic level of society causes energy demand also increase. Indonesia is still heavily dependent on fossil energy while this kind of energy reserves are declining [1,2]. The fact is also followed by the absence of new reserve discoveries and the transition to alternative renewable energy sources is not yet running [3] Therefore, alternative fuels instead of petroleum, especially diesel oil, can be used to reduce dependence of fossil fuels.

Biodiesel is an alternative fuel for diesel engines that are environmentally friendly and barely contain sulfur [4]. Biodiesel has advantages over diesel oils that is not emitting carbon monoxide, sulfur, hydrocarbons, and low burning fumes [5]. Biodiesel can be produced through the synthesis of vegetable oils such as palm oil or animal fats such as chicken fat or chicken oil.

Chicken is a relatively high-fat food in the body so that chicken fat is often separated and sold cheaply in the market even just thrown away as waste. This causes chicken fat has a low economic value. Chicken fat is in the form of liquid at room temperature. Chicken oil is a triglyceride composed of a series of saturated and unsaturated fatty acids where the largest fatty acid content is oleic acid which is unsaturated fatty acids [6]. The triglyceride could be reacted with an alcohol to form a potentially biodiesel alkyl ester by transesterification [7].

The biodiesel production through transesterification reaction is commonly used alcohol such as methanol or ethanol but methanol is more commonly used because it is more reactive and cheaper than that of ethanol [8]. The transesterification reaction is generally carried out using catalyst to speed up the reaction process. The most widely used catalysts are homogeneous catalysts such as NaOH and KOH [9]. The catalyst, however, has a weakness that is corrosive, hygroscopic, can cause saponification, and difficult to separate from the resulting product [10]. To overcome this problem, it has been developed the use of heterogeneous catalysts. This kind of catalyst has many advantages such as high activity, low reaction conditions, long life time, easily separated, reusable, and relatively cheap [11]. This fact leads to this study using heterogeneous catalysts. The transesterification reaction occurring between triglycerides and alcohols with the catalyst is shown in Figure 1

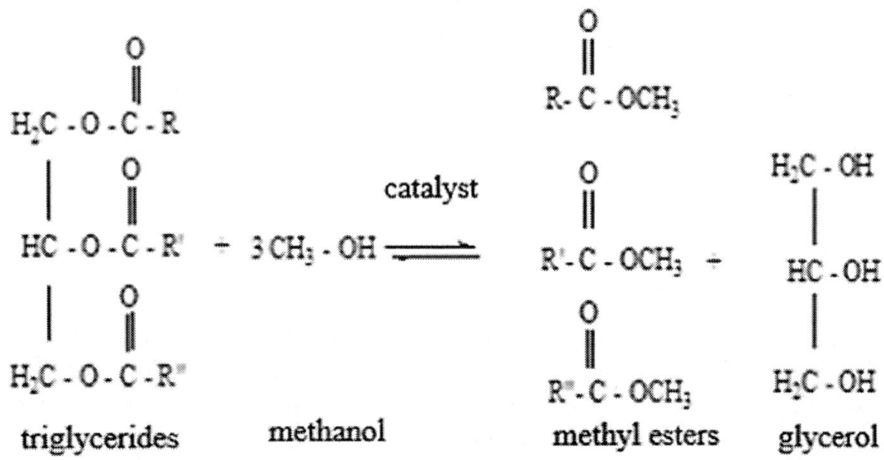

Figure 1. Transesterification reaction occurring between triglycerides and alcohols with a catalyst.

Heterogeneous catalysts like earth alkaline metal oxides such as CaO is often used. They are cheap, easily available, and less toxic [12] Calcium Oxide can be obtained from limestone, $CaCO_3$. When it is heated in a high temperature, the limestone limestone releases CO_2 and remaining CaO. The CaO catalyst has as a strong base but has a slightly weak activity when used as a catalyst to produce alkyl ester. One way to increase CaO activity is to combine it with magnesium oxide, MgO [7]. The MgO catalyst is obtained from $MgCO_3$ that is heated through a calcination process in a very high temperature to separate CO_2 from MgO. Therefore, the study of the coupling catalyst, CaO-MgO, in synthesis of methyl esters from chicken oil through transesterification reaction becomes interesting to do. Whats are the components resulted from transesterification dan its potential as a biodiesel are two interesting things to look at.

2. Materials and Methods

2.1 Preparation and Characterization of Chicken Oil

Chicken oil extraction was done by heating 450 g of chicken oil added 100 mL of distilled water in a Beaker glass. After heating for 3 to 4 hours will be obtained two layers. The top layer is chicken oil and the bottom layer is a water mixture with chicken fat impurities. The oil product is moved into a Beaker glass and heated to a temperature of 100 °C to remove residual water. Characterization of isolated chicken oil includes the properties of density, viscosity, refractive index, and acid number. Determination of free fatty acid content of chicken oil is done as the basis of whether transesterification can be done directly or need some stages of esterification.

2.2 Preparation of CaO-MgO Heterogeneous Catalysts
A total of 50.0 g of CaO and MgO catalysts were activated by calcination in the furnace for 2 hours at 825 °C. After cooling for a while, the heated catalyst was stored into the desiccator for ± 24 hours. A coupling catalyst of CaO and MgO are mixed with a weight ratio of 1: 1 to homogeneous before being used in methyl ester synthesis.

2.3 Synthesis of Methyl Ester with CaO-MgO Heterogeneous Catalyst
A total of 40.00 g of chicken oil is put into a three-neck flask. While heated slowly added methanol with a mole ratio of methanol: chicken oil (15: 1), which has added a catalyst mixture of CaO-MgO with a concentration of 5% of the sample weight. Heating with reflux at constant temperature between 60-65 °C while stirring at 300 rpm. The reaction process was followed by analysis with Thin Layer Chromatography (TLC) to obtain R_f of the product of synthesis and compared with the initial sample. Eluent used in the TLC analysis was mixture of n-hexane, diethyl ether and formic acid with a ratio of 70: 30: 2.

After the reaction process is stopped, the synthesis mixture was centrifuged for 20 minutes at a rate of 3000 rpm. The mixture was poured into a separating funnel and allowed for two hours to obtain two layers where the top layer is expected of methyl ester while the bottom layer is glycerol. Top layer was separated and washed with warm aquadest. The resulting methyl ester was dried by adding an anhydrous Na_2SO_4 crystals. The yield of synthesis is determined by comparing the weight of the synthesis product with the weight of the theoretical results. The identification of methyl esters of the synthesis results was done by using GC-MS instrument.

2.4 Characterization.
Characterization of chicken oil and methyl esters as the result of synthesis consist of density, viscosity, refractive index, and acid number. Determination of the density was carried out picnometer, determination of the viscosity was carried out using Oswald viscosimeter, determination of the refractive index using Abbe refractometer, and determination of the acid number through titration using a 0.1 N KOH solution.

3. Results and Discussion

3.1 Chicken Oil Extraction and Characterization
Extraction results with water solvent, as much as 450 g of chicken fat obtained 171.15 g of oil (38.3%). The physical properties of chicken oil are liquid, clear yellow, specific odor. The character of chicken oil obtained were the density of 0.91 g/mL, the viscosity of 60.36 cSt, refractive index of 1.465, and the acid number 1.58 mL KOH/g of chicken oil. The fatty acid content of chicken oil is 1.64%. The free fatty acid content is less than 2.5%, so it can be directly done through transesterification reaction [13].

3.2 Activation of CaO-MgO Heterogeneous Catalyst
Preparation of CaO-MgO catalyst refers to a study conducted by Indah et a. [12]. Catalyst of CaO and MgO respectively of 50.00 g were activated by calcination in the furnace for 2 hours at 825 °C. Calcination can improve the active side of the catalyst. Calcination is also done to reduce the water content in the catalyst. The constant weight of CaO and MgO catalysts after each heating was 36.60 g and 43.02 g. The CaO and MgO catalysts are stored into desiccator for 24 hours.

3.3 Synthesis of Methyl Ester with CaO-MgO Heterogeneous Catalyst
Oil is a triglyceride or fatty acid ester of glycerol. Through the transesterification reaction of three glycerides from chicken oil can be converted to another ester form. The optimum condition of methyl ester synthesis of chicken oil with methanol (1:15) with a heterogeneous CaO-MgO catalyst at a concentration of 5% by weight of the sample at 65 °C for 2 hours. The concentration of the catalyst

refers from a study conducted. The reaction of the transesterification reaction with the CaO-MgO catalyst occurring is revealed in Figure 2.

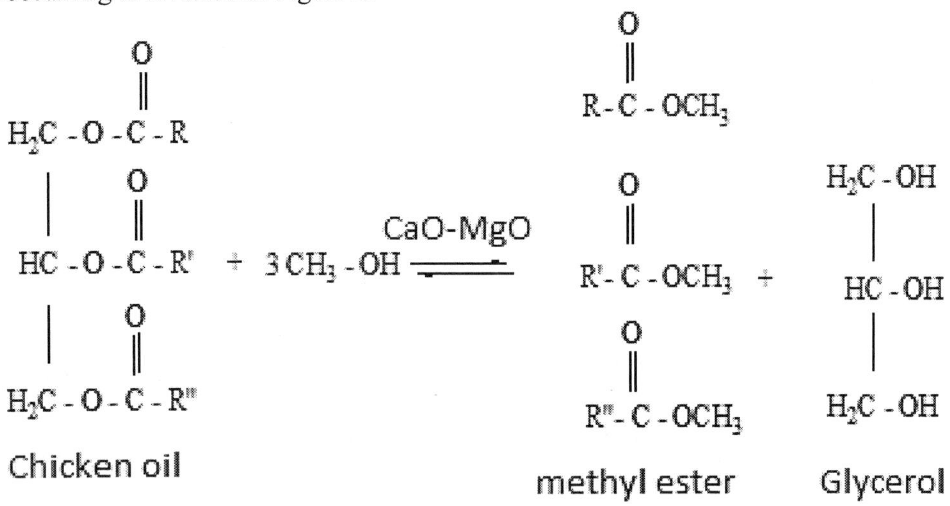

Figure 2. Transesterification reaction of threeglycerides

Based on the stoichiometric reaction the mole ratio of threeglyceride and alcohol was 1 : 3 and produced 3 moles of methyl ester. In this study excessive alcohols are used because of the reversible reaction properties to obtain product as much as posible. The transesterification reaction is reversible so that excess alcohol is required to obtain a maximum product [5]. The reaction is stopped after a change from oil to ester form can be tested by analysis with TLC showing difference Rf stain mixture of methyl ester with chicken oil. The result of synthesis of methyl ester by means of reflux can be seen in Figure 3.

Figure 3. Tormed two layers of liquid as a result of synthesis of methyl ester from chicken oil

The transesterification results form two layers where at the top is considered as a methylester and at the bottom is glycerol. Both layers were separated until the upper layer remains. The yield of methylester resulted at the synthesis using CaO-MgO catalyst was 81.39%.

IOP Conf. Series: Journal of Physics: Conf. Series **1093** (2018) 012035 doi:10.1088/1742-6596/1093/1/012035

3.4 Identification of Methyl Ester Using GC-MS

Methylesters synthesized from chicken oil in this research identified using GC-MS yields a well-defined chromatogram as shown at Figure 4.

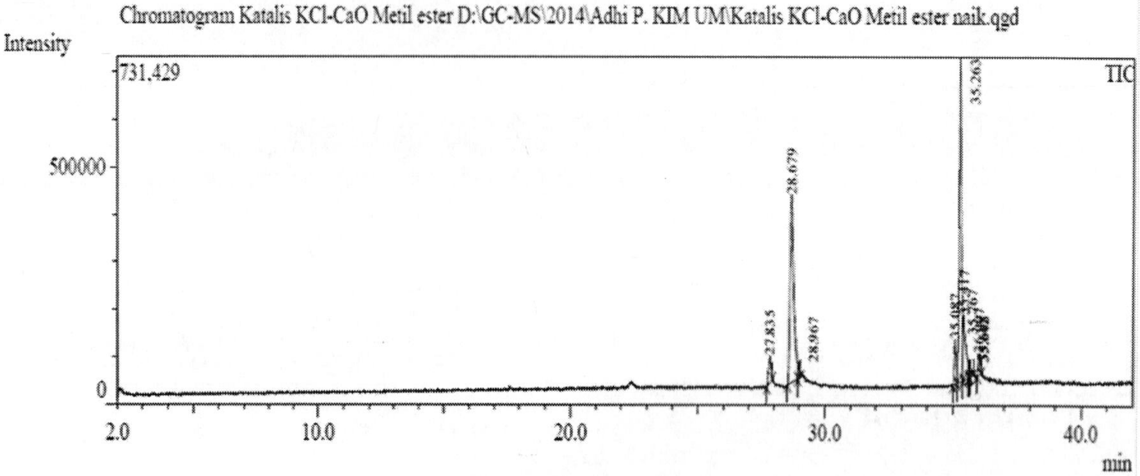

Figure 4. Chromatogram of Methylester Synthesized from Chicken Oil

Figure 2 shows there are 5 fairly large peaks that can be said there are five compounds making up the methylester mixture of the synthesis product. Each peak at chromatogram represents one compound. The total area shown in the chromatogram is proportional with percent composition of the compound in the mixture. The retention time and area of the five compounds making up the methyl ester mixture are listed in Table 1.

Table 1. Retention Time and Area of Five Compound Compounds of Methylester

Peak	Retention Time	Area	Percent Area (%)
1	27.833	391036	3.64
2	28.683	3980858	37.05
4	35.087	510951	4.75
5	35.263	4395149	40.90
10	36.007	245343	2.28

The compounds in the mixture of methylester as the product of synthesis can be identified using mass spectroscopy analysis. One Example of the result of mass spectroscopy at with retention time (tR) of 27.833 min is shown in Figure 5.

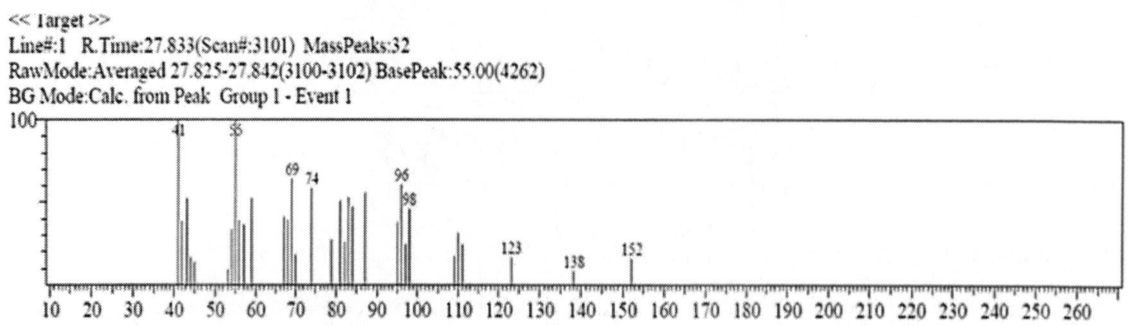

Figure 5. Spectrum Peaks of Chromatogram of Methyl ester from Chicken Oil with Retention Time (tR) 27,833 Minutes

The mass spectra obtained are compared with the mass spectrum found in the WILEY7.LIB Library and searched for peaks of relatively equal position. Based on observations of the WILEY7.LIB Library entry number of 199160, it is presumed that the compound is methyl palmitoleate or methyl 7-hexadecenoate or methyl 7-hexadecenoate. It is reinforced by a fragmentation pattern whose peaks have m/z 41, 55, 69, 74, 96 , 98, 123, 138 and 152. Peaks with m/z 152 and m/z 98 are thought to be obtained by breaking fragments is shown in Figure 6.

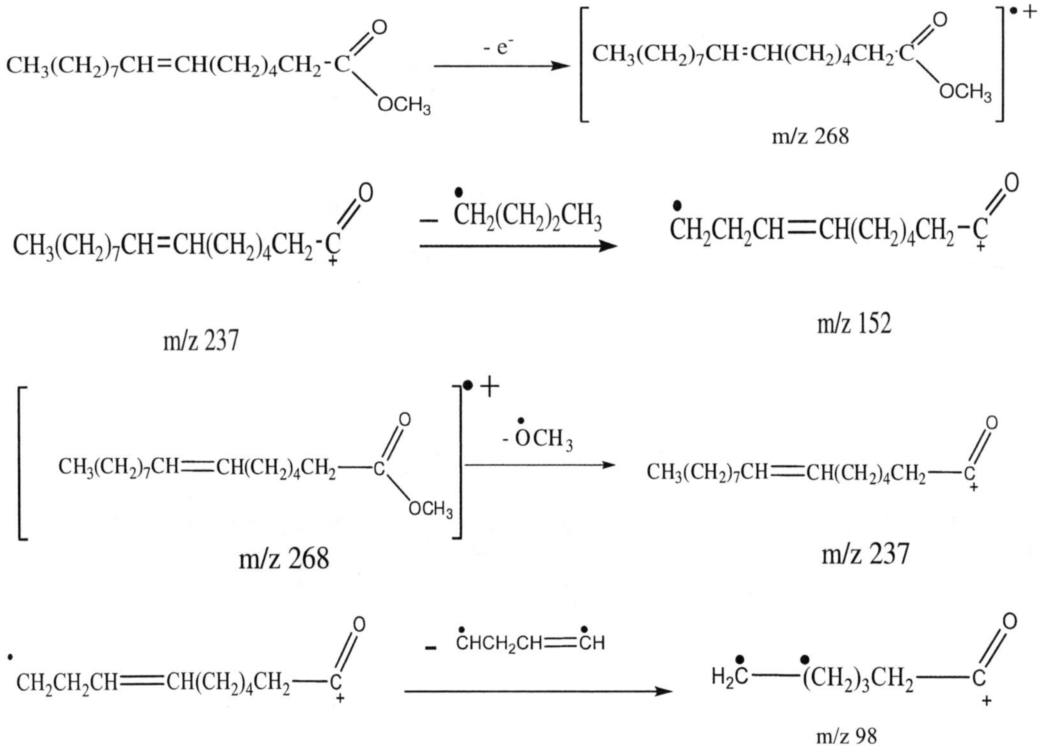

Peak with m / z 123 allegedly obtained from fragments with m/z 268 as follows:

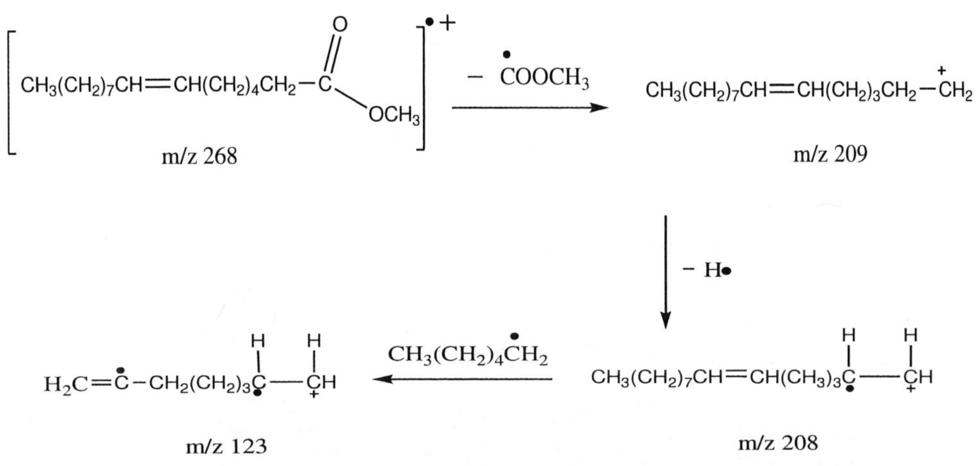

Figure 6. Fragmentation molecules of methyl ester of chicken oil.

IOP Conf. Series: Journal of Physics: Conf. Series **1093** (2018) 012035 doi:10.1088/1742-6596/1093/1/012035

Peaks with m/z 74 (base peak) and 55 are thought to be obtained via a molecular ion reconstitution of Mc Lafferty is shown in Figure 7.

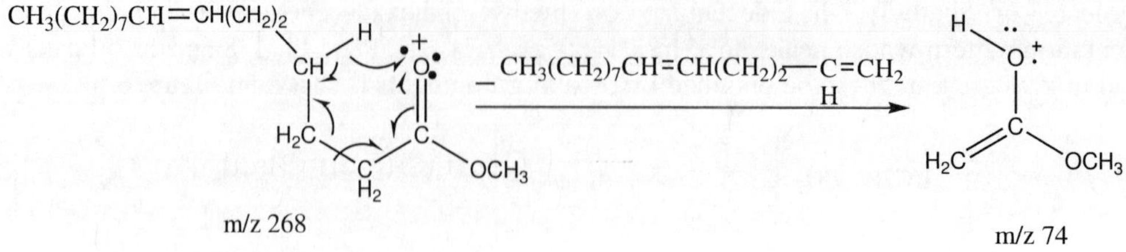

Figure 7. Molecular ion reconstitution of Mc Lafferty

Peaks with m/z 41 are thought to be obtained by breaking the bonds on the molecular ions is shown in Figure 8.

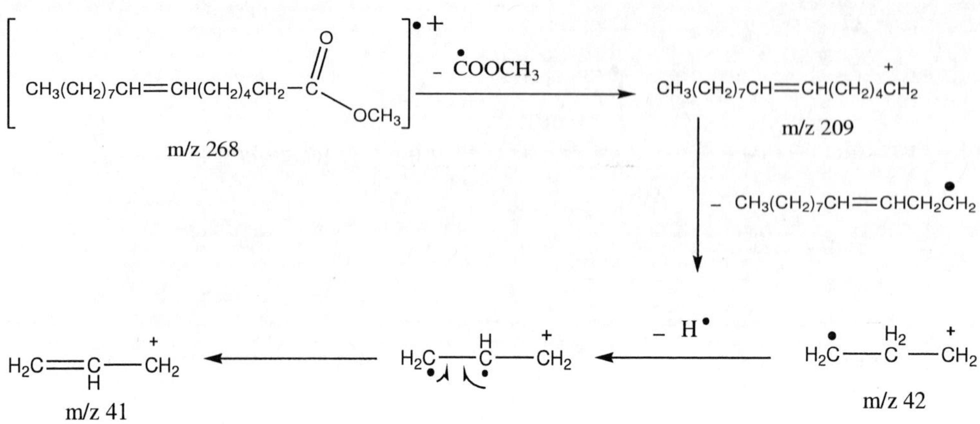

Figure 8. Breaking the bonds on the molecular ions at m/z 41.

Based on spectral fragmentation analysis with a retention time of 27.835 min, it can be concluded that the compound is methylpalmitoleate like shown in Figure 9.

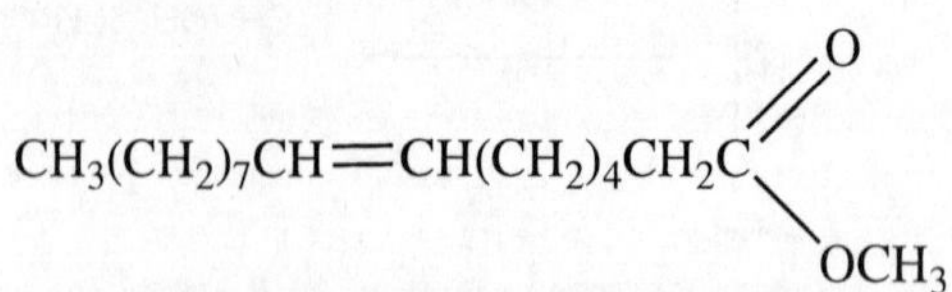

Figure 9. Methylpalmitoleate

Pattern analysis of mass spectrum fragmentation as a whole can be seen that the five main constituent components of methyl esters mixture of chicken oil are methyl palmitoleate (3.64%), methylpalmitate (37.05%), methyllinoleate (4.75%), methyloleate (40.90%), and methylstearate (2.28%).

3.5 Characterization of Methyl Ester as a Result of Synthesis
Characteristics of methyl esters of the synthesis include the density, viscosity. refractive index, and acid number.

Table 2. Comparison of MethylEster Characteristics of Synthesis and Chicken Oil

Parameter	Methylester as a product of synthesis	Chicken oil
Density (g/mL)	0.88	0.91
Viscosity (cSt)	5.84	60.36
Refractive Index	1.450	1.465
Acid number (mg of KOH/ g of oil)	0.69	1.58

Data at Table 2 shows that the characters of Methyl ester product of synthesis are different with the character of chicken oil. As an example, there is a change of viscosity from 60.36 to 5.84. it means that there is a decline of 55 cSt. This data indicates that methyl esters have been successfully synthesized. Characteristics of methyl esters synthesized when compared with biodiesel characteristics as shown in Table 3.

Table 3. Characteristics of Methylester as a Result of Synthesis from Chicken Oil and Biodiesel (SNI)

Parameter	Methyl ester as a product of synthesis	Biodiesel Standard of SNI
Density (g/mL)	0.88	0.850 – 0.890
Viscosity (cSt)	5.84	2.30 – 6.00
Refractive Index	1.450	Max 1.45
Acid number (mg of KOH/ g of oil)	0.69	Max 0.80

The data show that the character of methy lester as a result of synthesis approaches the biodiesel character according to SNI standard. Based on the results of the analysis, it can be concluded that the methyl ester of the synthesis has potential as biodiesel or fuel. This result is in line with the result of characterization of biodiesel produced from palm oil via base catalyzed transesterification [14].

4. Conclusion

Methyl esters can be synthesized from chicken oil using heterogeneous CaO-MgO catalyst at 5% concentration and obtained the yield of 81.32%. The selectivity each compound of methyl esters as the product of synthesis is methylpalmitoleate of 3.64%, methyl palmitate 37.05%, methyllinoleate of 4.75%, methyloleate of 40.90%, and methyl stearate of 2.28%. The character of methylester resulted from synthesis shows the density of 0.88 g / mL, viscosity of 5.45 cSt, refractive index of 1.45018, and acid number of 0.69 mg KOH/ g methyl ester, so that the methylester has potentially as a biodiesel.

References

[1] Sumari S, Roesyadi A and Sumarno S 2013 Effects of ultrasound on the morphology, particle size, crystallinity, and crystallite size of cellulose *Scientific Study & Research. Chemistry & Chemical Engineering, Biotechnology, Food Industry* **14** 229
[2] Budianto A., Sumari S. and Udyani K. ARPN Journal of Engineering and Applied Sciences
[3] Syah, A.N.A Biodiesel Jarak Pagar Bahan Alternatif yang Ramah Lingkungan.
[4] Suwarsono, WP., Gani, I.Y., Kusyanto Valensi, 1(2): 44-52.
[5] Kusmiyati K 2008 Reaksi katalitis esterifikasi asam oleat dan metanol menjadi biodiesel dengan metode distilasi reaktif *Reaktor* **12** 78–82
[6] Hermanto, S., Muawanah, A., Harahap, R. Profil dan Karakteristik Lemak Hewani (Ayam, Sapi dan Babi) Hasil Analisa FTIR dan GCMS.
[7] Tahvildari, K., Davari, N.,Ghasri. M. A., & Behrourzinavid, M. World Academy of Science,

Engineering and Technology, 55: 102-115
[8] Aziz, I. Valensi. 1(1) : 34-43
[9] Vyas, A.P., Jaswant, L.V., dan Subrahmanyam, N. Advances in Chemical Engineering and Science, 1: 45-50.
[10] Thanh L T, Okitsu K, Boi L V and Maeda Y 2012 Catalytic technologies for biodiesel fuel production and utilization of glycerol: a review *Catalysts* **2** 191–222
[11] Taufiq, Y.H., Fitriyah, N., Abdullah, Basri, M. Jurnal Sains Malaysiana, 40(6): 584-597.
[12] Lee, D., Park, Y., dan Lee, K. Catalysis Surveys From Asia, 13(2): 63-67.
[13] Padil, Wahyuningsih, S., Awaluddin, A. Jurnal Natur Indonesia, 13(1): 27-32.
[14] Eman N. A. & Tay C. I. Elsevier Procedia Engineering 53: 7 – 12

Performance of Activated Natural Zeolite/Cu as a catalyst on Degradation of Glycerol into Ethanol Assisted by Ultrasonic

Sumari*, Fauziatul Fajaroh, Aman Santoso, Riski Kusuma Wardani

Department of Chemistry, Faculty of Mathematics and Science, Universitas Negeri Malang, Jl. Semarang 5 Malang 65145, Indonesia

*Corresponding author's email: sumari.fmipa@um.ac.id

Abstract. The purpose of this study was to convert glycerol into ethanol using the catalyst of activated natural zeolite/Cu assisted by ultrasonic. The catalyst was characterized using X-Ray Diffraction and BET isotherm adsorption. The degradation process carried out at the temperature of 60 °C and at various time of 1, 2, and 4 hours. The product of glycerol degradation was analyzed using Gas Chromatography. The results showed that the structure of natural zeolite was a mordenite. The BET analysis showed that the surface area of natural zeolite is 99.096 m^2/g greater than that of natural zeolite impregnated Cu that is 6.597 m^2/g. The total pores, however, a natural zeolite is 0.026 cm^3 smaller than that of natural zeolite impregnated Cu that is 0.058 cm^3. The highest yield of ethanol as the product of glycerol degradation was 3.704% produced when the sonication times is 2 hours using activated natural zeolite catalyst. The activated zeolite has a prospect to be used as a degradation reaction catalyst for molecules that goes through the Bronsted Lowry acid mechanism.

Keywords: Catalysts, degradation of glycerol, ethanol production, *mordenite,* sonication

1. Introduction

Currently, new and renewable energy is an important issue according to the depletion of fossil fuels. The BPPT data in 2011 states that Indonesia's total oil reserves was 7.73 billion barrels and fell to 7.41 billion barrels in 2012 [1]. The rate of the consumption of petroleum increase along with the increase in population [2]. It is estimated that 15 to 20 years later the fossil energy source will run out. Considering that existing fossil material is limited and non-renewable resource, efforts should be made in the use of alternative energy considered technically, economically, and environmentally feasible [3,4]. This idea is in line with Presidential Instructions No.1 2006 on the supply and utilization of biofuel as an alternative fuel.

Biodiesel has become a renewable fuel that proves to be suitable as an alternative fuel for diesel engines. It is estimated that biodiesel can contribute as much as 20% to all transport fuels by 2020 [5]. The higher the production of biodiesel the higher the amount of glycerol as a by-product. Glycerol is obtained by the transesterification process of triglycerides/vegetable oils on biodiesel production [6,7]. A total of 1 mol of glycerol can be produced by every 3 moles of biodiesel production through transesterification reaction or about 10% of total weight [5]. The abundance of glycerol waste makes it have a low price of glycerol in the market, ranging from 1.28 to 1.50 US$ per kilogram [8]. The availability of glycerol that is inexpensive and has the potential to be utilized as a new type of energy that has high economic value like ethanol, methanol, acrolein, and allyl alcohol.

Buhler et al. [9] had conducted glycerol conversion through the glycerol molecular breaking reaction using aqueous media near the critical point. The reaction was run at a temperature of 622 to 748 K, pressures of 25, 35, and 45 MPa, the duration times of 32 to 165 seconds with varying concentrations. The products obtained were ethanol, acetaldehyde, propionaldehyde, acrolein, allyl alcohol, methanol, formaldehyde, carbon monoxide, carbon dioxide and hydrogen. Meanwhile, Yuniati et al. [10] stated that glycerol could be converted to ethanol and methanol through the glycerol molecule breaking reaction. This process was carried out at 400 °C and the pressure of 250 bars. The highest yield obtained was ethanol of 1.3 %wt and methanol of 16.7 %wt. This process, however, is less feasible because it is done at high temperature and pressure. Corma et al. [11] explain that glycerol can be converted through a series of reactions including dehydration, cracking, hydrogen transfer, and catalyzed using HZSM-5 to produce acrolein. The study of converting glycerol using H-Zeolite size 200 mesh at 350-500 °C was also obtained acrolein [6].

The previous studies described that the components of glycerol are potential to be utilized in a new type of energy. However, the degradation of glycerol still requires enormous energy in the use of high temperature and pressure. It is necessary to find an alternative glycerol degradation technology that requires cheaper and safer operational costs.

The utilization of ultrasonic began to be used for macromolecular degradation [12]. The degradation of glycerol using ɣ-alumina catalyst with a sonication technology yielded the highest yield methanol of 6.297 %wt when it was conducted at 50 °C for 70 minutes, allyl alcohol of 5.503 %wt when it was conducted at 30 °C for 70 minutes, and acrolein of 18.878% when it was conducted at 30 °C for 30 minutes [13]. This research used an activated natural zeolite/Cu catalyst with various sonication times for converting glycerol to ethanol. The expected product, ethanol, is one of the most widely used of fuels.

2. Materials and Methods

2.1. Materials

The materials used in this research were natural zeolite from South of Malang East Java Indonesia, glycerol, oxalic acid, NaOH p.a., Cu $(NO_3)_2.3H_2O$ p.a., an indicator of phenolphthalein, and demineralized water.

2.2. Methods

Natural zeolite was crushed and sieved into fine grains of 100 to 200 mesh. The zeolite was activated by calcination method at 500 °C for 5 hours. A part of the activated zeolite was impregnated with Cu so that the catalyst composes 15%wt of Cu. The impregnation was carried out using soaking the activated zeolite with a solution of Cu $(NO_3)_2$ then stirred using a shaker for 24 hours at the temperature of 60 to 70 °C. The impregnated of the activated zeolite then washed with demineralized water and dried at 110 °C for 24 hours. Both activated zeolites with or without impregnation were characterized using X-ray Diffraction (XRD) to determine the structure of zeolite and characterized using Brunauer-Emmett-Teller (BET) to determine the specific surface area and pore size. The degradation of glycerol to ethanol using a catalyst at the various time was assisted by ultrasonic of BRANSON 1510 at a frequency of 42 kHz. The products of glycerol degradation were analyzed using a Gas Chromatography (GC).

3. Results and Discussion

3.1. Calcination of natural zeolite

The aims of calcination is to release water molecules or hydrates attached to the pores or zeolite cavities. The loss of the water molecules can raise the cavity empty of the zeolite pores, so that expand the surface area of the zeolite. The physical change of zeolite because of calcination was a change in the zeolite color from gray to brown.

3.2. Characterization of Activated Natural Zeolite / Cu Using XRD

Impregnation was done by submersion activated natural zeolite into a solution of $Cu(NO_3)_2$ (aq). The impregnation of copper was expected that this metal can be dispersed evenly throughout the surface and the pores of the zeolite. The impregnated metal on the active site of the zeolite was considered to affect the character of the zeolite catalyst. The study of Bhatia et al. stated that with a more optimal dispersion of impregnated metal will increase the active site on the catalyst [14]. The structure of natural zeolite and natural zeolite-Cu were characterized using an XRD. The spectra generated from the XRD was analyzed and matched to the JCPDS (Joint Committee on Powder Diffraction Standards) data [15].

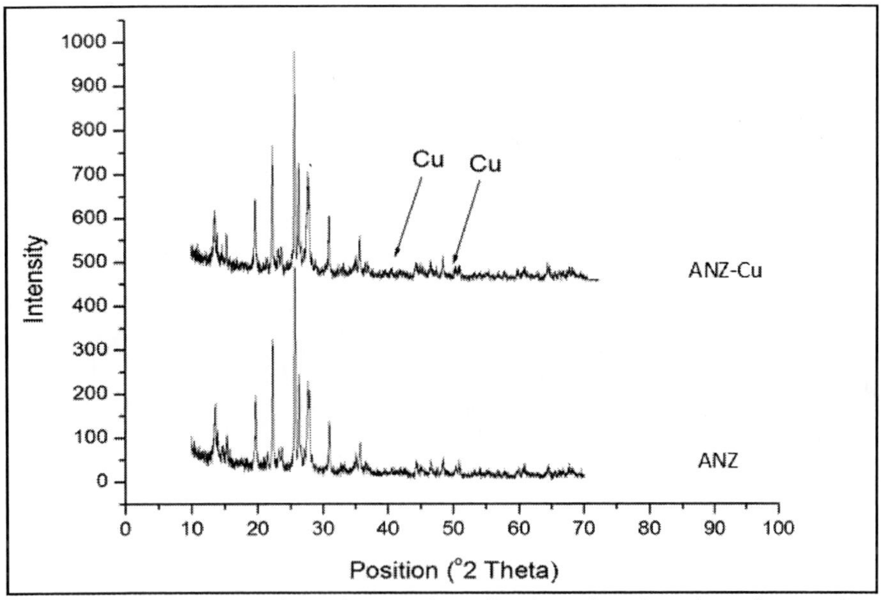

Figure 1. Diffractogram Activated Natural Zeolite (ANZ) and Activated Natural Zeolite ANZ-Cu.

The result of X-ray diffraction scanning on natural zeolite samples shows that the mineral was mordenite indicated by the emergence of peak region 2θ, while the level of component and crystallinity was indicated by the high peak intensity. Peaks pattern (Figure 1) shows that there is a similarity with peaks pattern of zeolite MOR according to JCPDS standard. In more detail can be seen in Table 1.

Based on the mapping of X-ray diffraction spectra between natural zeolite and JCPDS. There is a very high similarity in the appearance of the major peaks at the 2θ angle and also the intensity. The similarity of the diffractogram pattern of the main peaks in the same 2θ angles shows that the materials have the same structure [16,17]. Therefore it can be concluded that natural zeolite from Southern Malang is a Mordenite.

Meanwhile, XRD diffractogram of natural zeolite - Cu did not differ greatly with active natural zeolites as shown in Figure 1. The Diffractogram shows that zeolite impregnated of Cu only slightly change the number of peaks and its intensity compared to zeolite non-impregnated. The presence of Cu doped in the zeolite makes the crystallinity a slight increase. The impregnated Cu metal into the zeolite mostly forms CuO and Cu_2O, but there is also a small form of Cu. It can be seen in the diffractogram at 2θ emerging peaks around 44.3235°; 50,9708°; and 72.7535° which denotes the formation of Cu metal on zeolites corresponding to JCPDS Cu 04-0836. However, at the diffractogram also peak at 2θ around 35,6939°; 48,4564°; 60.7851° and 64.5100° which indicates CuO formed in the zeolite and peak 46,5855° which indicates the formation of Cu_2O in the zeolite.

Table 1. The data of main peaks at 2θ angles of Activated Natural Zeolite and zeolite MOR standard

Activated Natural Zeolite		**Zeolite MOR standard**	
2θ (°)	***d-spacing* (Å)**	**2θ (°)**	***d-spacing* (Å)**
13.4271	6.59454	13.45	6.584
15.2485	5.81067	15.30	5.791
19.5765	4.53472	19.61	4.527
21.3778	4.15653	21.45	4.142
22.2405	3.99722	22.20	4.004
23.3483	3.81001	23.64	3.764
25.6023	3.47946	25.63	3.476
26.2786	3.39165	26.25	3.395
27.6118	3.23063	27.67	3.223
30.8655	2.89709	30.89	2.894
35.6376	2.51934	35.61	2.521

3.3. Characterization of Activated Natural Zeolite/Cu Using BET

The analysis using the BET method was performed to find out the specific surface area and to know the pore volume of zeolite. There are two types of zeolites that are analyzed namely activated natural zeolite and activated natural zeolite impregnated Cu. The results of the BET analysis of activated natural zeolite and activated natural zeolite impregnated Cu are presented in Table 2.

Table 2. BET Analysis of Activated Natural Zeolite (ANZ) and Activated Natural Zeolite-Cu (ANZ-Cu)

Type of Catalyst	**Surface area (m^2/g)**	**Total of Pore Volume (cm^3/g)**
ANZ	99.096	0.026
ANZ-Cu	6.597	0.058

The results of the BET analysis shows that the ANZ had a surface area of 99.096 m^2/g and a pore volume of 0.026 cm^3/g. Meanwhile, ANZ-Cu had a surface area of 6.597 m^2/g and a pore volume of 0.026 cm^3/g. It means that the surface area of ANZ is higher than that of ANZ-Cu, but the pore volume of ANZ is smaller than that of ANZ-Cu. The surface area of these natural zeolites both ANZ and ANZ-Cu are smaller than that of synthetic zeolite, ie 374 m^2/g [18]. It can be possible because the natural zeolite still contains impurities which cover some of its pores thereby lowering the surface area. The decreasing surface area of ANZ-Cu can be considered that Cu closes several pores of the catalyst. Furthermore, it can be denoted that activated natural zeolite non-impregnated of Cu has a good adsorption power.

3.4. Degradation of Glycerol becomes Ethanol.

The catalyst of ANZ or ANZ-Cu here serves to decrease the intra-molecular bonding force within the glycerol molecule, so it is easy to react with the H^+ present in the mixture. The reaction can be seen in Figure 2.

3.5. Degradation Product as a function of the type of catalyst and sonication time

The product of degradation was analyzed using a Gas Chromatography. The analysis was proposed to find out qualitatively and quantitatively what the product is. The GC chromatogram data of the glycerol degradation products is shown in Table 3. The data showed that the product formed by the reaction of glycerol degradation is ethanol. The similar product was obtained on the degradation of glycerol that is ethanol through water supercritical conditions [19], using Ni/γ-Al2O3 Catalyst [9], [20]. However, Buhler et al. [9] state that ethanol is a product that occurs at high temperatures and through free radical mechanisms.

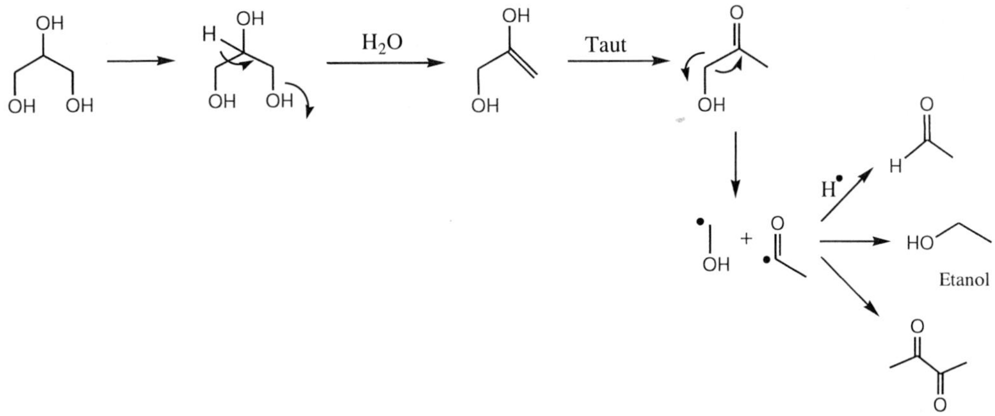

Figure 2. Mechanism of Glycerol Degradation Using a catalyst of ANZ or ANZ-Cu.

Based on the time variations performed on both catalysts, they showed that the highest yield of ethanol when it was conducted for 2 hours of sonication time with ANZ that is 3.704 %wt. The percentage of ethanol produced from the degradation of glycerol using ANZ catalyst was higher than that of using ANZ-Cu catalyst.

Table 3. Data of GC Analysis of Glycerol Degradation

Samples	Time (minute)	Area (pA*s)	Conc. EtOH (%wt)	Product
ANZ- 1 hour 60 °C	0.962	1.54400e5	0.849	EtOH
ZA- 2 hours 60 °C	0.953	6.73164e5	3.704	EtOH
ZA- 4 hours 60 °C	0.955	1.92004e5	1.056	EtOH
ZA-Cu- 1 hour 60 °C	1.066	2.27243e4	0.125	EtOH
ZA-Cu- 2 hours 60 °C	0.997	1.02768e5	0.565	EtOH
ZA-Cu- 4 hours 60 °C	1.030	6.50366e4	0.358	EtOH

According to sonication times, the highest yield of ethanol obtained when hydrolysis was conducted for 2 hours. The longer time of two hours sonication shows that the yield of ethanol produced was smaller. It can be stated that the optimal sonication time was 2 hours. At a longer time of sonication causes the ethanol to be degraded furthermore into smaller molecules. This product (ethanol) can be decomposed into CO, CO_2, and H_2 [20,21] so that the yield of ethanol becomes smaller at a longer time of sonication.

4. Conclusion

The structure of natural zeolite catalyst is a *mordenite*. The activated natural zeolite has the surface area of 99.096 m^2/g larger than that of activated natural zeolite-Cu that has a surface area of 6.597 m^2/g. However, the total pore volume activated natural zeolite of 0.026 cm^3/g is smaller than that of activated natural zeolite active-Cu of 0.058 cm^3/g. Both catalysts, ANZ and ANZ-Cu, provide a glycerol degradation to produce ethanol. The metal of Cu impregnated to zeolite catalyst decreases the yield production of ethanol. The optimum sonication time for glycerol degradation using both catalysts of ANZ and ANZ-Cu is 2 hours.

References

[1] Sugiyono A, Permana A ., Boedoyo M . and Adiarso *BPPT-Outlook Energi Indonesia 2013.* (JAkarta: PTPSE and BPPT)
[2] W K, C.B K, T.N K and A K 2012 Indonesia Energi Outlook (IEO).
[3] Sumari S, Achmad R and Sumarno S 2013 Effects Of Ultrasound On The Morphology, Particle

Size, Crystallinity, And Crystallite Size Of Cellulose *Chemistry & Chemical Engineering, Biotechnology, Food Industry* **14** 229–39

[4] A B, Sumari S and Udyani *ARPN Journal of Engineering and Applied Sciences.* **10** 10317–23

[5] Mohamed M, Sie Ting T, Amin N A S, Abdullah T A T and Mat R 2011 Conversion of glycerol to methanol in the presence of zeolite based catalysts (IEEE) pp 389–93

[6] Maulia U and Riana S 2011 PROSES KONVERSI GLISEROL MENJADI ACROLEIN DENGAN KATALIS HZEOLIT

[7] F B, I.S E and Setiadi T 2010 *Produksi Etanol dari Gliserol sebagai Hasil Samping Industri Biodiesel* (Bandung: Institut Teknologi Bandung)

[8] Pachauri N and He B 2006 Value-added utilization of crude glycerol from biodiesel production: A survey of current research activities. ASABE Paper No. 066223 *Am. Soc. Agric. Biol. Eng.: St. Joseph, MI*

[9] Bühler W, Dinjus E, Ederer H J, Kruse A and Mas C 2002 Ionic reactions and pyrolysis of glycerol as competing reaction pathways in near-and supercritical water *The Journal of supercritical fluids* **22** 37–53

[10] Y Y, Sumarno S and Mahfud M 2011 *The Journal for Technology and Science* **22** 36–40

[11] Corma A, Huber G W, Sauvanaud L and O'Connor P 2008 Biomass to chemicals: catalytic conversion of glycerol/water mixtures into acrolein, reaction network *Journal of Catalysis* **257** 163–71

[12] Suslick K S 1994 The yearbook of science and the future *Encyclopedia Britannica, Chicago* **138**

[13] R B, K K and Sumarno S 2011 *Jurnal Teknik Pomits* **2** 2–4

[14] Twaiq F A, Zabidi N A M and Bhatia S 1999 Catalytic Conversion of Palm Oil to Hydrocarbons: Performance of Various Zeolite Catalysts *Industrial & Engineering Chemistry Research* **38** 3230–7

[15] Treacy M M J and Higgins J B 2001 Collection of simulated XRD powder patterns for zeolites. Published on behalf of the Structure Commission of the 'International Zeolite Association' *Powder Patterns, 203* **204**

[16] Bhadauria J, Singh B ., Tomar A and Tomar R 2011 Synthesis and characterization of analogue of mordenite and its role as a catalyst for Friedel-Crafts acylation of anisole *Journal of Chemical and Pharmaceutical Research* **3**

[17] Kalla R, Sumarno S and Mahfud M 2015 Non-Catalytic and γ-Al2O3 Catalyst-based Degradation of Glycerol by Sonication Method *Bulletin of Chemical Reaction Engineering & Catalysis* **10** 304–12

[18] Selim M M, EL-Mekkawi D M, Aboelenin R M M, Sayed Ahmed S A and Mohamed G M 2017 Preparation and characterization of Na-A zeolite from aluminum scrub and commercial sodium silicate for the removal of Cd $^{2+}$ from water *Journal of the Association of Arab Universities for Basic and Applied Sciences* **24** 19–25

[19] Qadariyah L, Machmudah S, Sasaki M and Goto M 2011 Degradation of glycerol using hydrothermal process *Bioresource technology* **102** 9267–71

[20] Miranda B C, Chimentão R J, Santos J B O, Gispert-Guirado F, Llorca J, Medina F, Bonillo F L and Sueiras J E 2014 Conversion of glycerol over 10% Ni/γ-Al2O3 catalyst *Applied Catalysis B: Environmental* **147** 464–80

[21] Ott L, Bicker M and Vogel H 2006 Catalytic dehydration of glycerol in sub- and supercritical water: a new chemical process for acrolein production *Green Chem.* **8** 214–20

Effectivity of Black Tea Polyphenol in Adipogenesis Related IGF-1 and Its Receptor Pathway Through In Silico Based Study

Hendra Susanto[1,*], Viol Dhea Kharisma[2], Dwi Listyorini[1], Ahmad Taufiq[3], Sunaryono[3], Aulanni'am[4,5]

[1]Department of Biology, Faculty of Mathematics and Natural Sciences, Universitas Negeri Malang, Jl. Semarang 5 Malang 65145, Indonesia
[2]Department of Biology, Faculty of Sciences, Universitas Brawijaya, Indonesia
[3]Department of Physics, Faculty of Mathematics and Natural Sciences, Universitas Negeri Malang, Jl. Semarang 5 Malang 65145, Indonesia
[4]Department of Chemistry, Faculty of Mathematics and Natural Sciences, Universitas Brawijaya, Jl. Semarang 5 Malang 65145, Indonesia
[5]Faculty of Veterinary Medicine and Biosains Institute, Universitas Brawijaya, Indonesia

*Corresponding author's email: hendrabio@um.ac.id

Abstract. Obesity has been established as an emerging metabolic disease worldwide. The change of global lifestyle does not only become a significant health issue in the developed countries but also in the most deprived area around the world. The administration of obesity addressed to inhibit adipogenesis by using several potential herbal drugs. Recently, black tea treatment showed an essential result whereas it can decrease cell proliferation, differentiation, and induced apoptosis. The aim of this in Silico study was to provide a comprehensive data as the confirmation to in vitro and in vivo model. We compared the potential anti-obesity properties among black tea polyphenol on the activation of insulin-like growth factor 1 receptor (IGF-1R). The molecular interaction prediction between black tea polyphenol and IGF-1R was analyzed by docking model. Our in Silico data showed that theaflavin 3,3'-digallate (TF3) had the primary function to trigger cell apoptosis, differentiation inhibition, and improved global energy-linked binding affinity between TF3 and IGF-1R. The higher value of repulsive Van der Waals energy of TF3 enhanced the alteration of stability of IGF-1 and its receptor. As a result, the binding site dynamic of molecular complex between IGF-1 and IGF-1R significantly was changed by the molecular modification on autophosphorylation domain of IGF-1R. In conclusion, black tea polyphenol, TF3 tends to become a promising future candidate for herbal therapy combating obesity and metabolic syndrome. This preliminary bioinformatics data provides a hallmark for TF3 molecular activity during adipogenesis.

Keywords: Theaflavin, adipogenesis, obesity, IGF-1R, In Silico

1. Introduction

Obesity is a metabolic perturbation disease resulting from the imbalance between energy intake and energy dissipation [1]. The vast majority of obesity incidence was observed in the developed countries [2]. In contrary, the recent report showed that obesity is an emerging disease in the several developing countries particularly in the Asia Pacific [3]. Indeed, Indonesia and some Southeast Asia countries facing a global burden of obesity among different age-groups within their population. The gradual changes in obesity prevalence are closely and significantly associated with the common alteration of daily intake in the Asian community [4]. Modernization and westernization related to unhealthy food have strongly stimulated the progression of obesity in five Asian countries, including Indonesia [5].

The emerging problem of obesity in Indonesian urban area has suggested an origin from low socio-economic groups with different dietary pattern [6]. It was reported that high caloric or fast food consumption, lack of physical activities, sedentary lifestyle, and less fiber diet induce overweight and accelerate obesity development in the low-income and low-middle income (LMC) countries, especially in South and South East Asia region [7]. As a consequence, increased obesity prevalence showed a crucial contribution to several gastrointestinal and metabolic diseases development especially non-alcoholic fatty liver disease, gastrointestinal cancer and type 2 diabetes mellitus [8-10].

Several programs were developed to reduce the higher prevalence of obesity, including the regular monitoring of community health services through eating and daily activities [11]. Moreover, the exploration of novel and potential drug candidates from natural resources against obesity and its complication has been widely reported in the last two decades [12, 13]. The future goal of herbal medicine treatment will address to inhibit or reduce adipogenesis rate linked to adipocyte number and size. Previous studies in both animal and human trial have shown that herbal therapy including green tea and black tea administrations decreased adipogenesis. The metabolic intervention by using green tea reduced adipogenesis through angiogenesis and adipose tissue growth inhibition [14], lowering total cholesterol and LDL (low-density lipoprotein) concentration [15]. Moreover, it also induced endopeptidase neprilysin and AMPK (Adenosine Monophosphate-activated Protein Kinase) pathway activation [16, 17], regulate feeding behavior, energy homeostasis related to a circadian gene, and fat browning [18, 19]. In addition, this treatment was able to reduce fat deposition and inflammation [20], insulin resistance and blood pressure [21], and changing the serum level of ghrelin and leptin in the circulation [22].

Importantly, theaflavin (black tea polyphenol) has been suggested that it provides an opportunity to become a novel candidate for clinical obesity therapy. The polyphenol of black tea has been able to regulate lipid metabolism through upregulation of AMPK (Adenosine Monophosphate-activated Protein Kinase), inhibit pancreatic lipase by in silico modeling, protein phosphatase PTP1B (Tyrosine-protein phosphatase non-receptor type 1) inhibition, stimulate lipolysis via mitochondrial uncoupling activity, and inactivate PEPCK (phosphoenolpyruvate carboxykinase) promoter as the primary enzyme of gluconeogenesis through FOXO1a (Forkhead box protein O1a) phosphorylation [23-27]. Such data proved that theaflavin could control adipogenesis rate and indirectly could regulate lipid metabolism, glucose homeostasis, and adipocyte mass. Hence, a specific target molecule in adipogenesis and insulin signaling pathway combined with black tea polyphenol long-term treatment may offer as an alternative clinical target in the patient against obesity via down/up-regulation of the specific molecular pathway.

The primary pathway of preadipocyte differentiation into mature adipocyte and insulin signaling activity involve several growth factors, cytokines, kinase protein and critical transcription factors. Among those potential growth factors, it was well established that insulin-like growth factor-1 (IGF-1) and its receptor (IGF-1R) play an essential role during the adipogenesis program [28-30]. Interestingly, the expression of IGF-1 and IGF-1R were significantly increased in obese children [31]. Also, the activation of IGF-1R has a significant correlation with cancer-linked energy balance and obesity [32, 33]. This data imply that the activation of IGF-1R by insulin and mimics of insulin (IGF-1) is the critical target for combating obesity, metabolic syndrome and cancer-related obesity. Even though previous studies have been done to explore the potential role of herbal treatment model by using green and black tea polyphenol, however, there is a lack of information on how black tea polyphenol directly regulates

IGF-1R activity. Here, our In Silico-based study will provide the preliminary and comprehensive data related to the molecular interaction modeling between the black tea polyphenol and IGF-1R activity. We suggest that the potential polyphenol within black tea can regulate adipogenesis through the IGF-1R to control lipid deposition, pre-adipocyte differentiation, glucose and lipid metabolism, and indirectly decreased obesogenic program.

2. Experimental Methods

2.1. The preparation of target protein sample database, ligand chemical structure, and the prediction of potential molecular pathway

PassOnline web server (www.pharmaexpert.ru/passonline/), Swiss Target Prediction database (www.swisstargetprediction.ch) and STRING database (www.string-db.org) were used as the primary online dataset to predict the essential function of black tea polyphenol (theaflavin) on the target protein in the intracellular signaling pathway. The three-dimensional structure of black tea polyphenol and target protein were obtained from the molecular database (www.pubchem.ncbi.nlm.nih.gov). The detail of target protein sequence information for *insulin-like growth factor 1* (IGF-1) P05019 and *insulin-like growth factor 1 receptor* (IGF-1R) P08069IGF-1 were collected from UniProt database (www.uniprot.org). Furthermore, PassOnline web server also was utilized to predict the biological function of the ligand by submitting the SMILE query molecule to obtain the probability activity value (Pa) and probability inhibitor score (Pi). The probability activity value was expected > 0.7 (Pa > 0.7) that illustrates the similarity of the in Silico prediction and analysis compared to laboratory test/experiment.

2.2. Target protein modeling, virtual screening, and protein-protein docking

The tertiary structure modeling of IGF-1 and IGF-1R was done on protein modeling database *Swiss Model* (www.swissmodel.expasy.org) by using homology modeling. The homology modeling resulted in a homolog protein structure with conserved protein sequence with more than 20% similarity value. The validation of the previous protein structure was analyzed by 3D Ramachandran plot web server (www.mordred.bioc.cam.ac.uk/~rapper/rampage). To examine the binding affinity between black tea polyphenol and IGF-1R, virtual screening for molecular docking (reverse docking method) was done using PyRx software to build the simulation of molecular interaction among ligand to the target protein. In the additional test for ligand-protein interaction, to predict and develop a rigid molecular interaction model, PatchDock webserver (www.bioinfo3d.cs.tau.ac.il/PatchDock) was chosen to provide the structural conformation of binding complex and atomic contact energy (ACE). Then, the result of this step will be sent to FireDock web server for refinement docking to provide the flexible structure of protein complex/molecules complex, final binding energy, and Van der Waals binding data. All those data were used to know the difference of chemical interaction energy for every atom within the ligand-protein complex when IGF-1R bound to the tea polyphenol or in regular binding activity (IGF-1 and IGF-1R interaction).

2.3. Ligand-Protein Interaction and Visualization

The visualization of the three-dimensional structure of ligand and primary target protein related to it secondary structure were quantified and analyzed by PyMol software (www.pymol.org). The different types of amino acid residue position and ligand-protein interaction were shown on LigPlot software. The chemical interaction of both molecules related to hydrogen bond, hydrophobic bond, and the contribution of each amino acid residue on that binding complex was analyzed entirely using LigPlot software.

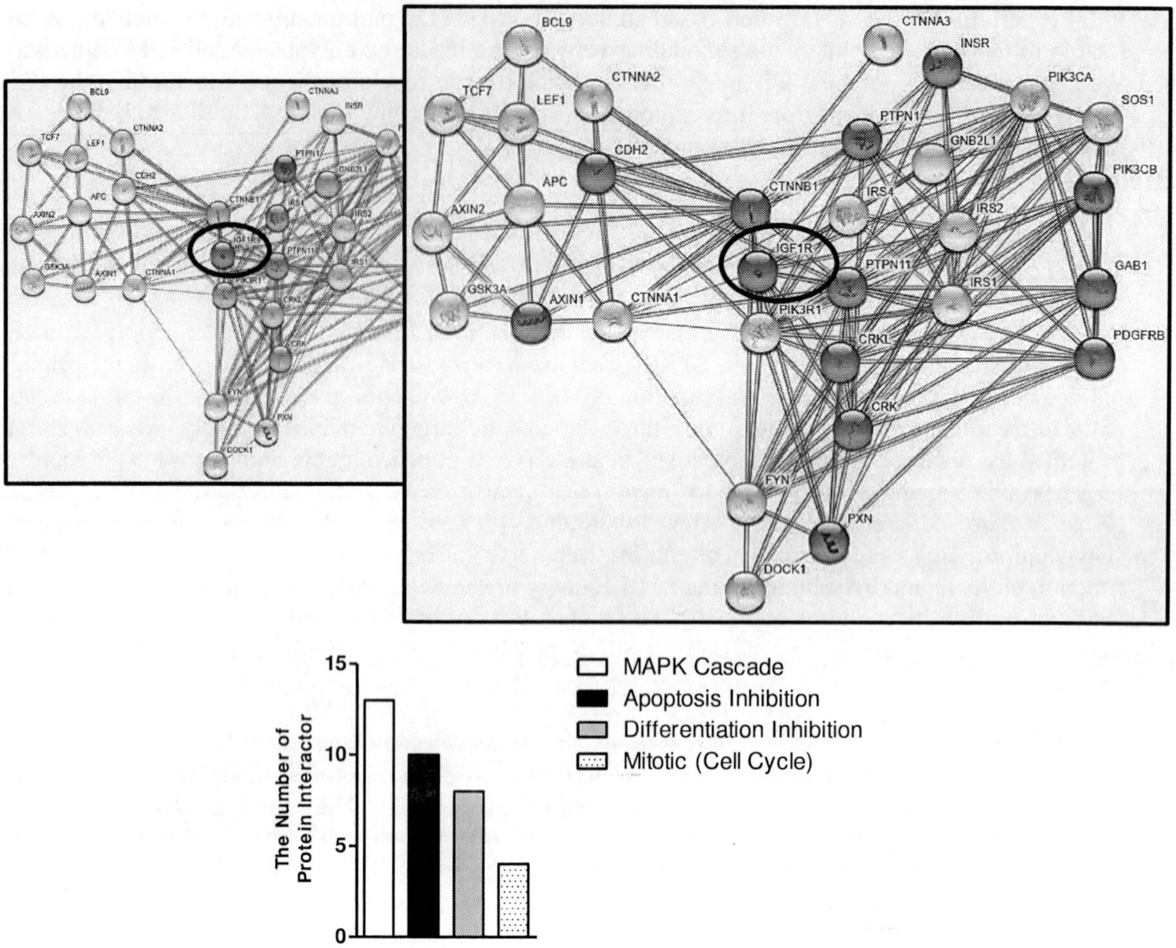

Figure 1. The Prediction of IGF-1R Signaling Pathway and its Protein Interactor. The red line shell shows protein query and the black circle illustrates the potential role of IGF-1R as the initiator protein on the primary signaling pathway.

3. Results and Discussion

The prediction of the potential molecular pathway of IGF-1R activation showed in Figure 1. The preliminary result of our In Silico study found that IGF-1R phosphorylation would induce the mitogen-activated protein kinase (MAPK) signaling pathway, the inhibition of apoptosis protein, and enhanced mitotic protein on cell cycle step. The predictive score or highest confidence value of this data was about 90% which meant the validity of those results was close to the real data from in vitro and in vivo database. The number of protein that contributed to that pathway was 35 protein involved in MAPK activity (13 protein), apoptosis silencing (10 protein), differentiation inhibition (8 protein), and mitotic cell cycle (4 protein). The majority of interactor protein for IGF-1R were contributed to the MAPK signaling pathway. The blue and purple lines implied the protein-protein interaction based on the database and experimental studies respectively (Figure 1).

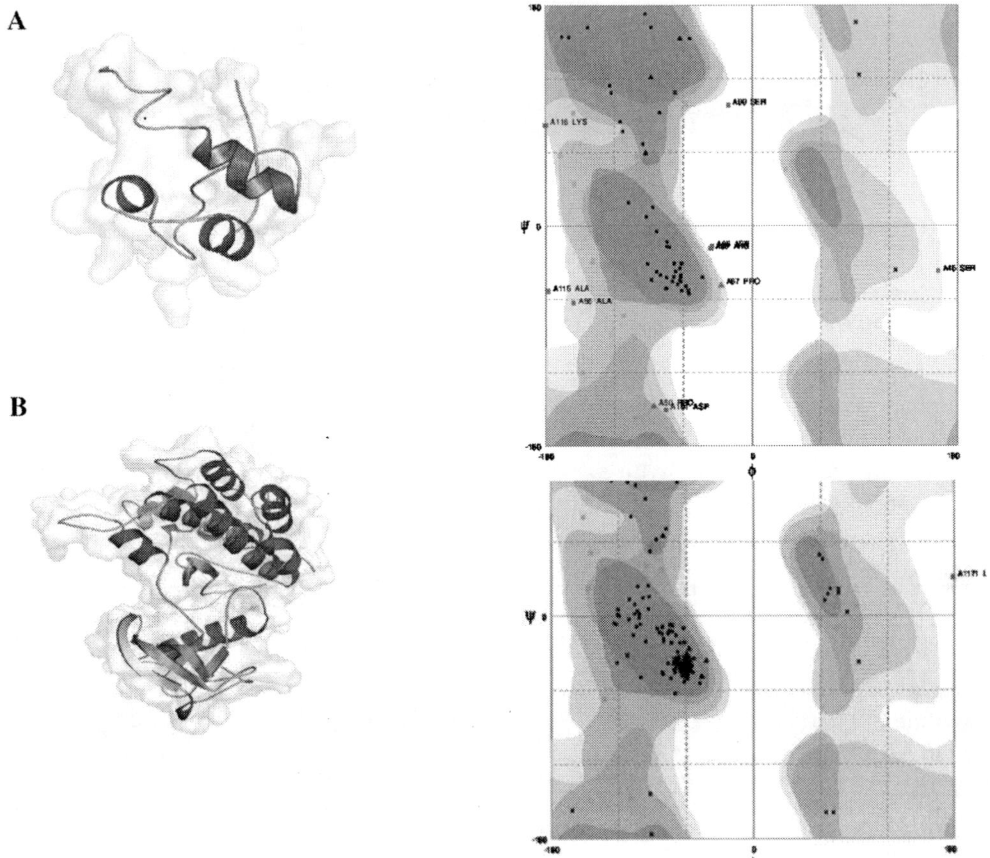

Figure 2. The Visualization of Protein Structure with Homology Modelling Method and Structure Validation. (A). The basic structure of *Insulin-like growth factor 1 (*IGF-1), (B). *Insulin-like growth factor 1 receptor* (IGF-1R)

The majority of essential chemical molecules consist of caffeine, theaflavin, catechin, theobromine, theophylline, thearubigin, gallic acid, chlorogenic acid, pectin, and theanine. For the next data analysis, the target protein homology modeling also described the conserved characteristic of IGF-1 and its receptor with a higher similarity score. The value of the favored region for IGF-1 and IGF-1R is more than 90% which mean both molecular structures have a more elevated validity model. The visualization of IGF-1 and IGF-1R is shown in Figure 2. The cartoon model with specific staining for the secondary structure of both proteins including alpha-helix, beta-sheet, and the coil is red, yellow and green respectively within the white transparent surface model. Despite the primary data from homology modeling, the researchers also tried to visualize the autophosphorylation domain of IGF-1R. The specific domain of protein phosphorylation for IGF-1R is shown in Figure 3. The phosphorylation domain of IGF-1R consists of tyrosine residue (Tyr-1161, Tyr-1165 and Tyr-1166). This specific domain provided in spheres and stick structure within the transparent surface model.

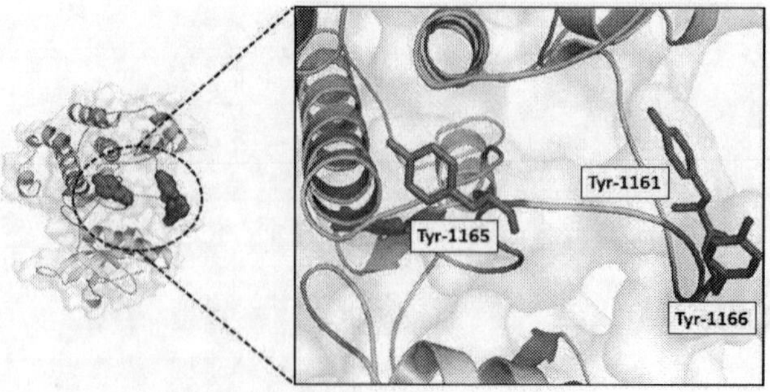

Figure 3. Autophosphorylation domain of IGF-1R at Tyrosine residues

Next, to trace the probability of black tea polyphenol binding activity to IGF-1R, reverse docking model was applied to our data. The result of this study proved that the lower binding affinity of black tea polyphenol was found in thearubigin (-10.0 kcal/mol) while another polyphenol showed a strong binding affinity to this receptor (theophylline = -7.0 kcal/mol, theaflavin 3,3'-digallate = -6.7 kcal/mol, and theaflavin = -6.7 kcal/mol). The detail information of reverse docking data analysis was shown in Table 1.

Table 1. The binding affinity of query molecules in black tea (*Camellia sinensis*).

Ligand	**PubChem ID**	**Target Protein**	***Binding affinity* (kcal/mol)**
Thearubigin	100945367	IGF-1R	-10.0
Theophylline	3055685	IGF-1R	-7.0
Theaflavin 3,3'-digallate	58252602	IGF-1R	-6.7
Theaflavins	114777	IGF-1R	-6.7
(-)-Epigallocatechin gallate	65064	IGF-1R	-6.6
(-) Epicatechin gallate	107905	IGF-1R	-6.5
Chlorogenic Acid	1794427	IGF-1R	-6,5
(-) Epicatechin gallate	107905	IGF-1R	-6,5
(-)-Epigallocatechin	72277	IGF-1R	-6,3
(-) Epicatechin	72276	IGF-1R	-6,3
Theaflavin-3-gallate	102115506	IGF-1R	-6.0
Pectin	441476	IGF-1R	-5.1
Theobromine	5429	IGF-1R	-5,0
Gallic Acid	370	IGF-1R	-4,8
Caffeine	2519	IGF-1R	-4,5
Theanine	228398	IGF-1R	-4,4

To classify the potential molecule with a strong binding affinity to autophosphorylation domain of IGF-1R, the computational analysis using LigPlot software was used. Interestingly, our data showed that theaflavin 3,3'-digallate (TF3) and theophylline had a significant interaction with the tyrosine residue at Tyr1611 (Table 2). However, there was no significant interaction between thearubigin and other polyphenol molecules.

Table 2. Ligand-Protein Interaction between Black tea Polyphenol and IGF-1R

Ligand	Interaction
Theaflavin 3,3'-digallate	Hydrogen bond : Glu1162, Asp1159, Asn1049, Arg1167, Glu1056 Hydrophobic : Val1053, Tyr1161, Ser1052, Arg1158, Phe1131
Theophylline	Hydrogen bond :- Hydrophobic : Ile1160, Thr1190, Lys1130, Tyr1161, Tyr1192, Ser1258, Asn1127, Phe1259, Ala1128
Theaflavins	Hydrogen bond : Ser1052, Glu1162, Lys1130, Asp1159 Hydrophobic : Glu1056, Lys1055, Asn1129, Val1053, Phe1131, Asn1049, Arg1158
Thearubigin	Hydrogen bond : Glu1056, Lys1055, Asn1129, Asn1058 Hydrophobic : Phe1131, Tyr1125, Ser1282, Phe1279, Cys1059
Autophosphorylation Domain	Hydrogen bond : Glu1056, Lys1055, Asn1129, Asn1058 Hydrophobic : Phe1131, Tyr1125, Ser1282, Phe1279, Cys1059

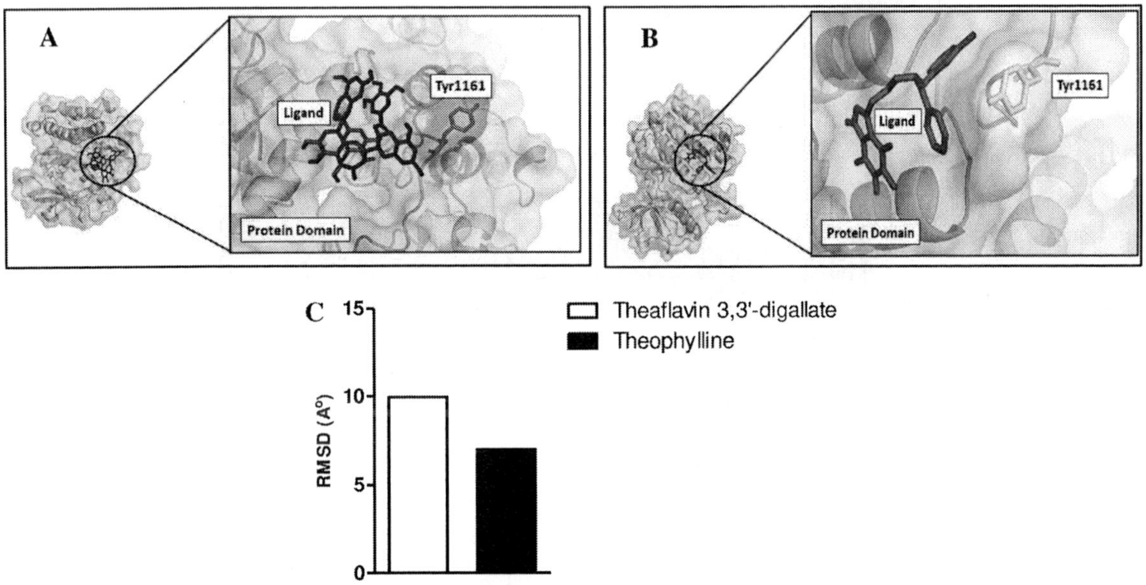

Figure 4. Ligand query interaction with amino acid residues Tyr1161 on IGF-1R autophosphorylation domain. (A). Theaflavin 3,3'-digallate (TF3); (B) Theophylline; (C) RMSD comparison of TF3 and theophylline.

The interaction of TF3 and theophylline on IGF-1R phosphorylation domain is shown in Figure 4. The clarification of the affinity profile of TF-3 and theophylline was supported by the distance of atomic binding according to root square deviation value (RMSD). This score obtained from the distance interaction between atoms from the ligand and amino acid residues at autophosphorylation domain. In the final analysis, we also tried to investigate the rigid model of ligand-protein interaction through refinement docking method. The binding energy resulting from atomic contact energy, repulsive and attractive Van der Waals bond, global energy and a hydrogen bond is shown in Figure 5. In the natural condition (green color), the absence of black tea polyphenols on IGF-1R autophosphorylation domain resulted in global energy about -46.16 kcal/mol, attractive and repulsive Van der Waals bond -42.86 and 17.30 kcal/mol respectively, and hydrogen bond -5.87 kcal/mol. However, the present of TF3 (blue color) caused the significant changes in global energy into -27.23 kcal/mol and -40.15 kcal/mol for theophylline (red color). Importantly, the repulsive Van der Waals energy of TF3 also showed more significantly compared to theophylline. Also, the binding energy from hydrogen bond formation in TF3 was more favorable (-2.46 kcal/mol) and more positive than theophylline (-4.44 kcal/mol). Due to the potential chemical properties of Theaflavin 3,3'-digallate (TF3) and theophylline on IGF-1R signaling, to complete our in silico based study the researcher finished with a biological function analysis using PassOnline database. The probability activity (Pa) score of TF3 was higher and significantly different in lipid peroxidation than apoptosis and phosphatase inhibition (Figure 6). TF3 was predicted to work as lipid peroxidase inhibitor, apoptosis agonist, and phosphatase inhibitor. Overall, the probability activity value (Pa) of TF3 was ranging from 70-92% while the probability score for phosphatase inhibitor (Pi) was in the lower range (2-13%).

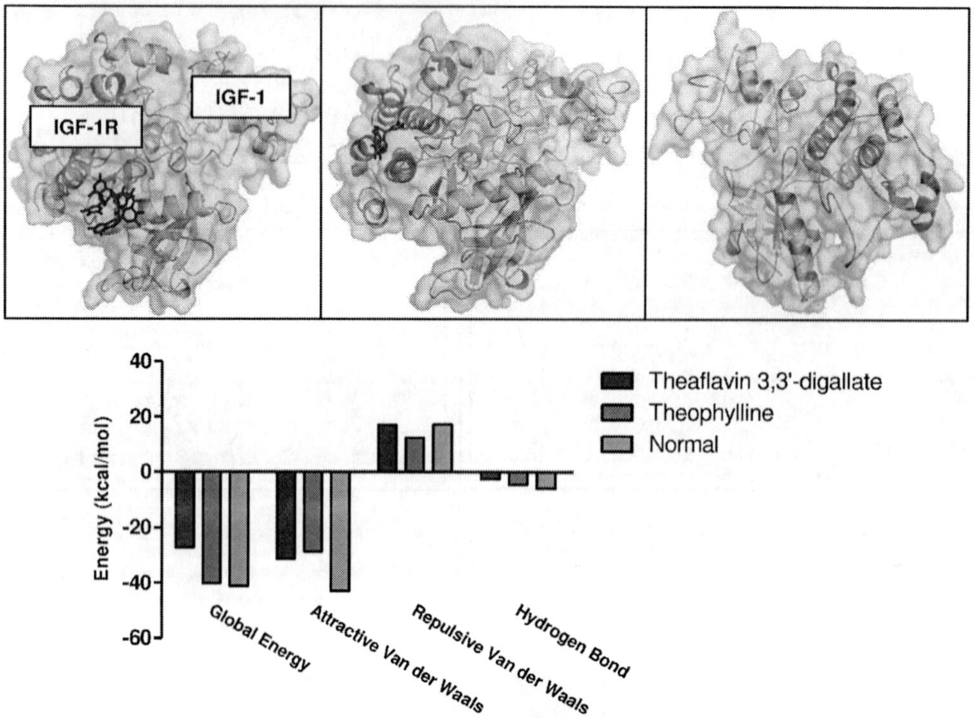

Figure 5. The comparison of binding energy between IGF-1, TF3, and Theophylline on IGF-1R

The primary function of IGF-1 and insulin was closely related to several aspects, including leading to protein synthesis (gene expression), cell proliferation or death, and lipid-glucose metabolism [34]. It was proven that the binding activity of IGF-1 to its receptor (IGF-1R) strongly associated with the up-regulation of pro-obesogenic gene expression. The differentiation of preadipocytes linked IGF-1 signaling correlated to MEK/ERK (*mitogen-activated protein kinase family*) activation to enhance the up-regulation of transcription factor PPARγ (*peroxisome proliferator-activated receptor gamma*) and C/EBPα (*CCAAT/enhancer-binding protein alpha*) gene expression [35]. The disruption of IGF-1 signaling suggested that the results in impaired adipogenic program and thermogenesis [36]. Moreover, the higher expression of IGF-1R was able to induce adipocyte differentiation from immature adipocytes or preadipocytes [37]. The recruitment of IGF-1R was reported to play a significant role in adipogenesis through regulating primary cilium elongation on preadipocytes during the differentiation process [38]. The IGF-1R was abundantly expressed in preadipocytes while IR predominantly found in mature adipocytes. Therefore, the primary treatment against obesogenic program was developed to inhibit IGF-1 and IGF-1R during preadipocyte differentiation.

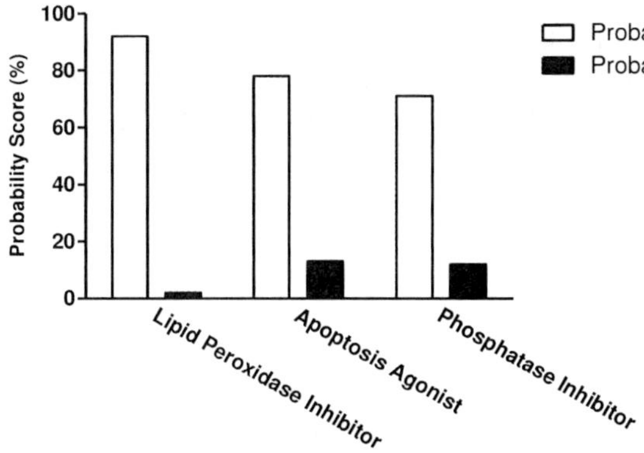

Figure 6. Probability activity (Pa) and probability inhibition (Pi) of Theaflavin 3,3'-digallate (TF3)

Some studies have proven that green tea and black tea polyphenol have a potential activity to reduce growth factor receptors and transcription factor activation. In silico study suggested that green tea polyphenol can inhibit the adipogenesis by blocking the primary transcription factor PPARγ (*peroxisome proliferator-activated receptor gamma*) activity [39]. Furthermore, theaflavin 3, 3'-digallate (TF3) showed an ability to downregulate EGFR (*epidermal growth factor receptor*) improved that anti-proliferative effect of black tea on cell proliferation [40]. TF3 was able to inhibit NIH3T3 cell proliferation more effectively than epigallocatechin-3-gallate (EGCG) corroborate, the likely impact of this black tea polyphenol against adipogenesis [41]. Also, theaflavins within black tea exhibited apoptosis and reduced inflammation through in vivo and in vitro experiments [42]. Thus, the essential polyphenol in black and green tea could be proposed as the future treatment against adipogenesis.

Importantly, the primary question that might have risen in our preliminary study was whether TF3 potentially decreased adipogenesis through the IGF-1R pathway. Linear to previous findings, in our both preliminary laboratory work through animal and cell culture experiment it was proven that black tea and its potential polyphenol could decrease adipogenesis significantly through downregulating the expression of IGF-1, ERK1/2 (*extracellular signal-regulated kinases*), and PPARγ (data not shown). Black tea polyphenol was suggested to suppress obesity, hyperglycemia, visceral fat mass, fatty acid synthesis related gene attenuation, and insulin resistance [20, 43, 44]. We hypothesized that the treatment of black tea extract on the animal model and TF3 on preadipocytes culture were disrupted IGF-1 expression and its receptor activation/phosphorylation. However, to elucidate our preliminary speculation the prediction of this biological pathways by using in silico model is required. Here, we tried to build a first molecular forecast on how black tea polyphenol can reduce obesogenic activity by using molecular docking the query component within black tea. The baseline data showed that IGF-1 phosphorylation modulates MAPK (*mitogen-activated protein kinase family*) signaling while the inhibition of this growth factor expression will decreasing the number of cell amount/induced apoptosis. It was supported by the highest confident score from in silico data on STRING database. The computational analysis has shown that black tea polyphenol may bind to IGF-1R binding domain particularly on tyrosine residue (Tyr-1161, Tyr-1165, and Tyr-1166). IGF-1R phosphorylation domain could be potential to become the primary target for some inhibitors including the query molecule of black tea (theaflavin).

Our study also delineated the binding performance by bioinformatics way through molecular docking of potential black tea polyphenol, theaflavin 3,3' digallate (TF3) with IGF-1R. In the drug development model, the general computational study suggested by reverse docking approach [45]. The results of our study showed that among potential black tea polyphenols, TF3 inhibit markedly IGF-1R activity. The binding affinity of TF3 was similar to theophylline with hydrophobic interaction on Tyrosin residue (Tyr1611). Molecular reverse docking model indicates that TF3 has a strong binding activity with IGF-1R compared to theophylline. The interaction of TF3 on autophosphorylation domain of IGF-1R results in significant positive global energy (-27.23 kcal/mol) than theophylline (-40.15 kcal/mol). This data imply the instability of IGF-1R phosphorylation domain due to TF3 binding activity induced the alteration of molecular interaction on IGF-1 and IGF-1R complex. The amount of energy for binding affinity can indicate some ability for active compound binding on the target protein, to form a protein-ligand complex stabilized and it requires the negative value of binding affinity. Even though the attractive Van der Waals of theophylline was much stronger than TF3, however, based on the global energy, repulsive Van der Waals, and hydrogen bond profile of TF3 improved that this black tea polyphenol has a more significant inhibitory effect than other polyphenols. The global energy value illustrates the total binding energy between the ligand and its receptor to form a complex during molecular interaction [46]. Also, the Van der Waals force is strongly associated with protein-protein interaction among ligand-receptor interaction [47] while thermodynamics protein association determine to contribute to complex binding stability [48]. As a consequence, it may stimulate the down-regulation of IGF-1 signaling pathway activation especially during adipogenesis program within adipose tissue.

Another additional data also showed that TF3 might act as the anti-obesity agent through inhibit cell proliferation (apoptosis agonist) and phosphatase inhibitor based on probability activity score (Pa score) analyzed by PassOnline measurement [49]. The higher score of TF3 probability activity (> 0.7 or 70%) emphasizes our theory which is the black tea polyphenol attenuate adipogenesis to induce apoptosis of immature adipocyte cells. We suggest that TF3 alleviated anti-apoptosis protein and enhanced cell death program. Nevertheless, the limitation of our study is that we cannot fully elucidate the blocking effect of TF3 to transcriptional molecules that play a pivotal role in the adipogenesis. Hence, the future pre-clinical investigation is crucial to recommend the administration of black tea polyphenol for a patient with metabolic syndrome symptom particularly overweight and obesity.

4. Conclusion

In summary, the black tea polyphenol Theaflavin 3,3'-digallate (TF3) is a promising and potential candidate as a herbal therapeutic agent for obesity prevention. TF3 may directly regulate adipose tissue growth via down regulate IGF-1 expression and IGF-1R activation. While our study employed fundamental data for further clinical applications, the extensive and comprehensive research are required to emphasize and clarify our findings. Importantly, the data presented here provide a foundation or hallmark for the molecular mechanism of black tea polyphenol in the translational investigation.

Acknowledgments

We thank Prof. Dr. M. Rasjad Indra, MS for his help and guidance for metabolic syndrome study. We also acknowledge his helpful discussion for the biochemistry and molecular biology suggestion during the manuscript preparation.

References

[1] Wang W and Seale P 2016 Control of brown and beige fat development *Nature Reviews Molecular Cell Biology* **17** 691

[2] Hossain P, Kawar B, and El Nahas M 2007 Obesity and diabetes in the developing world—a growing challenge *The New England journal of medicine* **2007** 213-5

[3] Ramachandran A and Snehalatha C 2010 Rising burden of obesity in Asia *Journal of obesity* **2010**

[4] Kelly M 2016 *Eating, Drinking: Surviving*: Springer) pp 83-90

[5] Worsley A, Wang W, Sarmugam R, Pham Q, Februhartanty J and Ridley S 2017 Family food providers' perceptions of the causes of obesity and effectiveness of weight control strategies in five countries in the Asia Pacific region: a cross-sectional survey *Nutrients* **9** 78

[6] Yulia, Khusun H and Fahmida U 2016 Dietary patterns of obese and normal-weight women of reproductive age in urban slum areas in Central Jakarta *The British journal of nutrition* **116 Suppl 1** S49-56

[7] Malik V S, Willett W C, and Hu F B 2013 Global obesity: trends, risk factors and policy implications *Nature Reviews Endocrinology* **9** 13

[8] Font-Burgada J, Sun B and Karin M 2016 Obesity and Cancer: The Oil that Feeds the Flame *Cell metabolism* **23** 48-62

[9] Pan A, Malik V S and Hu F B 2012 Exporting diabetes mellitus to Asia: the impact of Western-style fast food *Circulation* **126** 163-5

[10] Fan J-G, Kim S-U, and Wong V W-S 2017 New trends on obesity and NAFLD in Asia *Journal of Hepatology* **67** 862-73

[11] Wadden T A, Webb V L, Moran C H, and Bailer B A 2012 Lifestyle modification for obesity: new developments in diet, physical activity, and behavior therapy *Circulation* **125** 1157-70

[12] de Freitas Junior L M and de Almeida Jr E B 2017 Medicinal plants for the treatment of obesity: ethnopharmacological approach and chemical and biological studies *American journal of translational research* **9** 2050

[13] Liu Y, Sun M, Yao H, Liu Y, and Gao R 2017 Herbal Medicine for the Treatment of Obesity: An Overview of Scientific Evidence from 2007 to 2017 *Evidence-based complementary and alternative medicine: eCAM* **2017** 8943059

[14] Kim M-Y, Park B, Lee H, Park E, Hahm J, Lee J, Hong Y, Choi S, Park D and Lee H 2010 The anti-angiogenic herbal composition Ob-X inhibits adipose tissue growth in the obese mice *International journal of obesity* **34** 820

[15] Samavat H, Newman A R, Wang R, Yuan J M, Wu A H and Kurzer M S 2016 Effects of green tea catechin extract on serum lipids in postmenopausal women: a randomized, placebo-controlled clinical trial *Am J Clin Nutr* **104** 1671-82

[16] Muenzner M, Tappenbeck N, Gembardt F, Rulke R, Furkert J, Melzig M F, Siems W E, Brockmann G A and Walther T 2016 Green tea reduces body fat via upregulation of neprilysin *International journal of obesity (2005)* **40** 1850-5

[17] Yuan E, Duan X, Xiang L, Ren J, Lai X, Li Q, Sun L and Sun S 2018 Aged Oolong Tea Reduces High-Fat Diet-Induced Fat Accumulation and Dyslipidemia by Regulating the AMPK/ACC Signaling Pathway *Nutrients* **10**

[18] Li H, Kek H C, Lim J, Gelling R W and Han W 2016 Green tea (-)-epigallocatechin-3-gallate counteracts daytime overeating induced by high-fat diet in mice *Molecular nutrition & food research* **60** 2565-75

[19] Neyrinck A M, Bindels L B, Geurts L, Van Hul M, Cani P D and Delzenne N M 2017 A polyphenolic extract from green tea leaves activates fat browning in high-fat-diet-induced obese mice *The Journal of nutritional biochemistry* **49** 15-21

[20] Heber D, Zhang Y, Yang J, Ma J E, Henning S M and Li Z 2014 Green tea, black tea, and oolong tea polyphenols reduce visceral fat and inflammation in mice fed high-fat, high-sucrose obesogenic diets *The Journal of nutrition* **144** 1385-93

[21] Bogdanski P, Suliburska J, Szulinska M, Stepien M, Pupek-Musialik D and Jablecka A 2012 Green tea extract reduces blood pressure, inflammatory biomarkers, and oxidative stress and improves parameters associated with insulin resistance in obese, hypertensive patients *Nutrition research (New York, N.Y.)* **32** 421-7

[22] Haghighatdoost F, Nobakht M G B F and Hariri M 2018 Effect of green tea on plasma leptin and ghrelin levels: A systematic review and meta-analysis of randomized controlled clinical trials *Nutrition (Burbank, Los Angeles County, Calif.)* **45** 17-23

[23] Glisan S L, Grove K A, Yennawar N H and Lambert J D 2017 Inhibition of pancreatic lipase by black tea theaflavins: Comparative enzymology and in silico modeling studies *Food chemistry* **216** 296-300

[24] Lin C-L, Huang H-C, and Lin J-K 2007 Theaflavins attenuate hepatic lipid accumulation through activating AMPK in human HepG2 cells *Journal of lipid research* **48** 2334-43

[25] Ma J, Li Z, Xing S, Ho W-T T, Fu X and Zhao Z J 2011 Tea contains potent inhibitors of tyrosine phosphatase PTP1B *Biochemical and biophysical research communications* **407** 98-102

[26] Ko H-J, Lo C-Y, Wang B-J, Chiou R Y-Y and Lin S-M 2015 Theaflavin-3, 3′-digallate, a black tea polyphenol, stimulates lipolysis associated with the induction of mitochondrial uncoupling

proteins and AMPK–FoxO3A–MnSOD pathway in 3T3-L1 adipocytes *Journal of Functional Foods* **17** 271-82

[27] Pan H, Gao Y and Tu Y 2016 Mechanisms of body weight reduction by black tea polyphenols *Molecules* **21** 1659

[28] Rajkumar K, Modric T and Murphy L J 1999 Impaired adipogenesis in insulin-like growth factor binding protein-1 transgenic mice *The Journal of endocrinology* **162** 457-65

[29] Hu L, Yang G, Hagg D, Sun G, Ahn J M, Jiang N, Ricupero C L, Wu J, Rodhe C H, Ascherman J A, Chen L and Mao J J 2015 IGF1 Promotes Adipogenesis by a Lineage Bias of Endogenous Adipose Stem/Progenitor Cells *Stem cells (Dayton, Ohio)* **33** 2483-95

[30] Ali A T, Hochfeld W E, Myburgh R and Pepper M S 2013 Adipocyte and adipogenesis *European journal of cell biology* **92** 229-36

[31] Ricco R C, Ricco R G, Queluz M C, de Paula M T S, Atique P V, Custodio R J, Tourinho Filho H, Del Roio Liberatori R, Jr. and Martinelli C E, Jr. 2017 IGF-1R mRNA expression is increased in obese children *Growth hormone & IGF research : official journal of the Growth Hormone Research Society and the International IGF Research Society*

[32] Moore T, Checkley L A, and DiGiovanni J 2011 Dietary energy balance modulation of epithelial carcinogenesis: a role for IGF-1 receptor signaling and crosstalk *Annals of the New York Academy of Sciences* **1229** 7-17

[33] Doyle S L, Donohoe C L, Finn S P, Howard J M, Lithander F E, Reynolds J V, Pidgeon G P and Lysaght J 2012 IGF-1 and its receptor in esophageal cancer: association with adenocarcinoma and visceral obesity *The American journal of gastroenterology* **107** 196-204

[34] Garten A, Schuster S and Kiess W 2012 The insulin-like growth factors in adipogenesis and obesity *Endocrinology and metabolism clinics of North America* **41** 283-95, v-vi

[35] Prusty D, Park B-H, Davis K E and Farmer S R 2002 Activation of MEK/ERK signaling promotes adipogenesis by enhancing peroxisome proliferator-activated receptor γ (PPARγ) and C/EBPα gene expression during the differentiation of 3T3-L1 preadipocytes *Journal of Biological Chemistry* **277** 46226-32

[36] Boucher J, Mori M A, Lee K Y, Smyth G, Liew C W, Macotela Y, Rourk M, Bluher M, Russell S J and Kahn C R 2012 Impaired thermogenesis and adipose tissue development in mice with fat-specific disruption of insulin and IGF-1 signaling *Nature communications* **3** 902

[37] Boucher J, Tseng Y-H and Kahn C R 2010 Insulin and insulin-like growth factor-1 receptors act as ligand-specific amplitude modulators of a common pathway regulating gene transcription *Journal of Biological Chemistry* **285** 17235-45

[38] Dalbay M T, Thorpe S D, Connelly J T, Chapple J P and Knight M M 2015 Adipogenic Differentiation of hMSCs is Mediated by Recruitment of IGF-1r Onto the Primary Cilium Associated With Cilia Elongation *Stem cells (Dayton, Ohio)* **33** 1952-61

[39] Javaid M S, Latief N, Ijaz B and Ashfaq U A 2017 Epigallocatechin gallate as an anti-obesity therapeutic compound: An in silico approach for structure-based drug designing *Natural product research* 1-5

[40] Mizuno H, Cho Y Y, Zhu F, Ma W Y, Bode A M, Yang C S, Ho C T and Dong Z 2006 Theaflavin-3, 3′-digallate induces epidermal growth factor receptor downregulation *Molecular carcinogenesis* **45** 204-12

[41] Liang Y-C, Chen Y-C, Lin Y-L, Lin-Shiau S-Y, Ho C-T and Lin J-K 1999 Suppression of extracellular signals and cell proliferation by the black tea polyphenol, theaflavin-3, 3′-digallate *carcinogenesis* **20** 733-6

[42] Gosslau A, Jao E, Li D, Huang M T, Ho C T, Evans D, Rawson N E and Chen K Y 2011 Effects of the black tea polyphenol theaflavin-2 on apoptotic and inflammatory pathways in vitro and in vivo *Molecular nutrition & food research* **55** 198-208

[43] Nishiumi S, Bessyo H, Kubo M, Aoki Y, Tanaka A, Yoshida K-i and Ashida H 2010 Green and black tea suppress hyperglycemia and insulin resistance by retaining the expression of glucose transporter 4 in the muscle of high-fat diet-fed C57BL/6J mice *Journal of agricultural and*

food chemistry **58** 12916-23

[44] Lin J K and Lin-Shiau S Y 2006 Mechanisms of hypolipidemic and anti-obesity effects of tea and tea polyphenols *Molecular nutrition & food research* **50** 211-7

[45] Grinter S Z, Liang Y, Huang S-Y, Hyder S M and Zou X 2011 An inverse docking approach for identifying new potential anti-cancer targets *Journal of Molecular Graphics and Modelling* **29** 795-9

[46] Duhovny D, Nussinov R and Wolfson H J 2002 Efficient unbound docking of rigid molecules. In: *International workshop on algorithms in bioinformatics*: Springer) pp 185-200

[47] Roth C M, Neal B L and Lenhoff A M 1996 Van der Waals interactions involving proteins *Biophysical journal* **70** 977-87

[48] Ross P D and Subramanian S 1981 Thermodynamics of protein association reactions: forces contributing to stability *Biochemistry* **20** 3096-102

[49] Lagunin A, Zakharov A, Filimonov D and Poroikov V 2011 QSAR modeling of rat acute toxicity on the basis of PASS prediction *Molecular informatics* **30** 241-50

Analysis of Focal Mechanism Distribution in Northern Sulawesi

Muhammad Altin Massinai*, Muhammad Fawzy Ismullah M., Bambang Harimei, Fauziah Alimuddin

Geophysics Study Programs, Faculty of Mathematics and Natural Sciences, Hasanuddin University, Jl. Perintis Kemerdekaan Km. 10, Tamalanrea Indah, Tamalanrea, Tamalanrea Indah, Tamalanrea, Kota Makassar, Sulawesi Selatan 90245, Indonesia

*Corresponding author's email: altin@science.unhas.ac.id

Abstract. Sulawesi has been known to have the complex condition of the geology and tectonics. It is indicated by natural phenomena such as volcanoes and earthquakes in Sulawesi. This study aimed to determine the latest tectonic order in Sulawesi based on the distribution focal mechanism. The locations of the study were in the Sulawesi 2°22' N – 3°48' S and 119°22' – 124°22'. The earthquake data used were from BMKG since 2001 to 2015 with the criteria ≥ M4.5. The strike, dip, and rake data were through Global-CMT for the earthquake in Sulawesi at the same time. An AB cross-section lies along the neck of Sulawesi which has strike-slip fault type of focal mechanism with a shallow depth allegedly caused by Palu Koro Fault activity. CD longitudinal section elongated in the eastern arm of Sulawesi to the ocean, which is dominated by the strike-slip fault showing the activity of Batui Fault at the shallow depths. Matano Fault activity was observed in EF cross-section dominated by a strike-slip fault on the land and at sea by a normal fault. This activity was observed at the shallow depths which shows Matano Fault activity. GH cross section was an event with depth with various subtypes of reverse fault which is relatively uniform turning into an indication of the Sulawesi Sea subduction activity.

Keywords: Distribution, focal mechanism, northern, tectonic

1. Introduction

Sulawesi has been known to have the complex condition of the geology and tectonics. It is caused by main world plate & micro-continent movement around Sulawesi. It is indicated by natural phenomena such as volcanoes and faults in Sulawesi. Faults in Sulawesi such as Walanae Fault (South Sulawesi), Palu Koro Fault (Bone Gulf through Makassar Strait), Gorontalo Fault, Batui Fault (Central Sulawesi), Makassar Thrust (West Sulawesi), Matano Fault, Lawanopo Fault, and Kolaka Fault (Southeast Sulawesi) [1]. If Sulawesi is divided into two main regions, Northern and Southern Sulawesi, we saw more faults located in Northern Sulawesi. Earthquake information gives new perception about the tectonic process in Sulawesi. This perception can be used to explore the natural resource exploration, or for people in the earthquake hazard mitigation.

The research results reported that the hypocenter is on the fault zone or on the meeting area inter-plates. If this information is combined with focal mechanism data, it can visualize the tectonic pattern.

IOP Conf. Series: Journal of Physics: Conf. Series **1093** (2018) 012038 doi:10.1088/1742-6596/1093/1/012038

This study aimed to determine the latest tectonic order in Northern Sulawesi based on the focal mechanism distribution.

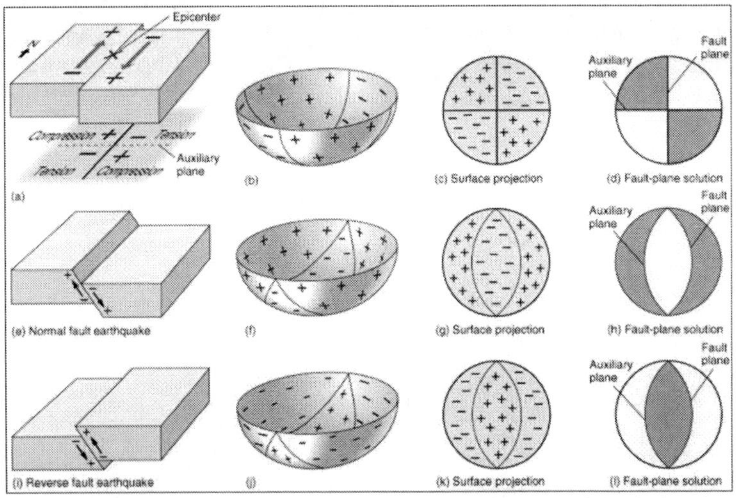

Figure 1. The relation between movement mechanism of faults and their focal mechanism [3].

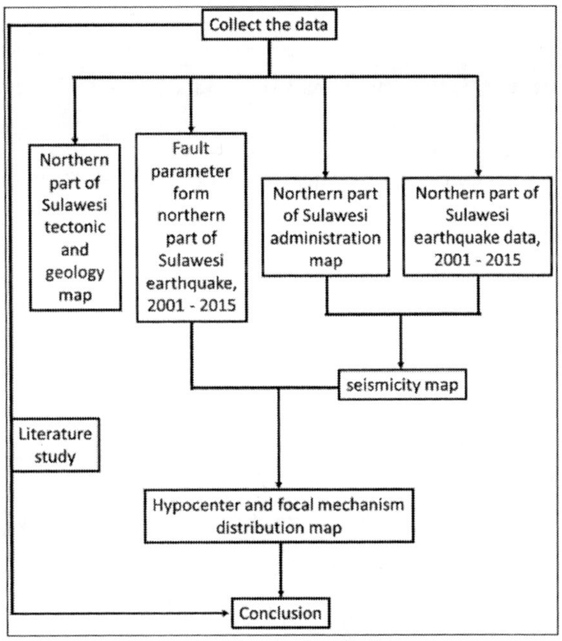

Figure 2. The research diagram

2. Experimental Methods

A focal mechanism solution is the result of waveforms generated by an earthquake and recorded by a number of seismographs. It usually takes at least 10 records to produce a reasonable focal mechanism diagram, and only the seismograph stations are well distributed geographically around the epicenter. The complete characterization of an earthquake focal mechanism provides essential information including the original time, epicenter location, focal depth, seismic moment (a direct measure of the energy radiated by an earthquake), and the magnitude and spatial orientation of the 9 components of the moment tensor. From the moment tensor, we could ultimately resolve the orientation and sense of the fault slip [2]. Figure 1 provides the information about the relationship between the movement mechanism of faults and their focal mechanism.

The research location covered Northern Sulawesi with 2⁰22' N – 3⁰48' S and 119⁰22' E – 124⁰22' E. The earthquake data were from Geophysics, Climatology, Meteorology Agency of Indonesia (BMKG on Indonesian) along 2001 – 2015 with criteria $\geq$ M4.5. The focal mechanism data were from Global-CMT for the earthquake in same location and time. The research method is explained by the diagram in Figure 2.

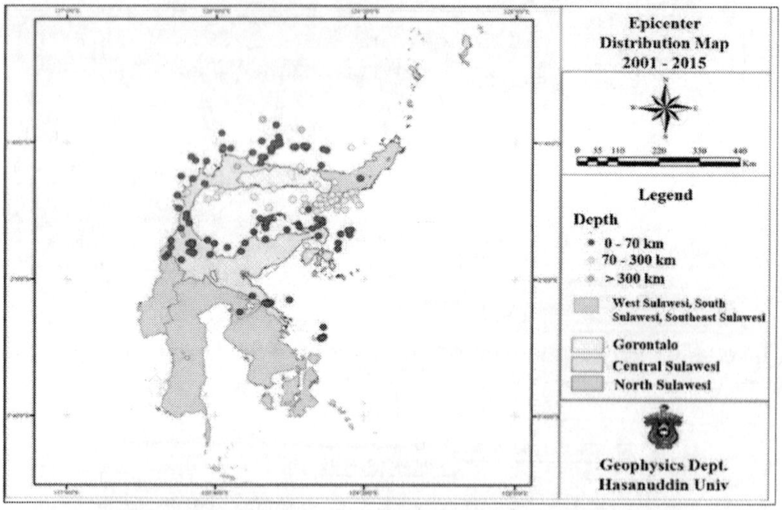

Figure 3. Epicenter distribution map based on depth

3. Results and Discussion

Figure 3 shows earthquake activity on the sea more than on the land. Besides, the shallow earthquake is more than others. The moderate earthquake occurred in southern Sulawesi North Arm (SNA), and the deep earthquake happened in northern SNA. The earthquake on the land is affected by the fault and the earthquake on the sea is influenced by subduction around SNA. This is caused by some structures such as Sulawesi Sea subduction and East Sangihe thrust in eastern and southern SNA [4]. The fault activity is shown from the shallow hypocenter, and the subduction activity is shown from the intermediate and deep hypocenter.

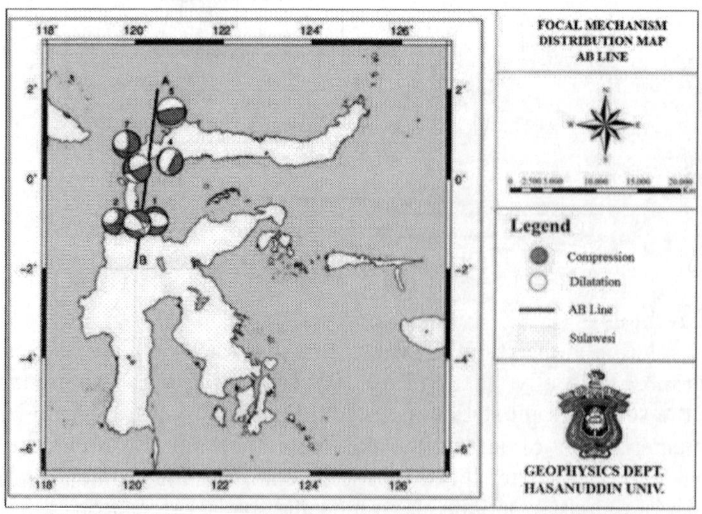

Figure 4. Focal mechanism distribution around AB line.

IOP Conf. Series: Journal of Physics: Conf. Series **1093** (2018) 012038 doi:10.1088/1742-6596/1093/1/012038

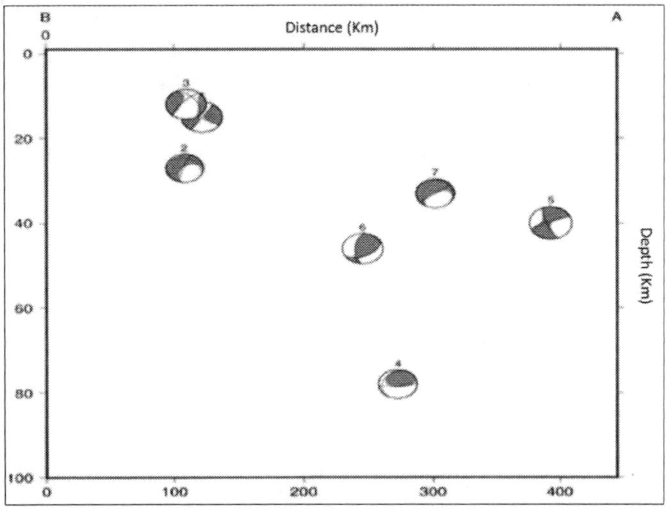

Figure 5. Focal mechanism cross-section around AB line.

Figure shows that many earthquakes occurred on northern Sulawesi. That is the indication for the complex tectonic pattern on northern Sulawesi. To find the depth effect of each earthquake in northern Sulawesi, the imaginary lines on several parts are drawn on the map to represent the earthquake in this area.

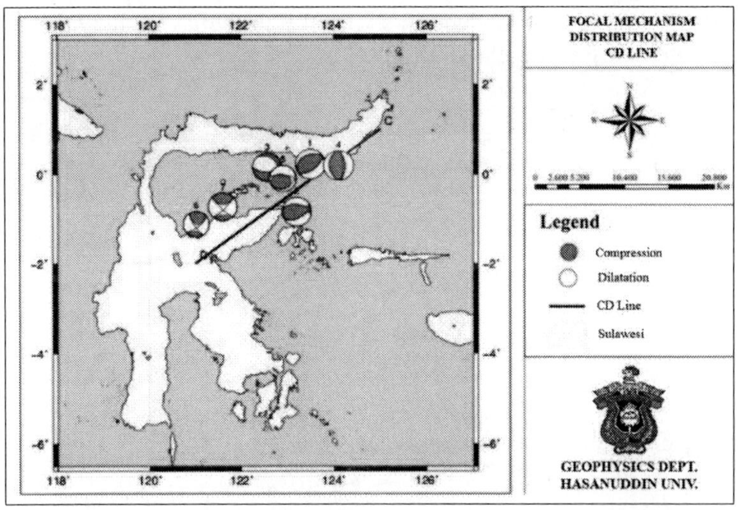

Figure 6. Focal mechanism distribution around CD line.

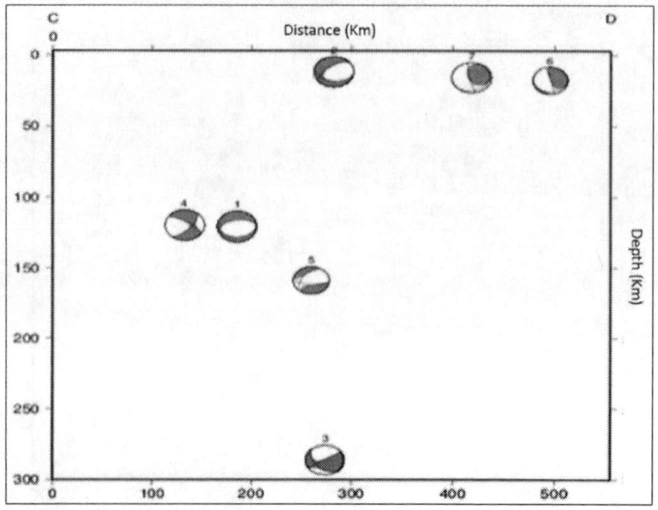

Figure 7. Focal mechanism cross-section around CD line.

CD line (Figure 6 & 7) is along Sulawesi East Arm to Sulawesi North Arm like AB line that shows 2 focal mechanism clusters. Meanwhile, Matano Fault activity is shown in EF line (Figure 8 & 9). The earthquake on the land is dominated by the strike-slip fault, and the earthquake on the sea is mainly caused by the normal fault. This activity is shown on the shallow depth assigned by Matano Fault activity. Furthermore, Sulawesi Sea Subduction shows the earthquakes along the GH line (Figure 10 & 11). The earthquake with various depths until 600 km and reverse fault type indicates this structured activity.

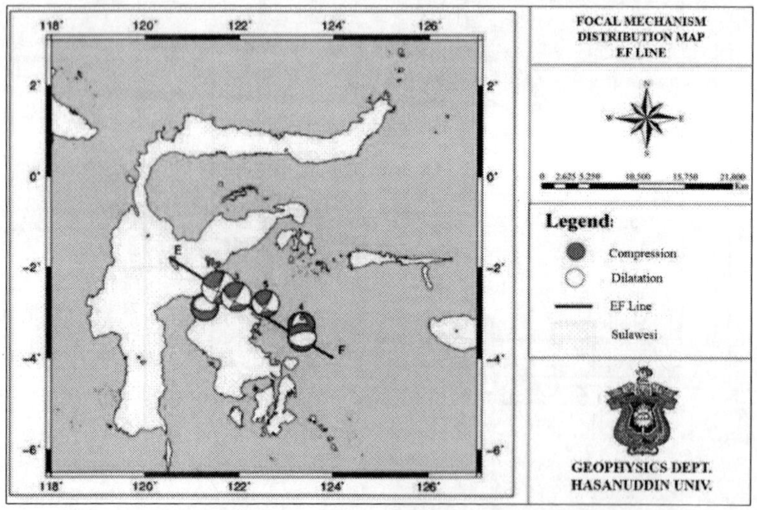

Figure 8. Focal mechanism distribution around EF line.

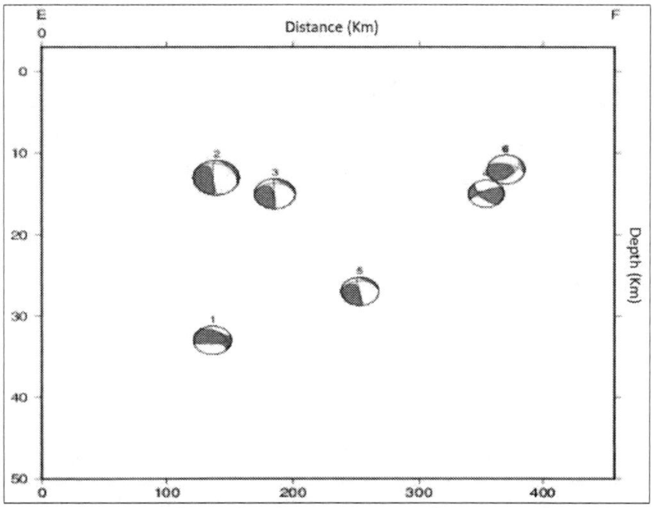

Figure 9. Focal mechanism cross-section around EF line.

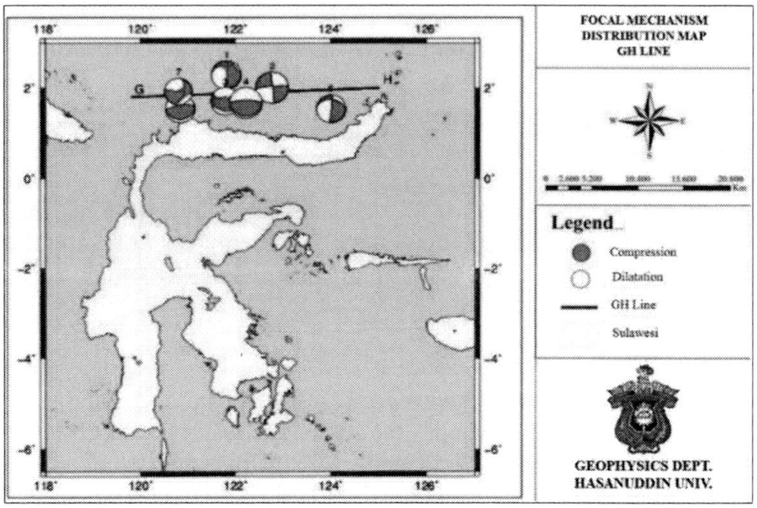

Figure 10. Focal mechanism distribution around GH line.

The four imaginary lines (AB, CD, EF, & GH) show focal mechanism distribution pattern on Northern Sulawesi. AB line lies on Palu – Koro Fault track on the southern cluster. It can be said that Palu – Koro Fault is the strike-slip fault. This result is different from [1, 5] in the other research. Previous work [1] reported that Palu – Koro Fault is a normal fault. The explanation provided by [5] using deformation analysis said that Palu – Koro Fault acts a transtensional deformation, normal faulting with a left lateral strike-slip component. CD line is located on Sulawesi East Arm. The western cluster is dominated by strike-slip fault indicating Batui Fault activity on the shallow depth. This is different from [1] who said Batui Fault is a reverse fault. Batui Fault, as known as Batui Thrust, separates the ophiolites from sedimentary formation [6] on this location. As thrust, the mechanism is a reverse fault, but the result shows the strike-slip. It may be affected by the activity of strike-slip Balantak Fault [7] located near Batui Fault. The eastern cluster lies on the sea with the profound depth, signed by the reverse fault and a normal fault earthquake. This indicates Moluccas Sea Double Subduction. Molucca Sea Double Subduction is generally a reverse fault and has a standard fault mechanism under 200 km.

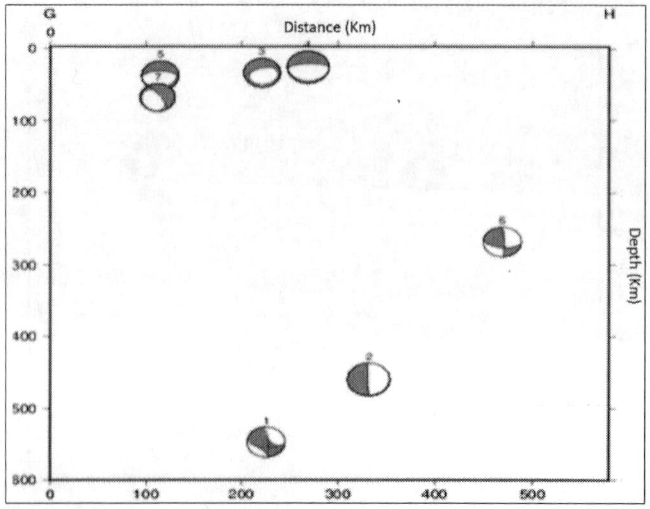

Figure 11. Focal mechanism cross-section around GH line.

EF line shows the activity of Matano Fault. Based on the focal mechanism distribution, that mechanism is a strike-slip fault. Matano Fault is a strike-slip fault with ESE – WNW orientation which cuts Southeast Sulawesi through Matano Lake. This fault is marked by a broad valley in the northern part of Southeast Sulawesi that disrupts local drainage pattern. Matano Fault is connected to Palu-Koro Fault on the western part and Sorong Fault on the eastern part [8]. Sulawesi Sea Subduction activity is shown in GH line. As known as North Sulawesi Subduction, this structure has a reverse fault on the shallow depth and strike-slip on deeper depth. This result agrees with [9] that earthquake focal mechanism on 0 – 200 km is a reverse fault and on 200 – 450 km is a strike-slip fault. The subduction has resulted in magmatism activities and a volcano that produces plutonic rocks and the volcano widespread in Sulawesi North Arm [10].

4. Conclusion

Some active structures in Sulawesi are shown by focal mechanism distribution. The strike-slip Palu-Koro Fault is shown on the shallow depth. Batui Fault is generally a strike-slip fault, with the segmentation of sea affected by Moluccas Sea Double Subduction activity and signed by a reverse fault type. Matano Fault is shown on the shallow depth with a strike-slip fault type. Sulawesi Sea Subduction activity is shown until 600 km depth dominated by the reverse fault type. This research needs to be developed in the future. Hypocenter determination becomes a vital thing for further interpretation. This result is expected to be a base in arranging the earthquake hazard mitigation.

Acknowledgments

We gratefully acknowledge BMKG and Global-CMT for their data used in this research. This research can be more informative if it is completed by the other data, like geology, tectonics, geodesy, etc. We thank Geophysics Dept., Faculty of Mathematics and Natural Sciences Hasanuddin University (UNHAS) for supporting this research.

References

[1] Massinai M F I, Lantu, Aswad S, and Massinai M A 2014 AIP Conf. Proc. 1658 030013-1 – 030013-10
[2] Cronin V S 2010 A Primer on Focal Mechanism Solutions for Geologists Baylor University.
[3] Gok R 2008 Earthquake Focal Mechanism and Waveform Modeling RELEMR Workshop Istanbul.
[4] Massinai M A, Amaliah K R, Lantu, Virman, Massinai M F I 2016 Prosiding Seminar Nasional Fisika UNJ 33 – 36.

[5] Bellier O, Sébrier M, Seward D, Beaudouin T, Villeneuve M, Putranto E 2006 Tectonophysics 413 201-220.
[6] Husein S, Novian M I, Barianto D H 2014 Proceedings Indonesia Petroleum Association IPA 14-G-137.
[7] Ferdian F, Hall R, Watkinson I 2010 Proceedings Indonesia Petroleum Association IPA 10-G-009.
[8] Lukman M N, Lamaliwa K A, Husein S, Setianto A, Samodra S B, Surjono S S 2016 Proceedings GEOSEA XIV Congress & 45th IAGI Annual Convention.
[9] Bailey I W, Alpert L A, Becker T W, Miller M S 2012 J. Geophys. Res. 117 B04404-1 – B04404-19.
[10] Lantu, Massinai M A, Amaliah K R 2016 ARPN Journal of Earth Sciences 5 64 – 69.
[11] Wessel P, Smith W H F, Scharroo R, Luis J, Wobbe F 2013 EOS Trans. American Geophys. Union 94 409 – 410.

Antibacterial Characteristic of Impregnated-Activated-Zeolite/Alumina/Glass-Ionomer-Cement Composites

Yudyanto*, **Isnindar Tandya Asri, Hartatiek and Nurul Hidayat**

Department of Physics, Faculty of Mathematics and Natural Sciences, Universitas Negeri Malang, Jl. Semarang 5 Malang 65145, Indonesia

*Corresponding author's email: yudyanto.fmipa@um.ac.id

Abstract. In this present paper, we introduce the beneficial use of Indonesian activated natural zeolite, precisely from Malang area, for a potential restorative material after combined with alumina and glass ionomer cement. The activated zeolite was previously impregnated by copper (II) chloride prior to prepare the composites of impregnated-activated-zeolite/alumina/glass-ionomer-cement composites. The X-ray fluorescence (XRF) test results support the X-ray diffraction (XRD) data in revealing the crystalline phase contents. The composites comprise of mordenite, sanidine, quartz, orthoferrosilite, and tolbachite. Scanning electron microscopy (SEM) photographs dictate the uniform $CuCl_2$ impregnation to the activated zeolite. The $CuCl_2$ impregnation, moreover the addition of alumina crystalline powder and GIC polymer, increase the density of activated zeolite. Finally, the impregnated-activated-zeolite/alumina/glass-ionomer-cement composites exhibit acceptable mechanical and antibacterial characteristics for dental restorative implants.

Keywords: Antibacterial, mechanical strength, phase compositions, restorative material.

1. Introduction

Bacterial activities in the mouth may lead to broken teeth issues [1], which is mostly driven by unwanted microorganisms [2]. The painful effect of those bacterial activities is permanently-damaged areas in teeth that may develop into holes, known as dental caries. It is one of the most common and injuring oral health disease. The best way to cure the disease is to use a restorative treatment [3]. A restorative treatment for dental caries is meant for removing insensitive and demineralized outer carious dentin; glass ionomer cement (GIC) is a harmless dental restorative material as filling and luting cement even for orthodontic attachment.

Though GIC is harmless, its crushing strength, compared to other restorative materials, is relatively a low crushing strength and is easy to dissolve in mouth liquids [4-5]. Additionally, GIC is often eroded by saliva giving accumulation of secondary caries [6]. Therefore, it is important to introduce another antibacterial material that is safe for a dentistry application. Nature provides such material in the form of calcium aluminum silicate hydrate, which can be obtained from the natural zeolite. It has good structure and porous stability for ion changes. On top of that, it shows a higher mechanical strength compared to GIC [7].

Prior to practical use, the natural zeolite needs to be activated by means of physical and chemical processes to eliminate impurities and enlarge its surface area [8]. To enhance the antibacterial properties of activated zeolite, $CuCl_2$ impregnation is required to control the microorganism growth. Another study reported that $CuCl_2$-impregnated zeolite has an excellent resistant to *streptococcus mutans* bacteria [9].

The desired applicable GIC-based restorative material can be obtained by not only introducing $CuCl_2$-impregnated activated zeolite but also adding extra metal oxides to enhance both mechanical strength and toughness. For this purpose, alumina meets all the requirements [8]. Therefore, impregnated-activated-zeolite/alumina/glass-ionomer-cement composites may contribute as secure dental restorative materials.

2. Materials and Methods

The natural zeolite sample was taken from Malang, Indonesia. The activation of the zeolite was done by washing zeolite powder and stirring for 4 h at 600 rpm in an ambient condition followed by heating at 120 °C for four h and immersing in 1% HF solution for 30 minutes and another heating at the same temperature and duration. Further treatment through 6 M HCl immersion for 30 minutes was completed, again, stirring at 600 rpm for four h. The residual powder after the second washing was immersed in 1 M NH_4Cl solution at a temperature of 90 °C for 3 h and finally sieved and heated at 120 °C for four h to obtain activated zeolite. The activated zeolite was impregnated by copper (II) ions via 0.3 M $CuCl_2$ solution at 600 rpm and 100 °C for one hour. The $CuCl_2$-impregnated activated zeolite was accomplished after washing and heating the powder at 100 °C for 24 h. Furthermore, GIC powder, impregnated activated zeolite, and alumina were mixed at various compositions by the following labelling convention: the weight fractions (%) of impregnated-activated-zeolite/alumina/GIC in sample A = 0/0/100, sample B = 50/0/50, sample C = 40/10/50, sample D = 30/20/50, sample E = 25/25/50, sample F = 20/30/50, sample G = 10/40/50, and sample H = 0/50/50. Finally, important characterizations by means of XRF, XRD, SEM, Vickers hardness, and antibacterial test.

3. Results and Discussion

Previous to XRD test, an elemental tracking of the zeolites, which can be seen in Table 1, was conducted by means of XRF. Table 1 shows elements (above 1%) within the prepared sample. Si is the most predominant atom, followed by Ca and Fe. As it is expected, Cu appears only after $CuCl_2$ impregnation. The atomic fractions for the constituents, except Al, tend to decrease because of the impregnation. Oxide compounds, detected by XRD, are based on these atoms.

Table 1. The XRF-based major elemental constituent in the zeolites.

Sample	Atomic fraction (%)					
	Si	Fe	Ca	Al	K	Cu
Activated zeolite	54.3	17.4	13.7	7.9	5.4	-
$CuCl_2$-impregnated activated zeolite	49.3	10.2	4.8	8.3	4.4	9.6

The XRD patterns of purified activated zeolite before and after $CuCl_2$ impregnation are shown in Figure 1. Changes in XRD patterns are detected in terms of diffractions peaks and intensities. Further analysis using RIR method was conducted and the results are represented in Figure 2. The crystalline constituents in the activated zeolite, in line with the XRF results, are 45% mordenite ($Ca_3Al_7Si_{40}O_{96}(H_2O)_{31}$), 23% quartz ($SiO_2$), 21% sanidine ($K(AlSi_3)O_8$), and 11% orthoferrosilite ($FeSiO_3$). Meanwhile, $CuCl_2$-impregnated activated zeolite contains 37% mordenite ($Ca_3Al_7Si_{40}O_{96}(H_2O)_{31}$), 32% sanidine ($K(AlSi_3)O_8$), 14% quartz ($SiO_2$), 13% orthoferrosilite ($FeSiO_3$), and 4% tolbachite ($CuCl_2$). Physically, $CuCl_2$ impregnation changed the color of the activated zeolite from yellowish-white to grey-white.

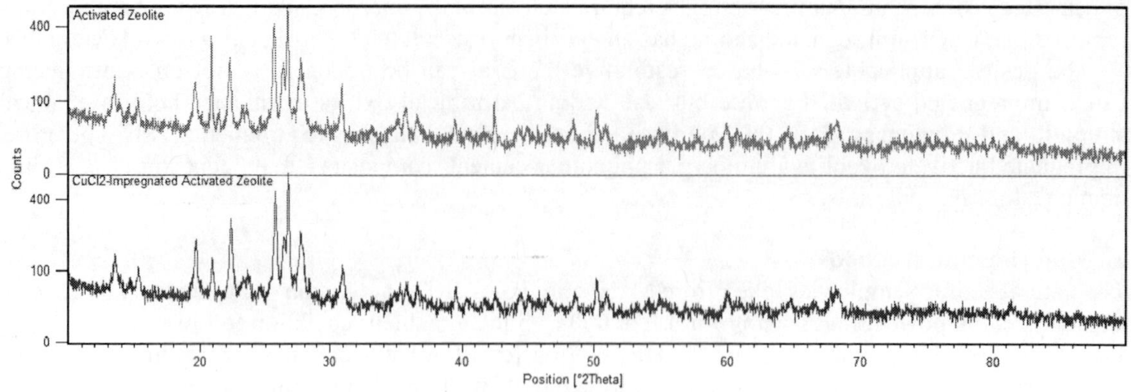

Figure 1. XRD patterns of activated zeolite and $CuCl_2$-impregnated activated zeolite.

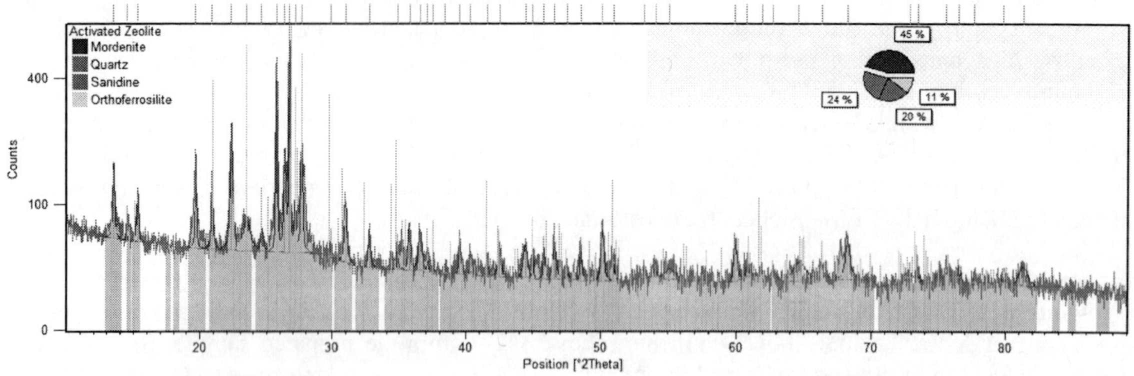

(a)

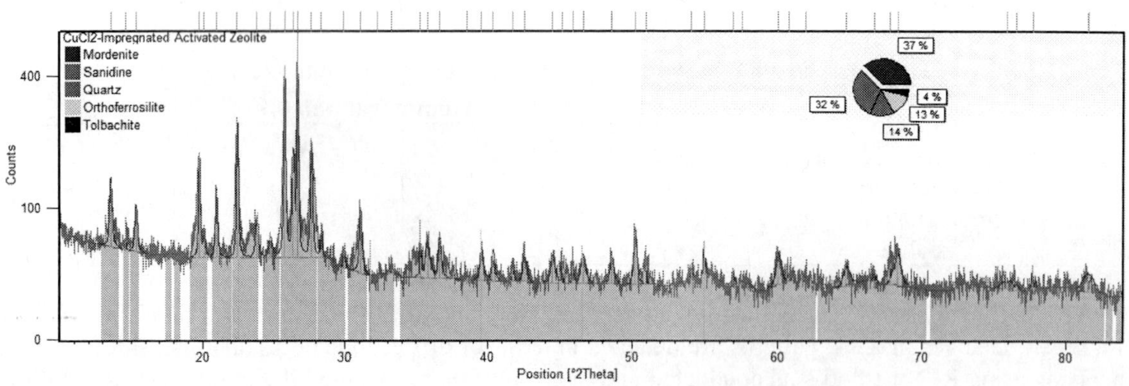

(b)

Figure 2. Phase identifications using RIR approach of (a) activated zeolite and (b) $CuCl_2$-impregnated activated zeolite.

IOP Conf. Series: Journal of Physics: Conf. Series **1093** (2018) 012039 doi:10.1088/1742-6596/1093/1/012039

Figure 3. Microstructure pictures of activated zeolite before (left) and after (right) $CuCl_2$ impregnation. The scale-bar represents 2 µm in length.

Figure 3 shows the SEM images of the zeolites. Both unimpregnated and impregnated zeolites perform similar particulates distribution, surface shape, and roughness. It is driven by the uniform $CuCl_2$ impregnation. Si/Al ratio, from SEM-EDX data, for activated zeolite and $CuCil_2$-impregnated activated zeolite, are respectively 4.63 and 5.31. Si/Al ratio for zeolite indicates the adsorption and ion transfer abilities. The larger Si/Al ratio of the impregnated sample provides better characteristics of zeolite. The particles of the impregnated activated zeolite are slightly denser than that of the pre-impregnated sample. It is due to impurities dissolution after acid pretreatment and $CuCl_2$ impregnation.

The composite of 25%-impregnated-activated-zeolite/25%-alumina/25%-GIC (sample E) and pure GIC polymer XRD profiles are given in Figure 4. Besides mordenite, sanidine, quartz, orthoferrosilite, and tolbachite, corundum-Al_2O_3 XRD peaks are found in the Bragg angles of 25.6°, 35.1°, 37.8°, 43.5°, 52.6°, 61.3°, 68.4°, and 77.6°. Broad XRD profiles of GIC represents that amorphous phase of the GIC polymer.

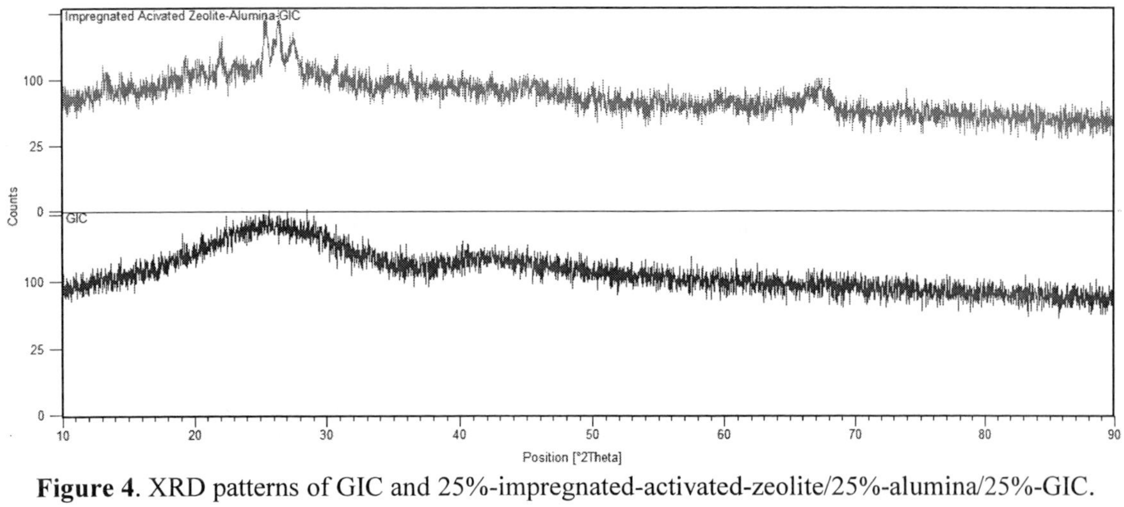

Figure 4. XRD patterns of GIC and 25%-impregnated-activated-zeolite/25%-alumina/25%-GIC.

Figure 5. SEM image of 25%-impregnated-activated-zeolite/25%-alumina/25%-GIC.

The corresponding SEM image of sample E is shown in Figure 5. Unlike the microstructures of impregnated activated zeolite, the composite of sample H shows even denser aggregates due to the presence of GIC polymer and alumina. This is the desired microstructural properties of the dental restorative material to fill the holes in human teeth.

Figure 6(a) represents the mechanical strength regarding Vicker's hardness number (VHN) and Figure 6(b) shows the antibacterial characteristic regarding the inhibition zone diameter. Note: the inhabitant of the composites was tested to *Streptococcus mutans* bacteria. To test the antibacterial performance, the composites were dissolved in water and incubated overnight. The incubation will provide an inhibition zone with which its diameter can easily be measured. A slow release agent mechanism due to copper (II) ions diffusion to the cell membrane of *Streptococcus mutans* bacteria. The oligodynamic properties of copper (II) ions within the composites reduce the growth of the microorganism.

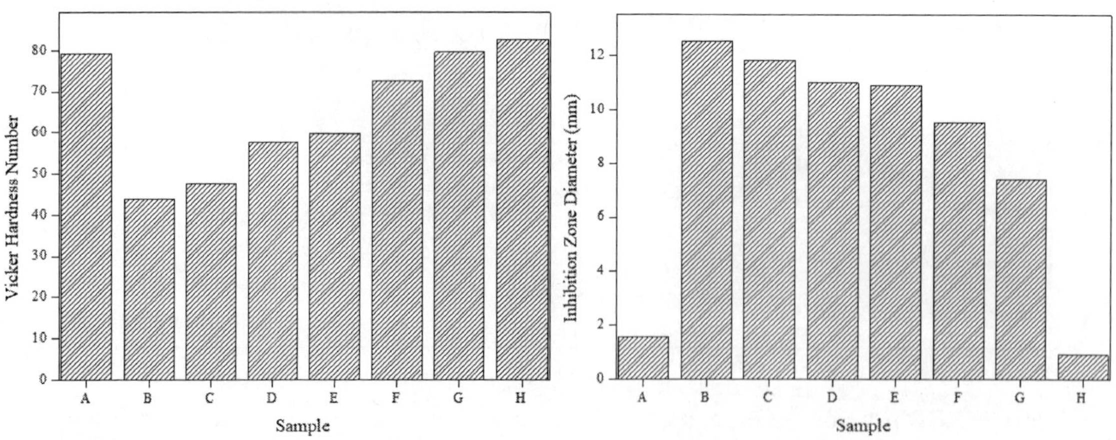

(a) (b)

Figure 6. (a) Mechanical strength and (b) antibacterial characteristic of impregnated-activated-zeolite/alumina/GIC composites.

Other reports showed that the concentration of silver nanoparticle reinforcement in the system of Ag/GIC increased the net setting time and the compressive strength over 30% larger than that of pure GIC. In terms of antibacterial performance, GIC reinforced Ag nanoparticle has significantly induced *escherichia coli* and *streptococcus mutans* growth inhibition zones due to the Ag^+ ions diffusion [10]. In general, all samples perform different mechanical and antibacterial characteristics. Fig 6 reveals the VHN of pure GIC is near 80, similar with sample G and H. However, the pure GIC shows a severe antibacterial property, its inhibition zone diameter is below 2 mm, identical to sample H. Therefore, sample H is not recommended as a proper dental restorative material. On the other hand, sample B and C perform a relatively excellent inhibitive feature, but their VHNs are relatively low. Sample D and E have HVN values above 50 (below sample F), and the inhibition zone diameter is around 10 mm, slightly higher than that of sample F. Hence, the incorporation of alumina and impregnated activated zeolite enhance the GIC characteristic, particularly the mechanical and antibacterial properties.

4. Conclusion

To conclude, a dental restorative material from impregnated activated zeolite/alumina/GIC composites have been successfully prepared. The prepared composites comprise of crystalline phases of mordenite ($Ca_3Al_7Si_{40}O_{96}(H_2O)_{31}$), quartz ($SiO_2$), sanidine ($K(AlSi_3)O_8$), orthoferrosilite ($FeSiO_3$) tolbachite ($CuCl_2$), and corundum-$Al_2O_3$; and amorphous phase of GIC polymer. The microstructural analysis reveals that the alumina and GIC lead to denser composite-particulates. In addition, the mechanical (in terms of Vickers hardness) and antibacterial (in terms of inhibitive performance) characteristics of GIC after $CuCl_2$ impregnated activated zeolite and alumina addition are feasible to be applied as in dentistry.

References

[1] Ozdemir D 2014 Dental Caries and Preventive Strategies Journal Of Educational And Instructional Studies In The World
[2] Peneva M 2007 Dental Caries–Disturbed Balance of the Risk Factors Journal of IMAB **13**
[3] Kidd E A M and Smith B G N 2000 Manual Konservasi Restoratif Menurut Pickard Alih bahasa: Narlan Sumawinata. Ed **6**
[4] De Caluwé T, Vercruysse C W J, Fraeyman S and Verbeeck R M H 2014 The influence of particle size and fluorine content of aluminosilicate glass on the glass ionomer cement properties Dental Materials **30** 1029–1038
[5] Bertenshaw B W and Combe E C 1972 Studies on polycarboxylates and related cements: I. Analysis of cement liquids Journal of dentistry **1** 13–16
[6] Lai G-Y, Zhu L-K, Li M-Y and Wang J 2013 An in vitro study on the secondary caries-prevention properties of three restorative materials The Journal of prosthetic dentistry **110** 363–368
[7] Iqbal N, Kadir M A, Iqbal S, Razak S I A, Rafique M S, Bakhsheshi-Rad H R, Idris M H, Khattak M A, Raghavendran H R B and Abbas A A 2016 Nano-hydroxyapatite reinforced zeolite ZSM composites: A comprehensive study on the structural and in vitro biological properties Ceramics International **42** 7175–7182
[8] Gultom F, Wirjosentono B,Thamrin, Nainggolan H, Eddiyanto 2015 Preparation and characterization from natural zeolite sarulla of North Sumatera polyurethane nanocomposite foams Chemistry and Materials Research **7**(10) 20-28
[9] Irnawati D, Agustiono P and Wardhani E H 2010 The Influence Of Cu Concentration On Cu-Zeolite To Antibacterial Power In Streptococcus Mutans Jurnal Zeolit Indonesia **9** 47–53
[10] Paiva L, Fidalgo T K S, da Costa LP, Maia L C, Balan L, Anselme K, Ploux L, Thiré R M S M 2018 Antibacterial properties and compressive strength of new one-step preparation silver nanoparticles in glass ionomer cements (NanoAg-GIC) Journal of Dentistry **69** 102-109

Preparation and Structural Characterization of Nanosized PVA/Fe_3O_4 Fibers

Maqrifatun Ainun Jannah[1], Ahmad Taufiq[1,*], Nurul Hidayat[1], Elsa Nuraini Usdiana[1], Sunaryono[1], Nur Aini Gama Lestari[1], Abdulloh Fuad[1], Sugiharto[2], Titik Istirohah[1]

[1]Department of Physics, Faculty of Mathematics and Natural Sciences, Universitas Negeri Malang, Jl. Semarang 5 Malang 65145, Indonesia
[2]Department of Sports Science, Faculty of Sports Science, Universitas Negeri Malang, Jl. Semarang 5 Malang 65145, Indonesia.

*Corresponding author's email: ahmad.taufiq.fmipa@um.ac.id

Abstract. Nowadays, nanosized Fe_3O_4/PVA fibers become promising materials in various advanced applications. Therefore, preparation and characterization become the essential aspects before applying the nanosized Fe_3O_4/PVA fibers in specific targets. In this work, we reported the preparation of the nanosized Fe_3O_4/PVA fibers from iron sand using co-precipitation and electrospinning techniques. The data analysis of the XRD data for the Fe_3O_4 particles presented a single phase of magnetite in cubic structure and crystallized in nanometric scale. The data of the SEM/EDAX characterizations revealed that the Fe_3O_4 nanoparticles have already been joined into the PVA fibers. In conclusion, the experimental result showed that the higher the Fe_3O_4 nanoparticles that enter into the PVA, the less the size of the fibers.

Keywords: Fe_3O_4, nanofiber; PVA, nanoparticles, Indonesian iron sand.

1. Introduction

Studies on magnetic nanomaterials have been exclusively conducted in various aspects, from the synthesis approaches to the functional applications. In particular, magnetic-based nanofibers preparation follows a line of investigation, either in laboratory scale or industrial scale, since they perform excellent characteristics in terms of thermo-mechanical and chemical stabilities [1]. Accordingly, numerous efforts have been completed to produce nanofiber materials with various shapes, sizes, and functions.

Considering the nanofibers-based nanomaterial applications, Fe_3O_4 nanofiber is one of the magnetic materials that has been extensively developed in the past few years, mainly related to its promising application as a smart magnetic material. The smart magnetic feature of Fe_3O_4 nanofiber is driven from its ability to be easily controlled under the influence of external magnetic fields and its adjusting superiority to external magnetic fields. At the same time, the selection of smoothly controlled biocompatible matrices is another research interest in producing nanosized Fe_3O_4 fibers. From many experimental results, polyvinyl alcohol (PVA) is found to meet the criteria as the biocompatible matrix for Fe_3O_4 nanofibers [2]. The demanding task is to reduce the synthesis expense of nanosized Fe_3O_4/PVA. Therefore, we introduce the use of Indonesian iron sand to produce highly crystallized magnetite-Fe_3O_4 as the reinforcement of Fe_3O_4/PVA nanometer-sized fibers.

There are numerous well-known synthesis approaches of Fe_3O_4/PVA composites such as template-assisted [3], self-assembly [4], phase separation [5], and electrospinning [6,7]. Above all, electrospinning has extra advantages because it is simple, practical, and efficient [7]. The electrospinning method can produce nanofibers on a large scale, and dimeter of the fiber can be controlled in various sizes [8]. Such method is also able to prepare materials directly from a polymer solution especially in the form of nonwoven fibers [9]. In this present study, we describe the electrospinning-based Fe_3O_4/PVA nanofibers from Indonesian iron sand. To complete our knowledge, fundamental investigation, particularly on their structural characteristics, is also reported.

2. Materials and Methods

The black and shiny Fe_3O_4 powder extraction from the natural iron sand taken from Pasur Beach Blitar, Indonesia was reported elsewhere [10]. The purified iron sand was initially immersed in HCl at room temperature to obtain a solution containing $FeCl_3$ and $FeCl_3$ prior to NH_4OH titration. The pH of the solution was controlled to normal condition to acquire pure Fe_3O_4 from the drying process for an hour. The pure Fe_3O_4 powders, with mass variations of 0.100 g (named as sample A), 0.125 g (named as sample B), and 0.150 g (named as sample C), were then homogeneously reacted with tetra-methyl ammonium hydroxide (TMAH) through a sonication process. PVA solution with 10% concentration was mixed with the Fe_3O_4 powders and sonicated for 3 hours. In the electrospinning method, PVA/Fe_3O_4 solution was put into an injection of 0.8 mm in diameter with a capacity of 10 mL at an injection rate of 10 μL/min. In this work, the collector was covered with aluminum foil set at an electric field at 20 kV/15 cm for 20 minutes to obtain nanofibers. The formed nanofibers were left for 24 hours to produce homogenous nanosized Fe_3O_4/PVA fibers. Finally, the samples were characterized by FTIR (Fourier transform infrared) spectrometer, EDX (Energy Dispersive X-Ray) spectroscopy, SEM (Scanning Electron Microscopy), and XRD (X-Ray Diffractometer) to respectively study the functional groups, size, the composition of constituent elements, morphology of particles, and their crystallographic characteristics.

3. Results and Discussion

3.1. Functional Groups of Nanosized Fe_3O_4/PVA Fibers

The characterization of functional groups of nanosized Fe_3O_4/PVA fibers was carried out using a Fourier transform infrared spectrometer with electromagnetic radiation in the range of 400 – 4000 cm^{-1}. The absorption peaks in various wavenumbers representing the functional groups of nanosized Fe_3O_4/PVA fibers are shown in Figure 1. The Fe^{3+}-O^{2-} and Fe^{2+}-O^{2-} ionic interactions, indicating the Fe_3O_4 formation, are detected at 424 cm^{-1} and 545 cm^{-1}, respectively. The absorption peaks at 934 cm^{-1} and 815 cm^{-1} indicate the O-H and C-C functional groups as reported by Shehap and Akil [11] and Marand et al. [12]. Furthermore, the C-O interaction is found at 1097 cm^{-1} [13]. Infrared absorption peaks at 1255 cm^{-1}, 1373 $cm^{-1,}$ and 1587 cm^{-1} refer to the mixture of CH and OH bindings. The C=O stretching is depicted by the absorption at 1732 cm^{-1} [11]. Meanwhile, the presence of CH_2 and OH molecules are respectively labeled by the absorptions near 2920 cm^{-1} and 3310 cm^{-1} [14].

3.2. Crystallographic Characteristics of Nanosized Fe_3O_4/PVA Fibers

Figure 2 shows the XRD patterns of the pure Fe_3O_4 nanoparticles and the composites of Fe_3O_4/PVA nanofibers. It reveals that for the XRD pattern of the pure Fe_3O_4 nanoparticles, only the magnetite spinel cubic structure is formed as the crystalline phase. The diffraction peaks of the magnetite refer to the Bragg planes of (220), (311), (400), (422), (511) and (440). The sequence of 2-theta positions and their corresponding intensities is perfectly matched with the Fe_3O_4 powder diffraction data (JCPDS code: 19-0629). Our Fe_3O_4-based XRD patterns show broader peaks than that of the JCPDS 19-0629 data, implying that we have a smaller crystalline size [15]. During the composite preparation, PVA plays an essential role in the formation of Fe_3O_4 by increasing the viscosity system and preventing the reactive species diffusion. Therefore, the decrease of crystallinity of Fe_3O_4 is reduced originating from PVA addition [16]. In this work, the particle size of the Fe_3O_4 powder is 21.54 nm, calculated from the full width at half maximum diffraction peak at (311) using the Scherrer equation [17].

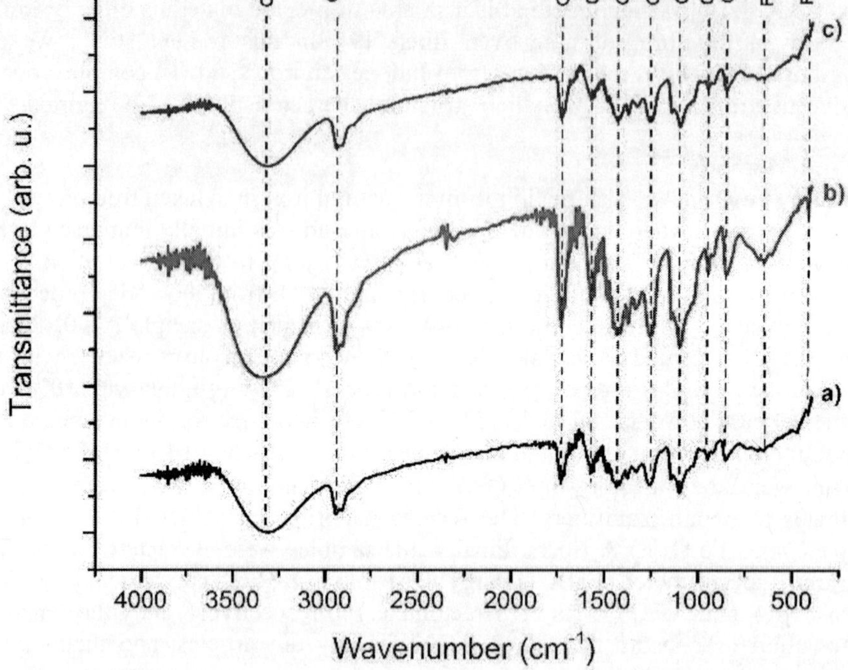

Figure 1. IR patterns of nanosized Fe_3O_4/PVA fibers: a) sample A, b) sample B, and c) sample C.

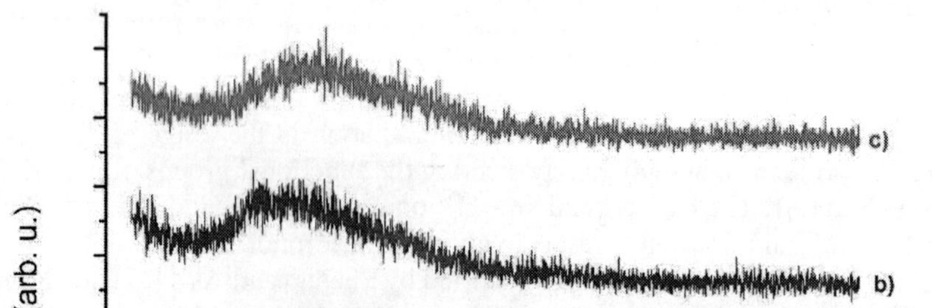

Figure 2. XRD patterns of pure Fe_3O_4 nanoparticle and nanosized Fe_3O_4/PVA fibers: a) sample A, b) sample B, and c) sample C.

In addition, PVA is a polymer showing an amorphous structure [18]. In the electrospinning process, polymer solutions are forced through the nozzle under a high electrical charge, and they are oriented towards the nanofiber axis. This leads to better molecular dispersion and consequently increases the molecular crystallization. The composites XRD patterns, in this study, show amorphous phases. There are no Fe_3O_4 diffraction peaks found in the XRD patterns. It is due to the domination of PVA solutions in the nanosized Fe_3O_4/PVA composites.

3.3. Morphological and Elemental Characteristics of Nanosized Fe_3O_4/PVA Fibers

The SEM-captured morphological characteristics of the produced composites are depicted in Figure 3. Although Fe_3O_4 agglomerations on the surface of the composites, all the SEM images have confirmed the formations of nano-sized Fe_3O_4/PVA fibers with the fiber sizes of 181.9 nm, 169.6 nm, and 151.1 nm respectively for sample A, B, and C. The nano-sized Fe_3O_4/PVA fibers produced in this study are relatively smaller than other reports. The electrospinning synthesis of nano-sized Fe_3O_4/PVA fibers prepared by other methods was reported to have an average diameter around 400 nm [1] and 230 nm [6]. In the electrospinning process, polymer concentration, applied voltage, and plate size are very crucial to prepare fine fibers. In addition, controlling morphology with the number of nanofiber granules is also vital [19, 20].

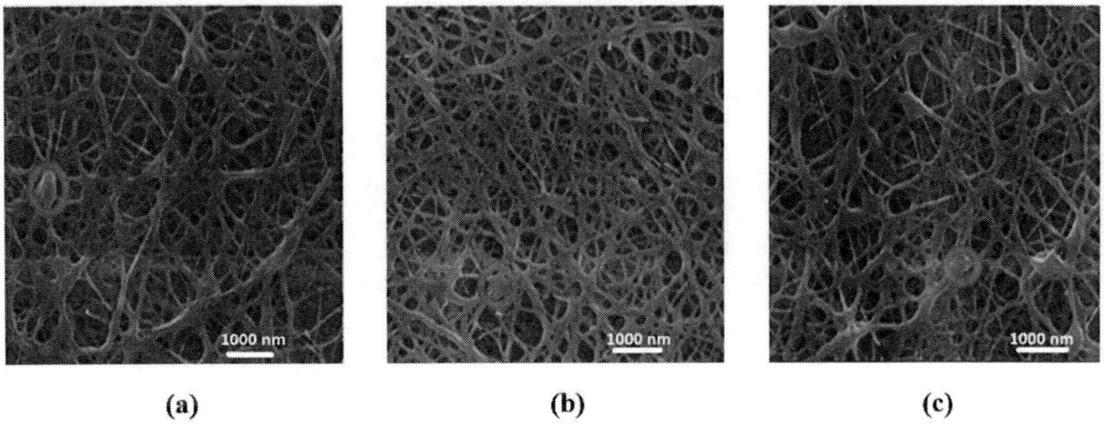

Figure 3. SEM images and selected area (as marked in the circles) for EDX characterization of (a) sample A, (b) sample B, and (c) sample C.

Table 1. EDX-based elemental content of nanosized Fe_3O_4/PVA fibers in the selected area of the SEM images (as marked in the circles).

Sample A			Sample B			Sample C		
Element	**wt%**	**at%**	**Element**	**wt%**	**at%**	**Element**	**wt%**	**at%**
C	13.73	21.93	C	27.95	42.84	C	22.08	33.10
O	36.18	43.39	O	20.22	23.27	O	33.76	38.00
Na	08.89	07.42	Na	06.97	05.58	Na	08.70	06.82
Mg	02.92	02.30	Mg	01.93	01.46	Mg	02.38	01.76
Si	32.43	22.15	Si	36.36	23.83	Si	28.39	18.20
Ca	05.84	02.80	Ca	06.58	03.02	Ca	04.70	02.11

From EDX characterization, it is known that there is no Fe content in the selected SEM images (red circles). However, from the FTIR data, Fe-O functional groups are detected. This "contradiction" is probably because (a) the PVA is too dominating so that Fe_3O_4 nanoparticles in the red circle area are found to be extremely small or (b) the grain size of Fe_3O_4 nanoparticles is too small [21]. Based on such result, it becomes a new challenge for developing a new method by employing natural iron sand as the main precursor in producing the nano-sized Fe_3O_4/PVA fibers with a high homogeneity to be applied in various fields.

4. Conclusion
In conclusion, reducing the synthesis expense of nanosized Fe_3O_4/PVA fibers can be acquired by extracting Fe_3O_4 nanoparticles from Indonesian iron sand, precisely from Pasur Beach Blitar. The particle size of the as-prepared Fe_3O_4 is 21.54 nm. Combining the Fe_3O_4 with PVA by means of electrospinning approach leads to diminishing the crystalline phases in the system, i.e., no diffraction peaks are detected in the nanosized Fe_3O_4/PVA fibers. The existence of Fe_3O_4 nanoparticles, though it was not revealed by the XRD data, was confirmed by the FTIR spectra especially from the Fe^{3+}-O^{2-} and Fe^{2+}-O^{2-} interactions at the respective wavenumber at 424 cm^{-1} and 545 cm^{-1}. Finally, the synthesized of nanosized Fe_3O_4/PVA fibers have an average diameter of 167.6 nm.

Acknowledgments
This work is supported by KEMENRISTEKDIKTI RI for AT.

References
[1] Beachley V and Wen X 2009 Effect of electrospinning parameters on the nanofiber diameter and length *Mater. Sci. Eng. C* **29** 663–668
[2] Wang S, Sun Z, Yan E, Yuan J, Gao Y, Bai Y, Chen Y, Wang C, Zheng Y and Jing T 2014 Magnetic composite nanofibers fabricated by electrospinning of Fe_3O_4/gelatin aqueous solutions *Mater. Sci. Eng. B* **190** 126–132
[3] Rambo C R, Recouvreux D O S, Carminatti C A, Pitlovanciv A K, Antônio R V and Porto L M 2008 Template assisted synthesis of porous nanofibrous cellulose membranes for tissue engineering *Mater. Sci. Eng. C* **28** 549–554
[4] Chang G, Zhu X, Li A, Kan W, Warren R, Zhao R, Wang X, Xue G, Shen J and Lin L 2016 Formation and self-assembly of 3D nanofibrous networks based on oppositely charged jets *Mater. Des.* **97** 126–130
[5] Zhang H, Liu X, Yang M and Zhu L 2015 Silk fibroin/sodium alginate composite nano-fibrous scaffold prepared through thermally induced phase-separation (TIPS) method for biomedical applications *Mater. Sci. Eng. C* **55** 8–13
[6] Wang S, Wang C, Zhang B, Sun Z, Li Z, Jiang X and Bai X 2010 Preparation of Fe_3O_4/PVA nanofibers via combining in-situ composite with electrospinning *Mater. Lett.* **64** 9–11
[7] Xingxing C, Xuebin Z, Hao S and Yi F 2014 Fabrication of Magnetic Fe_3O_4 Nanotubes by Electrospinning *Rare Met. Mater. Eng.* **43** 2330–2334
[8] Fang J, Wang X and Lin T 2011 *Functional applications of electrospun nanofibers* (InTech–Open Access Publisher)
[9] Tunák M, Antoch J, Kula J and Chvojka J 2014 Estimation of fiber system orientation for nonwoven and nanofibrous layers: local approach based on image analysis *Text. Res. J.* **84** 989–1006
[10] Taufiq A, Rachman Putra E G, Pratapa S and others 2015 Nano-structural studies on Fe_3O_4 particles dispersing in a magnetic fluid using X-ray diffractometry and small-angle neutron scattering *Materials Science Forum* vol 827 (Trans Tech Publ) pp 213–218
[11] Shehap A M and Akil D S 2016 Structural and optical properties of TiO_2 nanoparticles/PVA for different composites thin films *Int. J. Nanoelectronics and Materials* **9** 17-36
[12] Marand Z R, Farimani M H R and Shahtahmasebi N 2014 Study of magnetic and structural and optical properties of Zn doped Fe_3O_4 nanoparticles synthesized by co-precipitation method for biomedical application *Akush. Ginekol. (Sofiia)* **15** 238–47
[13] Zucchiatti P, Mitri E, Kenig S, Billè F, Kourousias G, Bedolla D E and Vaccari L 2016 Contribution of Ribonucleic Acid (RNA) to the Fourier Transform Infrared (FTIR) Spectrum of Eukaryotic Cells *Anal. Chem.* **88** 12090–8

[14] Khandanlou R, Ahmad M, Shameli K and Kalantari K 2013 Synthesis and Characterization of Rice Straw/ Fe_3O_4 Nanocomposites by a Quick Precipitation Method *Molecules* **18** 6597–607
[15] Yee W A, Nguyen A C, Lee P S, Kotaki M, Liu Y, Tan B T, Mhaisalkar S and Lu X 2008 Stress-induced structural changes in electrospun polyvinylidene difluoride nanofibers collected using a modified rotating disk *Polymer* **49** 4196–203
[16] Jayakrishnan P and Ramesan M T 2017 Synthesis, Characterization, Electrical Conductivity and Material Properties of Magnetite/Polyindole/Poly(vinyl alcohol) Blend Nanocomposites *J. Inorg. Organomet. Polym. Mater.* **27** 323–33
[17] Zhang P, Wang B, Williams G R, Branford-White C, Quan J, Nie H and Zhu L 2013 Self-assembled core-shell Fe_3O_4@SiO_2 nanoparticles from electrospun fibers *Mater. Res. Bull.* **48** 3058–64
[18] Fattah A A A, Elfadly A M, Naggar A Y, Ebiad M A and A. Mandor M 2013 Physicochemical caracterization of poly (ethylene glycol) with molecular weights (8000, 12000) by inverse gas chromatography *Int. J. Acad. Res.* **5** 226–31
[19] Nuryantini A Y, Ekaputra M P, Munir M M, Suciati T and others 2014 Sintesis nanoserat poli (vinil alkohol) dalam bentuk lembaran dengan pemintal elektrik multi nozel dan kolektor drum *J. Pendidik. Fis. Indones.* **10** 186–193
[20] Zheng J-Y, Zhuang M-F, Yu Z-J, Zheng G-F, Zhao Y, Wang H and Sun D-H 2014 The Effect of Surfactants on the Diameter and Morphology of Electrospun Ultrafine Nanofiber *J. Nanomater.* **2014** 1–9
[21] Singh G, Jalandhara D and Yadav K 2016 Effect of grain size on optical properties of iron oxide nanoparticles p 020409

Crystalline Phase Analysis and Thermal Expansion Coefficient Calculation of Quartz-Sand/Corundum Composites

Nurul Hidayat[1,*], Ahmad Taufiq[1], Sunaryono[1], Suminar Pratapa[2]

[1]Department of Physics, Faculty of Mathematics and Natural Sciences, Universitas Negeri Malang, Jl. Semarang 5 Malang 65145, Indonesia
[2]Department of Physics, Faculty of Science, Institut Teknologi Sepuluh Nopember, Jl. Arif Rachman Hakim, Surabaya 60111, Indonesia

*Corresponding author: nurul.hidayat.fmipa@um.ac.id

Abstract. Quartz-sand/corundum composites have been successfully produced by a simple solid-state reaction approach. The composites with corundum weight fractions of 10%, 20%, and 30% were prepared through grinding mixtures of purified Indonesian quartz sand and commercial high purity corundum for a half-hour with an addition of 3% polyvinyl alcohol. The composites were shaped into discs and pressed prior to the sintering at 950 °C and 1050 °C to form compact ceramic composites. The crystalline phase verification within the samples was conducted by means of X-ray diffraction (XRD). The coefficient of thermal expansion (CTE) of the composites was semi-theoretically evaluated, after being taken into consideration the XRD data analysis results, by means of Halpin-Tsai formula. The detected major crystalline phases within the sintered samples, based on the XRD profiles analysis, were quartz-SiO_2 and corundum-Al_2O_3. Meanwhile, the minor crystalline phases were wollastonite-$CaSiO_3$ (below 7%) for all samples and cristobalite-SiO_2 (around 1%) for the composite with 1050 °C sintering temperature. Generally, the addition of corundum into the quartz-sand/corundum composites reduced the CTE, having the value from 10.5 to 11.5 ppm/°C.

Keywords: Crystalline phase, coefficient of thermal expansion, quartz-sand, corundum.

1. Introduction

The advancement in materials science and engineering has encouraged researchers to investigate both fundamental and practical features of materials. The phase compositions and thermo-physical characteristics of materials play an essential role when designed as thermal environmentally responsive parts. In particular, SiO_2/Al_2O_3-based composites have been intensively studied and applied in various applications. SiO_2/Al_2O_3 nanocomposites have been combined with polyalkylene glycol lubricant and exposed to have excellent thermal conductivity and viscosity for the refrigeration system at the temperature range of 30 °C – 80 °C [1]. SiO_2/Al_2O_3 composites performed stronger acidity than that of pure SiO_2 or pure Al_2O_3 when they were applied as the support of Rh-Pd catalysts in working temperature of 250 °C – 450 °C [2]. SiO_2/Al_2O_3 was also reported as the primary factor on the effectivity of coal ash sintering and slagging at very high temperatures, i.e., 1500 °C [3] and 1550 °C [4]. In fuel cell technology, SiO_2/Al_2O_3-based sealants have been widely developed to optimize the fuel cell performance [5].

Herein, what makes it distinctive from those reported studies; we prepared SiO_2/Al_2O_3-based ceramic composites from naturally occurring minerals, i.e., Indonesian quartz sand. It was carried out to (1) reduce the synthesis expense of the beneficial composites and (2) escalate the added value of Indonesian quartz sand. We focused on the crystalline phase compositional analysis of the as-prepared quartz-sand/corundum (SiO_2/Al_2O_3) using a well-known Rietveld method. In addition, a theoretical model using the rule of mixtures based on Halpin-Tsai formula [6], by virtue of its significant consideration of constituent geometry and anisotropy, was preferred to use for calculating the composites CTE values rather than the Turner model [7] which assumes that only hydrostatic stresses occur in the crystalline phases or Kernel model [8] which accounts for phase isostatic and shear stresses.

2. Materials and Methods

The detail purification of Indonesian quartz sand (in which contains 97 97.8 (18)% quartz and 2.2(2)% calcite) was reported elsewhere [9]. The quartz-sand/corundum composites were prepared via sintering of quartz sand and corundum mixtures. 3% of polyvinyl alcohol was introduced to the mixtures to effectively bind the powder during pressing (with a force of 600 N to form a cylindrical sample with a diameter of 1.3(1) cm). The prepared quartz-sand/corundum pellets were then sintered at 950 °C and 1050 °C for 4 h to form ceramic composites. XRD test, using Cu-Kα radiation, was conducted to investigate the crystalline phase in the composites. After accurate phase identification, crystalline phase quantification was run by means of Rietica software [10] using the Rietveld method [11]. Furthermore, the CTE values of the sintered quartz-sand/corundum composites were calculated by means of Halpin-Tsai formula [6].

3. Results and Discussion

The XRD profiles of quartz-sand/corundum powders prior to sintering are depicted in Fig. 1. The major phases of the samples, after search-match analysis, are hexagonal quartz-SiO_2 (PDF reference code: 01-082-0511) and rhombohedral corundum-Al_2O_3 (PDF reference code: 01-081-2267). A very few and small peaks refer to rhombohedral calcite-$CaCO_3$ (PDF reference code: 00-001-0837) are also detected. The calcite phase is the remaining impurity in the purification of quartz from natural silica sand. Clearly, the Bragg peak intensities of corundum, particularly (102), (104), (110), and (113) are higher as the addition of corundum into the mixtures.

The phase weight fractions produced from the acceptable Rietica-assisted Rietveld refinements ($\chi^2 < 4$) are listed in Table 1. The table quantifies the exact values of the weight phase percentages. Calcite does not influence the corundum weight fractions since it already exists in the purified quartz-based silica sand. Phase quantification in the quartz-sand/corundum composites is crucial to predict the coefficient of thermal expansion values of the composites and to investigate the chemical reactions between quartz and corundum. The phase quantification was conducted by Rietveld analysis using Rietica software.

The successful XRD data refinements have been achieved for all XRD data, and the sample plots of the refinement outputs for QS20A950 (the composite with 20% corundum and sintering temperature of 950 °C) and QS20A1050 (the composite with 20% corundum and sintering temperature of 1050 °C) are respectively given in Fig. 2 and in Fig. 3. The measured and calculated XRD profiles are respectively visualized by the plus (+) sign and solid line. The difference of the refinement is represented by the green line.

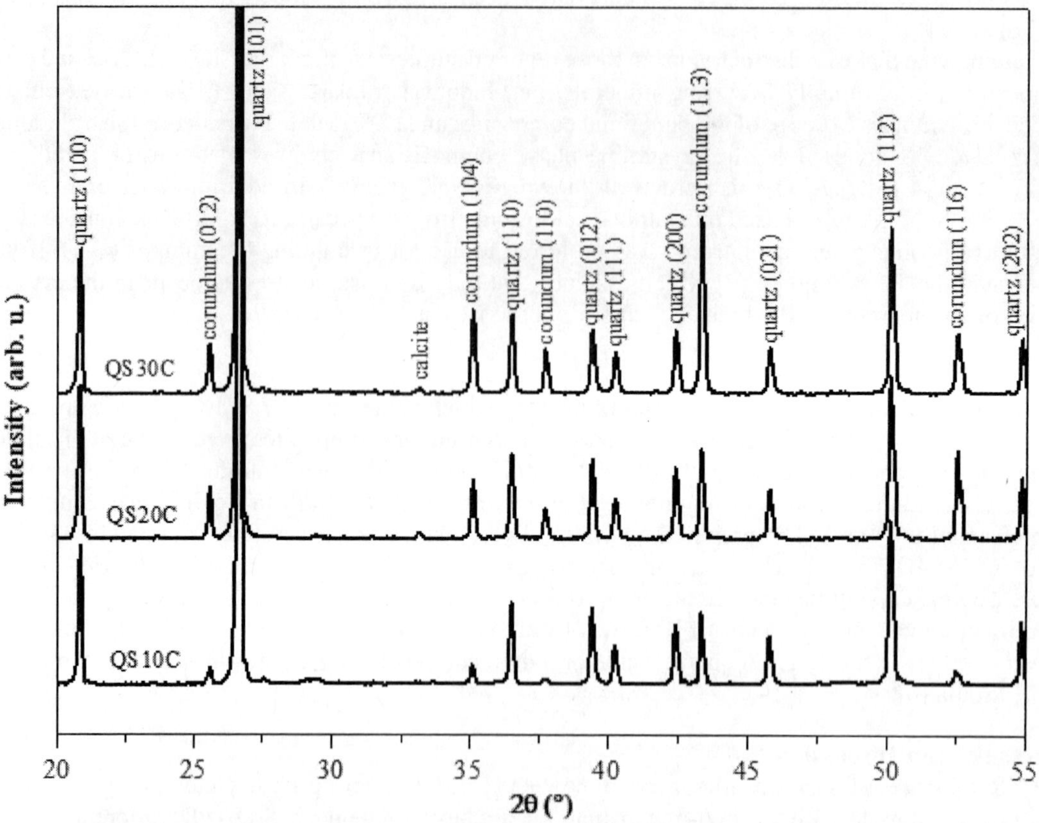

Figure 1. XRD profiles of pre-sintered quartz-sand/corundum mixtures. QS10C, QS20C, and QS30C represent quartz-sand/corundum mixtures with corundum weight fractions of 10%, 20%, and 30%.

Table 2 represents the crystalline phase quantification acquired from Rietveld analysis. From Table 2, the sintered composites contain quartz-SiO_2, corundum-Al_2O_3, and wollastonite-$CaSiO_3$. Cristobalite-SiO_2 grows at the sintering temperature of 1050 °C. The growth of cristobalite-SiO_2 in the 1050 °C sintered composites is due to the phase transformation of silica (from quartz to cristobalite) above 1000 °C [12]. Since the attachment of corundum reduces SiO_2 content, then the descent of cristobalite-SiO_2 becomes a chemical consequence. However, the drop of cristobalite content in this study is not drastic.

Table 1. Crystalline phase composition of pre-sintered quartz-sand/corundum mixtures.

Sample	Relative weight fraction (%)		
	Quartz (SiO_2)	Corundum (Al_2O_3)	Calcite ($CaCO_3$)
QS10C	86.9(29)	10.6(12)	2.5(7)
QS20C	79.2(15)	19.5(5)	1.3(3)
QS30C	68.3(17)	30.5(17)	1.2(3)

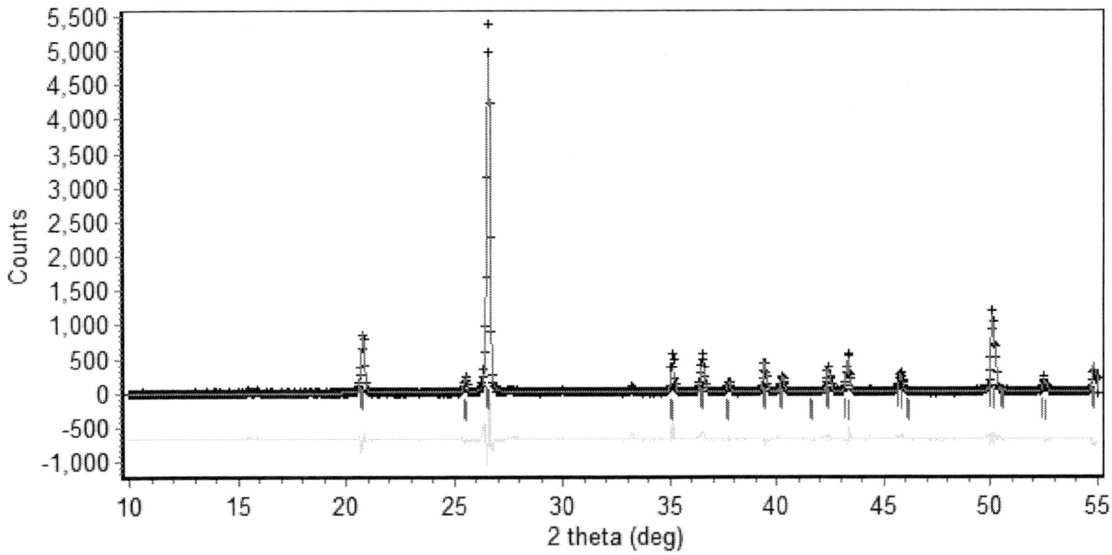

Figure 2. Plot of rietica-based rietveld refinement result for QS20C950.

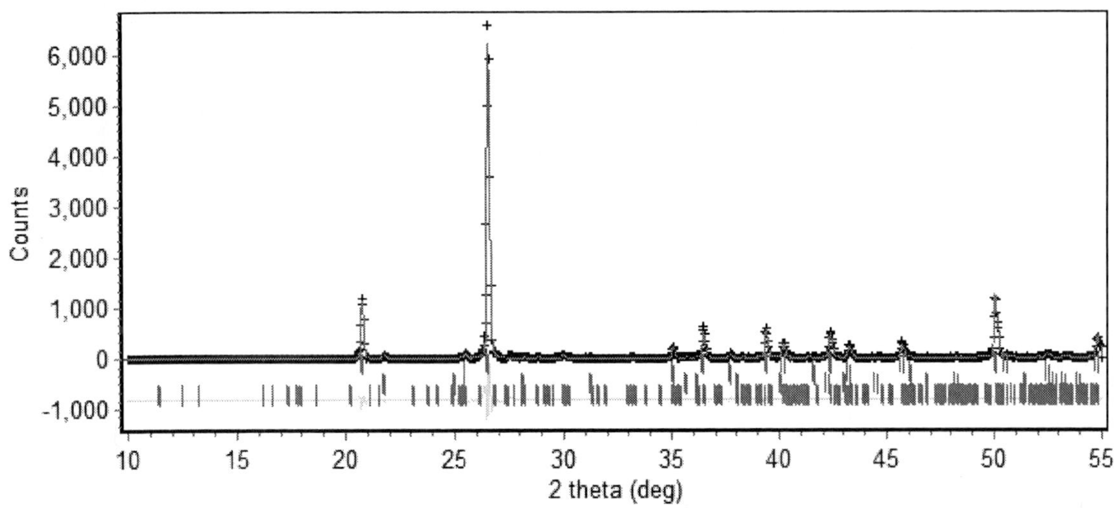

Figure 3. The plot of rietica-based rietveld refinement result for QS20C1050.

Another "new" phase that is possibly formed in the quartz-sand/corundum composites is mullite. It is a rare silicate mineral produced from a solid solution of SiO_2 and Al_2O_3 which can be in the form of $2Al_2O_3{\cdot}3SiO_2$ or $2Al_2O_3{\cdot}SiO_2$. From the SiO_2-Al_2O_3 phase diagram [13], the solidus line of mullite changes as the addition of Al_2O_3 above the silica-mullite eutectic temperature. The nucleation and growth of mullite occur above 1587 °C. On the other hand, quartz-SiO_2 is totally transformed to cristobalite-SiO_2 at 1570 °C [12]. Prior to mullite precipitation, the SiO_2-Al_2O_3 system may produce liquid phase formation after heating in the contact area of SiO_2 and Al_2O_3. Thus, it makes sense if the XRD data do not reveal the presence of mullite in our composites.

In addition, Table 2 shows the absence of calcite-$CaCO_3$ and the presence of wollastonite-$CaSiO_3$. It is important to note that $CaSiO_3$ is not the chemical reaction product between SiO_2 and $CaCO_3$, but between SiO_2 and CaO. The CaO was initiated from $CaCO_3$ decomposition around 900 °C [14]. SiO_2 that reacts with CaO is quartz because there is a decrease of quartz as the cristobalite phase grows.

Table 2. Crystalline phase composition of sintered quartz-sand/corundum ceramics.

No	Sample	Relative weight fraction (%)			
		Quartz (SiO_2)	Corundum (Al_2O_3)	Wollastonite ($CaSiO_3$)	Cristoballite (SiO_2)
1	QS950	95.2(26)	-	4.8(4)	-
2	QS10A950	83.7(24)	10.0(10)	6.3(5)	-
3	QS20A950	76.0(26)	20.4(14)	3.6(5)	-
4	QS30A950	66.1(17)	31.9(17)	2.0(4)	-
5	QS1050	93.2(27)	-	6.8(4)	-
6	QS10A1050	86.5(26)	9.2(11)	3.0(3)	1.2(1)
7	QS20A1050	72.0(22)	25.2(12)	1.7(2)	1.0(1)
8	QS30A1050	63.6(22)	34.4(19)	1.0(2)	1.0(1)

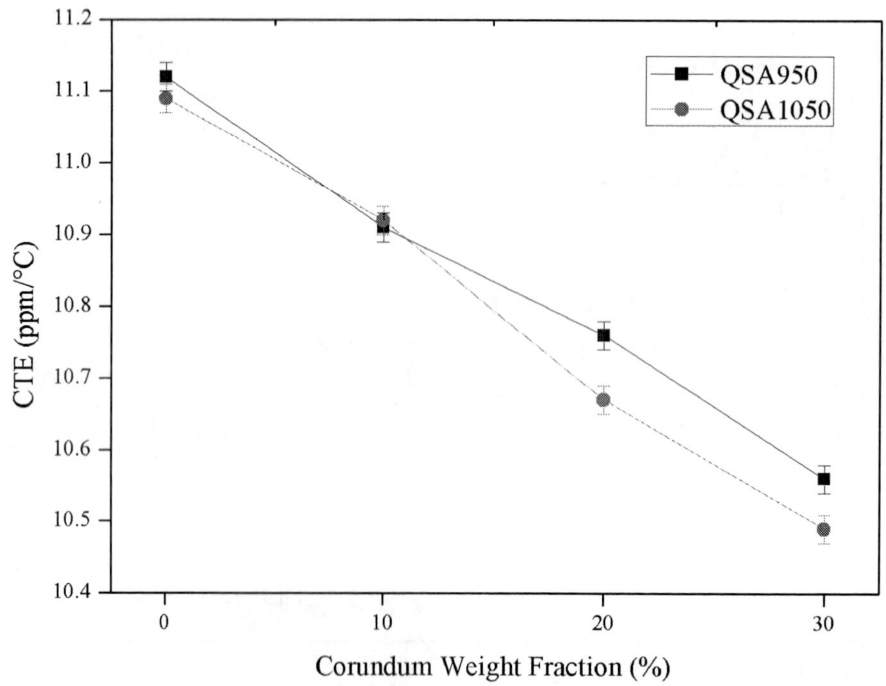

Figure 4. CTE values of quartz-sand/corundum composites calculated by Halpin-Tsai formula.

Furthermore, CTE calculation of the quartz sand and the composites are displayed in Fig. 4. Investigating crystalline phase composition in a composite is essential for predicting the composite characteristics, such as CTE, and for fundamental knowledge to develop new materials. Fig. 4 evidently displays that (1) the CTE values of the quartz-sand/corundum composites lie between 10.5-11.5 ppm/°C which is similar with quartz-sand/periclase/forsterite composites [15] and (2) addition of corundum, as well as sintering temperature, to the composites reduces the CTE values. When the temperature exceeds 1050 °C, the phase constituent within the quartz-sand/corundum-based composites become more complex, i.e., nearly 50 wt% of cristobalite-SiO_2 grows at 1250 °C [9]. The extra cristobalite-SiO_2 will decrease the CTE value of the composites since its CTE value is slightly smaller than that of quartz-SiO_2 or corundum-Al_2O_3 [16]. In terms of the possible application based on the CTE values, quartz-sand/corundum composites meet the criteria for fuel-cell seal materials.

4. Conclusion

A complete quantitative crystalline phase analysis on quartz-sand/corundum composites provides a feature to predict thermal coefficient expansion using the Halpin-Tsai equation. Quartz-SiO_2 and corundum-Al_2O_3 are found as the main phases in the composites, although minor wollastonite-$CaSiO_3$ and cristobalite-SiO_2 are also detected in the sintered composites, particularly at a sintering temperature of 1050 °C. The CTE values of the composites are between 10.5-11.5 ppm/°C and they decrease as the incorporation of corundum and higher sintering temperature.

References

[1] Zawawi N N M, Azmi W H, Redhwan A M M, Sharif M Z and Sharma K V 2017 Thermo-physical properties of Al_2O_3-SiO_2/PAG composite nanolubricant for refrigeration system Int. J. Refrigeration 80 1–10
[2] D'Ippolito S A, Especel C, Vivier L, Epron F and Pieck C L 2014 Influence of the Bronsted acidity, SiO_2/Al_2O_3 ratio and Rh-Pd content on the ring opening: part II. Selective ring opening of methycyclohexane Appl. Catal. A: Gen. 469 532–540
[3] Xuan W, Wang Q, Zhang J and Xia D 2017 Influence of silica and alumina (SiO_2 + Al_2O_3) on crystallization characteristics of synthetic coal slags Fuel 189 39–45
[4] Yan T, Bai J, Kong L, Bai Z, Li W and Xu J 2017 Effect of SiO_2/ Al_2O_3 on fusion behavior of coal ash at high temperature Fuel 193 275–283
[5] Dai Z, Pu J, Yan D, Chi B and Jian L 2011 Thermal cycle stability of Al_2O_3-based compressive seals for planar intermediate temperature solid oxide fuel cells Int. J. Hydrog. Energy 36(4) 3131–37
[6] Affdl J C H and Kardos J L 1976 The Halpin-Tsai equations: A review Polym. Eng. Sci. 16(5) 344–352
[7] Turner P S 1946 Thermal expansion stresses in reinforced plastics J. Res. NBS 37 239
[8] Kerner E H 1956 The elastic and thermoplastic properties of composite media Proc. Phys. Soc. B69 808-813
[9] Hidayat N, Istiqomah, Widianto M Y H, Taufiq A, Sunaryono, Triwikantoro, Zainuri M, Baqiya M A, Aristia G and Pratapa S 2017 Natural silica sand/alumina ceramic composites: promising candidates for fuel-cell sealants IOP Conf. Series: Materials Science and Engineering 202 012060
[10] Hunter B A 1998 Rietica-A visual Rietveld program Newsl. Int. Union Crystallogr. Comm. Powder Difraction
[11] Rietveld H M 1969 A profile refinement method for nuclear and magnetic structures J. Appl. Crystallogr. 2(2) 65–71
[12] Day A L and Shepherd E S 1906 The lime-silica series of minerals J. Am. Chem. Soc. 28(9) 1089–114
[13] Tse J S and Klug D D 1991 The structure and dynamics of silica polymorphs using a two-body effective potential model J. Chem. Phys. 95(12) 9176–85
[14] L'vov B V, Polzik L K and Ugolkov V L 2002 Decomposition kinetics of calcite: a new approach to the old problem Thermochim. Acta 390(1–2) 5–19
[15] Hidayat N, Triwikantoro, Baqiya M A and Pratapa S 2013 Thermal expansion coefficient prediction of fuel-cell seal materials from silica sand AIP Conf. Proc. 1555 99–101
[16] Carter C B and Norton M G 2007 Ceramic Materials: Science and Engineering (Springer)

Dielectric Properties of TiO_2 Nanoparticles Doped Flavonoid Extract of *Pterocarpus Indicus Willd* (PIW) Latex

M. Rynaldi Iqbal[1], Isnindar Tandya Asri[1], Harsipah[1], Sajida Rakhmah[1], Markus Diantoro[1,2,*], Yudyanto[1,2]

[1] Department of Physics, Faculty of Mathematics and Natural Science, Universitas Negeri Malang, Jl. Semarang 5 Malang 65145, Indonesia
[2] Central Laboratory of Minerals and Advanced Materials, Faculty of Mathematics and Natural Science, Universitas Negeri Malang, Jl. Semarang 5 Malang 65145, Indonesia

*Corresponding author's email: markus.diantoro.fmipa@um.ac.id

Abstract. Indonesia has many plants that contain flavonoids such as leaves or latex trunk from Angsana tree (*Pterocarpus indicus Willd*, PIW). To improve electrical properties, we can use several dopants such as protonation or introduce metal ions into the flavonoid of PIW extract. In our previous studies, we use silver nanoparticles in flavonoid films to increase the conductivity. It is rarely found the use of optical dielectric material such as TiO_2 induced in flavonoid. Flavonoid extraction from PIW latex was performed using the maceration method. Separately, we also prepared TiO_2 nanoparticle using the coprecipitation method. Meanwhile, the dissolving process of flavonoids and TiO_2 nanoparticle was carried out in an aquadest solvent with PEG 400. The FTIR spectrum shows the type of rutin. Analysis of XRD patterns showed an increase in the crystallinity of composite flavonoid-TiO_2 film along with the addition of TiO_2 nanoparticles. The SEM image shows that the TiO_2 nanoparticles are dispersed on the surface of the flavonoid-TiO_2 film. The dielectric constant of flavonoid film without TiO_2 nanoparticle dopant is 1.05×10^4. The dielectric constant with 0.05 grams of TiO_2 nanoparticle decreased to 75.8.

Keywords: Dielectric, TiO_2, Nanoparticles, Flavonoid, PIW latex

1. Introduction

Indonesia possesses various kind of plants which contain flavonoids such as latex from Angsana tree (*Pterocarpus indicus Willd*, PIW). In PIW latex includes flavones which are polyphenols of natural polymeric builders [1]. Flavonoids produced from PIW are mostly biologically derived only as a bioactive preparation [2], used in traditional medicine [3], biodiesel plants [4], and anti-microbial metabolism [5], [6]. Flavonoids possess a conjugated single, double bond on the functional group so that it can be used as a semiconductor or conducting polymer.

Conducting polymer is widely studied due to exhibits sizeable electrical conductivity, corrosion resistance, and is easy to obtain [7]. In principle, conducting polymers could be used as transistors, organic light emitters (OLED), photovoltaic devices, sensors, bioimplants, corrosion protection and electrical energy storage. To obtain good conducting polymer properties, metal ions can be added into the polymer. Metal doping is usually done to produce materials with excellent electrical properties. Inducing metal ions such as iron and copper would be an increase in crystallinity and conductivity [8].

The growth of the crystallinity will enhance the desired properties of the material [9]. Among the metal materials, Ag metal shows high electrical conductivity [10], it can be used to modify the polymer's properties to form conducting polymers [11]. The study of flavonoids of PIW latex with silver nanoparticles dopant exhibits a significant increase in its electrical conductivity from 3.22 to 542.85 S/cm [12]. In the case of silver-doped flavonoid, the film was extracted from Jatropha multifida Lin. (JML) latex showed an increase from 2.71 S/cm (0.1 M) to 745.25 S/cm (0.5 M) [12]. Thus, it possesses an excellent potential for solar cell applications, primarily as the electrode of the DSSC device [13]. We may infer that flavonoids or other types of polymers can be modified of its electrical properties by employing Ag or other metal elements. So far according to our knowledge, it is still challenging to find a report using TiO_2 as a dopant for natural flavonoids films. In the lower dimension, the dielectric constant of TiO_2 can be controlled by changing the particle size to the nanometer. It is well-known that nanometer size shows many advantages over the micrometer scale [14].

2. Materials and Methods

Flavonoid extraction was prepared using a raw material of PIW latex which was obtained from PIW trees around the campus of the Universitas Negeri Malang. We first put the latex into a petri dish then heat it over hotplate until dry (turned into powder) at a temperature of 100 °C. It was followed by weighing 5 grams of PIW latex powder subsequently, put it in a measuring cup and then mixed them with 10 mL of methanol (1: 2). It proceeded with the blending and homogenization using a magnetic stirrer for 30 minutes. The obtained solution was subjected to sonication for one hour followed by filtering the solution using filter paper and kept it for 24 hours.

2.1. Synthesis of TiO_2 Nanoparticles

The initial steps of TiO_2 nanoparticles synthesis were dissolved 10 mL of $TiCl_3$, 0.3 mL HCl, and 4.7 mL of aquadest then stirred at 360 rpm for 5 min. Furthermore, 30 mL of HCl 32% was added. During the stirring process, 180 mL of NH_3 was subjected to the purpose of precipitation until a purple solution was achieved and then turn to white. The results were kept for 24 hours. The precipitate has been formed for 24 hours, filtered using filter paper and washed using aquadest up to five times. After washing steps, the soft white sediments were obtained. The finished residue was filtered and subsequently annealing at 400 °C for 6 hours.

2.2. Synthesis of Flavonoid-TiO_2 NPs Solution

This synthesis was carried out by mixing the latex extraction and TiO_2 NPs into the aquadest solvent and PEG 400. Flavonoids were diluted with aquadest of 0.25 mL and then stirred for 15 minutes and subsequently added TiO_2 powder with PEG 400 of 0.02 mL. It was continued by a homogenizing the solution for 30 minutes at 500 rpm.

2.3. Fabrication of Flavonoid- TiO_2 NPs Film on ITO Substrate

The film of Flavonoid- TiO_2 NPs on ITO substrate was conducted by using a spin coating method. Firstly, we used the TiO_2 NPs-flavonoid solution which was dripped onto the ITO substrate. Initially, the ITO substrate was cleaned using acetone under sonication and washed using aquadest. The spin coating was performed at a rate of 1000 rpm for 30 seconds. The resulting film was dried in an oven at 230 °C for 15 minutes.

3. Results and Discussion

FTIR test results after PIW latex extraction are shown in Figure 1.

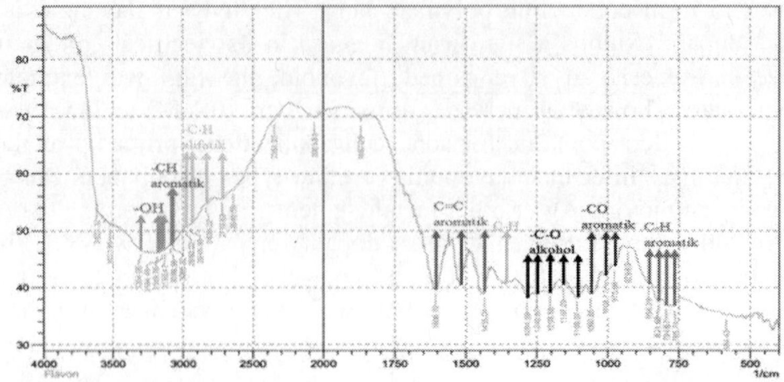

Figure 1. FTIR results from Angsana Latex Extraction

We found that from the FTIR spectra some aromatic of C-H functional groups existed, C-O aromatic, C-O alcohol, C-H, C = C aromatics, C-H aliphatic and –OH. It can be said that this flavonoid compound is included in the Rutin type. The aromatic C-H functional group at the wave number of 765.74, 794.67, 821.68, and 856.39 cm^{-1}. The peaks at wave numbers of 975.98, 1008.77, and 1060.85 cm^{-1} belong to the aromatic C-O functional group. The C-O alcohol functional groups occur at the wave numbers of 1109.07, 1157.29, 1203.58, 1249.87, and 1284.59 cm^{-1}. At a wave number of 1371.39 cm^{-1} is an indication of a functional group of C-H. For the functional group of C = C aromatics occurs at the wave numbers of 1435.04 and 1606.70 cm^{-1} whereas the functional group of CH aliphatic occurs at the wave number of 2719.63 cm^{-1}, 2839.22 cm^{-1}. The wave number 3130.47 cm^{-1} is a functional group - OH because these wave numbers appear in the range 3000-3500 cm^{-1} [15].

3.1. *XRD of Flavonoid-TiO_2*

The diffraction pattern flavonoid-TiO_2 films with a variation of TiO_2 NPs is shown in Figure 2.

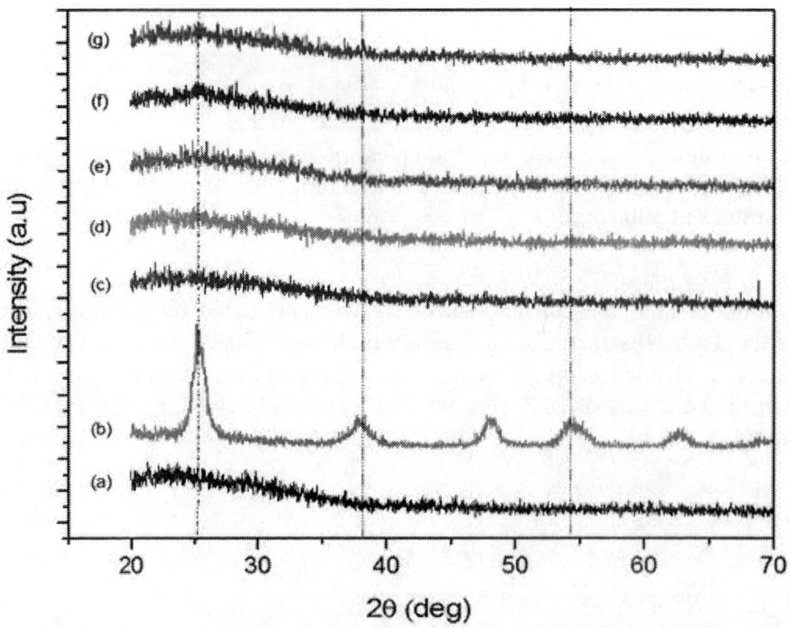

Figure 2. X-ray Diffraction Patterns of (a) Flavonoid, (b)TiO_2, and (c)-(g) Flavonoid doping TiO_2 from 0.01 gram to 0.05 gram.

Figure 2a shows the diffraction pattern of flavonoids. The diffraction pattern shows that flavonoids are included in the amorphous phase because there is no peak in the pattern [16]. Figure 2b shows the diffraction pattern of TiO_2. The phase shown is the anatase phase by matching the model data with TiO_2 data in this study. The grain size was calculated using Equation 1. The results of calculation using Equation 1 falls to 7.999 nm.

$$D = \frac{k\lambda}{\beta Cos\theta} \quad (1)$$

From XRD patterns of the flavonoid-TiO_2 film with variations of TiO_2 NPs is shown in Figure 2c-e, where no visible TiO_2 nanoparticles peak observed. Figure 2f shows that there is a weak peak of TiO_2 nanoparticles at the 2θ position of 25.356°. Figure 2(g) shows three small peaks of TiO_2 nanoparticles is observed at 2θ of 25.356°, 38.617°, and 54.010°.

3.2. *Crystallinity*
Crystallinity can be measured using XRD patterns data and analyzed using a winPLOTR of Fullprof item to create an amorphous model. We used Origin to calculate the amorphous and total area.

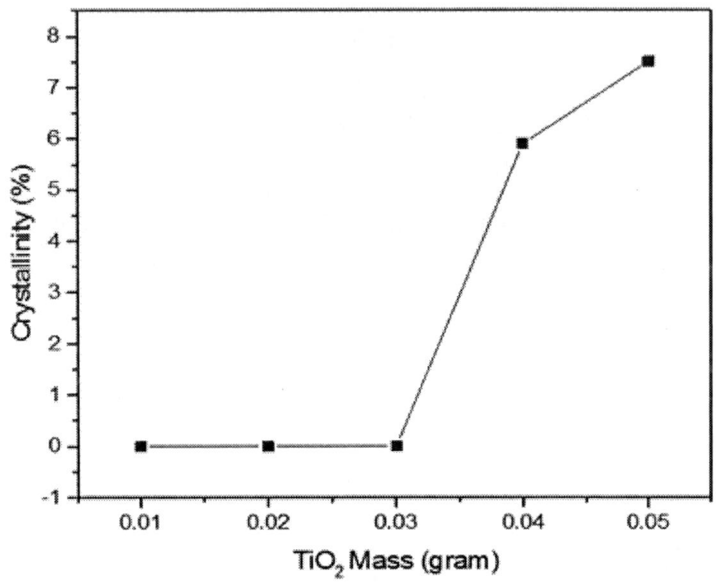

Figure 3. The crystallinity of TiO_2 Nanoparticles Mass Addition in Flavonoid-TiO_2 Films.

Figure 3 is a relationship between crystallinity and the addition of TiO_2 nanoparticles mass. The analysis results obtained crystallinity for TiO_2 mass of 0.01, 0.02, 0.03, 0.04, and 0.05 g respectively 0, 0, 0, 5.9%, and 7.5%. The 0% samples are associated with a no crystalline phase found in the patterns.

3.3. *SEM EDAX of Flavonoid-TiO_2 Films*
The morphology of flavonoid-TiO_2 films with a mass TiO_2 of 0.03 grams is shown in Figure 4. The SEM-EDX image shows that the white dots are scattered on the TiO_2 flavonoid film as in Fig. 4 a. with 20000x magnification is focused on white dots like Figure 4 b. The red box in Figure 4 b shows a selected area for EDAX test.

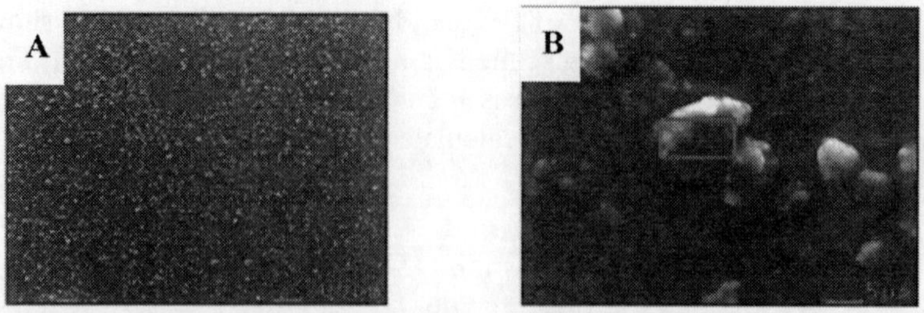

Figure 4. Morphology of the Flavonoid-TiO_2 Film (a) 500x of Magnification; and (b) 20,000x of Magnification.

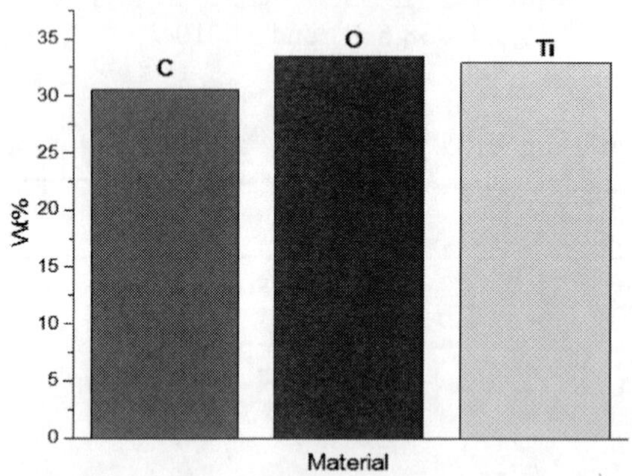

Figure 5. Elemental composition of the Flavonoid-TiO_2 film.

The EDAX result is presented in Figure 5. The content of C, O, and Ti content falls to 30.5, 33.52, and 32. %. It can be concluded that the white dot on the flavonoid-TiO_2 film is TiO_2. Thus, according to Fig. 4, a TiO_2 is evenly distributed on the surface of the flavonoid-TiO_2 film.

3.4. Dielectric Constant of Flavonoid-TiO_2

The relationship between the addition of TiO_2 mass and the dielectricity of the TiO_2 flavonoid film is shown in Figure 6.

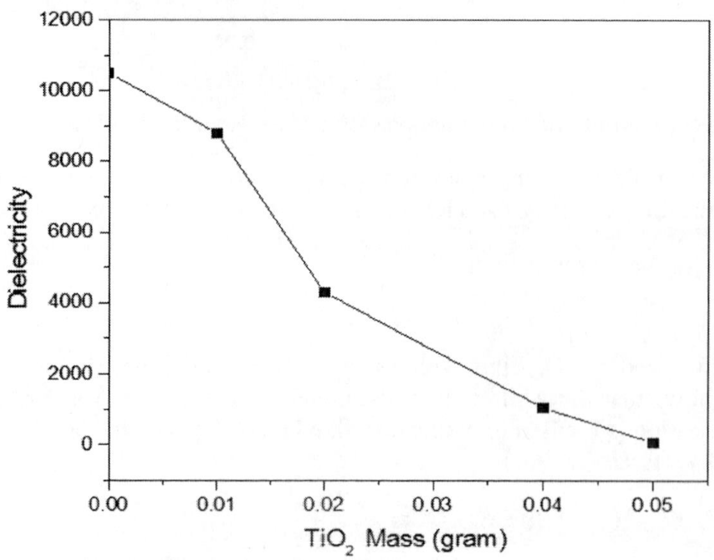

Figure 6. Dielectricity of Flavonoid-TiO_2 films with TiO_2 variation.

Based on Figure 6, it is found that the addition of TiO_2 affects the dielectric constant (dielectricity) of the flavonoid-TiO_2 film. The dielectricity of flavonoid film without doping of TiO_2 is 1.05×10^4. In flavonoid film with 0.01 gram of TiO_2 dielectricity obtained for 8.79×10^3. Dielectricity at 0.02 gram is TiO_2 of 4.3×10^3. The result of dielectricity obtained for the sample with 0.04 gram of TiO_2 is 1.04×10^3 while in 0.05 gram TiO_2 is 75.8. The dielectricity of flavonoid film without doping is TiO_2 is 1.05×10^4 whereas TiO_2 has a dielectricity of 53.1. Thus, the more nano mass of TiO_2 added, the dielectricity of the flavonoid- TiO_2 film decreases. This is consistent with previous research that TiO_2 nanoparticles decrease the dielectricity of PANI along with the mass of compounded TiO_2 nanoparticles sequentially dielectric constants is 3700, 3327, 1755, and 394 [17].

4. Conclusion

Based on the data analysis and discussion, it can be concluded that the extraction result from PIW latex (*Pterocarpus indicus Willd*) belongs to Rutin type flavonoid. The increase of TiO_2 gives rise to increase the crystallinity of the flavonoid-TiO_2 film. On the other hand, it decreases the dielectricity of the flavonoid-TiO_2 films. These results are supported by other similar reports.

Acknowledgments

Parts of this work were supported by the Ministry of Research and Higher Education of Indonesia through the DRPM of UM grant.

References

[1] Sabandar, C.W., Ahmat, N., Jaafar, F.M., Sahidin, I., Medicinal property, phytochemistry and pharmacology of several Jatropha species (Euphorbiaceae): A review. (Phytochemistry Vol. 85, 2013), pp. 7-29.

[2] Kanth, B.S., Kumar, A.S., Shinde, D.B., Babu, K.H., Raju, T.V., Kumar, C.G., Sujitha, P., Das, B., *New bioactive macrocyclic diterpenoids from Jatropha multifida* (Bioorganic & Medicinal Chemistry Letters vol. 21, 2011), pp. 6808-6810.

[3] Victorien, D.T., Jean Robert, K., Jacques, D.T., Julien, S., Jean-Marc, A., Aléodjrodo, E.P., Olufunkè, S., Ferdinand, D., Carlos, D., Frédéric, L., Karim, D., *Hemostatic activity screening and skin toxicity of sap of Jatropha multifida L. (Euphorbiaceae) used in traditional medicine (Benin).*(Asian Pac. J. Trop. Dis. Vol. 2, 2012), pp. S927-S932.

[4] Kumar, N., Singh, A.S., Kumari, S., Reddy, M.P., Biotechnological approaches for the genetic improvement of Jatropha curcas L.: A biodiesel plant (Ind. Crops Prod vol. 76, 2015), pp. 817-828.

[5] Falodun, A., Imieje, V., Erharuyi, O., Joy, A., Langer, P., Jacob, M., Khan, S., Abaldry, M., Hamann, M., *Isolation of antileishmanial, antimalarial and antimicrobial metabolites from Jatropha multifida* (Asian Pac. J. Trop. Biomed vol. 4, 2014), pp. 374-378.

[6] Rampadarath, S., Puchooa, D., Ranghoo-Sanmukhiya, V.M., *Antimicrobial, phytochemical and larvicidal properties of Jatropha multifida Linn* (Asian Pac. J. Trop. Med. Vol. 7, 2014), pp. S380-S383.

[7] Rangel-Vázquez, N.A., Salgado-Delgado, R., García-Hernández, E., Mendoza-Martínez, A.M., *Characterization of Copolymer Based in Polyurethane and Polyaniline (PU/PANI)* (J. Mex. Chem. Soc. Vol. 53, 2009), pp. 248-252.

[8] Markus Diantoro, Devy Purwaningtyas, Nazilah Muthoharoh, Arif Hidayat, Ahmad Taufiq, Abdulloh Fuad, The Influence of Iron- and Copper- doped of PANi Thin Film on their Structure and Dielectric Properties. American Institute of Physics, 1454. ISSN 0094-243X, p268-271. 2012.

[9] M. Diantoro, Kholid, A. A. Mustikasari, and Yudiyanto, The Influence of SnO2 Nanoparticles on Electrical Conductivity, and Transmittance of PANI-SnO2 Films IOP Conf. Series: Materials Science and Engineering 367 (2018) 012034 doi:10.1088/1757-899X/367/1/012034Mohanraj, V.J., Chen, Y., *Nanoparticles - A review* (Trop. J. Pharm.

Res.vol. 5, 2006), pp. 561-573.

[10] Markus Diantoro, Thathit Suprayogi, Ulwiyatus Sa'adah, Nandang Mufti, Abdulloh Fuad, Arif Hidayat and Hadi Nur (July 18th 2018). Modification of Electrical Properties of Silver Nanoparticle, Silver Nanoparticles Khan Maaz, IntechOpen, DOI: 10.5772/intechopen.75682.

[11] Ali, Y., Kumar, V., Dhaliwal, A.S., Sonkawade, R.G., Surface modification of polyaniline nanofiber using silver nanoparticles to enhance sensing properties (Adv Mater Lett Vol. 4, 2013), pp. 368–372.

[12] Diantoro, M., Hidayati, N.N.S., Latifah, R., Fuad, A., Nasikhudin, Sujito, Hidayat, A., Electrical conductivity modification using silver nanoparticles of Jatropha Multifida L. and Pterocarpus Indicus w. extracts films (AIP Conf. Proc. Vol. 1719), pp. 030034-1 – 030034-7.

[13] Markus Diantoro, Thathit Suprayogi, Arif Hidayat, Ahmad Taufiq, Abdulloh Fuad, and Risa Suryana, Shockley's Equation Fit Analyses for Solar Cell Parameters from I-V Curves, International Journal of Photoenergy, Volume 2018, Article ID 9214820, 7 pages https://doi.org/10.1155/2018/9214820

[14] Mohanraj, V.J., Chen, Y., *Nanoparticles - A review* (Trop. J. Pharm. Res.vol. 5, 2006), pp. 561-573.

[15] Sukadana, I.M., . Senyawa antibakteri golongan flavonoid dari buah belimbing manis (Averrhoa carambola Linn. L) (J. Kim. Vol. 3, 2009).

[16] Klug, H.P., Alexander, L.E., 1974. X-Ray Diffraction Procedures: For Polycrystalline and Amorphous Materials. Wiley.

[17] Dey, A., De, S., De, A., De, S.K., Characterization and dielectric properties of polyaniline–TiO_2 nanocomposites (Nanotechnology Vol.15, 2004), pp. 1277.

Extraction and Characterization of Crop Oil from Seed Kernels of Feun Kase (*Thevetia peruviana*)

Suwari[1,*], Herry Z Kotta[2], Yohanes Buang[1]

[1] Department of Chemistry, Faculty of Science and Natural Sciences, Nusa Cendana University Jl. Adisucipto Kampus Baru Penfui, Kupang, NTT 85361 Indonesia
[2] Department of Mining, Faculty of Science and Engineering, Nusa Cendana University

*Corresponding author's email: suwari_chem68@yahoo.com

Abstract. The seed kernel of *Thevetia peruviana* is a non-edible material but a potential resource to produce bulb lights in local society of West Timor. The light produced by the seeds makes it a probable feedstock for biodiesel production. In this report, the optimum condition for extraction of the crop oil from the seed kernel of Feun Kase (*Thevetia peruviana*) using conventional soxhlet technique was studied. The solvent used were chloroform, water, and a chloroform-water binary solvent. The parameters investigated effecting the oil yield involved various solvent polarities, extraction time, and temperature. Each experiment was conducted in 250 mL soxhlet apparatus. The extract was analyzed to examine the physicochemical characteristics including density, kinematic viscosity, acid value, iodine value, saponification value, and water content. The optimum conditions were found after 4.0 h extraction time, extraction temperature of 75 °C and a chloroform-water ratio of 100:0 (polarity index 4.10). The oil extract was found to be 52.43 ± 1.73%. These results revealed that the crop oil from the seed kernel of Feun Kase is a potential feedstock for biodiesel production.

Keywords: *Thevetia peruviana*, crop oil, seed kernel, biodiesel production.

1. Introduction

Thevetia peruviana is a small plant globally known as yellow oleander, lucky nut, bush milk, orange oleander or kolkaphul. In Indonesia, this plant is more popularly known as ginje, but in some regions, it is given different names, e.g. Nagasari (East Java) and Ki Hujan (West Java). On Timor Island, the community calls this plant as Feun Kase, meaning *Kemiri Kota* (City Candlenut). Traditionally, in West Timor Island, this plant was used as a fuel source of lamps, but this habit disappeared with the development of technology and modernization. This species of plant grows widely in the tropics and subtropics. All parts of this plant are toxic [1]. The toxic properties of these plants are reported to come from the apigenin-5-methyl-ether (flavonoid) compounds and several other compounds such as triterpenoids and glycosides [2]. Feun Kase stars flowering after one and a half year after planted and it blooms thrice every year. In a hectare, 3000 saplings can be planted and out of which 52.5 tons of seed can be collected [3]. It has been reported [4] that Feun Kase seed kernels have a high level of oil content (about 60-65%) and a valuable protein content (30-37%), without fat. It remains non-edible because of the presence of toxins [5]. The high oil content of Feun Kase seeds indicates potentiality of the plant as a raw material for a biodiesel production.

The problems related to raw materials and extracting methods are still the major challenges in a biodiesel production because 60-75% of the total cost of biodiesel production is the cost of providing raw materials and the remaining 25-40% processing costs [6]. Recently, the development and research on biodiesel is 95% statically directed to biofuel sources, such as edible oil or conventional sources, using a transesterification method [7]. This condition is very unfavorable because of the food competition and the high cost of processing problems which make the price of biodiesel more expensive than the fossil fuel. One of the potential plants that could be developed as a raw material for a biodiesel production is Feun Kase.

In addition, the production of biodiesel from vegetable materials is done through several stages, namely oil extraction, transesterification process, and purification of products. In obtaining the crop oil with high yield, the oil must be extracted using a high-performance extraction method. The choice of solvent type (polarity), extraction time, and temperature are also the main factors determining the number of isolation products [8-10]. The inability to control these factors can lead to failure to achieve the high yields and a good oil quality. Several studies of reported extraction methods also support the empirical fact of the importance of solvent selection [8,11]. The objective of this works was to investigate the effects of chloroform-water solvent with different polarities, extraction time, and temperature for the crop oil extraction from the seed kernels of Feun Kase using a conventional soxhlet technique to obtain maximum oil yield.

2. Materials and Methods

The seeds of Feun Kase that were used in this work was obtained from Soba Village, West Amarasi, Kupang District of East Nusa Tenggara Province in Indonesia. After the seeds of Feun Kase were collected, the seeds were carefully separated from the shells and removed from unwanted ingredients, such as remnants of the outer shell, twigs, and fine pebbles. The seeds of Feun Kase were opened manually to obtain the Feun Kase kernels. The kernels were ground using a sieve plate and shaker grinder to reduce it into a size of range 0.5 – 0.75 mm. The entire chemical used in this work, such as chloroform, iodine, ethanol, potassium hydroxide, hydrochloric acid, glacial acetic acid, sodium thiosulfate, and potassium iodide, were of analytical reagent grade obtained from Cica-Merck (Kantako Chemical).

The effects of solvent type (polarities), extraction time, and temperature were studied in order to obtain the optimum conditions for the extraction to achieve maximum oil yield. The crop oil from seed kernels of Feun Kase was extracted with a soxhlet method. Thirty-seven (37) extractions with different combinations of solvent types, extraction time, and temperature were performed. The extraction of the oils was carried out with a soxhlet apparatus of 250 cm^3 capacity using chloroform and water. A sample of 25 g of seed kernels was extracted with non-polar and polar solvents, with different ratios of chloroform to water (100:0, 90:10, 80:20, 70:30, 60:40, and 50:50 v/v) to give the solvent polarity of 4.10, 4.59, 5.08, 5.57, 6.06, and 6.55, respectively. Next, five different temperatures of 70 °C, 75 °C, 80 °C, 85 °C, and 90 °C were investigated to obtain the optimum extraction temperature, while the extraction time was varied between 3.0 to 5.5 h. At the end of the extraction process, the solvent was removed by a vacuum evaporator at 40 °C for 1 h. Furthermore, the oil obtained was filtered and then mixed with 2% of distilled water (degumming process). Subsequently, the mixture was heated at 70° C and stirred for about 30-60 minutes, added magnesium sulfate anhydrous and filtered to obtain the oil extract. The oil obtained after evaporation was weighed and the oil yield was calculated. All experiments were repeated at least twice and the final value was the average of all. The oil yield was calculated by using equation 1 [5].

$$Oil\,yield\,(\%) = \frac{W_o}{W_s}\,x\,100\% \qquad (1)$$

Where

W_o = mass of oil extracted (g)

W_s = mass of the seed kernels (g)

The physicochemical properties of the oil yield, namely: density, viscosity, acid value, iodine value, saponification value, and water content of the extracted oil were determined using the ASTM method.

3. Results and Discussion

The successful separation of oil components in Feun Kase seed kernels by the soxhletation technique is highly dependent on the level of solvent polarity and separated components. The difference in solubility of each component into the solvent, then the selectivity of the solvent is crucial to succeeding in the extraction or isolation process.

In principle, the extraction of Feun Kase seed kernels oil was done by weakening the attraction between the oil molecules in a material so that the oil molecules can separate from the material and assemble to form bonds or new interactions. The occurrence of this event can be done by dissolving the mixture in a suitable solvent. The interaction between the solvent molecule and the material molecule produces events such as the occurrence of penetration of solvent molecules into the material, on the contrary, the molecules of the ingredients penetrate into the solvent. The easier a solvent penetrates into the material or otherwise, the attraction between the host molecules gets faster so that the molecules of the material (oil) are more easily extracted from the material and drawn into the solvent phase.

The polarity of the solvent is measured from the dipole moment. Water is the most polar solvent and has a polarity index or Snyder index (SI) of 9.0, whereas n-hexane has a Zero-Snyder index, including very non-polar solvents [8]. The polarity of other solvents lies between the polarity of water and n-hexane. Chloroform classified as a semi-polar solvent with a Snyder index of 4.1. The physical properties of the solvents used are listed in Table 1. The similarity or closeness of the polarity index between the solvent molecules with the building block molecules of a material, allows the shift or displacement of building block molecules into the solvent, resulting in separation.

Table 1. Physical properties of chloroform and water [8,12]

Physical Properties	Unit	Name of Solvent	
		Chloroform	Water
Polarity (Snyder index)	-	4.1	9.0
Dipole moment	Debye	1.04	1.87
Boiling point	°C	61	100
Dielectric constant	-	4.71	78.36
Cohesive energy density	J mol / mL	332.00	2095.93
Viscosity	mPa	0.54	0.89
Surface tension	cal / mol A ^{2}	38.39	104.70

The separation of the oil components from the building molecules using the chloroform-water solvent was also influenced by the formation of hydrogen bonds between the oxygen atoms contained in the oil molecule with the hydrogen atom contained in a water molecule or vice versa. The preferential of the solvent to donate or accept electrons (nucleophilic) and to donate or accept H^+ (electrophilic) with the components in the material matrix also determines the ease of penetrating the oil component of the building block molecule [8]. The chloroform-water solvent composition of 100: 0; 90:10; 80:20; 70:30; 60:40; and 50:50 with a solvent polarity of 4.10 to 6.55 were tested for its ability to extract the oil from Feun Kase seed kernels. The amount of oil extraction was calculated in % of total oil present in the seed kernels of Feun Kase. The effect of the solvent composition on oil yield was shown in Figure 1.

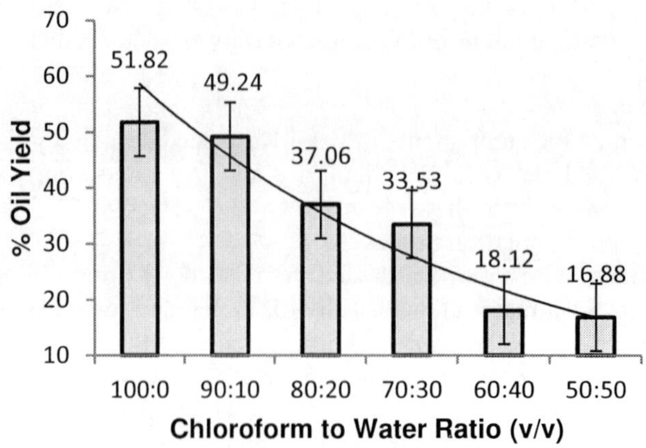

Figure 1. Effect of chloroform to water ratio on oil yield with 70 °C of extraction temperature and extraction time of 4 h

The combination of chloroform-water solvent actually decreases the percentage of oil extracted (Figure. 1). The maximum oil yield of 51.82% was obtained when using a single chloroform solvent. The increase of solvent polarity from 4.59 to 6.55 due to an increase in the water composition resulted in a decrease of oil yield by 4.9 - 67%. These results indicate that the percentage of non-polar lipid in Feun Kase seed kernels was greater than that of a polar lipid content. The solvent boiling point, surface tension, and viscosity decreased at lower temperatures, and then the solvent was able to reach the active sites in the matrix more easily. The water with higher viscosity and surface tension than that of chloroform (Table 1) was found to be inappropriate in oil extraction of Feun Kase seed kernels because it would hinder the solvent absorption into the active sites inside the matrix, however, the dielectric constant and cohesive energy are significantly higher, then the water molecules can strongly bond to the polar oil components in the matrix [8]. Therefore, the ratio of chloroform and water composition of 100: 0 (polarity 4.1) which gave the highest oil yield was considered the best solvent choice and to be studied further.

The extraction results are not only influenced by the type and polarity of the solvent, but also by extraction conditions such as extraction temperature and extraction time. The temperature may affect the rate of extraction and oil yield because the temperature change affects the kinetic energy of the solvent molecules and the collision frequency between the solvent molecules and the target molecules. The effect of the extraction temperature on the amount of extracted oil on Feun Kase seed kernels using the chloroform solvent was shown in Figure 2.

The increase in the extraction temperature can decrease the surface tension and solvent viscosity further increases the oil yield. The amount of extracted oil increased by 6.8% with an increase in the extraction temperatures from 70-75°C. Therefore, the selected extraction temperature was the temperature capable of producing the highest oil yield which was 75 °C. The results indicated that the increase of the extraction temperature will increase the kinetic energy of the solvent so that the penetration of the solvent into the material becomes larger in order the oil molecules are more easily extracted from the material and drawn into the solvent phase. The increase in temperature also affects the solvent's diffusivity into the inner part of the material and increases the solubility of the oil in the solvent, thus improve the extraction rate of oil from the seed kernels of Feun Kase. However, the extraction at 80-90 °C temperatures, the oil yield decreased by 9 to 22%. These results were in accordance with those reported by [13-15] as explained that the reduction of the oil yield would be at a high temperature. An increase in temperature reduced the oil kinematic viscosity with an increased mobility of biopolymers in cellular walls. Therefore, it can be deduced that 75 °C was adequate to give the highest yield of oil from the seed kernels of Feun Kase.

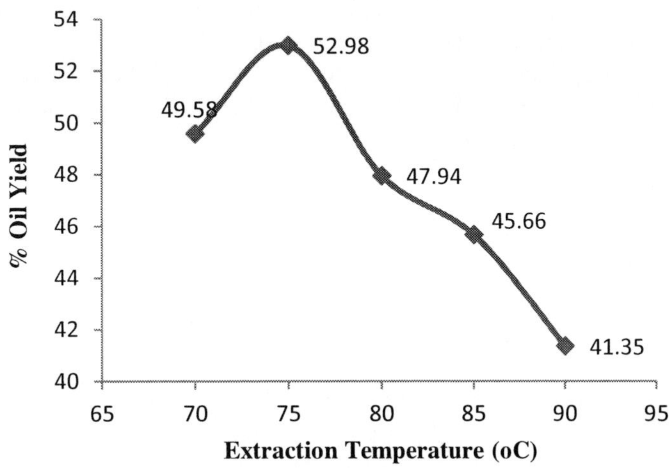

Figure 2. The effect of temperatures on oil yield at the reaction condition of chloroform to water ratio of 100:0 extraction time 4 h and various temperatures in °C

The extraction time is an important factor determining the efficiency of an oil extraction from vegetable materials. An appropriate timing of extraction directly affects to the oil yield, due to the maximum interaction between the solvent and the oil molecules when the contact time and residence time of the solvent in the extractor were optimum. Fig. 3 showed the total amount of oil extracted from the seed kernels of Feun Kase at different extraction times. The extraction times in this work were within the range of 3.0 h-5.5 h. The results showed that oil extracted from seed kernels of Feun Kase increased by 20% due to the increase of extraction time from 3.0 h to 4.0 h, but the subsequent increase in extraction time did not affect oil yield. It also observed that the rate of extraction was very high during the first 30 minutes of extraction and afterward it tapered off. The initial high rate of extraction may be due to the quick solubility of the oil present at the solid surface and higher mass transfer driving force provided by low oil concentration in the fresh solvent [16]. This shows that the optimum residence time or contact time required by the extractant (oil) in the chloroform solvent was 4.0 h. Hence, 4.0 h was chosen as the optimum time for Feun Kase seed kernels oil yield.

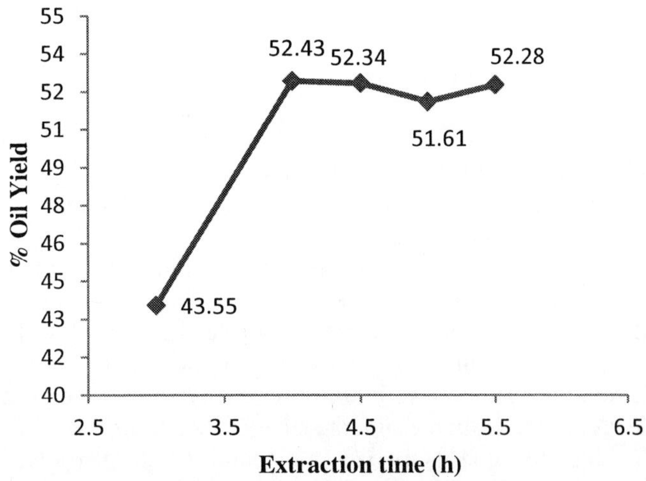

Figure 3. The effect of extraction time on oil yield at the reaction condition of chloroform to water ratio of 100:0, extraction temperature 75 °C

An oil quality analysis was performed to determine the physicochemical characteristics of the oil produced from the soxhletation extraction process using chloroform solvent at 75° C and 4.0 h of extraction time. The results of analysis of physicochemical properties of Feun Kase seed kernels oil are needed to determine the stage of transesterification process that would be used in the production of methyl ester and determining the required methanol and catalyst requirements. The parameters analyzed were density, viscosity, acid value, iodine value, saponification value, and water content. The physicochemical properties of the extracted oil at the optimum conditions are shown in Table 2. These properties were comparable to those of the National Standard of Indonesia (SNI) for biodiesel production.

Table 2. The physicochemical properties of crop oil from seed kernels of Feun Kase

Properties	Unit	Values	SNI Specification
Density	kg/m^3	1066	850 – 890
Viscosity	mm^2/s	2.18	2.3 – 6.0
Iodine value	g I_2/100 g oil	43.56	115 max
Acid value	mg KOH/g oil	1.86	0.8 max
Saponification value	mg KOH/g	140.52	500 max
Water content	%	0.032	0.05 max

Based on the analysis results, it was found that the density of oil before being processed into biodiesel was 1066 kg/m^3, this value was higher than the value of SNI (850-890 kg/m^3) and the value set by Pertamina for the diesel oil, which was 820 - 870 kg/m^3. The value of oil density is influenced by the fatty acid composition and the purity of the raw materials. A high-density value indicated that in the Feun Kase seed kernels oil contains many unsaturated fatty acids. Feun Kase seed oil contained predominantly 4 types of fatty acids, namely palmitic acid (17.02%), stearic acid (6.23%), oleic acid (41.91%) and linoleic acid (11.89%) [4,17]. Oleic acid (C18:1) and linoleic acid (C18:2) with the composition of 41.91% and 11.89%, respectively, included unsaturated fatty acids which predominantly affected the high density of Feun Kase seed oil. The more the number of double bonds with the same length of the carbon chain, the greater oil density. Since it contained 53.80% unsaturated fatty acids composition, its effect on the density was very dominant. The density will increase as the carbon chain length decreases and the increase in the number of double bonds in fatty acids. The high density of the extracted oil can be caused by the increasing number of unsaturated oil components in the extraction product. Therefore, before being used as a raw material for a biodiesel production, this extracted oil needs to be done degumming or bleaching process to decrease the oil density.

Viscosity is a measure of a fluid's resistance to flow. Overly high viscosity will cause smoke because the fuel is slow to flow and more difficult to atomize, as well as incomplete combustion and carbon deposition in the injector. The viscosity of a fuel is very influential on the performance of the engine so it becomes one of the important parameters. If the viscosity is high then the fuel will flow slowly and the degree of atomization of fuel will become more difficult, so the resulting energy becomes not optimal. The viscosity as a measure of oil resistance to fluid obtained in this works was 2.18 mm^2/s, so that the viscosity of oil from Feun Kase seed kernels fulfilled the standard of SNI. These results are in line with the findings [18] that the unsaturated fatty acids composition of the Feun Kase seed kernels oil would contribute mostly to the viscosity value. The more the number of double bonds with the same length of the carbon chain, the smaller the viscosity.

The saponification value indicated its average molecular weight of triglycerides in the oil. It was determined by the molecular weight of the constituent fatty acids. The saponification value obtained in this work was 140.52 mg KOH/g of oil. A low saponification value of the oil indicated that the triglyceride has a long carbon chain. The longer the carbon chain of fatty acid, the less fatty acid content in the oil. The oil composed of long-chain fatty acids indicated that a high molecular weight will have a small saponification value. On the contrary, low molecular weight oils will have high saponification value.

The acid value showed free fatty acid levels in the oil. It is indicated the amount of milligram KOH required to neutralize free fatty acids in one gram of oil. The presence of free fatty acids in biofuel products can lead to the formation of ash at the time of combustion, clogging the filter with precipitate and corroding the diesel engine. The acid value can be an indicator of damage occurring to the oil, due to oxidation activity. The results showed that the acid value in Feun Kase seed oil was 1.86 mg KOH/g oil. It was higher than the SNI specification for biodiesel product with the maximum of 0.8 mg KOH/g oil. A high acid value indicated a high fatty acid (FFA) derived from oil hydrolysis. The higher the acid value the lower the quality, because oils containing free fatty acids more than 1% will form emulsion formulations that make it difficult at the time of biodiesel separation. The water content is one important factor determining the oil quality. The high water content can lead to a blockage of fuel flow, the occurrence of hydrolysis reactions that will increase the acid value, decrease the pH and improve the corrosive properties. The water content of the Feun Kase oil was 0.032%. This value was relatively lower and met the specified quality standard. The iodine value indicates the degree of unsaturation or number double bonds contained in oils expressed in I_2/100 g oil. The iodine value obtained in this work was 34.56 g I_2/100 g of oil and fulfilled the standard of SNI. This indicates that the iodine value attached to the double bond was relatively low so that the degree of unsaturation of the fatty acids was low. The amount of unsaturated fatty acid compounds in the oil allows the compound to react with oxygen at the time of combustion, forming a chain with a very large molecular weight. The iodine value has a direct relationship to viscosity and cetane numbers. The higher the iodine value the lower the cetane number and vice versa. The decrease in viscosity and cetane numbers will lead to an increase in oil unsaturation. The maximum iodine value for biodiesel feedstock was 70-100. The physical and chemical properties of crop oil from seed kernels of Feun Kase in this study compared with previous studies to determine its suitability as a feedstock for biodiesel production was shown in Table 3.

Table 3. The comparison of physicochemical properties of the oil in this study to other studies

Properties	**Unit**	**Other research**					**This study**
		[17]	[19]	[20]	[21]	[22]	
Density	kg/m^3	-	940	899	921	-	1066
Viscosity	mm^2/s	40.42	39.84	-	47,0	-	2.18
Iodine value	g I_2/100 g oil	-	-	71.20	12.60	71.40	43.56
Acid value	mg KOH/g oil	-	-	0.568	4.70	0.658	1.86
Saponification value	mg KOH/g	-	-	-	412.30	121.00	140.52
Oil yield	%	-	48.82	-	-	-	52.43

The viscosity obtained in this study was lower than the previous research by [17, 19, 21]. The lower the viscosity of the oil, the better the oil to be used as a feedstock for biodiesel production. The density of the oil was higher than the previous study by [19, 20, 21]. The high content of oleic acid and linoleic acid with a double bond and unsaturation level contributed to the high of the density in this study. The iodine values obtained in this study was better than the previous studies by [20, 22], it fulfilled the standard of SNI. The saponification value of the oil met with the standard of SNI indicated that the oil can be used as an effective feedstock for biodiesel production. The non-edible property and oil content of a seed are the first factor to be considered among others when large-scale extraction of the oil is to be considered [7, 21]. Any non-conventional and non-edible seeds with their oil contents more than 40% are the most as suitable feedstock to produce biodiesel. The oil yield obtained in this study (52.43%) was higher than that reported by [19] (48.82%), and exceeded the value for *Pongamia glabra* seed oil (33%)[23], *Moringa oleifera* seed oil (35%)[24], as well as

Jatropha curcas seed oil (30 – 50%) [25,26]. Prospectively, based on the physicochemical properties and oil yield, Feun Kase seed oil has a potential for a feedstock in biodiesel production

4. Conclusion
The crop oil from seed kernels of Feun Kase can potentially be utilized as a feedstock of biodiesel production. Therefore, it can reduce the dependency on using edible oil feedstock. The optimum conditions were obtained at chloroform to water ratio of 100:0 (polarity 4.10), extraction temperature of 75 °C and extraction time of 4.0 h. The optimized oil yield was found to be 52.43 ± 1.73% (weight basis). It had been observed that polarity of the solvent and extraction time was the most crucial parameters that affecting the oil yield. The oil yield and physicochemical analysis of the extracted oil further suggests its potential application as a feedstock for a biodiesel production.

Acknowledgments
The authors would like to thank the Directorate of Research and Community Service, the Directorate General for Research and Development, the Ministry of Research, Technology and Higher Education of the Republic Indonesia who provided the funding for the research through Applied Research of Superior University competition scheme with grant No. 195/UN15.19/LT/2017.

References
[1] S. Kishan, K.K. Agrawal, V. Mishra, S.M. Uddin and A. Shukla, International research Journal of Pharmacy 3, 74-77 (2012).
[2] D. Patel, S. Shukla and S. Gupta, International Journalof Oncology 30, 233-245 (2007).
[3] T. Balusamy and R. Marappan. Journal of Scientific and Industrial Research 66, 1035-1040 (2007).
[4] S.A. Ibiyemi, V.O. Fadipe, O.O. Akinremi and S.S. Bako, Journal Applied Science and Environment 6, 61-65 (2002).
[5] Ana Godson R.E.E and G.B. Udofia, International Journal of Sustainable and Green Energy 4, 150-158(2015).
[6] Ma Fangrui, A. H. Milford. Bioresource Technology 70, 1 – 115 (1999).
[7] S. Basumatary, Research Journal of Chemical Sciences 3, 99-103 (2013).
[8] M. Markom, M. Hasan, W.R.W. Daud, H. Singh and J.M. Jahim, Separation Purification Technoloy 52, 487-496 (2007).
[9] M.C. Straccia, F. Siano, R. Coppola, F.L. Cara and M.G. Volpe, Chemical Engineering Transactions 27, 391-396 (2012).
[10] A.G. Sicaire, M. Vian, F. Fine, F. Joffre, P. Carre, S. Tostain and F. Chemat, International Journal of Molecular Sciences 16, 8430-8453 (2015).
[11] T.P.L. Ferraz, M.C. Fiuza, M.L.A. Dos Santos, L.P. de Carvalho and N.M. Soares, Journal of Biochemical and Biophysical Methods 58, 187-193 (2004).
[12] C.H. Gu, H. Li, R.B. Gandhi, K. Raghavan, International Journal of Pharmaceutics 283, 117-125 (2004).
[13] S. Sayyar, Z.Z. Abidin, R. Yunus and A. Muhammad, American Journal of Applied Sciences 6, 1390-1395 (2009).
[14] O.S. Lawson, A. Oyewumi, F.O. Ologunagba and A.O. Ojomo, Journal of Engineering and Applied Sciences 5, 51-55 (2010).
[15] B. Panchal, S. Deshmukh and M. Sharma, International Journal of Oil, Gas and Coal Engineering 2, 1-6 (2014).
[16] D.K. Saxena, S. K. Sharma and S. S. Sambi, Journal of Engineering and Applied Sciences 6, 84-89 (2011).
[17] M.I. Oseni, S.E. Obetta and F.V. Orukotan, Journal of Scientific and Industrial Research 3, 62-68 (2012).
[18] Joeliangsih, A.H. Tambunan, T.H. Soerawidjaya, S. Yasuyuki and A. Kamaruddin, Jurnal

keteknikan Pertanian 22, 1-6 (2008).
[19] S.B. Dhoot, D.R. Jaju, S.S. Deshmukh, B.M. Panchal and M.R. Sharma, Journal of Alternative Energy Sources & Technologies 2, 8-16 (2011).
[20] D.C. Deka and S. Basumatary, Biomas Bioenergy 35, 1797-1803 (2011).
[21] C.I. Yarkasuwa, D. Wilson and E. Michael, Journal of the Korean Chemical Society 57, 377-381 (2013).
[22] M.M. Bora, D.K. Kakati, P. Gogoib and D.C. Deka, Industrial Crops and Products 52, 721-728 (2014).
[23] A.K. sarma, D. Konwer and P.K. Bordoloi, Energy Fuels 19, 656-657 (2005).
[24] U. Rashid, F. Anwar, B.R. Moser and G. Knothe, Bioresource Technology 99, 8175-8179 (2008).
[25] Y.H.T. Yap, M.Z. Hussein and R. Yunus, Biomass Bioenergy 35, 827-834 (2011).
[26] K. Permanik, Renewable Energy 28, 239-248 (2003).

Performance of Solar Cells under Total Solar Eclipse in Central Sulawesi of Indonesia

Darsikin, Nurgan, Sahrul Saehana*

Departement of Physics Education, Tadulako University, Jl. Soekarno-Hatta km. 09 Tondo, Palu, Indonesia, 94118

*Corresponding author's email: sahrul_saehana@untad.ac.id

Abstract. This paper reports the observation of performance solar cells when the total solar eclipse occurs in central Sulawesi, Indonesia. The observation was done by measuring some parameter and calculating the performance of solar cells. It was found that performance of solar cell was decreased linearly before total solar eclipses happen and its performance increase after the total solar eclipse. The performance of a solar cell when a total solar eclipse occurs is nearly zero because the intensity of the light measured by the instrument is equal to zero. It was concluded that the performance of solar cells was influenced by the intensity of light.

Keywords: Solar cell, total solar eclipse, Sulawesi

1. Introduction

Solar cells are electronic devices that serve to convert photons into electrical energy [1-5]. This device is one of the leading in the conversion of renewable energy into electricity. The performance of this device depends on the magnitude of the intensity of photons received. Electricity generated at low intensity is smaller when compared to high intensity [6-10]. The efficiency of solar cells is influenced by light intensity [11]. At high light intensity, solar cell efficiency tends to decrease. Therefore, the effect of light intensity on the performance of solar cells is significant to investigate.

A total solar eclipse is an event where the sun position parallel to the moon so that within a few minutes sunlight is not transmitted to the earth (dark earth). A total solar eclipse is a rare event [12-20]. This event occurs over a span of decades in different places. In 2016, total solar eclipses can only be observed in several regions of Indonesia such as Palembang, Palu, Ternate, Palangkaraya, and Balikpapan.

Some papers reported observations on several physical parameters when a solar eclipse. This phenomenon affects the meteorological parameters, such as solar radiation, temperature, and humidity [12]. Solar total eclipse also related to meteorological parameters and environment [13]. According to their reports, it was found that some physics parameter has specific behavior when the total solar eclipse occurs. Thus, in this paper, we report the behavior of solar cells when the solar eclipse occurred Palu (Central Sulawesi, Indonesia) on March 9th, 2016. The effect of another parameter that changed when the total solar eclipse occurs in the performance of solar cells was also discussed.

2. Experimental Methods

The location of the observation is Tondo at Palu District (Central Sulawesi, Indonesia). Observations were performed on March 9'h, 2016 from 07.28 to 08.47 WITA, where measurements of all variables were performed on every two minutes. The intensity of the sunlight was measured with a solar power meter (TM206). To obtain data on the performance of solar cells when a solar eclipse occurs we measured the I-V characteristics of silicon solar cells. Current and voltage measurements were examined using multimeters while measuring the intensity of light using a solar power meter.

3. Results and Discussion

Total solar eclipse is a condition where sunlight is covered by the moon so that no sunlight is received on the surface of the earth. On solar total eclipse phenomenon, sunlight reduces to minimum condition and in this condition is very interesting to reveal the performance of solar cells considering there is no reference to report it. Photos, when the total solar eclipse occurs, are shown in Figure 1 (a). The observed silicon solar cell performance is shown in Figure 1 (b).

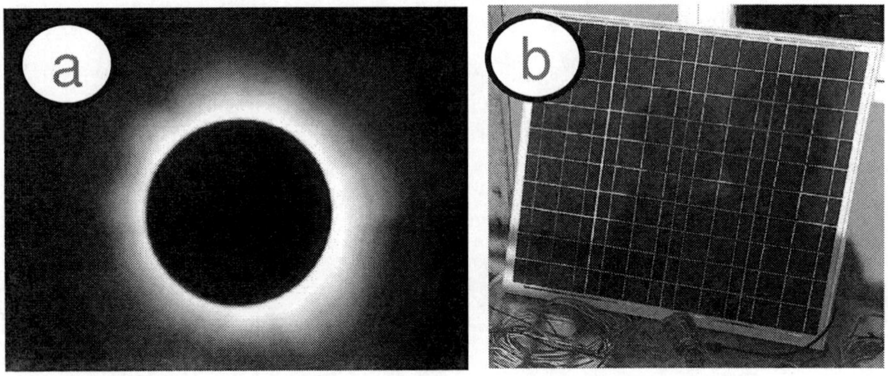

Figure 1. (a) Solar Total Eclipse; (b) Silicon Solar Cells

The performance of solar cells was depended on solar radiation as formulated in Equation (1).

$$E = \varepsilon \langle A \rangle \Delta t \quad (1)$$

where E is the efficiency of solar cells; ε is the average intensity of sunlight on the cell; $\langle A \rangle$ is the cross-sectional area of the solar cell panel, and Δt is the time radiation of the sun.

In this study, the intensity of sunlight was observed every two minutes, and it was started at 07.28 until 08.47 WITA as shown by the graph in Figure 2. It can be seen that there is a decrease in the intensity of sunlight between 07.28 until 08.36 and the minimum intensity was found between 08.38 sd 08.41 WITA. Furthermore, the intensity of sunlight has increased again. In relation to Equation 1 that the electric power generated by solar cells is proportional to the intensity of sunlight. It is evidenced by observations on electric current and the voltage generated when a solar eclipse occurs.

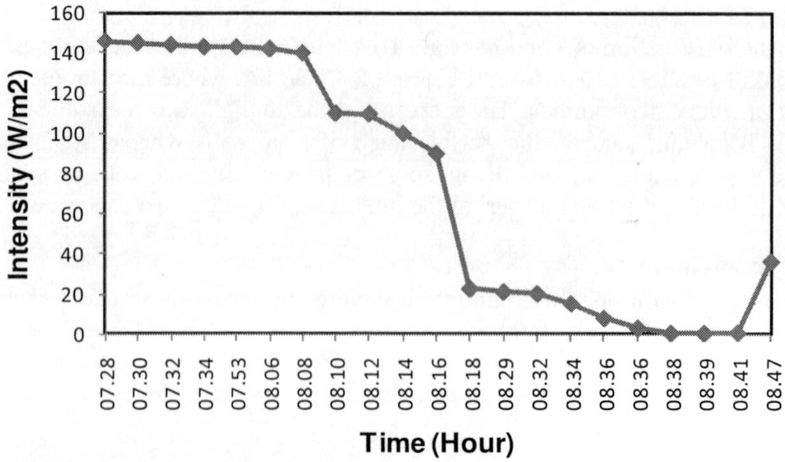

Figure 2. Observation of the intensity of sunlight during a total solar eclipse

The observation of the electric voltage generated by the solar cells is shown in Figure 3. The reduction of the voltage occurs with the change in the intensity of the sun. The minimum voltage that a solar cell possesses occurs when the intensity is minimum. This data is in line with the observed light intensity shown in Figure 2.

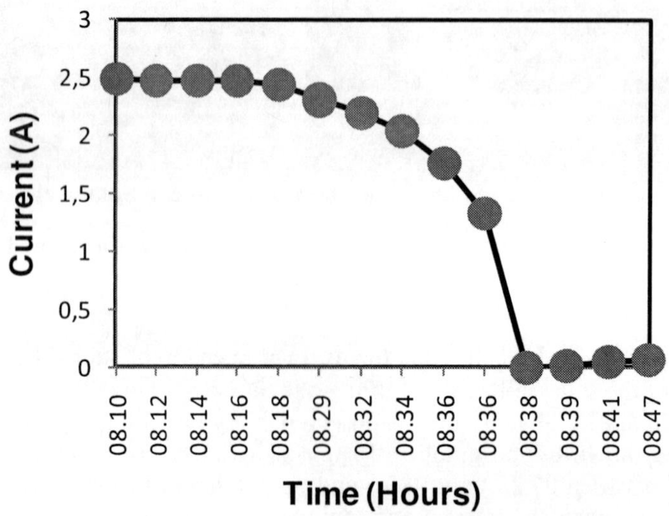

Figure 3. Observation of the solar cells voltage during a total solar eclipse

Observation of the resulting electric current was also done. The measurement results are illustrated in Figure 4. The data show that the electric current generated by the solar cell is proportional to the intensity of the sunlight. It can be seen from the decrease in electric current when the intensity of sunlight decreases and increases.

Temperature change is one factor that affects the performance of solar cells. Ref. [21] reported that with a significant increase in temperature give rise to the decrease of the electric voltage and fill factor. The reduction of voltage and fill factor causes a decrease in solar cell performance. From observations during the total solar eclipse, it is known that the temperature change is only about three degrees Celsius. Thus, no significant effect of temperature change during the total solar eclipse on the performance of silicon solar cells.

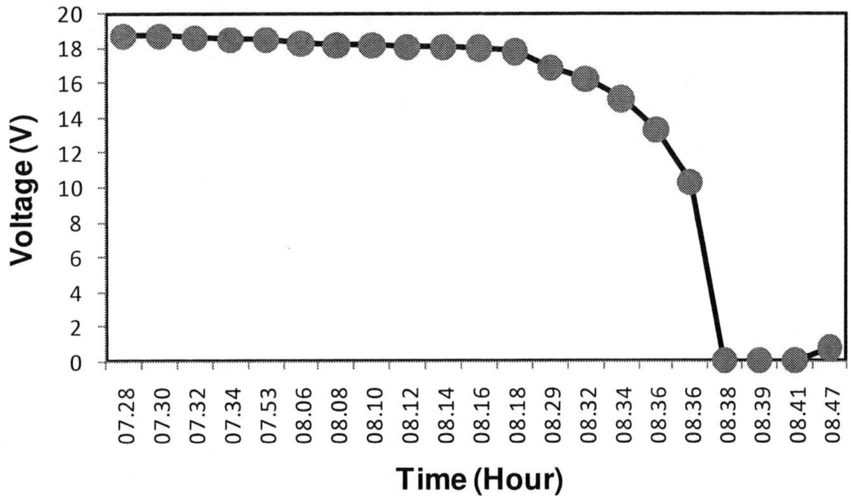

Figure 4. Observation of the solar cells current during a total solar eclipse

The presence of magnetic fields also affects the performance of solar cells. Ref. [22] reported that the presence of a high magnetic field causes a decrease in the efficiency of solar cells. The magnetic field may relate to the mobility of the carriers (electrons and holes) in the solar cells. From the observations during the total solar eclipse occurs it is known that no significant magnetic field changes [13].

The observations as reported in Ref. [14] also indicate that there is no significant gravitational change in Poso (Central Sulawesi) and Lembang (West Java) areas when the total solar eclipse occurs. Thus, there is no gravitational influence on the performance of solar cells at the time of total solar eclipse occurred in Central Sulawesi [12,23]. However, other environmental effects such as winds are expected to affect the dust flying around the observation site [11,17]. We suspect that the presence of dust during a total solar eclipse also resulted in decreased solar cell performance [24].

4. Conclusion

The performance of solar cells under the solar eclipse has been successfully observed. It is known that when a solar eclipse occurs, the intensity of sunlight decreases before eclipse and increases again after the eclipse occurs. It was found that performance of solar cell was decreased linearly before total solar eclipses happen and its performance increase after the total solar eclipse. The performance of a solar cell when a total solar eclipse occurs is nearly zero because the intensity of the light measured by the instrument is equal to zero. It was concluded that the performance of solar cells was influenced by the intensity of light.

Acknowledgments

This work was supported by Hibah Bersaing Research Grant from the Research, Technology and Higher Education of the Ministry of the Republic of Indonesia in the fiscal years of 2016.

References

[1] P. Joshi, Y. Xie, M. Ropp, D. Galipeau, S. Bailey, and Q. Qiao, Energy Environ. Sci., 2, 426 (2009).

[2] M. Grätzel, J. Photochem. Photobiol. C: Photochem. Rev., 4, 145 (2003).

[3] B. Lee, D.-K. Hwang, P. Guo, S.-T. Ho, D.B. Buchholtz, C.-Y. Wang, and R.P.H. Chang, J. Phys. Chem. B, 114, 14582 (2010).

[4] A. Islam, S. P. Singh, M. Yanagida, M. R. Karim, and L. Han, Int. J. Photoen., 2011, 1 (2011).

[5] S. P. Singh, A. Islam, M. Yanagida, and L. Han, Int. J. Photoen., 2011, 1 (2011).

[6] S. Saehana, P. Arifin, Khairurrijal, and M. Abdullah, J. Appl. Phys., 111, 123109 (2012).

[7] S. Saehana, Muslimin, and M. Abdulah, J. Renew. Sustain. Energy, 6, 023109-1 - 023109-6 (2014).
[8] K.K. Woo, F.H. Ngou, L.S. Ngo, W.K. Soong, and P.Y. Tang, Am. J. Food Tech., 6 (2), 140-148 (2011).
[9] G. Nair, S. Ramli, M. Irwanto, M.I. Yusoff, M. Fitra, U. Hashim, and N. Mariun, Appl. Mech. Mat., 699,516-521 (2015).
[10] M-J. Jeng, Y-L. Wung, L-B. Chang, and L. Chow, Int. J. Photoenergy, 2013, 1-9 (2013).
[11] R.A. Elgani, M.H.M. Hilo, M.D.A. Allah, A.A. Hassan, and R.A. Elhai, Natural Science, 5 (10), 1069-1077 (2013).
[12] B. Paramitha, R. Zaen and A.B.D Nandiyanto, "Changes in Meteorological Parameters (i.e., UV and Solar Radiation, Air Temperature, Humidity and Wind Condition) during the Partial Solar Eclipse of 9 March 2016" in 1st Annual Applied Science and Engineering Conference (AASEC), IOP Conf. Series: Materials Science and Engineering 180, edited by A.G. Abdulah *et al.* (American Institute of Physics, Melville, NY, 2017), pp. 1–6.
[13] S.S. Kumar, and R. Rengaiyan, Natural Science 3, 69-74 (2011).
[14] A. Laesanpura, T. Hidayat, D. Abdurachman, P. Mahasena, P.W. Premadi, H. Wulandari, Y. Suharyadi, and A. Sjarmidi, "Micro-gravity measurements during the total solar eclipse of 9 March 2016 in Indonesia" in International Symposium on Sun, Earth, and Life (ISSEL), IOP Conference Proceedings 771, edited by H. Gunawan *et. al.* (American Institute of Physics, Melville, NY, 2016), pp. 1–4.
[15] E.P. Raharja and Y. Pramudya, "Outreach activities in anticipation of the 2016 solar eclipse in Sorong" in International Symposium on Sun, Earth, and Life (ISSEL), IOP Conference Proceedings 771, edited by H. Gunawan et. al. (American Institute of Physics, Melville, NY, 2016), pp. 1–4.
[16] J.M. Pasachoff, Research in Astron. Astrophys. 9 (6), 613–634 (2009).
[17] A.P. Rachmadian, C. Kunjaya, W. Wahono, and A.A. Anugrah, "Public outreach and education during the 2016 total solar eclipse in Palu and Malang" in International Symposium on Sun, Earth, and Life (ISSEL), IOP Conference Proceedings 771, edited by H. Gunawan *et. al.* (American Institute of Physics, Melville, NY, 2016), pp. 1–4.
[18] L.S. Riza, Y. Wihardi, E.A. Nurdin, N.D. Ardi, C.P. Asmoro, A.F.C. Wijaya, J.A. Utama, and A.B.D. Nandiyanto, "Analysis on atmospheric pressure, temperature, and wind speed profiles during total solar eclipse 9 March 2016 using time series clustering" in International Symposium on Sun, Earth, and Life (ISSEL), IOP Conference Proceedings 771, edited by H. Gunawan et. al. (American Institute of Physics, Melville, NY, 2016), pp. 1–4.
[19] S. Fatima, Widyanita, H. Fahriyah, A.K. Rhodiyah, C.D. Satrya, M. Hilmi, G.E. Ramadhania, L. Naufal, F.A.M. Mulki and D. Herdiwijaya, "Total solar eclipse education for young generation at Palangkaraya, Central Kalimantan" in International Symposium on Sun, Earth, and Life (ISSEL), IOP Conference Proceedings 771, edited by H. Gunawan *et. al.* (American Institute of Physics, Melville, NY, 2016), pp. 1–4.
[20] A.T. Handini, Y. Yulianty, P.W. Premadi, and A. Annafi, "UNAWE Indonesia project: raising total solar eclipse 2016 awareness through educational packages" in International Symposium on Sun, Earth, and Life (ISSEL), IOP Conference Proceedings 771, edited by H. Gunawan *et. al.* (American Institute of Physics, Melville, NY, 2016), pp. 1–4.
[21] P. Singh, N.M. Ravindra, Solar Energy Materials & Solar Cells, 101, 36–45 (2012).
[22] M. Zoungrana, I. Zerbo, B. Soro, M. Savadogo, S. Tiedrebeogo, D.J. Bathiebo, Adv. Sci. Tech. Res. J, 11(2), 133–138 (2017).
[23] E. Urrejola, K. Peter, H. Plagwitz, and G. Schubert, J. Appl. Phys. 110, 056104 (2011).
[24] S.A. Sulaiman, H.H. Hussain, N.S.H.N. Leh, and M.S.I. Razali, Int. J. Mech. Mechatron. Eng. 5(10), 2011.

Distribution of Silver (Ag) Nanoparticle in PVA/Ag Nanofiber Fabricated by Electrospinning Method

Sunaryono[1,2,*], Aulia Rachmawati [1], Elya Khunazatus Shima[3], Ahmad Taufiq[1,2], Hendra Susanto[3], Nandang Mufti[1,2], Sujito[1], Sugiharto[4], Munasir[5,6]

[1]Department of Physics, Faculty of Mathematics and Natural Sciences, Universitas Negeri Malang, Jalan Semarang 5 Malang 65145, Indonesia
[2]Research Center of Minerals and Advanced Materials, Faculty of Mathematics and Natural Sciences, Universitas Negeri Malang, Jl. Semarang 5 Malang 65145
[3]Department of Biology, Faculty of Mathematics and Natural Sciences, Universitas Negeri Malang, Jl. Semarang 5 Malang 65451, Indonesia
[4]Department of Sports Science, Faculty of Sports Science, Universitas Negeri Malang, Jl. Semarang 5 Malang 65145, Indonesia.
[5]Department of Physics, Faculty of Mathematics and Natural Science, Universitas Negeri Surabaya, Kampus Ketintang, Jl. Ketintang, Surabaya, 60231.
[6]Research Center of Advanced Materials, Universitas Negeri Surabaya, Kampus Ketintang, Jl. Ketintang, Surabaya 60231.

*Corresponding author's email: sunaryono.fmipa@um.ac.id

Abstract. This research has successfully fabricated Ag nanoparticles and PVA/Ag nanofiber as antibacterial agents. The silver nanoparticles were synthesized using a chemical reduction method, while PVA/Ag nanofiber was successfully fabricated with silver nanoparticles and PVA polymer materials using electrospinning method. The chemical reduction method is an effective method for obtaining nanometer-sized silver particles, while electrospinning is one of the most efficient and inexpensive methods to produce nanofiber products. The silver nanoparticles and PVA/Ag nanofiber were characterized using instruments of XRD, SEM, EDX, and antimicrobial test for PVA/Ag nanofiber. In addition, through an X-ray diffraction characterization, the average size of silver nanoparticles was approximately 19.12 nm. Meanwhile, the SEM morphological analysis of PVA/Ag nanofiber has an average fiber diameter of 449.11 nm. The result of EDX analysis of PVA/Ag nanofiber showed the aggregation of silver nanoparticles in the range of polymer fibers shown by white clumps. In addition, the antibacterial test results on PVA/Ag nanofiber within 24 hours of incubation showed that the largest nanofiber zone of inhibition was in the diameter range of 0.9 cm. This case shows that PVA/Ag nanofiber has a relatively low antimicrobial power. This condition can be increased by reducing the possible impact of contamination from PVA/Ag nanofiber by the surrounding environment. Thus, the PVA/Ag nanofiber resulted in this study has a potential to be used as a wound dressing under optimal conditions.

Keywords: Silver, PVA, electrospinning, nanofiber.

1. Introduction

Nanoscience is a science that studies the nature of matters with the size range of 1-100 nm [1]. Meanwhile, nanotechnology is a technique of manipulating atoms and molecules to create new materials and structures at the nanoscale (1-100 nm) with new properties [2]. Nanotechnology involves the process, separation, deformation of the material by one atom or one molecule [3]. The material in the form of nanoparticles has unique properties, which can be controlled and modified in size, shape, chemical properties, and surface functionalization [4]. Nanomaterials have diverse applications in areas such as electronics, medicine, health, energy, biotechnology, information technology, agriculture, environmental food, and others [2]. Metal nanoparticles, such as gold, silver, iron, zinc, and metal oxides as well as multi-walled carbon nanotube (MWCNT)-magnetite nanocomposites have great opportunities in biomedical applications due to their large surface area to volume ratio [5,6].

The studies on one-dimensional nanostructured materials such as nanofiber, nanorod, nanobelt, and nanotubes have been reported by Xinxing *et al.* [7]. Nanofiber is identified as a very fine fiber with a very small diameter (below 100 nm) [8]. Producing nanofiber can be done with several techniques or methods, one of them is electrospinning method which is a relatively easy, simple, and effective method to produce nanofiber [9] starting from the submicron diameter to nanometer diameter by using a high electric field potential [9], [10]. Each type of nanofiber can be produced from various types of polymers including natural and synthetic polymers. The polymers used to produce the nanofiber generally have good characteristics thus have a potential as filtration, optical fiber, pharmaceutical delivery systems, medical scaffolding network, and protective clothing [11].

Silver nanoparticles have been extensively studied for their unique physical, chemical, and microbial properties, especially in the fields of optics, catalysts, and biomedicine [12]. The silver nanoparticles synthesized by using the Tollens method and dextrin as the stabilizers have a uniform morphology and particle size distribution as well as having the function as antimicrobials [13]. The Ag/PVA nanofiber composites produced by electrospinning technique have an average diameter of 13.8 nm as an antimicrobial agent [14]. The PVA/AgNPs nanofibers produced by electrospinning technique have a nanofiber range size of 135-160 nm, while the size of silver nanoparticles (AgNPs) are ranged between 15 to 27 nm, in which the results can be used for wound dressing components, protective coatings, biomedical devices, and water purification [15]. However, research on the distribution of silver nanostructures in PVA nanofiber is rarely conducted, this study focused on the characterization of silver nanoparticles in PVA nanofiber and antibacterial characterization that have potential applications as wound dressings.

2. Materials and Methods

The materials used for this research were silver nitrate ($AgNO_3$), sodium borohydride ($NaBH_4$), polyvinyl alcohol (PVA), polyvinyl pyroledone (PVP), 96% alcohol, and aquades. In the initial stage, the oxygen PVP (first solution) was reacted with 475 g of PVP and then put into 475 ml of aquades and $NaBH_4$ solution (second solution) by adding 0.433 g $NaBH_4$ and incorporated into the 25 mL of PVP solution. And then, 0.196 gram of $AgNO_3$ was incorporated into the first solution, stirred with a magnetic stirrer until became a homogeneous solution. The temperature was kept between 5 to10° C. Then, the deposition of the solution was put in the oven at a temperature of 100 ° C. After that, the characterization was done by using XRD.

After the Ag nanoparticles were formed, the next stage to be conducted was to produce a PVA/Ag nanofiber with an electrospinning technique. 0.125 gr of Ag nano was incorporated into 1 ml of aquades stirred with a magnetic stirrer until homogeneous. Next, the 0.5 gr of PVA was incorporated into 3.375 ml of aquades stirred with a magnetic stirrer at a speed of 700 rpm and a temperature of 120° C. Subsequently, the Ag nano solution, and PVA solution were then mixed, stirred with a magnetic stirrer. Furthermore, the formed solution was inserted into the ultrasonic cleaner for 2-3 hours. The solution was put into the syringe and then the process of electrospinning was conducted. After the nanofiber was formed, it was followed by morphological characterization using SEM and antibacterial activity test.

3. Results and Discussion

3.1. Silver (Ag) Nanoparticles

X-ray diffraction results of Ag nanoparticles are shown in Figure 1.

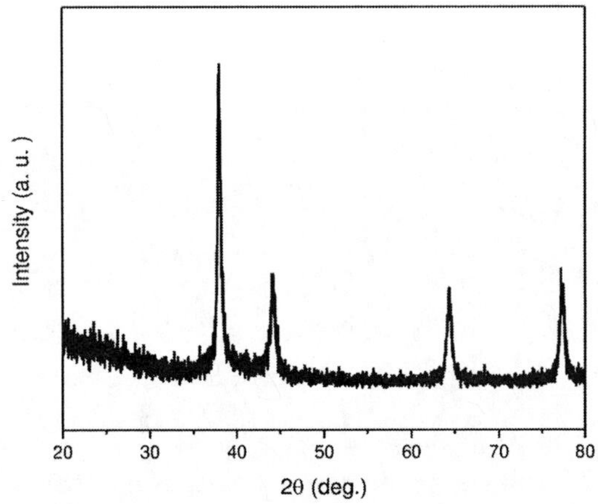

Figure 1. The Peak of X-Ray Diffraction of Ag Nanoparticles

Based on the diffraction pattern of Figure 1, the peaks that appear are located in the planes 111, 200, 220, and 311. The highest peak is in the (111) plane with the diffraction angle (2θ) between 36.76 °-39.02 °. This diffraction pattern is in accordance with previous research which has been done by Zhang *et al.* [14] which states that the highest peak lies at an angle of 38.2°. In order to find out the particle size of the silver nanoparticle sample, the fittings were conducted on the highest peaks of Figure 1. The parameters obtained from the fitting results are calculated into the Scherrer equation to obtain the value of the crystal size. The following Scherrer equations are used.

$$D = \frac{k\lambda}{\beta \cos\theta} \tag{1}$$

Where:
k = constant of *Scherrer* (0.9)
λ = length of radiation wave of Cu Kα (1.5406 Å)
β = FWHM (radian)
θ = Bragg Angle (radian)

From the highest peak analysis results from X-ray diffraction data of silver nanoparticles using equation 1, a particle size of 19.12 nm was obtained. The results of different particle sizes shown in other studies were the obtained size of nanosilver particles which were 13.8 nm [14] and 15-27 nm [15].

IOP Conf. Series: Journal of Physics: Conf. Series **1093** (2018) 012045 doi:10.1088/1742-6596/1093/1/012045

3.2. Results of SEM-EDX characterization on PVA/Ag Nanofiber

The morphology of the PVA/Ag nanofiber characterized by SEM instrument is shown in Figure 2.

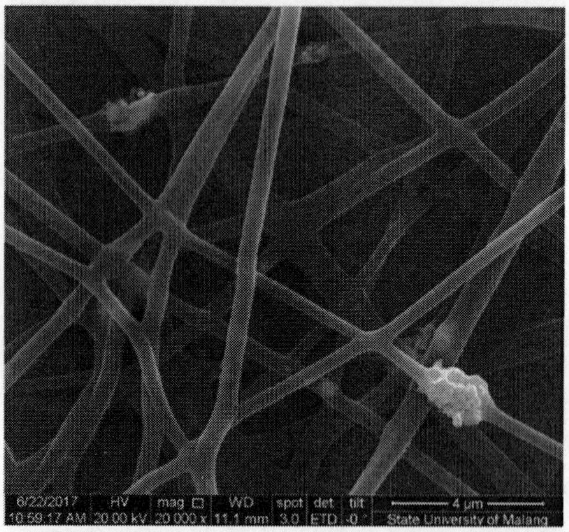

Figure 2. SEM Morphology of PVA/Ag Nanofiber (20.000 x)

As seen from the result of SEM characterization with a 20,000x magnification, the morphology of the PVA/Ag nanofiber produced has an average diameter size of 449.11 nm. The results obtained in this study were quite large compared to the previous research that has been done by Tan *et al.* [15] which obtained a nanofiber diameter of 135-160 nm. The difference of the diameter resulted depends on several parameters during the electrospinning process, among which were the properties of polymer solutions such as a solution concentration, molecular weight, viscosity, surface tension, dielectric constant, and conductivity [14,15].

The EDX characterization of the PVA/Ag nanofiber sample was performed to determine the percentage of silver nanoparticles contained. The result of EDX PVA/Ag nanofiber characterization is shown in Figure 3.

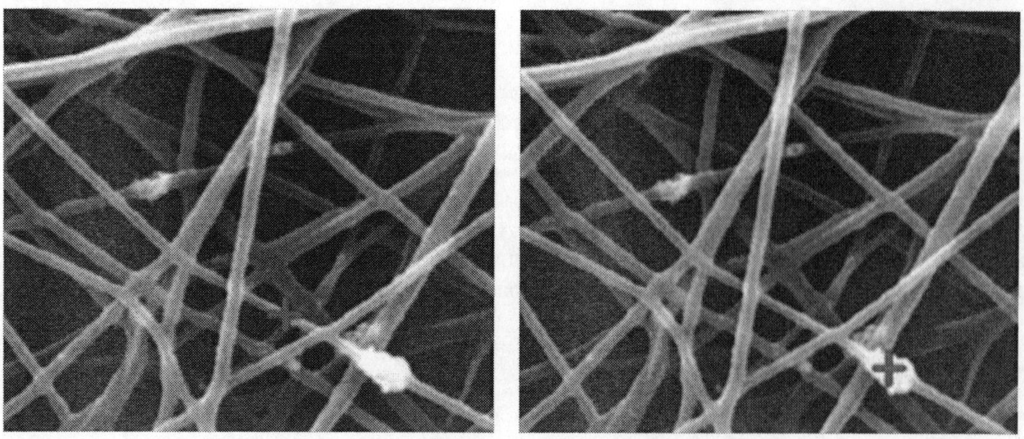

a) b)

Figure 3. The Result of SEM-EDAX PVA/Ag Nanofiber

Figure 3 shows an EDX characterization of a PVA/Ag nanofiber sample at a particular point marked with a plus sign (+) of red. The percentage of silver nanoparticle content obtained in Figure 3a is 5.68 wt% while in Figure 3b the percentage is 12.24 wt%. The result in Figure 3b was larger because at that point the silver nanoparticles underwent agglomeration indicated by white clumps. The similar study conducted by Lin *et al.* [16] showed an EDX analysis of several samples of nanofiber PVA/Ag with a nanoparticle content about 2.6 wt% and 12.4 wt%. The higher particle content of Ag in the PVA/Ag nanofiber causes the resulted diameter to be greater [17,18].

3.3. The Result of the Antibacterial Test of PVA/Ag Nanofiber

The results of the antibacterial test performed on the PVA/Ag nanofiber shown in Table 1 with the bacteria used for testing were *Escherichia coli* and *Staphylococcus aureus*.

Table 1. The Result of Antimicrobial Potential in PVA/Ag Nanofiber

Bacteria	Repetition	Diameter of Resistor Zone (cm)
	1	0.80
Escherichia coli	2	0.90
	3	0.70
	1	0.70
Staphyllococcus aureus	2	0.70
	3	0.70

The antibacterial test was conducted in 24 hours of incubation time. The antibacterial activity test on PVA/Ag nanofiber was performed on two types of bacteria, *Escherichia coli* (gram-negative) and *Staphylococcus aureus* (gram-positive). Based on the results obtained, PVA/Ag nanofiber could cause the zone of inhibition against the growth of colonies of *E. coli* and *S. aureus* bacteria. The zone of inhibition diameter resulted varied and was based on the measurement results, the largest zone of inhibition was present in the test of *Escherichia coli* (see Table 1).

Table 1 shows that the PVA/Ag nanofiber produces larger inhibitory diameters on *Escherichia coli* bacteria compared to the inhibitory diameters of *Staphylococcus aureus* bacteria. In the 24-hour incubation time, the largest zone of inhibition was only 0.9 cm in diameter, it can be said that the PVA/Ag nanofiber has low rates of antimicrobial resistance.

Different results of study are presented by Kong and Jang [19] in their research on the bioactivity of nanocomposite antibacterial performance against E. Coli bacteria showed that the presence of silver nanoparticles in the composite was effective in killing and inhibiting the growth of E. coli bacteria, besides the composites studied using methyl acrylate and Ag nanoparticles have been proven to have antibacterial power. The antibacterial characteristics of the silver nanoparticle composites are closely related to the properties of silver nanoparticles that have antibacterial properties [20,21]. The differences in the results of this study with previous studies are due to the time interval of PVA/Ag nanofiber synthesis with antimicrobial tests was too long, allowing the PVA/Ag nanofiber to be contaminated and reduce its antibacterial power. Thus, it needs to be re-tested to ensure the fact about the antibacterial power of PVA/Ag nanofiber.

4. Conclusion

This research concluded that the result of crystal size analysis from the Ag nanoparticles obtained was 19.12 nm. The morphological analysis of PVA/Ag nanofiber using SEM has an average diameter of 449.11 nm. Meanwhile, EDAX analysis of PVA/Ag nanofiber indicated that the aggregation of silver nanoparticles in the range of polymer fibers was characterized by white clumps. The results of the antimicrobial testing on the PVA/Ag nanofiber in the 24-hours of incubation period resulted in a 0.9 cm diameter zone of inhibition, it can be said that the PVA/Ag nanofiber has low rates of antimicrobial power. This is due to the time interval between producing the PVA/Ag nanofiber with the antimicrobial test was too long, thus allowing PVA/Ag nanofiber to be contaminated and reduce its antibacterial power. In sum, the potential of the PVA/Ag nanofiber produced is as an antibacterial agent which can be applied as a wound dressing.

Acknowledgments
The authors would like to express our deep gratitude for the financial support given by PSNI of KEMENRISTEKDIKTI for SN.

References

[1] de Mello Donegá C 2014 *Nanoparticles: Workhorses of Nanoscience* (Springer)
[2] Prasad T N V K ., Subba Rao Kambala V and Naidu R 2011 A critical review on biogenic silver nanoparticles and their antimicrobial activity *Curr. Nanosci.* **7** 531–544
[3] Elumalai E K, TNVKV P, Nagajyothi P C and David E 2011 A bird's eye view on biogenic silver nanoparticles and their applications *Chem Sin* **2** 88–97
[4] Nagarajan R and Hatton T A 2008 *Nanoparticles: synthesis, stabilization, passivation, and functionalization* (ACS Publications)
[5] Prasad S B, Aeri V and others 2013 Current Understanding of Synthesis and Pharmacological Aspects of Silver Nanoparticles *Am. J. Phytomedicine Clin. Ther.* **1** 536–547
[6] Rahmawati R, Melati A, Taufiq A, Sunaryono, Diantoro M, Yuliarto B, Suyatman S, Nugraha N and Kurniadi D 2017 Preparation of MWCNT-Fe_3O_4 Nanocomposites from Iron Sand Using Sonochemical Route *IOP Conf. Ser. Mater. Sci. Eng.* **202** 012013
[7] Xingxing C, Xuebin Z, Hao S and Yi F 2014 Fabrication of Magnetic Fe_3O_4 Nanotubes by Electrospinning *Rare Met. Mater. Eng.* **43** 2330–2334
[8] Frenot A and Chronakis I S 2003 Polymer nanofibers assembled by electrospinning *Curr. Opin. Colloid Interface Sci.* **8** 64–75
[9] Bhardwaj N and Kundu S C 2010 Electrospinning: a fascinating fiber fabrication technique *Biotechnol. Adv.* **28** 325–347
[10] Sill T J and von Recum H A 2008 Electrospinning: applications in drug delivery and tissue engineering *Biomaterials* **29** 1989–2006
[11] Wahyudi T and Sugiyana D 2011 Pembuatan Serat Nano Menggunakan Metode Electrospinning *Arena Tekst.* **26**
[12] Korbekandi H and Iravani S 2012 Silver nanoparticles *The delivery of nanoparticles* (InTech)
[13] Zhang Z-Q, Higgins J, Kim C L and Chang C 2013 Synthesis and antimicrobial properties of nanosilver *J Mater Env. Sci* **4** 139–142
[14] Zhang Z, Wu Y, Wang Z, Zou X, Zhao Y and Sun L 2016 Fabrication of silver nanoparticles embedded into polyvinyl alcohol (Ag/PVA) composite nanofibrous films through electrospinning for antibacterial and surface-enhanced Raman scattering (SERS) activities *Mater. Sci. Eng. C* **69** 462–469
[15] Tan G, Sağlam S, Emül E, Erdönmez D and Sağlam N 2016 Synthesis and characterization of silver nanoparticles integrated in polyvinyl alcohol nanofibers for bionanotechnological applications *Turk. J. Biol.* **40** 643–651
[16] Lin S, Wang R-Z, Yi Y, Wang Z, Hao L-M, Wu J-H, Hu G-H and He H 2014 Facile and green fabrication of electrospun poly (vinyl alcohol) nanofibrous mats doped with narrowly dispersed silver nanoparticles *Int. J. Nanomedicine* **9** 3937
[17] Cui W, Zhou Y and Chang J 2010 Electrospun nanofibrous materials for tissue engineering and drug delivery *Sci. Technol. Adv. Mater.* **11** 014108
[18] Thompson C J, Chase G G, Yarin A L and Reneker D H 2007 Effects of parameters on nanofiber diameter determined from electrospinning model *Polymer* **48** 6913–6922
[19] Kong H and Jang J 2008 Antibacterial Properties of Novel Poly(methyl methacrylate) Nanofiber Containing Silver Nanoparticles *Langmuir* **24** 2051–6
[20] Saquing C D, Manasco J L and Khan S A 2009 Electrospun nanoparticle–nanofiber composites via a one-step synthesis *Small* **5** 944–951

[21] Ahamed M, Khan M M, Siddiqui M K J, AlSalhi M S and Alrokayan S A 2011 Green synthesis, characterization and evaluation of biocompatibility of silver nanoparticles Phys. E Low-Dimens. Syst. Nanostructures 43 1266–1271

UV Irradiation Enhanced In-Vitro Cytotoxic Effects of ZnO Nanoparticle on Human Breast Cancer

U Lestari[1,*], N Mufti[2], D A Lutfiyah[1], U Fitriyah[1], Y Annisa[1]

[1]Department of Biology, Faculty of Mathematics and Natural Sciences, Universitas Negeri Malang, Jl. Semarang 5 Malang 65145, Indonesia

[2]Department of Physics, Faculty of Mathematics and Natural Sciences, Universitas Negeri Malang, Jl. Semarang 5 Malang 65145, Indonesia

*Corresponding author's email: umie.lestari.fmipa@um.ac.id

Abstract. Cancer is a disease characterized by abnormal cell growth in body tissues. Generally, cancer can be treated with surgery, chemotherapy, radiation therapy, and many others. However, these treatments have side effects on healthy cells. ZnO nanoparticle is one of considerable interests in targeting pharmaceutical for the treatment of cancer by inhibiting the cell proliferation. Therefore, in this study, we investigated *In vitro* cytotoxic effects of ZnO nanoparticles on the growth of MCF-7 breast cancer cells. The cytotoxic test was performed using varying concentration of ZnO nanoparticles (0, 25, 50, 75, 100 µg/mL) with different times of UV irradiation (0, 90, 180, and 270 seconds). The cytotoxic effect was evaluated by measuring the absorption of formazan crystal at wavelength of 490 nm using ELISA reader and converted to determine the percentage of living cells. The surviving cancer cells are characterized by their ability to convert MTT salts into formazan crystals. The concentration of ZnO nanoparticles caused the decrease in viability of MCF-7 cells, with the lowest viability at ZnO 100 µg/ml concentration of 57.5%. UV radiation increased the toxicity effect of ZnO nanoparticles on MCF-7 cells. A combination of ZnO 100 µg/ml nanoparticles with UV radiation exposure for 180 seconds had the lowest viability of the MCF-7 cells of 13.58%.

Keywords: Cancer, ZnO, UV irradiation, cytotoxic

1. Introduction

Cancer on the reproductive system, such as the mammary glands, is the leading cause of death among women worldwide. The female reproductive organ that possibly gets cancer is breast [1]. Breast cancer is the one of cancers widely studied and it becomes the most common cause of cancer-related deaths in women worldwide [2]. Based on GLOBOCAN (IARC) data in 2012, it is known that breast cancer is a cancer with the highest percentage of new cases in the female population in the world, that was 43.3%, and the percentage of deaths due to breast cancer was 12.9%. In Indonesia, the estimated incidence of breast cancer is 40 per 100,000 women [3].

The most common cancer treatments include surgery, radiation, and chemotherapy [4]. However, these treatments have adverse effects with high levels of morbidity in the form of normal tissue damage that can kill and damage healthy cells [5], and have toxic effects on healthy cells [4]. Photodynamic therapy (PDT) is a new method in the treatment of cancer. PDT has been proposed as a minimally invasive therapeutic procedure, utilizing three important elements to induce cell death, namely photosensitizer (PS), light with a certain wavelength, and oxygen molecules [6]. PDT has no long-term side effects when compared to surgical procedures, chemotherapy or radiotherapy [7]. ZnO nanoparticle as a photosensitizer has an excellent chance to be utilized in the photodynamic (PDT) therapy with its ability as an anticancer agent. The previous study showed that, for example, Li et al. (2010) used UV radiation to increase the strength of ZnO nanoparticles to suppress the growth of hepatocellular carcinoma SMMC-7221 cancer cells [8]. A ZnO nanoparticle is a semiconductor material with a wide band-gap, that can readily absorb UV rays [9].

ZnO nanoparticles have been used in various applications in cancer therapy, biosensing, drug/ gene delivery, nanomachines, and biomedical application [9]. Akhtar et al. (2012) reported that ZnO nanoparticles have toxic effects on cancer cells (human hepatocellular carcinoma HepG2, human lung adenocarcinoma A549, and human epithelial bronchiolus BEAS-2B), but not toxic to healthy cells (astrocytes and hepatocytes) [10]. This study aimed to determine the effect of ZnO nanoparticles with UV radiation exposure on breast cancer cells (MCF-7) evaluated with the test of cytotoxicity by using 3- (4,5-dimethylimidazole-2-yl) reagents is 2,5-diphenyl tetrazolium bromide (MTT).

2. Materials and Methods

2.1. Cell Culture

Michigan Cancer Foundation-7 (MCF-7) cells (purchased from Sentral Ilmu Hayati Laboratory Universitas Brawijaya) were maintained in Roswell Park Memorial Institute -1640 medium (Gibco, USA) supplemented with 10% fetal bovine serum (Sigma, USA), 10% penicillin (Sigma, USA)-streptomycin (Sigma, USA) and were incubated in 5% CO_2 at 37^{0}C. When the cells were confluent, the experiment could be started.

2.2. ZnO Nanoparticles

ZnO nanoparticles were purchased from Mineral and Advanced Material Laboratory FMIPA Universitas Negeri Malang, and dissolved in culture medium with the concentration of 0, 25, 75, and 100 μg/mL.

2.3. MTT-Assay

The effect of different ZnO nanoparticles concentrations on MCF-7 cancer cells was observed by MTT assay. The ZnO nanoparticles were at the concentrations of 0, 25, 75, and 100 μg/mL with different times of UV irradiation at 0, 90, 180, and 270 seconds. Initially, $1x10^4$ cells were seeded into each well containing 200 μL cell culture mediums in 96-well plate and incubated for 24 hours. On the next day, the cell culture was exposed to ZnO nanoparticle with serial concentration and followed by UV irradiation at different times exposure. The cells were incubated in 5% CO_2 at 37 ^{0}C for 24 hours and 48 hours. At the end of the incubation, the cells were washed using PBS. Then, 200 μL MTT solution (5 mg/mL) was added into the well and were incubated for 4 hours until forming formazan crystals. The formation of formazan crystals was observed under an inverted microscope (Olympus 1X-71, Japan), and absorption of formazan crystal was measured at wavelength of 490 nm using ELISA reader (Biorad), then the absorbance value was converted to determine the percentage of living cells. The placebo was cultivated under the same conditions without addition of ZnO nanoparticles and UV irradiation.

3. Results and Discussion

The cytotoxic effects of ZnO nanoparticles were determined by the MTT-assay method. The formation of formazan crystal in the cell culture is shown in Figure 1. The cell viability number was equal to the amount of formazan crystal within the cell and associated with the higher metabolic rate of the cell. [11].

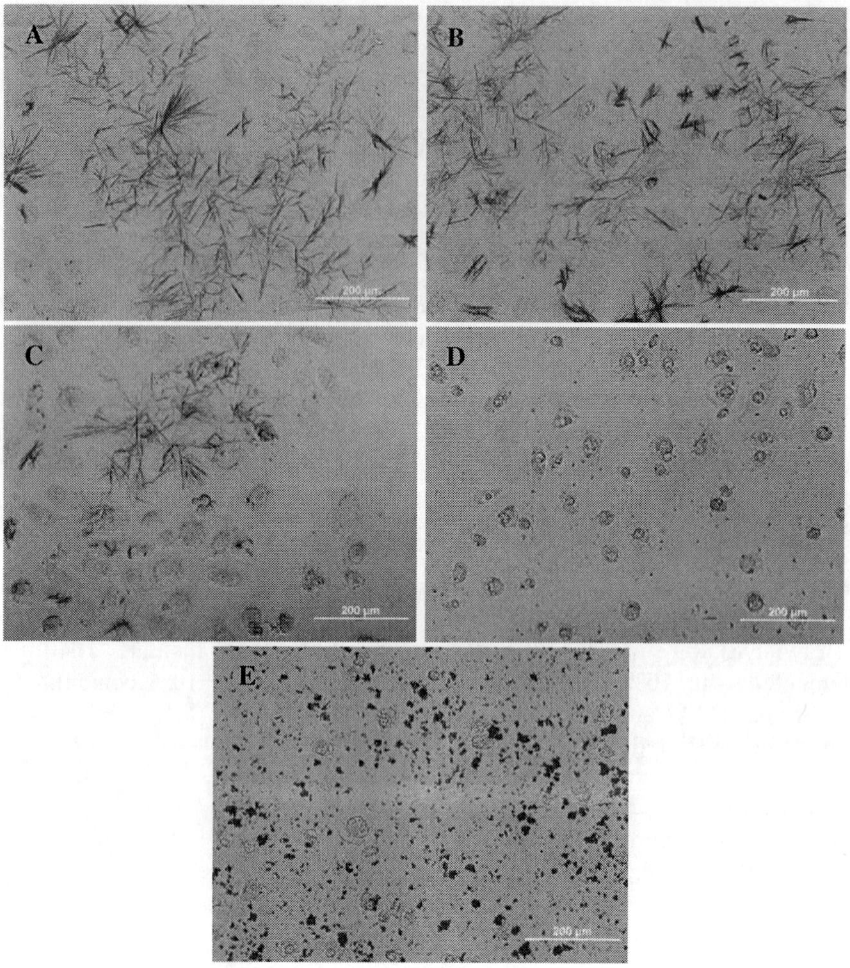

Figure 1. Formation of formazan crystal on MCF-7 cells after treatment with ZnO nanoparticles and UV radiation for 90 s at 48 hours incubation; (A) ZnO nanoparticles with the concentration of 0 μg/ml, (B) ZnO nanoparticles with the concentration of 25 μg/ml, (C) ZnO nanoparticles with the concentration of 50 μg/ml, (D) ZnO nanoparticles with the concentration of 75 μg/ml, (E) ZnO nanoparticles with the concentration of 100 μg/ml; 400x.

The data analysis in Figure 2 shows an increase in the concentration of ZnO nanoparticles (without UV irradiation) causing the decrease in viability of MCF-7 cells, with the lowest viability at ZnO 100 μg / ml concentration of 57.5%. This result corresponds the study of Taccola et al. (2011), who reported that ZnO nanoparticles had selective cytotoxic properties against cell proliferation in benign and malignant tumors [12]. Mathuraman et al. (2014) also reported that ZnO nanoparticles could alter 3T3-L1 cell morphology, and increase cell death at the highest concentration of ZnO [13]. Elevated levels of intracellular ZnO can enhance cytotoxicity through zinc-mediated protein activity disequilibrium and oxidative stress. ZnO nanoparticles can induce oxidative stress in cancer cells, due to the semiconductor nature of ZnO induces ROS generation [9]. Production of reactive oxygen species (ROS) results in oxidative stress and cellular toxicity, which leads to DNA damage and cell death when the anti-oxidative capacity of the cell is exceeded [14].

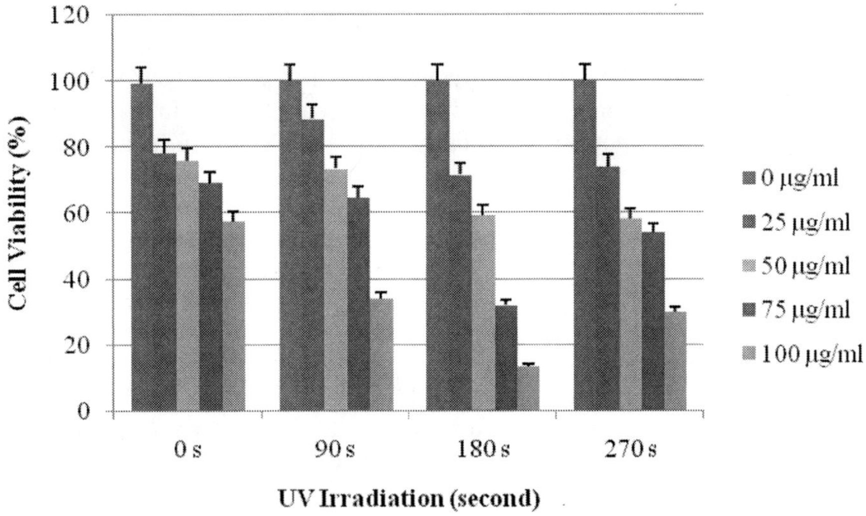

Figure 2. The level of cell MCF-7 viability after treatment with ZnO nanoparticles and UV irradiation at 24 hours incubation; the values are presented as mean ± SD.

UV radiation increased the toxicity effect of ZnO nanoparticles on MCF-7 cells, as indicated by the decreased viability of MCF-7 cells with exposure to UV radiation for 90, 180, and 270 s (Figures 2 and Figure 3). On 24-hour incubation, a combination of ZnO 100 µg / ml nanoparticles with UV radiation exposure for 180 seconds had the lowest viability of the MCF-7 cells of 13.58%. While at 48 hours incubation, the lowest cell viability of 13.61% was demonstrated at the same dose of ZnO nanoparticles, but with exposure to UV radiation for 90 seconds.

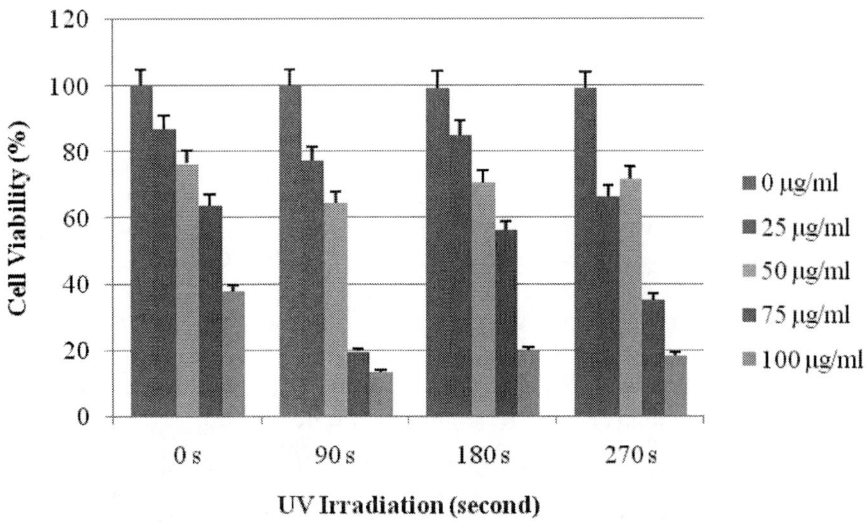

Figure 3. The level of cell MCF-7 viability after treatment with ZnO nanoparticles and UV irradiation at 48 hours incubation; the values are presented as mean ± SD.

UV light is a form of electromagnetic radiation that can induce apoptosis through various molecular pathways [15]. UV radiation exposure can produce oxidative free radicals that are capable of interacting with macromolecules such as proteins, lipids, RNAs, and DNAs that alter the structure and disrupt the function of the macromolecule. UV light is efficient in generating ROS that can damage DNA through indirect photosensitization reactions, and can also be absorbed directly by the DNA that causes the mutation [16].

ZnO has a light-capturing group called chromophores, which will generate excessive energy when absorbing UV light at specific wavelengths, resulting in the unstable ZnO structure. Hence, there will be electron transfer (e^-) from the valence band to the ZnO conduction band, simultaneously producing a hole (h^+) in the valence band. There is a charge imbalance so that ZnO is immediately neutralized by the H_2O molecule to form a hydroxyl radical (OH·) which is free radical. According to Konopka and Goslinski (2007), the formed free radicals are very active and can damage proteins, lipids, and other cellular components, which can destroy cancer cells [17].

ZnO nanostructures are well potential photosensitizers based on their phototoxic action upon UV irradiation. ZnO nanoparticles are reported to exhibit a robust preferential ability to kill cancerous cells compared to healthy cells, and able to induce excessive production of reactive oxygen species (ROS) in various cell lines, resulting in oxidative stress and cellular toxicity [14]. The ZnO nanoparticles were effective to decrease the viability of cancer cells, and UV irradiation could significantly enhance the cytotoxic effect on cancer.

4. Conclusion

ZnO nanoparticles were useful to decrease the viability of MCF-7 cells, in the amount of 57.5% at ZnO 100 µg / ml concentration. UV radiation could enhance the ability of ZnO nanoparticles against MCF-7 cells of equal level, with reduced viability to 13.58%. The combination of ZnO nanoparticles and UV radiation can be developed as a method of cancer therapy on an in-vivo scale. ZnO nanoparticles potentially becomes an anti-cancer agent in the reproductive system, especially breast cancer cells.

Acknowledgments

We would like to thank Helly Nurul Karima, S.Pt., M.P. at LSIH-UB who has provided support for the development of this research as well as to other technicians.

References

[1] Toss A, Grandi G, Cagnacci A, Marcheselli L, Pavesi S, De Matteis E, Razzaboni E, Tomasello C, Cascinu S and Cortesi L 2017 The impact of reproductive life on breast cancer risk in women with family history or BRCA mutation. *Oncotarget* **8** 9144–54

[2] Costa E C, Gaspar V M, Marques J G, Coutinho P and Correia I J 2013 Evaluation of Nanoparticle Uptake in Co-culture Cancer Models *PLoS ONE* **8** 1–13

[3] Kementrian Kesehatan RI Pusat Data dan Informasi Kesehatan 2015 Stop Kanker *Infodatin-Kanker* hal 3

[4] Wang S, Fan W, Kim G, Hah H J, Lee Y K, Kopelman R, Ethirajan M, Goswami L N, Pera P, Morgan J and Ravindra K 2013 NIH Public Access **43** 686–95

[5] Yang Y and Wang H 2013 Perspectives of Nanotechnology in Minimally Invasive Therapy of Breast Cancer *J. Healthc. Eng.* **4** 67–86

[6] Debele T A, Peng S and Tsai H C 2015 *Drug carrier for photodynamic cancer therapy* vol 16

[7] Calixto G, Bernegossi J, de Freitas L, Fontana C and Chorilli M 2016 Nanotechnology-Based Drug Delivery Systems for Photodynamic Therapy of Cancer: A Review *Molecules* **21** 342

[8] Li J, Guo D, Wang X, Wang H, Jiang H and Chen B 2010 The photodynamic effect of different size ZnO nanoparticles on cancer cell proliferation in vitro *Nanoscale Res. Lett.* **5** 1063–71

[9] Bisht G and Rayamajhi S 2016 ZnO Nanoparticles: A Promising Anticancer Agent *Nanobiomedicine* **3** 9

[10] Javed Akhtar M, Ahamed M, Kumar S, Majeed Khan M, Ahmad J and Alrokayan S A 2012 Zinc oxide nanoparticles selectively induce apoptosis in human cancer cells through reactive oxygen species *Int. J. Nanomedicine* **7** 845–57

[11] Riss T L, Moravec R A, Niles A L, Duellman S, Benink H A, Worzella T J and Minor L 2013 Cell Viability Assays *Assay Guid. Man. Internet* **114** 785–96

[12] Taccola L, Raffa V, Riggio C, Vittorio O, Iorio M C, Vanacore R, Pietrabissa A and Cuschieri A 2011 Zinc oxide nanoparticles as selective killers of proliferating cells. *Int. J. Nanomedicine* **6** 1129–40

[13] Muthuraman P, Ramkumar K and Kim D H 2014 Analysis of Dose-Dependent Effect of Zinc Oxide Nanoparticles on the Oxidative Stress and Antioxidant Enzyme Activity in Adipocytes *Appl. Biochem. Biotechnol.* **174** 2851–63

[14] Tripathy N, Ahmad R, Ah Ko H, Khang G and Hahn Y B 2015 Multi-synergetic ZnO platform for high performance cancer therapy *Chem. Commun.* **51** 2585–8

[15] Pronin S, Koh C H and Hughes M 2017 Effects of Ultraviolet Radiation on Glioma: Systematic Review *J. Cell. Biochem.* **118** 4063–71

[16] D'Orazio J, Jarrett S, Amaro-Ortiz A and Scott T 2013 UV radiation and the skin *Int. J. Mol. Sci.* **14** 12222–48

[17] Konopka K and Goslinski T 2007 Photodynamic therapy in dentistry *J. Dent. Res.* **86** 694–707

Optimization of Synthesis of Biodiesel from *Jatropha curcas* L. with Heterogeneous Catalyst of CaO and MgO by Trans-esterification Reaction Using Microwave

Aman Santoso*, Sumari, Dedek Sukarianingsih, Rosi Mayang Sari

Departement of Chemistry, Faculty of Mathematics and Natural Science, Universitas Negeri Malang, Jl. Semarang 5 Malang 65145, Indonesia

*Corresponding author's email: aman.santoso.fmipa@um.ac.id

Abstract. The purpose of this study was to find the optimum conditions in synthesis of Biodiesel from *Jatropha* Curcas with heterogeneous catalyst of CaO and MgO by trans-esterification reaction using microwave. The research was conducted in five stages, that is determining acid number of *Jatropha* oil; esterification process; preparation of CaO and MgO catalysts; synthesis of biodiesel via transesterification reaction in various catalyst concentration and various times; identification and characterization of biodiesel involves the synthesis, testing TLC, GC-MS analysis, the index test bias, the density test, test the viscosity and acid number test. The results showed that optimum conditions the synthesis of biodiesel from *Jatropha* oil through transesterification reaction using microwaves produce yield 92.39% ± 0.04, with a catalyst concentration of 7.5% CaO-MgO. Furthermore, identification of compounds contained in biodiesel using GC-MS showed that biodiesel contains palmitoleic methyl compound (2.40%), methyl palmitate (21.20%), methyl oleate (62.40%), and methyl stearate (14.00%). We found also that the refractive index of the biodiesel is 1.465, the density of 0.888 g/mL, viscosity 4.23 cSt, and acid number is 0.58 mg KOH/g.

Keywords: Biodiesel, *Jatropha curcas*, heterogen catalyst, microwave.

1. Introduction

Petroleum fuels are one of the most important sources of energy for human life. Fuel needs continue to increase, along with the development of industry in various fields. Sustainable petroleum exploration (nonrenewable resources) on an ongoing basis will lead to an energy crisis in the world. Utilization of renewable alternative fuels has great importance for the sustainability of a nation [1,2]. The superiority of renewable energy is that it can be produced in quick time, availability (a lot or renewable source), does not produce harmful, cheap and easy to get emissions.

Alternative fuels, one of which is biodiesel that can be made from raw materials of palm oil, non-edible *Jatropha* and used as fuel for diesel engines [3,4]. Vegetable oils are triglycerides and can be used as raw materials for biodiesel, such as *nyamplung* seed oil, palm oil, and castor oil fence. The use of palm oil as a raw material for biodiesel production collides with the needs of community food production. The selection of biodiesel feedstock is very important to prevent competition between food needs and production needs. Indonesian people recognize *Jatropha curcas* as a bush fence plant that has not been fully utilized, whereas *Jatropha curcas* L. is one of the oil producing plants that can be used as raw material for biodiesel production. *Jatropha curcas* has developed as a raw material in the manufacture of biodiesel [4], [5] by pressing oil from *Jatropha* seeds.

Biodiesel from vegetable oils is made by converting triglycerides to fatty acid alkyl esters by utilizing the catalyst through a transesterification reaction [1,2]. Transesterification reaction is generally done conventionally, so it requires a long reaction time. The biodiesel synthesis time of *Jatropha* oil [1] was 2 hours, using a heterogeneous CaO catalyst of 1.5%. The catalyst was activated by immersing the CaO in an ammonium carbonate solution then dried in an oven at 1100 °C and calcined at 9000 °C for 1.5 hours. Comparison of oil mol: methanol used is 1:12. The castor oil converted to biodiesel was refluxed at 700 °C and a yield of 95% was obtained. The conventional method of biodiesel synthesis of rapeseed oil [6] takes 3.5 hours at 64.50 °C, and the ratio of mole oil to methanol is 1:18. The catalyst used was CaO and MgO by 10%. CaO and MgO catalysts were each calcined at 7000 °C for 2 hours then mixed and stirred. The yield of biodiesel produced is 92%.

Long reaction time on conventional methods can be overcome by utilizing microwaves, so the reaction time goes faster. Utilization of microwaves in the synthesis of biodiesel from fish oil ever done [6]. The ratio of oil and methanol ratio is 1:18, the catalyst used is NaOH, the optimum reaction time required is 10 minutes and yields the yield of 84.5% biodiesel. Synthesis of biodiesel using microwave has also been done [6,7]. The raw material used is kapok seed oil, reaction time required is 5 minutes. The catalyst used is KOH with ratio of oil and methanol that is 1: 7, and yield of biodiesel produced is 98.52%.

In this study used *Jatropha* oil as the raw material of biodiesel production because *Jatropha* oil does not compete with the food needs of the community. The transesterification reaction of *Jatropha* oil and methanol is supported by heterogeneous catalysts of CaO and MgO and microwaves. The purpose of this research is to find the optimum condition of biodiesel synthesis from *Jatropha* oil using microwave wave through transesterification reaction with time variation and CaO and MgO catalyst concentration. The synthesized biodiesel is identified and characterized as, TLC, GC-MS analysis, refractive index test, type density test, viscosity test and acid number test.

Transesterification is the reaction between triglycerides and alcohols to form a new ester and a new alcohol. Transesterification is usually referred to as an alcoholic reaction, since the ester is reacted with an alcohol with the aid of an acid or base catalyst to a new ester form and a by-product of glycerol [8]. The equation of the transesterification reaction is shown at Figure 1.

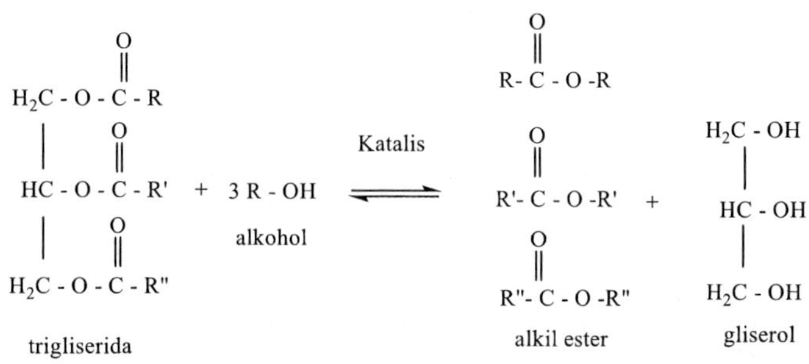

Figure 1. The transesterification reaction occurring between triglycerides and alcohols with the catalyst

The transesterification reaction is reversible that occurs at room temperature and runs slowly in the absence of a catalyst, so as to push the reaction to the right it can be used excess alcohol. The commonly used alcohol in the transesterification reaction is methanol because it is cheap and more reactive than the longer chain alcohol [2,9]. Transesterification requires the catalyst in its reaction, because in the absence of a catalyst, the resulting conversion is maximum but the reaction proceeds slowly. The catalyst commonly used in the transesterification reaction is an alkaline catalyst [8,10].

One type of heterogeneous catalyst that has been studied by some researchers on the manufacture of biodiesel is a group of heterogeneous alkaline base group catalysts, such as magnesium oxide [8]

and calcium oxide [9]. The use of CaO as an alkaline catalyst has many advantages, due to its high activity, low reaction conditions, long catalyst period, and low catalyst cost [1,11]. For MgO catalysts to have optimum activity, MgO catalysts need to be combined with other catalysts such as in combination with CaO catalysts as in a study conducted by Liang *et al.* [3], the yield of biodiesel from rapeseed oil obtained was 92%, using a CaO mixed catalyst and MgO.

The biodiesel properties are similar to those of diesel oil, so biodiesel can be used as diesel fuel. Conversion of triglycerides into methyl esters through a transesterification reaction reduces the molecular weight of triglycerides by one-third, reduces viscosity by one-eighth, and slightly increases the flash point [4,15]. The use of the Microwave oven itself works so quickly and efficiently because the electromagnetic waves penetrate the food and excite the water and fat molecules evenly (not just the surface) [7]. Waves at a frequency of 2.5 GHz is absorbed by water, fat, and sugar. When absorbed, the atoms are excited and generate heat. This process does not require a heat conduction like an ordinary oven, because that's how the process can be done very quickly.

2. Materials and Methods

This research was a laboratory experimental study that aims to obtain optimum conditions of biodiesel synthesis utilizing microwave. The alcohol used for the transesterification reaction is methanol, the catalyst used is heterogeneous CaO and MgO catalysts. The study consisted of 5 stages including determination of *Jatropha* oil number, esterification process, CaO and MgO catalyst preparation, optimization of biodiesel synthesis through transesterification reaction of *Jatropha* oil of esterification using microwaves with variation of time and concentration of CaO and MgO, identification and characterization of synthesis biodiesel includes, TLC test, GC-MS analysis, refractive index test, type density test, viscosity test and acid number test.

2.1 Determination of Jatropha Oil Number

Determination of *Jatropha* oil number is by weighing ± 1 gram of oil and put into Erlenmeyer. Then plus 20 mL of n-hexane and 20 mL of alcohol. Added 3 drops indicator pp then titrated with 0.1 N KOH solution until the color of the solution turns into pink.

2.2 Esterification

A total of 200 g of oil was introduced into a three-necked flask, heated to the temperature of 70 °C, carefully added 12 g of methanol and 1.7 mL of concentrated H_2SO_4, then the mixture was refluxed for 2 hours at temperature of 69 to 71 °C while stirring. Added 6 g of methanol and 0.8 mL of concentrated H_2SO_4, continued reflux for 1 hour at temperature of 69 to 71 °C with stirring, cooled and then separated in separating funnel. The top layer was washed with a little water and heated with a hotplate at a temperature of 90 to 100 °C.

2.3 Preparation of CaO and MgO Catalyst

CaO powder is calcined for 2 hours at temperature of 700 °C with the aim of activating the active side of the catalyst. Similarly, the calcined MgO powder for 2 hours at temperature of 700 °C for the purpose of activating the active side of the catalyst, then stored in a desiccator 24 hours. After CaO and MgO are obtained, the CaO and MgO catalysts are prepared by mixing CaO and MgO with a ratio of 1: 1.

2.4 Biodiesel Synthesis Optimization Through Transesterification Reactions Using Microwaves

A total of 25 g of esterified oil was introduced into the beaker, then carefully added CaO and MgO catalysts with appropriate concentrations (5%, 7.5% and 10%) and the mole oil ratio: methanol 1:18 was then stirred with mechanical stirrer with a speed of 200 rpm. The mixture was introduced into the microwave with time variation (5; 7.5; 10; 12.5; 15) minutes, then centrifuged for 20 minutes at a speed of 3000 rpm. The mixture is then poured into the separating funnel and left for ± 24 hours. The esters in the top layer are separated from the glycerol on the bottom layer by opening the separating funnel valve. The top layer is then washed to remove residual catalyst, water and glycerol with warm water until neutral pH and heated at temperature of 90 to 100 ° C to remove water leftover laundry.

3. Results and Discusion

3.1. Determination of Figures of Jatropha Oil

Jatropha oil which will be converted into biodiesel must be determined its acid number first. The determination of *Jatropha* oil is important because it can affect the amount of biodiesel product. The maximum acceptable acid number in a system using an alkaline catalyst is less than 2.5%. The transesterification reaction of a vegetable oil containing high levels of fatty acids and using an alkaline catalyst will lead to the formation of soap. Soaps will decrease the yield of biodiesel and will complicate the separation of biodiesel by glycerol [5]. The free fatty acid saving reaction is as follows:

$$\underset{\text{Free fatty acid}}{R_1\text{-COOH}} + \underset{\text{sodium hydroxide}}{NaOH} \longrightarrow \underset{\text{fatty acid salls}}{R_1\text{-COONa}} + \underset{\text{water}}{H_2O}$$

The number of *Jatropha* oil acids after analyzed is 14.39% ± 0.07. The acid number is still high and has not met the standard acid number less than 2.5% [2], this will cause the reaction of saponification. *Jatropha* oil must be lowered by acidification by esterification process. *Jatropha* oil must be esterified in advance for transesterification reaction to produce maximum biodiesel product. The reaction occurring in *Jatropha* oil which is esterified using methanol and concentrated sulfuric acid catalyst is as follows.

$$R-\overset{\overset{\displaystyle O}{\|}}{C}-OH + CH_3-OH \xrightleftharpoons{H^+} R-\overset{\overset{\displaystyle O}{\|}}{C}-OCH_3 + H_2O$$

The esterification result is then separated in a separating funnel. The result of esterification reaction is formed two layers as shown in Figure 2.

Figure 2. Esterification Reaction Generate Two Layers

The esterification process takes place twice because in the first esterification, the acid number is still above 2.5% so that the esterification is done again to decrease the acid number to <2.5%. The second esterification resulted in an acid number of 2.26% ± 0.

3.2. Biodiesel Synthesized from Jatropha Oil

The process of activating CaO and MgO catalysts refers to [4,6] study, ie by heating the catalyst in the furnace (calcination) at temperature of 700 ^{0}C for 2 hours. After CaO and MgO are obtained, the CaO and MgO catalysts are prepared by mixing CaO and MgO with a ratio of 1: 1.

The optimum condition of biodiesel synthesis from *Jatropha* oil through transesterification by inhomogeneous catalyst [4,9] using microwaves based on the highest yield is shown in Figure 3.

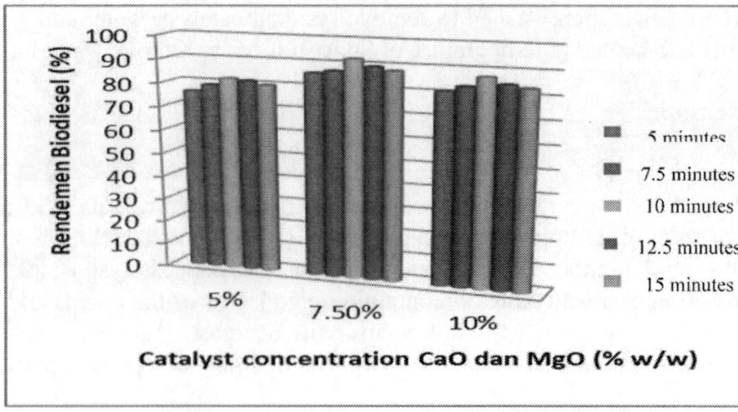

Figure 3. The Yield of Biodiesel Product

Figure 4 shows that the yield of biodiesel increases with increasing catalyst concentration and reaction time, reaching maximum at 7.5% concentration at 10 minutes with a yield of 92.39% ± 0.04. The yield increases as the catalyst concentration increases and the duration of the reaction. This indicates that the longer the reaction time and the greater the concentration of the catalyst, it will increase the yield of biodiesel, but if the reaction has reached equilibrium then the reaction time and the magnitude of the catalyst concentration will not increase the yield of the catalyst. The results showed that at 10% CaO and MgO catalyst concentration, the yield of biodiesel began to decrease. This happens because the long reaction of biodiesel synthesis causes the biodiesel product to react with the soap produced from the transesterification reaction, so the product becomes reduced resulting in reduced biodiesel yield.

The optimum condition of biodiesel synthesis from *Jatropha* oil through transesterification reaction using microwave was at 10 minutes with CaO and MgO concentration of 7.5%. The highest yield (92.39% ± 0.04) obtained from *Jatropha* oil synthesis with CaO and MgO catalysts using microwaves has a yield value that is not much different from that of [1,4] that is 92%.

3.3. Identification of Biodiesel Components from Jatropha Oil

Transesterification biodiesel from *Jatropha* oil is a mixture of methyl esters which can be analyzed using GC-MS. The purpose of the analysis is to know the composition of the compounds that make up the biodiesel. The analysis is based on structural approaches, compositions and molecular weights of biodiesel or methyl esters which will produce chromatograms of typical peaks. The GC-MS biodiesel chromatogram is shown in Figure 4.

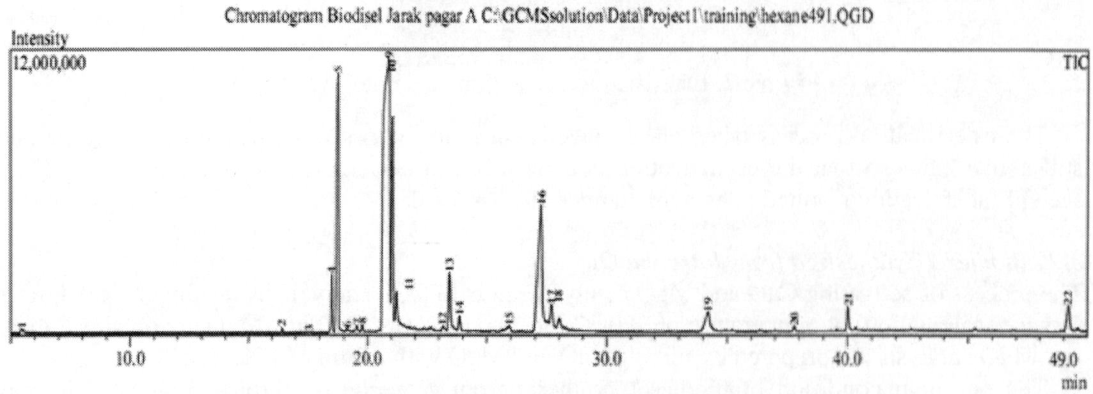

Figure 4. Chromatogram of Biodiesel from *Jatropha* oil

According to Figure 4 is shown that Methylesters or biodiesel from *Jatropha* oil contain four major constituent components. Retention time and the peak area of the chromatogram of the four main components of the biodiesel constituent of *Jatropha* oil is shown in Table 1

Table 1. Time Retention (tR) and of Four Major Components of Biodiesel from *Jatropha* Oil

Peaks	Retention (minutes)	Time Area (a. u.)	Percent Area (%)
4	18.502	7798897	2.40
5	18.759	68252263	21.20
9	20.908	201368020	62.40
10	21.058	45192916	14.00

The chromatogram peaks of mass spectrum of biodiesel at retention time of 18.502 minutes is shown in Figure 5.

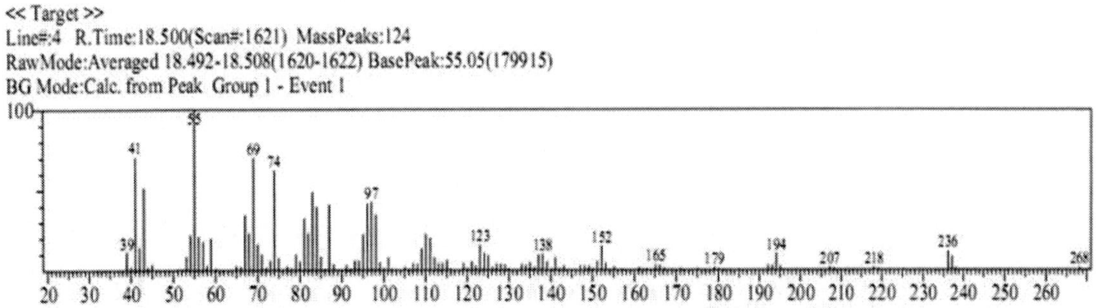

Figure 5. Peaks of Chromatogram of Biodiesel from *Jatropha* oil at tR of 18.759 Minutes.

The mass spectrum obtained is then compared to the mass spectrum found in the WILEY7.LIB library and searched for peaks of relatively equal position. Based on the observation, the peaks are positioned relatively similar to those contained in the library WILEY7.LIB entry number 178119, it is suspected that the compound is methyl palmitoleate. The possibility of the methyl ester compound is methyl palmitoleate, it is reinforced by a possible fragmentation pattern, particularly the major peaks with m/z of 41, 55, 97, 123, 152, and 236.

Peak with m/z 55 are thought to be obtained from the peak with m/z 268 with carbon-carbon and hydrogen bond (I) termination bonds as follows:

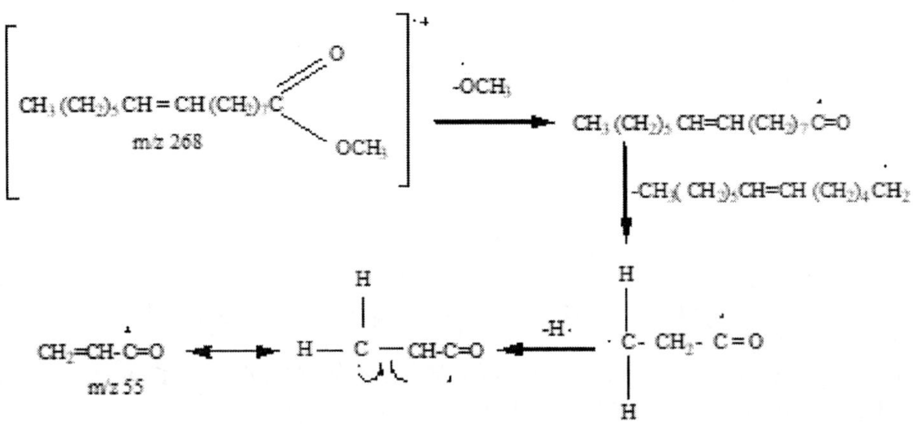

IOP Conf. Series: Journal of Physics: Conf. Series **1093** (2018) 012047 doi:10.1088/1742-6596/1093/1/012047

Based on the analysis of peak fragmentation pattern of biodiesel chromatogram with retention time of 18.502 min with m/z of 41, 55, 97, 123, 152, and 236, it can be concluded that the compound is methyl palmitoleate with $C_{17}H_{32}O_2$ molecular formula and structure formula as follows:

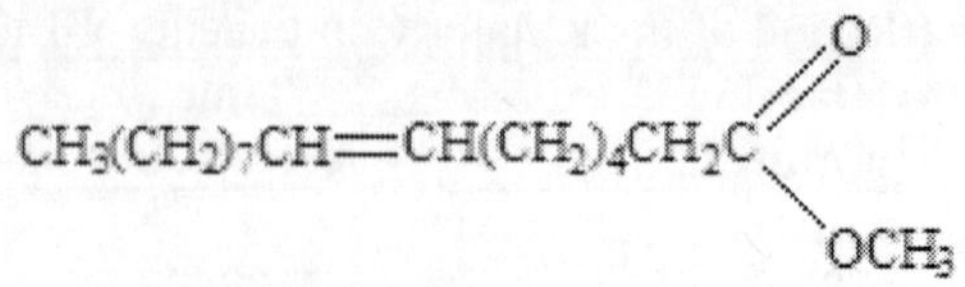

Methyl palmitoleate

Based on the analysis result of biodiesel character from synthesis of *Jatropha* oil, it can be concluded that biodiesel from synthesis has potential as fuel because its character fulfill the standard range of biodiesel quality according to National Standard of Indonesia (SNI). The synthesis biodiesel character of *Jatropha* oil and biodiesel SNI are listed in Table 2. Biodiesel Character Result of Synthesis from *Jatropha* Oil and Biodiesel (SNI)

Table 2. The Character of Biodiesel from *Jatropha* Oil and and Biodiesel according to SNI

Parameter	Biodiesel from *Jatropha* Oil	Biodiesel according to SNI
Refractive index	1.46445	Max 1.45
Density (g/mL)	0.888	0.850 – 0.890
Viscosity (cSt)	4.23	2.30 – 6.00
Acid number (mg KOH/g biodiesel)	0.58	Max 0.80

Biodiesel synthesized from *Jatropha* oil has character similar to the character of biodiesel which has been done by Shah *et al.* [12]. The values of the density, viscosity, and acid number of the previous research are 0.88 g/mL, 3.89 cSt, and 0.48 mg KOH /g biodiesel respectively.

4. Conclusion

The optimum condition of biodiesel synthesis from *Jatropha* oil using microwaves was using catalyst of CaO and MgO with concentration of 7.5% during reaction time of 10 minutes. The yield of biodiesel produced was 92.39% ± 0.04. The synthesized biodiesel from *Jatropha* oil was prepared using CaO and MgO catalysts assisted by microwaves yielded a methyl palmitoleate of 2.40%, methylpalmitate of 21.20%, methyloleate of 62.40%, and methylstearate of 14.00%. The properties of synthesized methyl ester possess refractive index of 1.465, density of 0.888 g/mL, viscosity of 4,23 cSt, and acid number of 0.58 mg KOH/g in accordance with national standards of Indonesian biodiesel.

References

[1] Hawash S, El Diwani G, Abdel Kader E. Optimization of biodiesel production from *Jatropha* oil by heterogeneous base catalysed transesterification. International Journal of Engineering Science and Technology 2011;3:5242–5251.

[2] Leung DY, Wu X, Leung MKH. A review on biodiesel production using catalyzed transesterification. Applied Energy 2010;87:1083–1095.

[3] Yan S, Lu H, Liang B. Supported CaO Catalysts Used in the Transesterification of Rapeseed Oil for the Purpose of Biodiesel Production. Energy & Fuels 2008;22:646–51. doi:10.1021/ef070105o.

[4] Hwei Voon Lee, Joon Ching Juan, Nurul Fitriyah Binti Abdullah, Rabiah Nizah MF and Yun Hin Taufiq-Yap. Heterogeneous base catalysts for edible palm and non-edible *Jatropha*-based biodiesel production n.d.

[5] Munishwar N Gupta and Smita Raghava,. Relevance of Chemistry to white technology,

Chemistry Central Journal 2007 1:17 n.d.

[6] Quitain, Armando T. Katoh, Shunsaku, Goto, Motonobu. Microwave-assisted synthesis of biofuels. In: Biofuel Production-Recent Developments and Prospects. InTech, 2011.

[7] Sameh M Osman, Mohamed H El-Newehy, Salem S Al-Deyab and Ayman El-Faham. Microwave synthesis and thermal properties of polyacrylate derivatives containing itaconic anhydride moieties, Chemistry Central Journa l20126:85 n.d.

[8] Betiku E, Adepoju TF. Methanolysis optimization of sesame (Sesamum indicum) oil to biodiesel and fuel quality characterization. International Journal of Energy and Environmental Engineering 2013;4:9.

[9] Inkollu Sreedhar and Yandapalli Kirti Kishan,. Process standardization and kinetics of ethanol driven biodiesel production by transesterification of ricebran oil, International Journal of Industrial Chemistry 2016 7:74 n.d.

[10] Cantrell DG, Gillie LJ, Lee AF, Wilson K. Structure-reactivity correlations in MgAl hydrotalcite catalysts for biodiesel synthesis. Applied Catalysis A: General 2005;287:183–190.

[11] Kouzu M, Kasuno T, Tajika M, Sugimoto Y, Yamanaka S, Hidaka J. Calcium oxide as a solid base catalyst for transesterification of soybean oil and its application to biodiesel production. Fuel 2008;87:2798–2806.

[12] Shah, Shweta, Sharma. Shweta. Gupta, M. N. Biodiesel preparation by lipase-catalyzed transesterification of Jatropha oil. *Energy & Fuels*, 2004, 18.1: 154-159.

Volatiles Dynamics on Solid Interface to Model Chemical Cues Release in Tritrophic Interaction

Surjani Wonorahardjo[1,*], Nurindah[2], Dwi A. Sunarto[2], Sujak[2], Endang Budiasih[1], M. Sodiq Ibnu[1], Dedek Sukarianingsih[1], Mahardika E. Prawiro[1], Meilah Handayani[1], Rizky S. Amalia[1]

[1]Department of Chemistry, Faculty of Mathematics and Science, Universitas Negeri Malang Jl. Semarang 5 Malang 65145, Indonesia.
[2]Indonesian Sweetener and Fibre Crops Research Institute, Jl. Raya Karangploso, PO Box 199 Malang, Indonesia

*Corresponding author's email: surjani.wonorahardjo@um.ac.id

Abstract. Phenomena of triptrophic interaction are well known in the agricultural area. Some specific volatiles is released from host plants after the infestation of pest insects, to attract parasitoids. Volatiles are formed during the biochemical process before released to open air. The phenomena of volatiles adsorption and desorption were investigated using spectroscopic methods. The model material was biosilica-nanocellulose which is porous and made by sol-gel processing and represented the wounded tissue plant in the tritrophic interaction. Some similar hydrocarbons from the previous study were adsorbed onto the surface and investigated. Some UV-visible and infrared spectra of adsorbed chemicals were recorded while the released gasses were captured and separated in the gas chromatography column. The result showed qualitatively molecular attachment into polar surfaces, and gas chromatography-mass spectroscopy indirectly determines the molecule desorbed from the solid interface and its pattern.

Keywords: Volatile dynamics, solid interface, triptrophic interaction, polar surface.

1. Introduction

Tritrophic interaction is a common mechanism used by plants in a self-defense situation [1–4]. The interaction between host-plant – herbivore- parasitoid in such condition involves certain chemical cues, which are biosynthesized and released as some volatile compounds. One model of this tritrophic interaction has been explored and rice plant attacked by brown plant-hopper (BPH) was explored. The chemical cues released attracted Anangrus Nilaparvateae parasitoid, and they consisted of mostly long-chained hydrocarbons as well as alcohols and esters [4]. The infested paddy plant was undergoing soxhlet extraction using polar and non-polar solvents and the results were analyzed by gas chromatography and mass spectrometry. Some chemicals were increased in dominance, but there are also new compounds appear and some disappear in the extract after infestation. The result was quite persistent and also in good agreement with other exploration of rice from the other region of Asia [3], [5,6].

The behavior of insects wounding plant leaves-tissue for egg deposition was already explored and described [1]. The insects would choose the best place for egg-laying but this process was also correlated with the chemical cues from the plants as well as with climate and working hour of the insects. The egg-induced changes to the biochemistry of the plant and as the result, primary and secondary metabolites of the plants would also change. The chemicals from the plants act also to preserve the eggs and influence the quality of insect egg as well as embryo development. However, the real biochemistry routes investigation would need more depth in laboratory experiments. In fact, this phenomenon is becoming more interesting as human being need to imitate the nature to take the control in agriculture work.

To make the story shorter, investigation on chemical cues involved, some extraction method was done and the content was determined by gas chromatography and mass spectrometry [2,3]. This job was clear and in accordance with another result from the Asian region. On the other hand, for bio-control purposes, there were some attempts to make such system for fabrication. A system with similar wounded-moist leaves releasing chemicals is under consideration. A biomaterial from silica and cellulose was under investigation since the structure is porous and can be made humid and then some similar hydrocarbon or other volatile compounds can be made released from the interface. In this case, similar wounded leaves are on the attempt.

The study of the released volatiles from wounded leaves has not been investigated so far. The material of the leaves can be cellulose, lignin and many other organic compounds covered by kind of paraffin to adjust the moist besides other functions for defense. When it is wounded by herbivores, some holes are created and volatiles is released due to abiotic stress. This type of situation is to be imitated while the material themselves are designed from the laboratory. Combination of silica and cellulose were chosen due to their properties and the similarity with leaves system. Raw materials were chosen from biomaterial too, those are rice husk ash as the source of silica and hydrolyzed *nata de coco* as the source of cellulose.

Silica was extracted from rice husk ash and gelled by sol-gel processing, during which nano-cellulose was added to alter the surface [4]. Originally the silica was extracted from rice husk ash by sol-gel processing [5]. Nano-cellulose was obtained by hydrolysis in the sulfuric acid catalyst. The cellulose itself was taken from bacterial cellulose "nata de coco" which was made with the aid of *Acetobacter xylinum* [6,7]. This cellulose is relatively pure and ready to hydrolyze to cut the long carbon chains. Working with cellulose, there are more possibilities to explore since modification with the cellulose can be done for various applications [8–11]. This type of material has more applications as adsorbent and still more exploration would be done related to solid-gas and solid-liquid interface. Therefore some applications of cellulose surface are for adsorbents and in this opportunity the silica and hydrolyzed cellulose during gelling in the sol-gel processing. This method of material for adsorbent was registered as Indonesian patent in 2017 with the register number P00201508272. The imitated wounded leaves from silica-cellulose were then used to observe hydrocarbon release. Some similar chemical compounds were adsorbent onto the surface and the amount released as well as the retention time were tried in the laboratory as well as in olfactometer systems. This type of formulation was also registered as an Indonesian patent. The big aim of this project was to make similar attractant from the similar adsorbent (silica and cellulose undergo gelling together) and similar synthetic components of volatiles added intentionally. The study in this experiments covers the dynamics of those molecules in the solid-gas interface as can be approached using FTIR and GCMS instruments.

2. Experimental Methods

In this part, there were two sections, the formulation of bio-attractant from silica-cellulose material and the spectroscopy study using infrared spectroscopy for the adsorbed attractant and gas-chromatography-mass spectrometry for the released gas separation and identification. One series of mass measurements were done to get information about attractant given off the surface to the air especially during the earlier stages of the making. The third part is about bioassay, to prove the volatiles given off by attracting Anangrus Nilaparvateae parasitoid to the source of a synthetic odor.

The chemical used in this study were obtained from Sigma Aldrich and E. Merck. Most of the solutions needed were made using distilled water. FT-IR measurement used IRPrestige-21 from

Shimadzu and GC/MS experiments were done in QP2010Plus from Shimadzu. Scanning Electron Microscope used was the one from FEI Type Inspec S50. All bench activities used pyrex glassware.

Materials for adsorbent was made of rice-husk ash as the source of silica, according to Kalapathy [12] with some modification. The extraction was by sol-gel processing and some hydrolized cellulose were added during the gelation process for the *in-situ* incorporation. The method has been registered in Indonesian Patent (*2nd December 2015, No.* P00201508272). The bio-attractant formulation was done using some hydrocarbon compounds namely heptane, hexadecane, and n-hexane was used as a standard. The other system used 1-octanol, butyl acetate, and hexatriacontane. The third system used the same compounds but the adsorbent contained one surfactant (*n*-CTMABr) to shift the surface properties and to make the material homogeneous in the pore system. The real formulation method of bio-attractant was also registered in Indonesian Patent with registration number P00201603970, 16th June 2016. The bio-attractant were put in closed vials for spectroscopy measurement. The mass-reduction measurement was done immediately after formulation, for the second and third systems, and the vials were kept open, the same as in the real situation with wounded leaves. FTIR experiments were run at the attractant formula together with the adsorbent as a solid sample, and GC/MS were done for the released gasses after formulation using a gas-tight syringe. The gasses in closed vials were taken by tight syringe gas prior to injection. Bioassay for the bio-attractant was done but would be reported elsewhere.

3. Results and Discussion

There are several sections on this result and discussion. The first section will be discussed the materials for adsorbent, which were used for the bio-attractant formulation by adding some chemicals similar to kairomones compounds extracted before [2]. The appearance of this adsorbent can be found in Figure 1 (a) and the SEM picture can be seen from the magnification of 100.000 times (Figure 1.b). The white powder of silica-cellulose is actually a porous medium, and according to the SEM picture below, the diameter of particles is around 50 nm. This material can be categorized nanomaterial since the diameter, as well as the dimension of porosity, is below 100 nm. This material is also hydrophilic due to polar functional groups on the surface, which come from the silanols and hydroxyl groups of the cellulose. The polar porous surface can retain small molecules, and therefore they can release the gaseous particles slowly and step by step. This is the desirable property of a good releaser so that the bio-attractant can be effective after sometimes. The silica surface itself tends to be too polar that retains the molecules stronger. The presence of cellulose somehow reduces the polarity and the material can be a good and effective releaser.

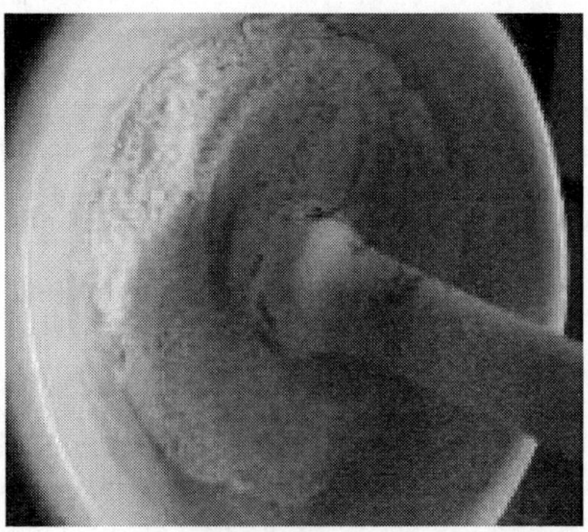

(a)

(b)

Figure 1. The appearance of silica-cellulose adsorbent after drying (a) and the 100.000x magnification of the material (b).

The white powdered material in Figure 1(a) shows no sign of porosity. Only by big enough magnification can reveal the voids in between particles. The texture of the surface cannot be assigned clearly from the SEM picture. However, the homogeneous particles and porosity can be claimed from the rough observation. The silica particles mixed well with cellulose as globular particles altogether. The system itself is a non-conducting material so that the measurement can be done after a relatively thick coating.

The formulation of bio-attractant was actually done to mimic the wounded leaves due to oviposition of insect pest. Under abiotic stress, some volatiles as developed by biochemistry processes of the plants is released to air to attract the parasitoid of the pest. The volatiles is specific and undergoes complex biochemistry routes of synthesis from jasmonates and its derivates [13,14]. The composition of volatiles was already done in similar ways, namely extraction and identification in the past [2,3,15]. In this case, silica-cellulose adsorbent resembles the leaves tissue, which is also humid and the similar volatile compounds were added to the adsorbent and were released and observed (Figure 2a). The mechanism for releasing volatiles was investigated by FTIR and GC/MS.

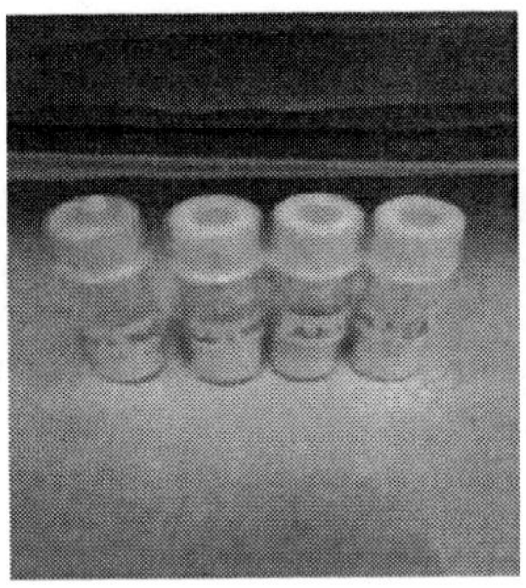

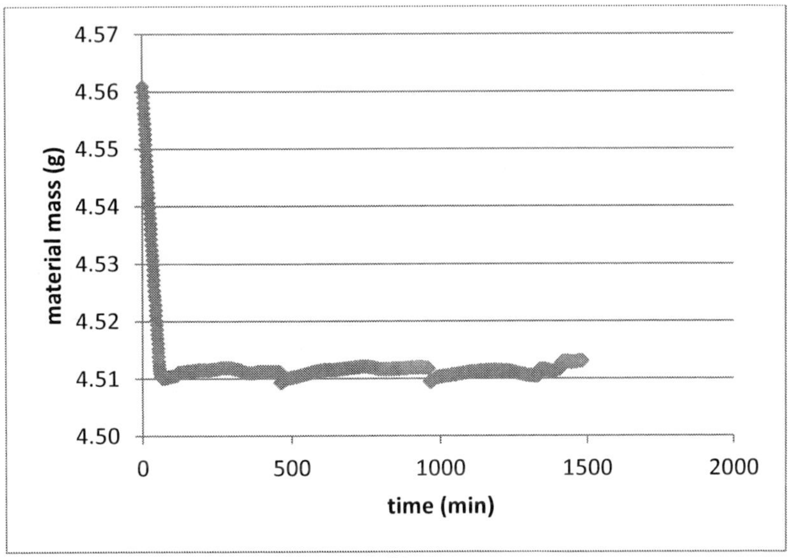

(a) (b)

Figure 2. Formulation of bio-attractant with a silica-cellulose adsorbent (a) and mass reduction measurement (b) for 0-1500 minutes.

The mass measurement was done to observe the volatiles release process (Figure 2b). The original mass dropped significantly during the early hour and leveling off after sometimes. However, not all the adsorbed molecules went away, there were some on the surface in equilibrium with outer water molecules as well as other gasses on air. The mass did not change significantly anymore, the up and down in a long time showed the equilibrium on the surface. Moreover, in the FTIR spectra, up to the day-8, the sign of functional groups of the volatiles were still present. The gas released also appeared in GC/MS analysis up to day -10. The observation and analysis of adsorbates, as well as gasses released, would be discussed below.

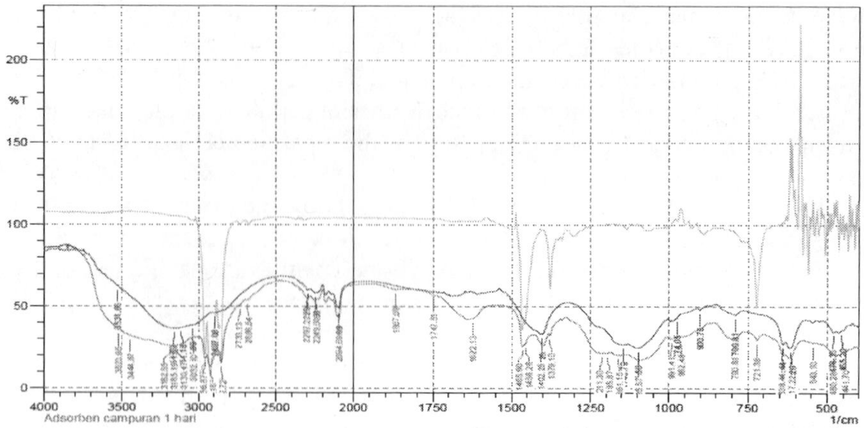

Figure 3. Combined IR spectra of silica-cellulose adsorbent (dark blue), (1:1:1) mixture of *n*-hexane, *n*-heptane, *n*-hexadecane (light blue), adsorbed mixture onto adsorbent (red) after one-day exposure.

From the figure above it can be seen that adsorbent material, both with and without adsorbed hydrocarbons give different patterns compared to the liquid hydrocarbons (light blue spectrum). The broad signals indicate restricted vibrations on surfaces and the peaks are broader. However, it gives the signal of hydroxyl groups from silanols (≡Si-OH) as well as much hydroxyl of cellulose from the stretching vibration around 3600-3200 cm^{-1} (dark blue spectrum in Figure 3). The spectrum of hydrocarbon mixture indicates clearly the symmetric and asymmetric –C-H stretching (around averaged at 2900 cm^{-1}), $-CH_2-$ and $C-CH_3$ bending (1471 cm^{-1} and 1381 cm^{-1}) of all hydrocarbons. The band around 725 cm^{-1} looks strong for the bending vibration of the $-(CH_2)_{4-8}-$ units for all hexane, heptane, and hexadecane. The character of this hydrocarbons still appears after they are absorbed in the surface of silica-cellulose. The detail inside the circles indicates the same components on the surface, as they remain on the surface even though the molecules release themselves after sometimes. The bands broadened themselves since the vibrational modes are now more restricted on polar surface. The signals weaken after 8 days of exposure. This is the first system in this report.

The second system used one long-chained primary alcohol (*1*-octanol), one branched ester (butyl-acetate) as one fresh odor source in plants, and also one long chained hydrocarbon (hexatriacontane). These three types of organic compounds have different polarity and also the tendency of retention and interaction on the polar surface. However, it is difficult to distinguish which interaction is which and what appears in the spectrum would be the net result of so many possibilities in interactions.

The additional functional groups on this system can be seen in the blue spectrum above. The primary hydroxyl group from *1-octanol* can be seen from the stretching vibration which appears broad due to hydrogen bonding and also together with hydrogen bonding from the surface water after adsorption. Additional small peaks appear in the fingerprint area are from the character of acetate esters, which in turn are not seen in the adsorbed phase. However, the long-chained hexatriacontane cannot be avoided even after adsorption. This is only a qualitative assignment, there is no way to calculate the adsorbates quantitatively using the simple infrared spectrometer.

The third system in this series used surfactant on the surface. *n*-Cetyl Trimethyl Ammonium Bromide (Figure 5 a-b) is one cationic surfactant that is used mainly to alter the surface property. In solution, it is readily ionized. In making materials, surfactants are used as templates in the gelation stage [16]. The gelling process to make porous material can be followed in situ by NMR method[17] besides FTIR discussed previously. The negative and positive ends of the long chained compound can make micelle according to the surface and space-configuration so that the main charge, as well as its distribution, shifted. However, the appearance of FTIR spectrum of silica-cellulose is more or less the same as the previous adsorbents since after the gelation the adsorbents were washed with distilled water. The small amount of surfactant is not revealed in the spectrum even though it might alter the surface properties and make the porous system homogeneous.

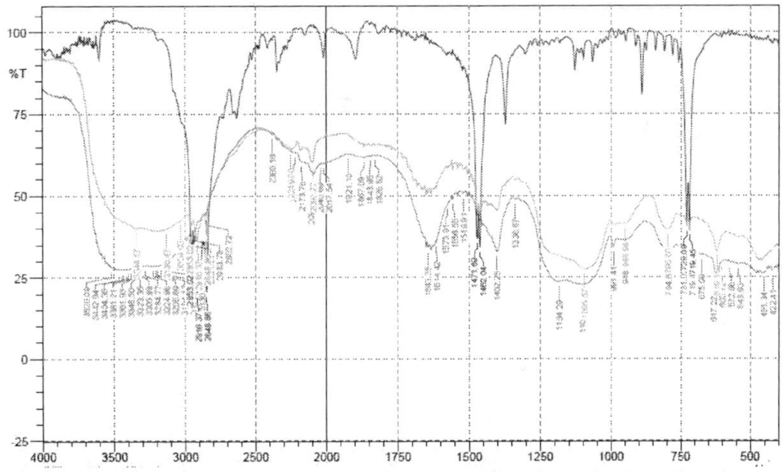

Figure 4. Combined IR spectrum of silica-cellulose adsorbent (grey), (1:1:1) mixture of *n*-hexane, *n*-heptane, *n*-hexadecane (light blue), adsorbed mixture onto adsorbent (green) after twelve days exposure.

Within the porous medium, the complexity increases and possible interaction with adsorbate compounds cannot be assign from FTIR spectrum. When chemical bonding with the surface is not occurring, the interactions between anything on the surface would be reversible. That is why the volatile compounds can be released into the air and compounds in solution undergo surface dynamics together with surface solvent molecules.

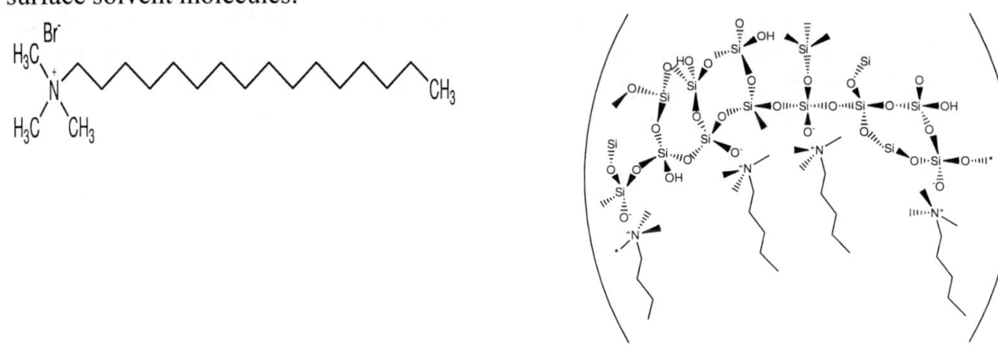

(a) (b)

Figure 5. The molecular structure of n-CTAB (a) and one possible interaction in silica surface (b) as given in Holmes et.al. [16].

Gas analysis attempt was done by separating in silica derivate column and analyze the gas by mass spectrometry. The original compounds were done before the adsorbed on the silica-cellulose surface. In Figure 6 we can see the appearance of adsorbed hexane, heptane, and hexadecane cannot be separated well since they are strongly bonded to the surface. Since the three peaks are so broad, they cannot be resolved well in the similar retention time. This, however, cannot explain the retention mechanism on the surface.

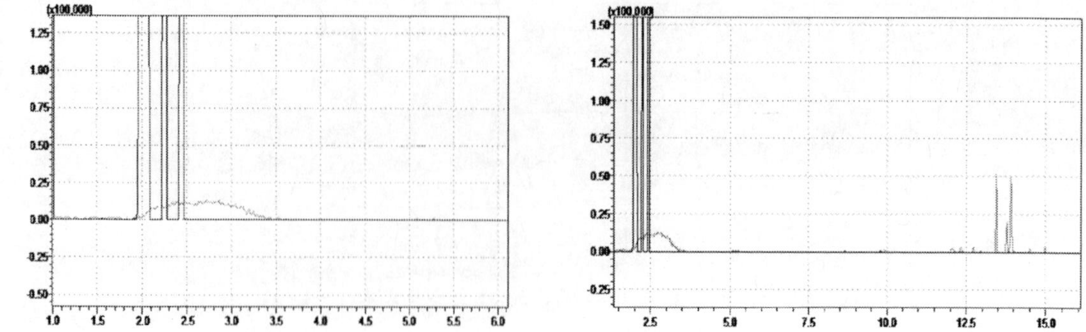

(a) (b)

Figure 6. The chromatogram of free (black) and adsorbed *n*-hexane, *n*-heptane, and *n*-hexadecane mixture (pink) from RTX-5 column, immediately after formulation (a) and after 1 day of exposure (b).

Equilibration in porous medium takes sometimes. There was something interesting in the chromatogram of the formulation after one day of exposure. There were some gasses appear in longer retention time even though the assignment from Wiley library is not really clear. However, due to the inhomogeneity of the sample, some analysis must be done. The similar hydrocarbons have almost the same properties since the elements are only carbon and hydrogen. That is how the mass spectrum of the longer retained components was not clear. The unresolved components in earlier retention time (around 2.5 minutes) shows the strong interaction with the surface in the column. Certainly, there was a similar interaction with the surface column which is made of more silica. As we know silica retains more and during elution occurred non-equilibrium mass transfer[18]. Some part of the gas would still retain but due to strong eluent gas flow then some part of the components appeared later (Figure 6.b). Of course, it is not the best condition for the separation and analysis. Also as a consequence, Wiley 8 library still suggested some long-chained hydrocarbons for the peak but only similar with hydrocarbon used, not exactly matched the same as supposed to be.

Analogous to this separation and longer retention time, the investigation of small molecule mobility is a porous medium for the partially filled porous system has been ling investigated using NMR relaxation and diffusion experiments [19–21]. A strange kinetic occurring as a consequence of surface diffusion has been thoroughly discussed in some paper about NMR methods[22]. Before appeared in longer retention time, the molecules undergo complex behavior related to the affinity of molecules unto the surface. Silica cellulose adsorbent is less polar than silica only so that the gasses were released easier. During separation, the gasses behaved differently.

For the second and third system, the presence of *n*-CTABr surfactant was also in the adsorbent system, there came the same tendency. In the similar place, comes some unresolved peaks, for the newly made formulation. After some time of equilibration, the unresolved peaks decrease in intensity but appeared more peaks with longer retention times (Figure 7). There was interesting that the number of gasses from the adsorbent made using surfactant appeared more than silica-cellulose only. Since more molecules released from the interface, the amount of volatiles after 5 days of exposure remained less from the surfactant adsorbent. This might due to the more homogeneous matrix of silica while it was gelled with surfactant template [23]. However, after 10 days appeared more volatiles from the surfactant adsorbent compared to the second system. In this case, the system of surfactant adsorbent retained more volatiles in equilibrium, might be related to polar functional groups they have (alcohol, esters, and also hydrocarbon).

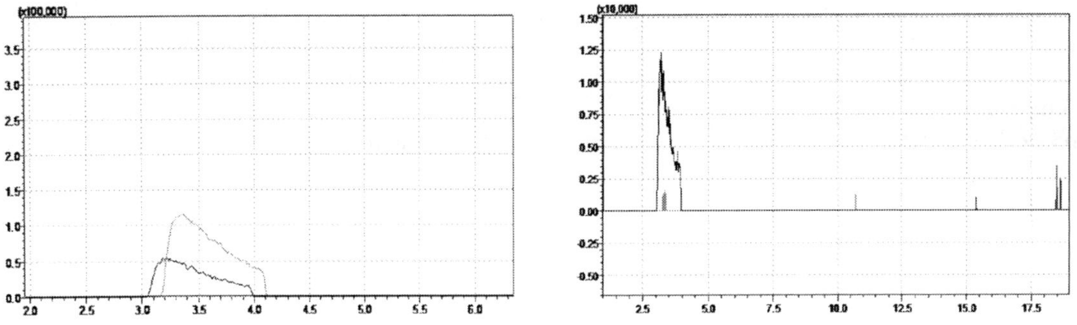

(a) (b)

Figure 7. The chromatogram of adsorbed *1*-octanol, hexatriacontane, and butyl acetate in silica-cellulose made without surfactant (black) and adsorbed mixture on silica-cellulose made with surfactant (pink) from RTX-5 column, immediately after formulation (a) and after 5 days of exposure (b).

After long days of exposure, the unresolved peaks disappeared but the peaks in longer retention time appear clearly (Figure 8). This indicated the ability of the interface to hold the volatile compounds to be released after some time. This is a good property of a bio-attractant so that it still shows effective in use after a long time.

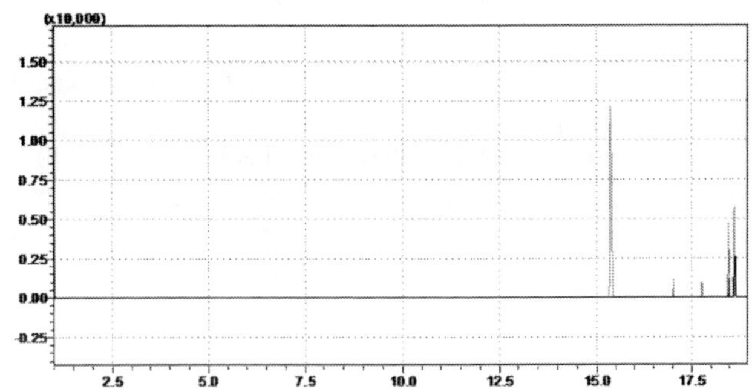

Figure 8. The chromatogram of adsorbed *1*-octanol, hexatriacontane, and butyl acetate in silica-cellulose made without surfactant (black) and adsorbed mixture on silica-cellulose made with surfactant (pink) from RTX-5 column, 10 days of exposure.

The presence of gasses must be studied further using bioassay, to see the effect as bioattractant for the parasitoid. The assignment using Wiley 8 database would not explain the real gasses appear in the longer retention times since the similarity index (SI) would not be 100%. Further mass spectrometry must be done to ensure the assignment using the MS database. However, the best method to further investigate the gasses is bioassay using the real parasitoid in olfactometry or in the field, as already did for the extract of infested paddy plant [2,3].

4. Conclusion

The dynamics on the solid-gas interface was followed by mass-reduction experiments, and gas-chromatography-mass spectrometry FTIR spectra described the presence of bio-attractant components in the surface. GC/MS shows the released volatiles from the surface after some certain time retention on surfaces. Polar compounds behave different compared to non-polar hydrocarbons as qualitatively seen in the IR spectra and gas chromatograms. Mass reduction showed the release of gasses from the beginning until equilibration. However, since the original intention of this research is for modeling the tritrophic interaction, the further attempt has to be bioassay using real parasitoid insects. Further research would also deal with surfactant effect in releasing the volatiles or other adsorbates on interfaces in the more comprehensive discussion. The changing gasses released in longer retention time must be clearly explained.

Acknowledgments
This work was supported by Ministry of Research, Technology, and Higher Education, Indonesia, partly under the National Innovation Reseach Grant (INSINAS, 2015-2016) and also Fundamental Research Grant (2015-2016).

References

[1] L. A. Dyer, "Tropical Tritrophic Interactions : Nasty Hosts and Ubiquitous Cascades," in Tropical forest community ecology. Blackwell Science, Oxford, 275-293 2005.
[2] Y. Lou, M. Du, T. E. D. C. J. Turlings, J. Cheng, and W. Shan, J. Chem. Ecol., vol. 31, no. 9, pp. 1985–2002, 2005.
[3] Y.-G. Lou, B. Ma, and J.-A. Cheng, J. Chem. Ecol., vol. 31, no. 10, pp. 2357–72, Oct. 2005.
[4] N. Nurindah, S. Wonorahardjo, D. Adi Sunarto, and S. Sujak, J. Pure Appl. Chem. Res., vol. 6, no. 1, pp. 49–56, 2017.
[5] Y.-G. Lou, G.-R. Zhang, W.-Q. Zhang, Y. Hu, and J. Zhang, Biol. Control, vol. 68, pp. 103–116, Jan. 2014.
[6] Y. Lou, G. Zhang, W. Zhang, Y. Hu, and J. Zhang, Biol. Control, vol. 67, no. 1, pp. 8–20, 2013.
[7] M. Hilker and T. Meiners, Phytochemistry, vol. 72, pp. 1612–1623, 2011.
[8] S. Wonorahardjo, Nurindah, D. A. Sunarto, Sujak, and N. Zakia, J. Entomolgi Indones., vol. 12, no. 1, pp. 48–57, 2015.
[9] S. Wonorahardjo, A. R. Wijaya, and S. Suharti, The J. of Pure and App. Chem. Res. 5 (2), 48-54 , 2016.
[10] U. Kalapathy, A. Proctor, and J. Shultz, Bioresourse Tech. vol. 73, 2000.
[11] A. Budhiono, B. Rosidi, H. Taher, and M. Iguchi, Carbohydr. Polym., vol. 40, pp. 137–143, 1999.
[12] M. Iguchi, S. Yamanaka, and a. Budhiono, J. Mater. Sci., vol. 35, pp. 261–270, 2000.
[13] T. Heinze and A. Koschella, Polimeros, vol. 15, pp. 84–90, 2005.
[14] A. Stark, M. Sellin, B. Ondruschka, and K. Massonne, Sci. China Chem., 2012.
[15] S. Wonorahardjo, M. S. Ibnu, and E. Budiasih, Scientific Study & Research. Chemistry & Chemical Engineering, Biotechnology, vol. 17, no. 2, pp. 179–188, 2016.
[16] H. H. Bi, R. Sen Zeng, L. M. Su, M. An, and S. M. Luo, J. Chem. Ecol., vol. 33, no. 5, pp. 1089–1103, 2007.
[17] T. Xu, Q. Zhou, W. Chen, G. Zhang, G. He, D. Gu, and W. Zhang, Chinese Sci. Bull., vol. 48, no. 18, pp. 1982–1987, 2003.
[18] S. M. Holmes, V. Zholobenko, A. Thursfield, R. Plaisted, C. Cundy, and J. Dwyer, J. Chem. Soc. Faraday Trans., vol. 84, pp. 2025–2032, 1998.
[19] S. Wonorahardjo, G. E. Ball, J. Hook, and G. Moran, J. Noncrystalline Solids, vol. 271, no. 1–2, pp. 137–146, 2000.
[20] S. Wonorahardjo, Metode-Metode Pemisahan Kimia, Sebuah Pengantar (Separation Chemistry Methods, An Introduction), I. Jakarta: Indeks Akademia, 2013.
[21] C. Mattea, R. Kimmich, I. Ardelean, S. Wonorahardjo, and G. Farrher, J. Chem. Phys., vol. 121, no. 21, pp. 10648–10656, 2004.
[22] A. Bede, A. Bede, A. Scurtu, and I. Ardelean, Cement and Concrete Research. no 89. 56-62, 2016.
[23] I. Ardelean, C. Mattea, G. Farrher, S. Wonorahardjo, and R. Kimmich, J. Chem. Phys., vol. 119, no. 19, p. 10358, 2003.
[24] R. Kimmich, Chem. Phys. vol. 284, no. 253-285, 2002.
[25] V. L. Zholobenko, S. M. Holmes, C. S. Cundy, and J. Dwyer, Microporous Mater., vol. 11, no. 1–2, pp. 83–86, 1997.

Effect of SnO_2 Nanoparticles on Band Gap Energy of $x(SnO_2)$-$y(Ag)$-β-Carotene/FTO Thin Film

Solehudin[1], Markus Diantoro [1,2,*], Ulwiyatus Sa'adah[1], Arif Hidayat[1,3]

[1]Department of Physics, Faculty of Mathematics and Natural Sciences, Universitas Negeri Malang, Jl. Semarang 5 Malang 65145, Indonesia
[2]Center of Advanced Materials for Renewable Energy, Universitas Negeri Malang, Jl. Semarang 5 Malang 65145, Indonesia
[3]Optics and Computational Physics, Faculty of Mathematics and Natural Sciences, Universitas Negeri Malang, Jl. Semarang 5 Malang 65145, Indonesia

*Corresponding author's email: markus.diantoro.fmipa@um.ac.id

Abstract. SnO_2 shows higher electron mobility (100–200 $cm^2V^{-1}s^{-1}$) than TiO_2 (0.1-1.0 $cm^2V^{-1}s^{-1}$), indicated that it has higher electron diffusion transport. As a result, it may minimize interfacial charge recombination losses to oxidized redox species in the electrolyte or solid-state hole transporter, thus enhancing solar cell device performance. Many researchers have done the synthesis of SnO_2 by controlling morphologies and architectures to optimize solar cells performance. Unfortunately, it's just view of them investigated the properties of the composite SnO_2 with metallic materials and its potentials for solar cell devices. In this work, SnO_2 nanoparticles, Ag nanoparticles, and β-Carotene were composited with a various mass portion of SnO_2 nanoparticles to control the band gap energy and the other optoelectronic properties. The composites subsequently were deposited on FTO substrates using spin coater. The thin film of SnO_2-Ag-β-Carotene/FTO composites were characterized by UV-Vis, FTIR spectroscopy, and XRD measurements. We found that the maximum direct band gap energy of 3.74 eV was reached by the sample 20, while the minimum direct band gap energy of 3.54 eV was achieved by sample 15. The maximum indirect band gap of 1.59 eV resulted in sample 20. On the other hand, the minimum value of the indirect band gap is 1.27 eV achieved by sample 25. Both direct and indirect Tauc plot analysis showed that there was no significant influence of SnO_2 mass portion on the band gap energy of SnO_2-Ag-β Carotene composite thin film. But it was a fact that each mass fraction composites had different band gap energy. This fact can be used to control the band gap energy to optimize the use of SnO_2 composite as DSSC electron transporting electrode. The FTIR spectroscopy and XRD result were also detail examined in this work.

Keywords: SnO_2-Ag-β-Carotene/FTO composites, UV-Vis spectroscopy, band gap energy, DSSC

1. Introduction

In the past several decades, Dye-Sensitized Solar Cells (DSSCs) emerged as a generation of cost-effective solar cells with the highest performance have been those fabricated using TiO_2 nanoporous films as the electron transporting electrode [1]. However, further performance of the TiO_2-based DSSCs was limited due to TiO_2 low electron mobility. SnO_2 has over two orders of magnitude higher electron mobility (~100 – 200 $cm^2/(V.s)$ than TiO_2 (0.1 - 1.0 $cm^2/(V.s)$) in their bulk single crystal phases [2], indicated that it has higher electron diffusion transport. As a result, it may minimize interfacial charge recombination losses to oxidized redox species in the electrolyte or solid-state hole transporter, thus enhancing solar cell device performance [3]. SnO_2 is also has a larger band gap (3.6 eV) than TiO_2 (3.2 eV), which would create fewer oxidative holes in the valence band (VB), thereby minimizing the dye-degradation rate and facilitate the long-term stability of DSSCs [4]. On the other hand, larger band gap also means that higher energy was necessary to exciting the electron of SnO_2.

Many researchers have reported the synthesis of SnO_2 with different morphologies and architectures ranging from zero-dimensional (0D) [5,6], one-dimensional (1D) [7–10], and two-dimensional (2D) [11,12] structures to complex three-dimensional (3D) architectures assembled from nanoparticles (0D) [13], nanorods (1D) [14], nanosheets (2D) [15] and core-shell as building blocks [16]. Unfortunately, it's just view of them investigated the properties of the composite SnO_2 with other materials such as polymer [17], metallic and its potentials for solar cell devices. In this work, SnO_2 nanoparticles, Ag nanoparticles, and β-Carotene were composited with the various mass fraction of SnO_2 to control the band gap energy to optimize its application in Dye-Sensitized Solar Cells (DSSC). This work is closely related to our previous research on the synthesis of Ag NPs [18] and its usage in electrical properties modification [19,20]. The competition of SnO_2 NPs and Ag NPs is studied related to its structure as well as band gap.

2. Materials and Methods

Silver nanoparticles (Ag NPs) was prepared according to the chemical reduction method. The materials used are silver nitrate ($AgNO_3$) as a precursor, sodium borohydride ($NaBH_4$) as a reducing agent, and Mercaptosuccinic acid (MSA) as a stabilizer to prevent the nanoparticle from agglomerating. Tin oxide nanoparticles (SnO_2NPs) have been obtained by a sol-gel method from stannic chloride dihydrate ($SnCl_2.2H_2O$). In a typical synthesis procedure, 8 g $SnCl_2.2H_2O$ was dissolved in 25 ml methanol under continuous stirring at 400 rpm. After 30 minutes, 0.1 M polyethylene glycol (PEG, Mw=6000) added into the solution. The next step was refluxed at 200 rpm for 5 hours till and the temperature maintained at 120 °C. After the solution transformed into amber one, kept it to obtain the precipitate. Further, the yellowish precipitate was filtered and annealed at 100 °C for 3 hours. Finally, the precipitate calcined for 1 hour at 350 °C resulting the SnO_2 nanoparticles.

The composite of x(SnO_2)-y(Ag)-β-Carotene solutions were prepared by adding 0.04 Ag NPs into one mL H_3PO_4 stirred for 1 hour at 1000 rpm. 0, 0.005, 0.010, 0.015, 0.02 and 0.025 g of SnO_2 was added into the solutions then stirred for 2 hours at 1000 rpm. The fraction of SnO2 NPs also followed by AgNPs of 0.04, 0.045, 0.050, 0.055, and 0.060 g. Subsequently, about 0.013 g of β-Carotene in 0.1 M solutions were added into the SnO_2-Ag solutions and then stirred and sonication each for 2 hours. The fraction of β-Carotene was kept constant. The composite thin films were deposited on FTO substrate by using a spin coating method for 10 seconds at the speed of 1500 rpm. The produced film was then annealed at 100 °C for 4 hours. These samples are labelled as 0, 5, 10, 15, 20, and 25 or associated to the fraction of SnO_2 respectively as 0, 0.09, 0.16, 0.23, 0.28, and 0.33 to the total mass.

The functional chemical bonds that were appeared in the sample was characterized using spectrophotometer Shimadzu IR Prestige-21. The crystallinity, grain size, and phase that was formed in the samples. Characterized using diffractometer x-ray PanAnalytical XPERT-type Pro with Cu-Kα radiation sources (λ_1 = 1.54060 Å). The optoelectronic property (band gap energy) was characterized using spectrophotometer UV-1700 Series. The UV-Vis spectrophotometer results were then analyzed using Tauc plot linear fitting with both direct and indirect approach. The fitting processes were a done using Microcal Origin software.

3. Results and Discussion
Figure 1 represents the XRD patterns of the (Ag-SnO_2) NPs-β-Carotene thin films. Generally, it could be seen that four peaks appear at 37.80°, 44.87°, 64.53° and 77.53° belong to silver with low intensity. There are peaks appear at 26.60°, 33.90°, 37.94°, 51.75°, and 64.77° belong to SnO_2 NPs detected in the sample 0, 5, 10, and 15. The 26.60°, 33.90° peaks on 0 g SnO_2 shows that the SnO_2 of the FTO substrate was detected due to the thickness of the film was very thin.

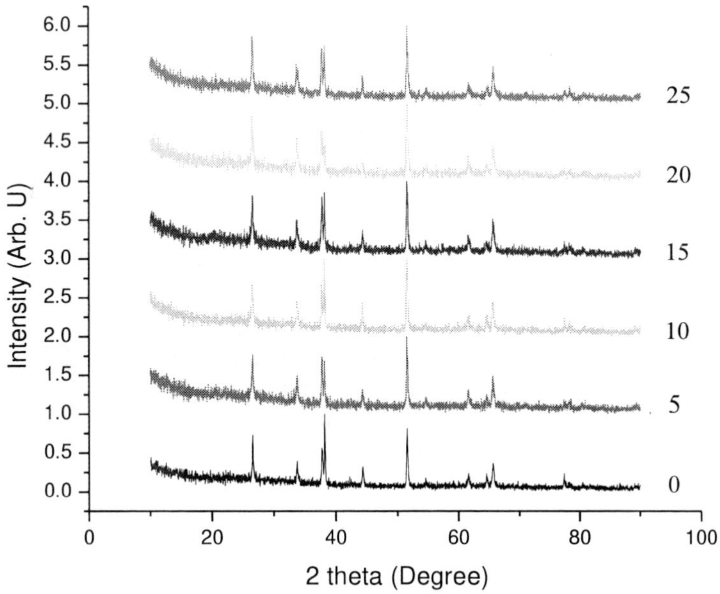

Figure 1. X-ray diffraction patterns of SnO_2-Ag-β-Carotene/FTO composite thin films with various SnO_2.

FTIR measurements were carried out to find the functional groups of films. Some peaks according to the functional groups of chemical bonds are drawn in Figure 2. The results of the analysis show that some peaks identified as C-H, O-H, N-H, O-Sn-O, and M-O. The peaks at a wave number of 592.15 cm^{-1} belong to the type of (M-O) metallic oxygen bond; it assigns to silver which is bond the oxygen. Stretching modes of the O-Sn-O functional group shown by existing at wave number 827.46 cm^{-1}. The wave number about 3496.94 cm^{-1}, which is in the range of 3349-3409 cm^{-1} related to stretching vibrations of the O-H bond. Beside, O-H stretching vibration also identified at 359 cm^{-1} indicating the presence of hydroxyl group. The peak at 1319.31 cm^{-1} suggests the appearance of a C-H bond corresponds to the β-carotene functional group. At wave number 2833.43 cm^{-1}, investigated as N-H functional group. There are no remarkable differences since SnO_2 NPs added in a tiny portion.

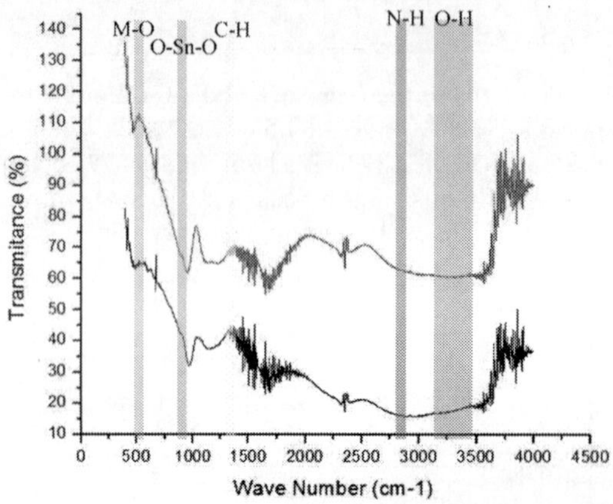

Figure 2. FTIR spectra of SnO_2-Ag-β-Carotene/FTO thin films with 0 g (black), and 25 g of SnO_2 (red).

The optoelectronic property represented by band gap energy of SnO_2-Ag-β-Carotene/FTO composite thin films was characterized using UV-Vis spectrometer. The UV-Vis Spectra of SnO_2-Ag-β-Carotene/FTO composites thin film with various SnO_2 mass portions were shown in Figure 3.The band gap energy data can be extracted from UV-Vis spectra using Equation (2) based on linear fitting.

$$\alpha = A\frac{(hv - E_g)^n}{hv} \quad (1)$$

$$(\alpha hv)^2 = A^2 hv - A^2 E_g \quad (2)$$

$$y = bx + a \quad (3)$$

where α is the absorption coefficient, h is Planck constant, v is UV-Vis electromagnetic frequency, and E_g is the band gap energy. Equation (1) can be rearranged into Equation (2) which is analog to a linear of Equation 3. From Equation (2) we can make $(\alpha hv)^2$–hv plot called the Tauc plot from which we can get the band gap energy information. The band gap energy then is

$$E_g = b/m \quad (4)$$

where b is $(\alpha hv)^2$ value when hv equal to zero and m is the gradient of the linear fit of the Tauc plot. The Tauc plot (direct), Tauc plot (indirect), band gap energy (direct), and band gap energy (indirect) of SnO_2-Ag-β-Carotene/FTO composite thin films are shown in Figure 4, 5, 6, and 7 respectively.

IOP Conf. Series: Journal of Physics: Conf. Series **1093** (2018) 012049 doi:10.1088/1742-6596/1093/1/012049

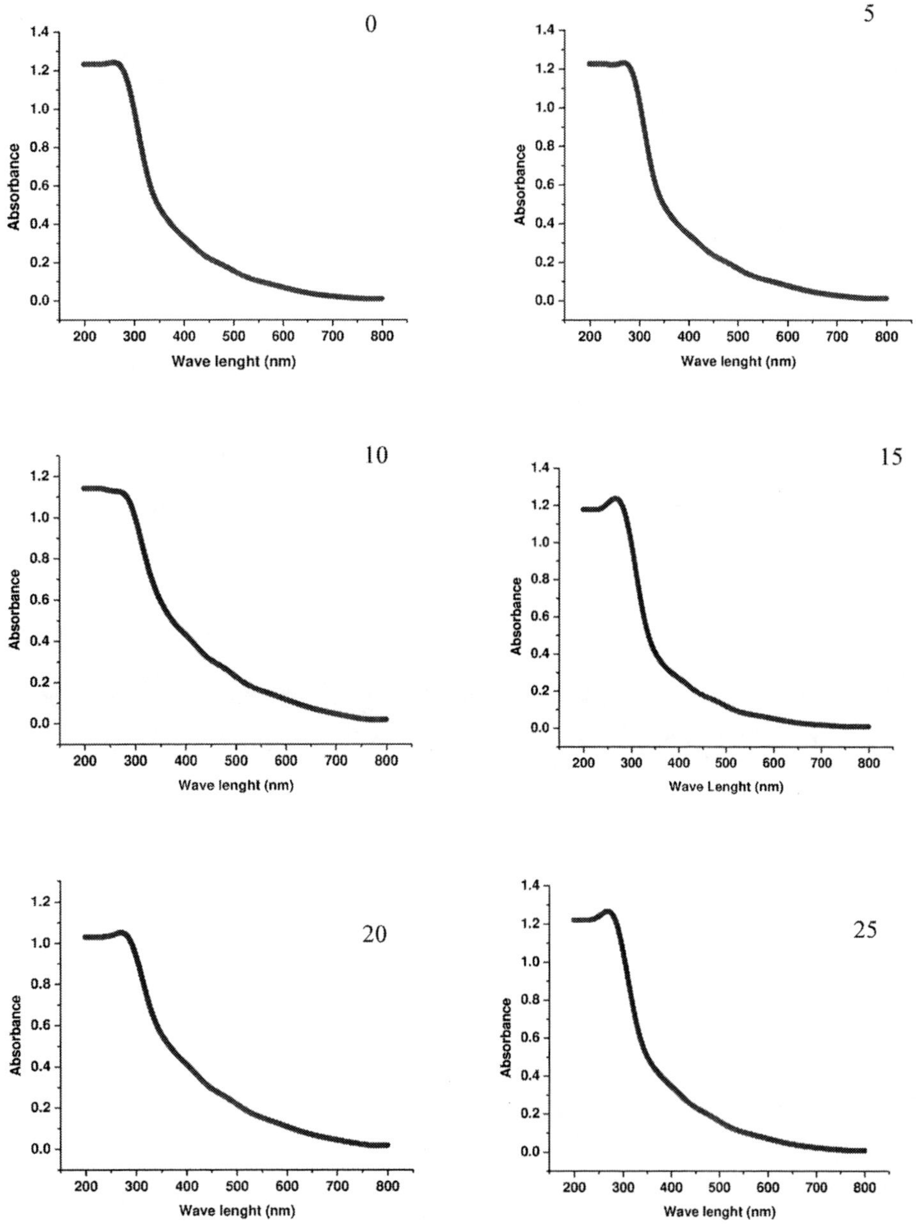

Figure 3. UV-Vis data plot of various SnO_2 in SnO_2-Ag-β Carotene/FTO thin films

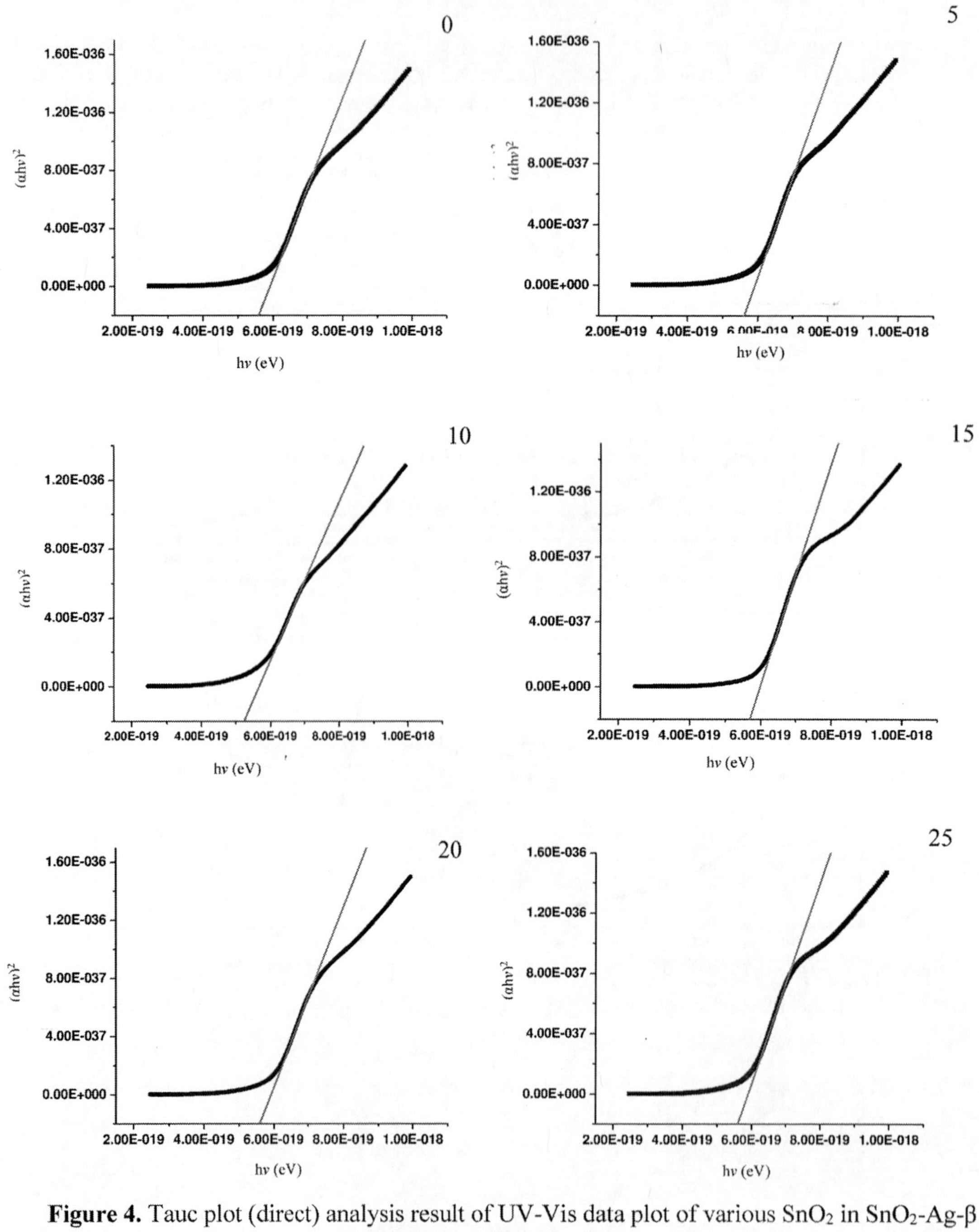

Figure 4. Tauc plot (direct) analysis result of UV-Vis data plot of various SnO_2 in SnO_2-Ag-β Carotene/FTO thin films.

IOP Conf. Series: Journal of Physics: Conf. Series **1093** (2018) 012049 doi:10.1088/1742-6596/1093/1/012049

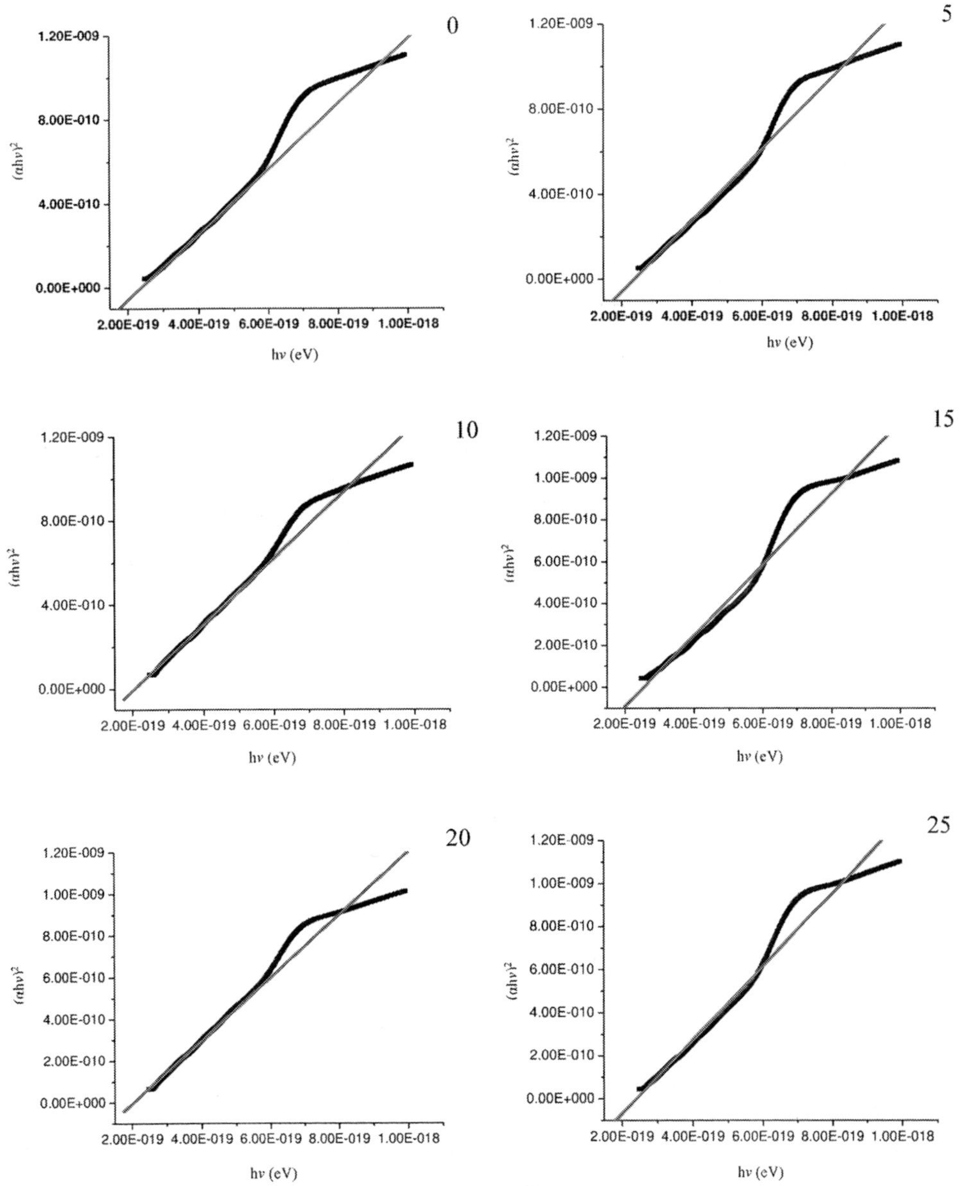

Figure 5. Tauc plot (indirect) analysis result of UV-Vis data plot of various SnO_2 in SnO_2-Ag-β Carotene/FTO thin films

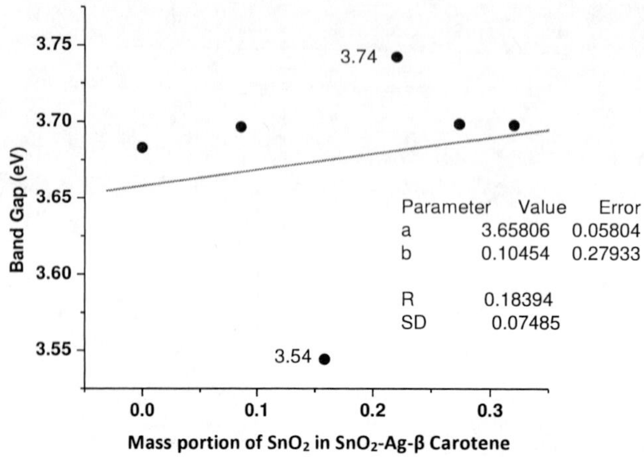

Figure 6. Band gap energy (direct) as the function of mass portion of SnO_2 in SnO_2-Ag-β Carotene

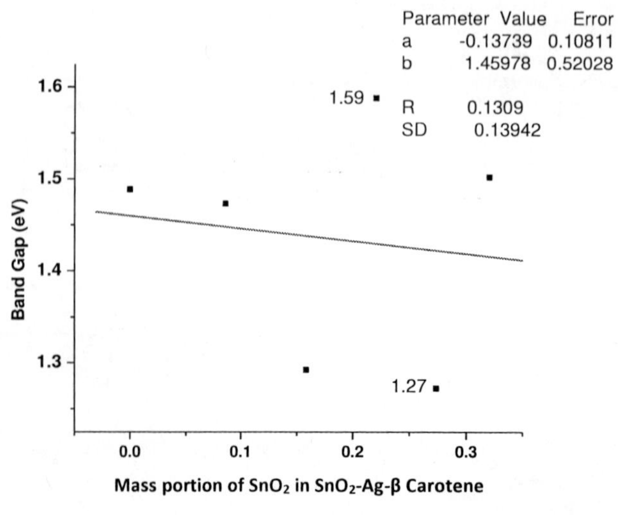

Figure 7. The band gap energy (indirect) as the function of the mass portion of SnO_2 in SnO_2-Ag-β Carotene

Figure 6. shows that the maximum direct band gap energy of 3.74 eV was achieved by sample 20 of a SnO_2 fraction, while it showed a minimum value of direct bandgap energy of 3.54 eV by sample 15. Figure 7. shows that the SnO_2:Ag:β-Carotene mass portion gave maximum indirect band gap energy of 1.59 eV yielded by sample 20, while a minimum value of indirect band gap energy of 1.27 eV was reached by sample 25. For direct Tauc plot analysis, the slope of the linear fitting curve was 0,10454 and -0.13739 for indirect Tauc plot analysis. These results showed that there was no significant influence of SnO_2 mass portion on the band gap energy of SnO_2-Ag-β Carotene composite thin film for both direct and indirect Tauc plot analysis. But it was still an important fact that each mass variation composites had different band gap energy. In the case of indirect analysis, the SnO_2

fraction tends to decrease the band gap. This fact can be used to control the band gap energy to optimize the usage of SnO_2 composite as DSSC electron transporting electrode.

4. Conclusion

We have successfully fabricated a thin film of x(SnO_2)-y(Ag)-β Carotene/ ITO PET at the different SnO_2 portion. The maximum direct band gap energy at 3.74 eV was reached by the sample 20, while the minimum direct band gap energy of 3.54 eV was achieved by sample 15. The maximum indirect band gap of 1.59 eV resulted in sample 20. On the other hand, the minimum value of the indirect band gap is 1.27 eV achieved by sample 25. Both direct and indirect Tauc plot analysis showed that there was no significant influence of SnO_2 mass portion on the band gap energy of SnO_2-Ag-β Carotene composite thin film. But it was still a remarkable fact that each mass variation composites had different band gap energy. This fact can be used to control the band gap energy to optimize the use of SnO_2 composite as DSSC electron transporting electrode. The competition between SnO_2 NPs and Ag NPs may take a role in this composite film.

Acknowledgments
The author's thanks to the Ministry of Research and Higher Education of Indonesia through the DRPM for supporting this research.

References

[1] Green M A, Emery K, Hishikawa Y and Warta W 2010 Solar cell efficiency tables (version 35) *Prog. Photovolt. Res. Appl.* **18** 144–50
[2] Fonstad C G and Rediker R H 1971 Electrical Properties of High-Quality Stannic Oxide Crystals *J. Appl. Phys.* **42** 2911–8
[3] Li Z, Zhou Y, Sun R, Xiong Y, Xie H and Zou Z 2014 Nanostructured SnO_2 photoanode-based dye-sensitized solar cells *Chin. Sci. Bull.* **59** 2122–34
[4] Senevirathna M, Pitigala P, Premalal E, Tennakone K, Kumara G and Konno A 2007 Stability of the SnO_2/MgO dye-sensitized photoelectrochemical solar cell *Sol. Energy Mater. Sol. Cells* **91** 544–7
[5] Li L-L, Zhang W-M, Yuan Q, Li Z-X, Fang C-J, Sun L-D, Wan L-J and Yan C-H 2008 Room Temperature Ionic Liquids Assisted Green Synthesis of Nanocrystalline Porous SnO_2 and Their Gas Sensor Behaviors *Cryst. Growth Des.* **8** 4165–72
[6] Shi L and Lin H 2010 Facile Fabrication and Optical Property of Hollow SnO_2 Spheres and Their Application in Water Treatment *Langmuir* **26** 18718–22
[7] Cheng B, Russell J M, Shi, Zhang L and Samulski E T 2004 Large-Scale, Solution-Phase Growth of Single-Crystalline SnO_2 Nanorods *J. Am. Chem. Soc.* **126** 5972–3
[8] Vayssieres L and Graetzel M 2004 Highly Ordered SnO_2 Nanorod Arrays from Controlled Aqueous Growth *Angew. Chem. Int. Ed.* **43** 3666–70
[9] Huang H, Lim C K, Tse M S, Guo J and Tan O K 2012 SnO_2 nanorod arrays: low temperature growth, surface modification and field emission properties *Nanoscale* **4** 1491–6
[10] Gubbala S, Chakrapani V, Kumar V and Sunkara M K 2008 Band-Edge Engineered Hybrid Structures for Dye-Sensitized Solar Cells Based on SnO_2 Nanowires *Adv. Funct. Mater.* **18** 2411–8
[11] Her Y-C, Wu J-Y, Lin Y-R and Tsai S-Y 2006 Low-temperature growth and blue luminescence of SnO_2 nanoblades *Appl. Phys. Lett.* **89** 043115
[12] Wang C, Zhou Y, Ge M, Xu X, Zhang Z and Jiang J Z 2010 Large-Scale Synthesis of SnO_2 Nanosheets with High Lithium Storage Capacity *J. Am. Chem. Soc.* **132** 46–7
[13] Deng D and Lee J Y 2008 Hollow Core–Shell Mesospheres of Crystalline SnO_2 Nanoparticle Aggregates for High Capacity Li^+ Ion Storage *Chem. Mater.* **20** 1841–6
[14] Ohgi H, Maeda T, Hosono E, Fujihara S and Imai H 2005 Evolution of Nanoscale SnO_2 Grains, Flakes, and Plates into Versatile Particles and Films through Crystal Growth in Aqueous

Solutions *Cryst. Growth Des.* **5** 1079–83

[15] Sun P, Zhao W, Cao Y, Guan Y, Sun Y and Lu G 2011 Porous SnO2 hierarchical nanosheets: hydrothermal preparation, growth mechanism, and gas sensing properties *CrystEngComm* **13** 3718

[16] Zhang D-F, Sun L-D, Jia C-J, Yan Z-G, You L-P and Yan C-H 2005 Hierarchical Assembly of SnO_2 Nanorod Arrays on α-Fe_2O_3 Nanotubes: A Case of Interfacial Lattice Compatibility *J. Am. Chem. Soc.* **127** 13492–3

[17] Diantoro M, Kholid, Mustikasari A A and Yudiyanto 2018 The Influence of SnO_2 Nanoparticles on Electrical Conductivity, and Transmittance of PANI-SnO_2 Films *IOP Conf. Ser. Mater. Sci. Eng.* **367** 012034

[18] Diantoro M, Fitrianingsih R, Mufti N and Fuad A 2014 Synthesis of silver nanoparticles by chemical reduction at various fraction of MSA and their structure characterization pp 257–61

[19] Diantoro M, Suprayogi T, Sa'adah U, Mufti N, Fuad A, Hidayat A and Nur H 2018 Modification of Electrical Properties of Silver Nanoparticle *Silver Nanoparticles - Fabrication, Characterization and Applications* ed K Maaz (InTech)

[20] Diantoro M, Hidayati N N S, Latifah R, Fuad A, Nasikhudin, Sujito and Hidayat A 2016 Electrical conductivity modification using silver nano particles of Jatropha Multifida L. and Pterocarpus Indicus w. extracts films p 030034

IOP Conf. Series: Journal of Physics: Conf. Series **1093** (2018) 012050 doi:10.1088/1742-6596/1093/1/012050

Synthesis of Silica Xerogel based Bagasse Ash as a Methylene Blue Adsorbent on Textile Waste

Lukluil Maknun, Nazriati*, Ida Farida, Nur Kholila, Ratu Baiqis Muyas Syufa

Department of Chemistry, Faculty of Mathematics and Natural Sciences, Universitas Negeri Malang, Jl. Semarang 5 Malang 65145, Indonesia

*Corresponding author's email: nazriati.fmipa@um.ac.id

Abstract. The textile industry is one of the fastest growing industries in Indonesia. One of the important components in the textile industry is methylene blue dye. In the process of dying, methylene blue is only tied at 5%, while 95% as waste. The non-biodegradable methylene blue properties make it as a contaminant in the water. Therefore, it is necessary to handle methylene blue in the textile industry waste to reduce the level of contamination in the waters. One of material that can be used as an adsorbent is bagasse ash. The high silica content in bagasse ash allows bagasse as a raw material for the manufacture of silica xerogel which acts as a blue methylene adsorbent. The synthesis of silica xerogel was done through two stages of extraction and sol-gel. The silica xerogel obtained was a white-colored powder to be used for adsorption of a methylene blue solution. The initial step for the adsorption process was to test the methylene blue solution using UV-vis spectroscopy. The results showed that the maximum λ of methylene blue solution was at 665 nm and the correlation coefficient (R^2) was 0.9980. The adsorption capacity of the xerogel silica increases with the increase of adsorption contact time and the optimum contact time at 60 minutes. The increased initial concentration of methylene blue further increases the adsorption capacity. Under the equilibrium conditions, the adsorption capacity increased 2.18 mg / g to 10.53 mg / g. Adsorption of blue methylene by silica xerogel following Langmuir adsorption isotherm model, takes place spontaneously and exothermically.

Keywords: Silica xerogel, bagasse ash, *methylene blue,* adsorbent, textile waste.

1. Introduction

The development of fashion trends has pushed the textile industry to grow rapidly. Around 15% of the total dye production in the world is used by the textile industry and its waste is discharged into the environment [1]. Textile dyes in the form of organic compounds that are nonbiodegradable. One of the common dyes is methylene blue ($C_{16}H_{18}N_3SCl$). Methylene blue has several advantages such as the price is relatively cheap, easy to obtain, good solubility, and is a basic dye. In the coloring process, methylene blue is bonded about 5% while the remaining 95% is wasted as waste, so the methylene blue waste in the aquatic environment is very large [2]. Methylene blue waste in addition to damaging aquatic ecosystems can disrupt human health as well [3].

Various ways have been applied to reduce the level of contamination of dye waste in waters, such as chemical oxidation, coagulation-flocculation, biological treatment and filtration [4]. Another effective method of reducing dye contamination in waters is adsorption. Types of adsorbents used include active carbon and silica. Silica adsorbents can be obtained from ash waste bagasse. Bagasse is a solid sugar cane obtained as waste from sugar cane processing in the sugar processing industry. The results show that bagasse ash contains more than 50% silica and the silica has an amorphous phase. The amorphous silica in bagasse ash can be taken by alkaline extraction to form sodium silicate solution followed by sol-gel method [5]. The content of silica in a high bagasse ash has the potential to become the raw material for the manufacture of silica xerogel used as a blue methylene adsorbent.

2. Materials and Methods

The materials used in this research are bagasse ashes from Krebet II New Sugar Factory Malang; Standard methylene blue solution; NaOH; HCl; Water demineralization; Whattman-42 filter paper; And universal indicators. Tools used include balance sheet; Pumpkin powder; beaker; Erlenmeyer; pipette; funnel; Stirring rod; Hotplate; Magnetic stirring bar; oven; Crucial; A crucial clamp; And spray bottles.

The silica extraction step is carried out by inserting 10 g of bagasse ash into 60 mL sodium hydroxide 2 M. The mixture was heated at its boiling temperature for 1 hour while constantly stirring using a magnetic stirrer to dissolves the silica. The mixture was cooled and filtered using Whatman-42 ashless filter paper and produce the sodium silicate solution. The next step was the synthesis of xerogel silica, which was 10 mL of the extracted filtrate fed into 30 mL of a pH 2 cation exchange resin with stirring for 30 min. The mixture was further filtered and the filtrate was produced in the form of silicic acid. The 10 mL silicate acid was added dropwise ammonium hydroxide 2 M to form a gel. Hydrogel formed then aging at room temperature for 18 hours. The gel was washed with demineralized water until it was clean and the pH neutral. The resulting gel was dried at 80 °C for 24 hours. Characterization of the silica xerogel was done using FTIR (8400 S, Shimadzu) and analysis methylene blue was using spectrophotometer UV-Vis (BIO-RAD).

3. Results and Discussion

Silica xerogel obtained in the form of white powder. The following is the result of the xerogel silica formed on the sol-gel process.

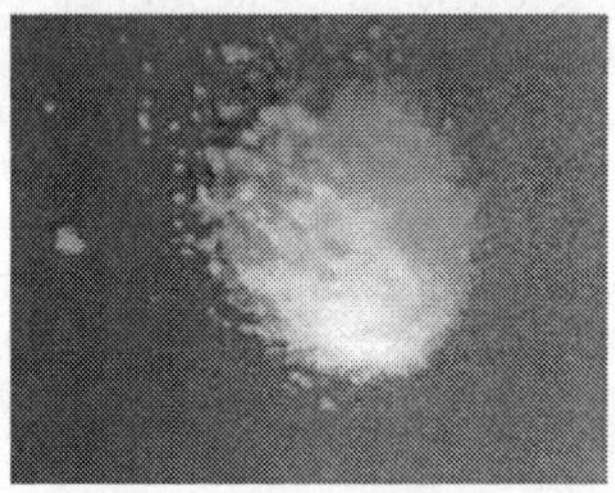

Figure 1. Silica Xerogel

Silica xerogel based bagasse ash synthesis was characterized by X-Ray Fluorescence (XRF) and Fourier Transform Infrared (FT-IR) spectroscopy. The results of the XRF test is shown in Table 1. Based on the table, it can be seen that the most elemental content in silica xerogel is silicon with a percentage of 92.8%. This indicates that the most compound in silica xerogel is a silicon compound. To know what functional group is contained in silica xerogel, further characterization was done in the form of FT-IR characterization as shown in Figure 2.

The silica xerogel spectrum shows the absorption bands at wave numbers 3444,87 cm^{-1} to 3224,98 cm^{-1} and 1633,71 cm^{-1} indicating the O-H vibration of the silanol (Si-OH) group. In the wave number 1085, 92 cm^{-1} appears an absorption band that indicates the vibration of the Si-O-Si bond. This signal indicates that the silica xerogel has been successfully prepared. The silica xerogel is further used as a blue methylene adsorbent. The following is the result of the xerogel silica application as a methylene blue adsorbent.

Table 1. The Results of *XRF* Silica Xerogel

Conponent	Mass (%)
Si	92.8
Ca	1.97
Ti	0.23
Cr	0.11
Fe	0.52
Ni	2.83
Cu	0.39
Zn	0.06
Ba	0.2
Re	0.35

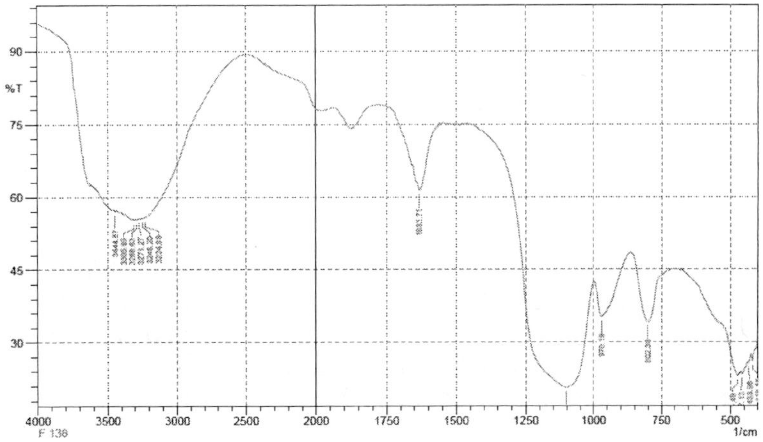

Figure 2. The FT-IR Silica Xerogel Spectrum

The application of xerogel silica as a blue methylene adsorbent is carried out by studying the effect of contact time variation on adsorption capacity. Based on Table 2, the xerogel silica adsorption capacity increases with the increase of adsorption contact time. At the contact time of 60 minutes, the adsorption capacity increase is not very significant, so it can be stated that the optimum contact time is 60 minutes.

Table 3 shows that with the increased initial concentration of methylene blue, the adsorption capacity also increases. An increase in the initial concentration of methylene blue leads to increased competition between the adsorbate molecules to be absorbed into the adsorbent bed, thereby increasing the interaction between blue methylene with silica xerogel which will result in a higher adsorption capacity.

Table 2. Methylene Blue adsorbed on Contact Time Variation

Time (min)	Concentration (mg/L)		Methylene Blue adsorbed (%)	qt (mg/g)
	C_0	C_t		
10	6	2.804	98.668	7.99
20	6	2.789	98.662	8.03
30	6	2.411	98.505	8.97
40	6	1.780	98.242	10.55
50	6	1.517	98.132	11.21
60	6	1.368	98.070	11.58
90	6	1.368	98.070	11.58
120	6	1.187	97.994	12.03

Table 3. Methylene Blue adsorbed on Adsorbent Concentration Variation

Concentration (mg/L)		Methylene Blue adsorbed (%)	qt (mg/g)
C_0	C_t		
1	0.129	87.081	2.18
2	0.239	88.038	4.40
3	0.268	91.069	6.83
4	0.598	85.048	8.50
5	0.789	84.211	10.53

Thermodynamics that can be calculated during the adsorption process are the free energy changes Gibbs (ΔG°), enthalpy changes (ΔH°), and entropy changes (ΔS°). Based on the calculation, the values obtained at ΔG °, ΔH °, and ΔS ° as presented in Table 4.

Table 4. The Thermodynamic Parameters

T (K)	*ΔG°* (kJ/mol)	*ΔH°* (kJ/mol)	*ΔS°* (J/mol)
308	-4.805		
318	-4.013	-18.848	-45.960
328	-3.900		

In Table 4 an increasingly negative ΔG ° value was obtained with a decrease in temperature, indicating that the methylene blue adsorption by silica xerogel runs spontaneously and is better at low temperatures. The negative value of ΔH ° indicates that the adsorption is exothermic, while the ΔH ° value is -18.848 kJ / mol (<80 kJ / mol) indicating that the adsorption type is physical adsorption. The value of the ΔS° entropy change shows a negative value indicating that there is a decrease in irregularity in the adsorption process [6]. The adsorption isotherm describes the relationship between adsorption capacity and the concentration of adsorbate at a constant temperature. In summary, the Langmuir and Freundlich adsorption isotherm values of the calculations are presented in Table 5.

Table 5. The Parameters of Langmuir and Freundlich Isotherms Model

Temperature	Isoterm Langmuir			Isoterm Freundlich		
(°C)	R^2	Q_0 (mg/g)	b_L (L/mg)	R^2	k_f	n
35	0.937	14.085	4.4375	0.924	16.319	2.053
45	0.977	14.493	2.0294	0.915	15.578	2.278
55	0.875	15.625	1.1852	0.652	14.481	2.924

Based on Table 5, it shows that the Langmuir isotherm model gives a correlation coefficient value (R^2) better than the Freundlich isotherm model. Therefore, the Langmuir isotherm model is better used in explaining the blue methylene adsorption process by silica xerogel. It also proves that the surface of the xerogel silica is a homogeneous side.

The kinetic parameter is used to determine the absorption rate in the adsorption process through the influence of contact time and this parameter to explain the efficiency characteristics of the adsorption occurring. The data in the adsorption kinetics study were taken from the influence of contact time data on the percentage of methylene blue adsorbed, the data used at 10-60 minutes. The following Table 6 is presented to compare two possibilities of a pseudo-first order and pseudo-second order adsorption kinetics model.

Table 6. The Parameters of Pseudo-First Order and Pseudo-Second Order

qe (mg/g)	Pseudo-first order			Pseudo-second order		
	qe cal (mg/g)	k_1 (min^{-1})	R^2	*qe cal* (mg/g)	k_2 (g/mg.min)	R^2
12.03	1.142	0.058	0.866	12.897	0.051	0.965

Based on Table 6 it can be seen that the value of the adsorption capacity of the pseudo order (*qe cal*) approaches the value of the experimental adsorption capacity (*qe*) of 12.897 mg/g. This indicates that the methylene blue adsorption process with silica xerogel can be well explained using a pseudo-second order model. The statement is also proved by the value of the pseudo-second order coefficient (R^2) which is closer to 1 than the pseudo-first order one [7].

4. Conclusion

Silica xerogel in the form of white-colored powder is used for adsorption of methylene blue solution. The adsorption capacity of silica xerogel increased with increasing of adsorption contact time and optimum contact time at 60 minutes. The more amount of the initial concentration of methylene blue, increases the more the adsorption capacity also increases. Under equilibrium conditions, the adsorption capacity increased 2.18 mg/g to 10.53 mg/g. The adsorption of blue methylene by silica xerogel following Langmuir adsorption isotherm model takes place spontaneously and exothermically.

Acknowledgments

The authors would like to thank the Directorate General of Higher Education (DGHE), the Ministry of Education and Culture, Indonesia for *Hibah* PKM 2017.

References

[1] Houas A, Lachheb H, Ksibi M, Elaloui E, Guillard C and Herrmann J-M 2001 Photocatalytic degradation pathway of methylene blue in water *Applied Catalysis B: Environmental* **31** 145–157

[2] Gong J-L, Zhang Y-L, Jiang Y, Zeng G-M, Cui Z-H, Liu K, Deng C-H, Niu Q-Y, Deng J-H and Huan S-Y 2015 Continuous adsorption of Pb (II) and methylene blue by engineered graphite oxide coated sand in fixed-bed column *Applied Surface Science* **330** 148–157

[3] Gürses A, Hassani A, Kıranşan M, Açışlı Ö and Karaca S 2014 Removal of methylene blue from aqueous solution using by untreated lignite as potential low-cost adsorbent: kinetic, thermodynamic and equilibrium approach *Journal of Water Process Engineering* **2** 10–21

[4] Valix M, Cheung W H and McKay G 2004 Preparation of activated carbon using low temperature carbonisation and physical activation of high ash raw bagasse for acid dye adsorption *Chemosphere* **56** 493–501

[5] Affandi S, Setyawan H, Winardi S, Purwanto A and Balgis R 2009 A facile method for production of high-purity silica xerogels from bagasse ash *Advanced Powder Technology* **20** 468–472

[6] Nandi B K, Goswami A and Purkait M K 2009 Adsorption characteristics of brilliant green dye on kaolin *Journal of hazardous materials* **161** 387–395

[7] Tan K A, Morad N, Teng T T, Norli I and Panneerselvam P 2012 Removal of cationic dye by magnetic nanoparticle (Fe3O4) impregnated onto activated maize cob powder and kinetic study of dye waste adsorption *APCBEE Procedia* **1** 83–89